STATISTICAL ANALYSIS

STATISTICAL ANALYSIS

An Interdisciplinary Introduction to Univariate & Multivariate Methods

Sam Kash Kachigan

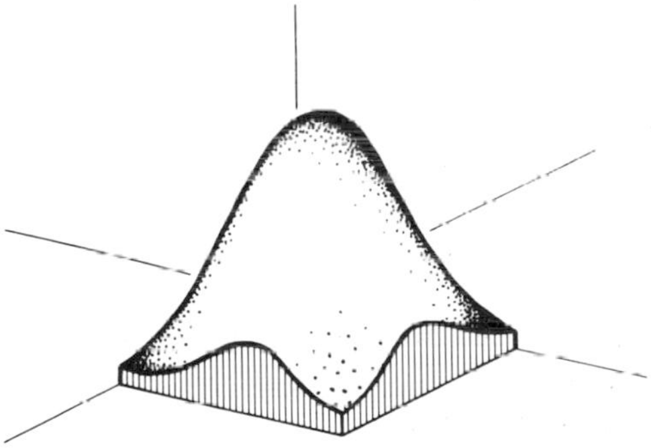

Radius Press

New York

STATISTICAL ANALYSIS

Type by Science Typographers
Graphics by Karl Kachigan and Suzanne Roberts

Manufactured in the United States of America

Library of Congress Cataloging-in-Publication Data

Kachigan, Sam Kash.
Statistical analysis.

Expanded version of the author's Multivariate statistical analysis.
Bibliography: p.
Includes index.
1. Multivariate analysis. I. Title.
QA278.K323 1986 519.5 85-61811

ISBN 0-942154-99-1

15 14 13 12 11 10 9 8 7 6 5 4 3 2

To my students and teachers

Preface

This book is designed for a *first course* in both univariate and multivariate methods of statistical analysis, for research-oriented students with a typical mathematics background.

Overall, the most unique feature of the book is that it recognizes and addresses the needs and abilities of a wider range of students than the conventional introductory statistics textbook, both in approach and content. A fundamental principle of education is that individuals differ in their native aptitudes for various subject matter. While relatively few students have a strong aptitude for manipulating mathematical symbols—the essential audience for many statistics texts—many more sharp students are capable of grasping concepts and arguments that are effectively presented *verbally*, *numerically*, *geometrically*, *graphically*, and by *example*, *repetition*, *logic*, and *analogy*; methods which are used extensively in this text. Mathematical symbols are reserved for introducing the basic univariate and bivariate concepts, while multivariate topics are treated conceptually as extensions of the bivariate relation.

The emphasis in the presentation is on the *rationales*, *interpretations*, and *applications* of the most commonly used statistical methods, rather than on their computational aspects. This approach has been dictated by the increasing availability of computing hardware and software, a technological innovation that has had the twofold consequence of (1) increasing the user base for statistical analysis, and (2) increasing the variety and power of readily available analytical techniques. Historically, use of the advanced multivariate methods introduced in the text were time-consuming and were only available to a relatively small handful of researchers with advanced training, but now with the proliferation of computers they are accessible to everyone and are commonplace in contemporary research. As a result, more students must learn more statistical techniques than in the past, and instruction must become more efficient.

Throughout the presentation a *building-block* approach is used, in which each new concept is shown to be a logical extension, variation, or combination of previously developed ones. Also, the position is taken that the methods of statistical analysis fill one or more of three basic objectives, including *data reduction*, *inference*, and the *identification of associations* among variables. With this and the building-block approach, the student will more easily see statistics as a unified subject area based on a handful of concepts and principles, rather than as a collection of disjointed techniques. This philosophy is reflected in the overall organization of the book's 21 chapters:

Part I: FUNDAMENTAL CONCEPTS
1. The Nature of Statistical Analysis
2. Objects, Variables, and Scales

Part II: DATA REDUCTION
3. Frequency Distributions
4. Central Tendency
5. Variation

Part III: INFERENCE
6. Basic Probability
7. Sampling Distributions
8. Parameter Estimation
9. Hypothesis Testing

Part IV: ASSOCIATION (Multivariate Analysis)
10. Correlation Analysis
11. Regression Analysis
12. Analysis of Variance
13. Analysis of Category Data
14. Discriminant Analysis
15. Factor Analysis
16. Cluster Analysis
17. Multidimensional Scaling

Part V: SELECTED SUBJECTS
18. Time Series Analysis
19. Nonparametric Analysis
20. Advanced Probability Topics
21. Decision Analysis

Because of the wide range of topics, the text can fill a wide variety of curriculum needs at both the undergraduate and graduate level.

The treatment of the multivariate methods in Part IV is conceptual and practical in nature, requiring no knowledge of computer programming, calculus, or matrix algebra, although the interested instructor can easily introduce these aspects during lectures. In their professional lives, the majority of students will be engaged in the design, interpretation, and application of multivariate research, and that is the focus of the presentation. For the student who will specialize in statistical methods, the text will serve as a springboard for advanced mathematical study and as a communications tool to be used in their eventual consulting capacity with naïve users of statistical methods of data analysis.

Since the principles of statistical analysis are perfectly general, cutting across all academic disciplines, students in all curriculums can use the text. The examples are drawn primarily from the *behavioral*, *biological*, *environmental*, and *monetary* sciences, with each illustration relevant to several disciplines. This approach is aimed at emphasizing the interdisciplinary nature of many research problems and the absolute generality of statistical theory.

Exercises for each chapter are included at the back of the text in the Appendix, their location simulating real life in that the test of our knowledge is typically removed from the point of the learning experience. Aside from assessing basic definitional and computational skills, the exercises test for the conceptual understanding of the techniques, including their interpretations and applications.

The many users of my earlier book *Multivariate Statistical Analysis: A Conceptual Introduction* (1982)—an abbreviated version of the present text—were instrumental in the development of this expanded volume, both through their acceptance of the overall approach and their desire for wider coverage. I have tried my best to accommodate as many of the often conflicting instructor needs as possible, while still maintaining an evenness of presentation and not losing sight of the primary introductory audience.

I am grateful to the Literary Executor of the late Sir Ronald A. Fisher, F.R.S., to Dr. Frank Yates, F.R.S., and to Longman Group Ltd., London, for permission to use Tables II, III, and IV from their book *Statistical Tables for Biological, Agricultural and Medical Research* (6th edition, 1974). And special thanks to Myrna Tenenbaum for her excellent management of the manuscript.

S.K.K.

ABOUT THE AUTHOR

Sam Kash Kachigan received his education at the University of Wisconsin-Milwaukee, the University of Washington, and Columbia University; and has held a variety of university, corporate, and consulting positions. He lives in New York and Paris and divides his time between teaching, writing, and research projects. This is his third book. He is also the author of *The Game: A Lesson* (1981), a demonstration of operational writing; and *Multivariate Statistical Analysis: A Conceptual Introduction* (1982), the original shorter version of this book.

Contents

Part I: FUNDAMENTAL CONCEPTS

Part II: DATA REDUCTION

Part IV: ASSOCIATION (Multivariate Analysis)

GREEK ALPHABET

A	α	alpha	N	ν	nu
B	β	beta	Ξ	ξ	xi
Γ	γ	gamma	O	o	omicron
Δ	δ	delta	Π	π	pi
E	ϵ	epsilon	P	ρ	rho
Z	ζ	zeta	Σ	σ	sigma
H	η	eta	T	τ	tau
Θ	θ	theta	Υ	υ	upsilon
I	ι	iota	Φ	ϕ, φ	phi
K	κ	kappa	X	χ	chi
Λ	λ	lambda	Ψ	ψ	psi
M	μ	mu	Ω	ω	omega

Part I.

FUNDAMENTAL CONCEPTS

Chapter 1

The Nature of Statistical Analysis

1. Introduction

The field of *Statistical Analysis* is concerned with the *collection, organization, and interpretation of data according to well-defined procedures.*

As such, it is the heart of every scientific discipline, for whether we are studying the behavioral and biological characteristics of human beings, or aspects of plant and animal life, or the dynamics of our physical environment all of which interact with one another—we need a methodology to systematically analyze and draw conclusions about the things that we study.

We will discover that the basic principles and concepts of statistical analysis are relatively simple and few in number, yet they generate a wide variety of specific analytical techniques to cope with different types of data problems. The biggest challenge is to be able to recognize which types of analysis are most appropriate for a given situation, and how to interpret and apply the results.

Although the actual computations involved in the execution of the various statistical techniques are often complicated, the rapid growth in computer technology has removed this time-consuming and often tedious aspect of the analyses. As a result, we can better concentrate on the basic underlying principles of statistical analysis and the logic behind the various methods, as well as their interpretations and applications. This emphasis will help us to both formulate and solve research problems, for, as is so often the case, a problem properly defined is half the solution.

We have defined statistical analysis as the *collection, organization, and interpretation of data according to well-defined procedures.* There are many terms in this statement that need further definition, and ultimately it will take the balance of the book to accomplish the task. We can begin with the concept of *data.*

2. Data

By *data* we mean observations made upon our environment—observations which are the result of measurements using clocks, balances, measuring rods, counting operations, or other objectively defined measuring instruments or procedures. These observations answer questions such as *how much*, *how many*, *how long*, *how often*, *how fast*, *where*, and *what kind*, and have the characteristic that they can be represented by numbers, since numbers have the distinct advantage over words in that there is less ambiguity to their meaning. It is one thing, for example, to be told that *most* patients survive a particular surgical operation, and quite another to be told that 58 out of 100 survive the operation; and this type of distinction is especially important if we are at the receiving end of the surgeon's scalpel. So while words such as *many*, *most*, *few*, *alot*, *usually*, *majority*, *several*, and *substantial* convey some information, they are not likely to be interpreted as precisely as statements using numbers based on standardized operations.

As might be expected, data can be found everywhere we look. There is no part of our environment that is not a source of data—ourselves, other individuals, family units, societies, cultures, races, cities, countries, planets, volcanoes, DNA strands, atomic particles, soils, medicines, molecules, plant and animal species, organs, cells, schools, governments, religions, etc.—in short, every aspect of our existence.

Just as there are myriad sources of data, there are countless measurements we can make, limited only by our imaginations. With regard to individuals we can measure such characteristics as income, blood pressure, weight, annual consumption of candy, amount of education, or scores on achievement, aptitude, or attitude tests. With regard to cities we can measure population, unemployment rate, incidence of crime, sales of Brand *A* toothpaste, extent of air pollution, vote on referendum *A*, snowfall, tax rate, etc. Foods can be measured in terms of vitamin, mineral, fat, protein, and carbohydrate content, as well as their price, or rated with respect to taste or quality. Metals can be measured with regard to corrosiveness, tensile strength, or hardness. We can try to measure the stars in the sky with respect to their age, temperature, distance, and number.

Since everything in our environment is a source of data, and with respect to any number of variable characteristics, it is no wonder, that given our natural curiosity and quest for order and understanding of our surroundings, that a field of study such as statistical analysis has developed which is concerned with methods of dealing with the everpresent data.

3. Univariate vs. Multivariate Analysis

When we are solely interested in an isolated characteristic of a set of objects, irrespective of other variable characteristics possessed by the objects, we are involved in a *univariate* analysis—i.e., one variable characteristic. Examples would include the study of the achievement test scores of students, the white blood cell counts of patients, the salaries of workers, the gold content of ore samples, etc.

Multivariate statistical analysis, on the other hand, is that branch of statistical analysis which is concerned with the simultaneous investigation of *two or more variable characteristics* which are measured over a set of objects. For example, interest may be in the relationships that exist among career performance, academic achievement, and aptitude test scores; or among crop yields, fertilizer levels, rainfall, and soil pH levels; or among crime, unemployment, and interest rates; or among product sales, pricing, and advertising levels; or among brain chemistry, diet, attitude, and behavior measures; or among any of a number of other sets of variables.

Of the two, multivariate analysis is the more interesting and scientifically productive. Therefore, the bulk of our attention will be devoted to it. However, before we can study the procedures for analyzing the relationships that exist among two or more variables, we need to know the basics of analyzing the individual isolated variables. Consequently, the early parts of the book are devoted to fundamental univariate topics. They will provide the basis for studying the multivariate methods in the remaining chapters.

4. Objectives of Statistical Analysis

We can think of the overall objective of statistical analysis as portrayed in Figure 1. Observations of the world are converted into numbers, the numbers are manipulated and organized, and then the results are interpreted and translated back to a world that is now hopefully more orderly and understandable than prior to the data analysis. This process of *drawing conclusions and understanding more about the sources of our data* is the goal of statistical analysis in its broadest sense.

More specifically, we can view the data manipulation and organization as achieving one or more of three basic objectives: (1) *data reduction*, (2) service as an *inferential measuring tool*, and (3) the *identification of associations or relationships* between and among sets of data. The following paragraphs provide an overview of these three basic objectives.

Data reduction. The *reduction of data* is perhaps the most pervasive role of statistical analysis. As the name implies, it involves the *summary* of data. Large masses of unorganized numbers are beyond anyone's comprehension. Statistical analysis is concerned with techniques that will reduce large sets of data into smaller sets that describe the original observations without sacrificing critical information. As with any summary, something is lost but something is gained: *We sacrifice detail and nuance for parsimony and salience.* The specific

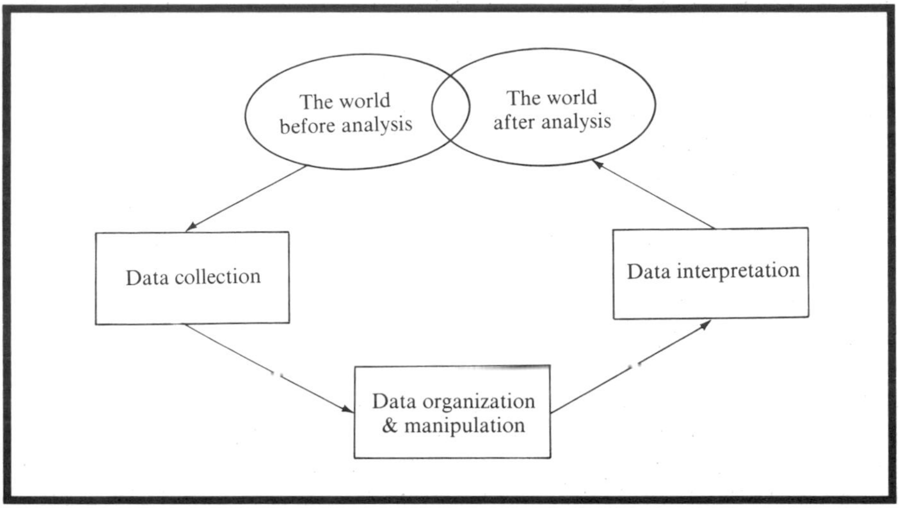

Figure 1 A schematic diagram of the discipline of statistical analysis.

techniques for accomplishing this task will be among the subjects of the remainder of the book.

Inference. The second major role of statistical analysis lies in its use as an *inferential measuring tool*, in the sense that it provides procedures for stating the degree of confidence we have in the accuracy of the measurements we make on our environment—whether it is the effectiveness of a vaccine, public attitudes toward a social issue, or the sales effectiveness of a TV commercial.

All measurements are subject to error. Ten individuals measuring the height of this page with a finely calibrated measuring rule would likely come up with several different readings. Components produced by the highest precision machinery and destined to be used in an aircraft engine will nonetheless vary in size. Statistical analysis provides the methods for stating the degree of precision of our measurements, when those measurements represent an estimate of the "true" but unknown value of a characteristic.

The same statistical principles apply to the evaluation of observed *differences* between sets of data. For example, is there a difference in the crop yield produced by two different fertilizers? Is one drug more effective than another in treating a disease? Is one toothpaste formulation better tasting than another? Is one teaching method better than another? Do males differ from females in mathematical aptitude? Is one advertisement more persuasive than another? All of these questions call for a procedure that will help to make the "right" inference, since we usually want to generalize our conclusions beyond the sample of observations upon which the conclusions are made. The field of statistics provides the necessary techniques for making statements of our certainty that there are real as opposed to chance differences between sets of observations.

This theme of inference, like that of data reduction, will appear throughout our discussion of the various key analytical techniques.

Identification of relationships. The third major objective of statistical analysis is the identification of *associations* or *relationships* that exist between and among sets of observations. In other words, does knowledge about one set of data allow us to infer or predict characteristics about another set of data? Such information is very useful: scores on a paper and pencil test that predict performance in various occupations; blood test results that predict an incipient disease; answers to a questionnaire that separate good credit risks from bad; economic policies that are related to rates of inflation or unemployment. There is virtually no limit to the types of relationships that we might identify. Prior to the development of many of today's statistical methods many associations existing in the world went unnoticed, and other associations which were thought to exist, upon closer scrutiny proved to be nonexistent.

The associations uncovered by statistical analysis are basically of two types. In one, the relationship identified between two sets of observations is purely *descriptive* or "correlational" in nature, and no conclusions about causality can safely be made. For example, while the characteristics of a blood sample may accompany or predict the onset of a disease, it does not follow that these features of the blood *caused* the disease, nor provide an explanation of its occurrence. Or, as another example, consider the personnel test that identifies a relationship between interest in certain hobbies and job success in a managerial position. Now interest in the hobby does not in itself cause the job success, as would be evident to anyone who tried to achieve such success by cultivating an interest in the selected hobbies. Rather, the reasons for such relationships are extremely complicated and virtually impossible to disentangle. Still, the identification of such descriptive relationships does have very important *practical* value, as we will learn in later chapters.

There is a second type of relationship, however, in which we can be relatively confident that the variables are related in a *causal* manner. These are *experimental* relationships (i.e., experimentally based) in which we as researchers manipulate the levels of one variable and observe changes in another. This is in contrast to the descriptive or correlational relationships which are simply observed as they occur in the natural environment. Examples of experimentally based relationships include differences in crop yield as a function of the amount of irrigation; differences in product sales as a function of advertising levels; differences in achievement as a function of hours of instruction, differences in disease duration as a function of drug dosage; etc.

These types of causal relationships are especially important to us, since they often allow us to bring our environment and its consequences under our control. This is not the case with descriptive or correlational relationships—e.g., school grades and aptitude test scores—in which the variables are not under our control. This distinction between correlational and experimental relationships will be one of the recurrent themes throughout our study of multivariate analysis in the later chapters.

Not only can statistical analysis be used to identify relationships, but processes can be discovered that actually *function according to statistical principles*. For example, notions of statistical variability help to explain the evolution of various species' characteristics. Consider the classical example of the giraffe's long neck. During periods of food scarcity, those giraffes with the longer than average necks were more able to reach the upper branches of trees and were therefore more likely to survive and pass on their genetic characteristics to succeeding generations. And in the field of experimental psychology, theories of learning have been formulated which are based on principles of statistical sampling of information. In quantum physics, the behavior of particles is seen as probabilistic in nature. In marketing, the brand switching behavior of consumers is also modelled on notions of probability, one of the central concepts of statistical analysis.

So, more and more, researchers are discovering that the principles of statistics are not only useful for identifying relationships between sets of *empirical* data—i.e., worldly observations—but also for the *theoretical modelling* of the process or processes that *generate* the data. In other words, instead of the application of statistical principles as a means to an end, the principles become an end in themselves, as they serve to explain aspects of our existence.

The above three roles of statistical analysis—*data reduction*, *inference*, and *the identification of relationships*—will be the central topics of the text; and we will discover that each of the specific statistical techniques that we study, can, and most often do, serve more than one of these three key functions. Since our primary concern is with multivariate statistical analysis, we will focus most sharply on the third stated objective—the study of the relationships among variables—and at the same time discover that the other two issues of data reduction and inference are unavoidable.

5. Descriptive vs. Inferential Analysis

Any particular statistical analysis can be characterized as *descriptive* or *inferential*, depending upon the nature of the data collection procedure and the objective of the data analysis. To understand the distinction between the two it is necessary to consider the concepts of *population* and *sample*.

A *population*, also referred to as a *universe*, is any well-defined collection of things. Again, by well-defined we mean that the members of the population are spelled out, or an unequivocal statement is made as to which things belong in it and which things do not. Populations can be as large or as small as we wish them to be, consisting of anything we specify; e.g., "stocks listed on the New York Stock Exchange," "college students in the state of California," "European cities with a population over 100,000," "users of Brand A toothpaste," "apples in a particular orchard," "voters for candidate B," etc.

A *sample*, on the other hand, is a subset or "portion" of a population. Samples are extremely important in the field of statistical analysis, since due to economic and practical constraints we usually cannot make measurements on every single member of the particular population in which we are interested. Rather, we will make observations on a sample from that population and then make inferences about the population as a whole. The techniques for drawing samples from populations in order that our inferences are valid is a field of study in itself. However, as we will learn later, one of the most useful methods of drawing a sample which allows us to make inferences about the population is referred to as random sampling. By a *random sample* we mean that every member of the population has an equal chance of being included in the sample; more strictly, every possible sample of the specified size has an equal chance of being selected from the population. It is equivalent to putting all the members of the population into a hat, shaking thoroughly, and then drawing a sample of the desired size. While there are more sophisticated ways of drawing a random sample from a population, drawing blindly from a hat captures the essence of the technique. It is because of the importance of this type of sampling procedure that the "collection" of data was included in our definition of statistical analysis.

Now that we have the distinction between the population and sample in hand, we can define the difference between descriptive and inferential statistical analysis. In the case of *descriptive* analysis we are concerned with the *direct exhaustive measurement* of population characteristics. These defining characteristics of populations are called *parameters*. Observations must be made on every single member of the population in question in order to precisely state the value of the parameter. For example, if our population of interest consists of the subscribers to a particular magazine, and the parameter of interest is the average age of those subscribers, a descriptive statistical analysis would call for the measurement of every single subscriber's age in order to arrive at the average value.

In contrast to the above descriptive approach, *inferential* statistical analysis is concerned with measuring the characteristics of only a *sample* from the population and then making inferences, or estimates, about the corresponding value of the characteristics in the population from which the sample was drawn. For instance, in the above magazine subscriber example, rather than measuring the age of every subscriber, we would draw a random sample of individuals, determine their average age, and then make an inference about the average age of the population of subscribers as a whole. Whereas a defining characteristic of a population is referred to as a parameter, that characteristic when measured only on a sample is referred to as a *statistic*. This use of the word "statistic" should not be confused with the lay person's loose usage of the term to represent virtually any number. Also, its plural form, "statistics," should be distinguished from its usage as a shorthand expression for the discipline of statistical analysis.

In summary, we can state that *descriptive* statistical analysis is concerned with the *direct measurement of population parameters*; while *inferential* statistical analysis is concerned with the measurement of *sample statistics* which in turn are treated as *estimates of population parameters*. During the course of our study we will be concentrating primarily on inferential techniques, since in most cases we will want to generalize the results of our data analyses beyond the samples of observations upon which they are based. We will discover along the way that many statistical techniques are applicable to both descriptive and inferential analyses, while others have meaning only for inferential purposes.

6. Statistical Analysis vs. Art and the Occult

Our working definition stated that statistical analysis is concerned with the collection, organization, and interpretation of data according to well-defined procedures. The reason for the last phrase in the definition, *well-defined procedures*, is to rule out from the field such "organizers and interpreters" of data as crystal ball gazers, tea leaf readers, psychic forecasters, astrologers, palm readers, and artists.

While these individuals are all concerned with organizing and interpreting data, their methods are not exactly well-defined. That is, the rules of organization and interpretation are not unequivocally spelled out so that other individuals studying the same data will perform the same analysis and arrive at the same conclusions. The rules of statistical analysis, on the other hand, are spelled out so that given the same set of data any investigator applying the same analytical technique will arrive at the same result.

However, it must be admitted that while the rules of statistical analysis are well-defined, there is not always agreement as to interpretation, once the results of the analysis must be reapplied to the world from which the data originated. This is especially true for some of the more advanced computer-based techniques that process unimaginable amounts of data, usually in a number of different stages. Referring back to Figure 1, it is the final link in the chain of statistical analysis that has the greatest weakness—interpreting the analyzed data with respect to the world from which it was drawn: *The translation of numbers back into words and actions.*

While in most analyses the implications are quite clear, there are instances in which the interpretation may vary from individual to individual. At this point the analysis becomes an art as well as a science. So it is best to think of the interpretation issue as belonging on a continuum, with statistical analysis on one end and art, tea leaf reading, astrology, and crystal ball gazing at the other extreme.

Although it has its limitations, one of the beautiful features of statistical analysis is that its principles and concepts are perfectly general, applicable to data arising from any field of study, whether from the physical, biological, or behavioral sciences, or from education, business, and government. To appreciate this generality is truly one of the most important things to be learned from the discussions that follow.

7. Concluding Comments

While we have learned that the three basic roles of statistical analysis include data reduction, inference, and the identification of relationships, we will also discover that the specific statistical techniques used to serve these roles, from the simplest to the most complex, are all based on a handful of relatively simple basic building block concepts. As a result we will come to recognize the field of statistical analysis as a unified discipline aimed at a better understanding of the world around us, with the ultimate goal that we can all lead more fulfilling lives.

Chapter 2

Objects, Variables, and Scales

1. Introduction

In the preceding chapter we learned that data is the central subject matter of statistical analysis, and that data can be defined as observations made upon our environment. Further, these observations were said to be converted into numbers in order to avoid the ambiguity of words. But numbers in and of themselves mean nothing. We must somehow relate the numbers to the worldly observations that gave rise to them. To do this, we need to develop a conceptual bridge or *construct system* to join the world of our senses with the world of numbers, to allow easy access back and forth between the two.

2. Objects

To begin to understand the component parts of this conceptual bridge, and eventually the entire bridge itself, we can ask ourselves what exactly represents the *source* of our observations? In the preceding chapter we named several specific sources; sources such as individuals, plants, animals, schools, families, tornadoes, time periods, fossils, cities, stocks, retail outlets, hospitals, etc. We found that nearly everything is a potential source of data. So, we can begin our definition of the conceptual bridge by giving these sources a general collective name.

We could call these data sources *things*, as we have been doing; or we might call them *entities*; or we could call them *objects*; even though our popular notion of concepts such as things, entities, and objects tend to exclude from their meaning such non-physical or intangible data sources as human beings, geographic locations, time periods, or events such as auctions or elections. An alternative is to call them by the more generally descriptive, though cumbersome, name of *observational units* or *analytical units*.

The point is that is does not really matter what we call these data sources, so long as we understand among ourselves what we mean by the term we choose. While unfortunately there is no universally accepted expression for data sources, *objects* seems to be a favored term among many workers in the field; but even then it is not always used as a totally comprehensive term, since it often excludes geographic locations, time periods, or events. Nevertheless, we will adopt its usage in a perfectly general sense: Its reference could be to *any* data source, whether *individuals*, *physical* or *biological things*, *geographic locations*, *time periods*, or *events*; that is, anything upon which observations can be made.

Every field of study—the sciences, education, industry, medicine, etc.—has its own set of objects of interest. In the world of business, for instance, the objects of interest are especially wide in range; whether employees or customers (individuals), products or branch offices (physical things), sales territories or store sites (geographic locations), workdays or sales periods (time periods), auctions or stock offerings (events). Analogous objects can be identified in other disciplines as well, and so they are, in more than one sense of the word, the *objects* of our analysis.

3. Variables

Moving toward a further development of the conceptual bridge between the observable world and the world of numbers, we need to probe the nature of the observations made upon the objects in our environment. The very fact that we are *measuring* objects with respect to some characteristic implies that the objects *differ* in that characteristic; or, stated another way, that the characteristic can take on a number of different values. And this is the case; objects *do* differ in their characteristics.

Indeed, an object can be thought of as nothing but a bundle of characteristics. An object does not have an existence independent of its characteristics. We see a child playing with a red ball, but upon closer inspection we discover that it is really an object with a variety of characteristics; an object spherical in *shape*, red in *color*, rubber in *composition*, having a particular *size* and *weight*. It is these characteristics of shape, color, composition, size, and weight that define the object. In each case they come to our attention because *alternative values* of the characteristic are possible: Its shape could have been elliptical, its color could have been blue, its composition could have been plastic, its size and weight could have been greater or less.

These properties or characteristics of an object that can assume *two or more* different values are referred to as *variables*. Just as objects are defined in terms of their values on variables, the variables have meaning only with respect to objects. Neither an object nor a variable can exist in a vacuum. The variable of height, for example, would have no meaning were it not for objects which have this property.

As with objects, every discipline has its own variables of interest. Education is concerned with scholastic achievement and teaching methods. Sociology studies family characteristics and crime rates, for example. In the worlds of business and economics the relevant variables include product quality, prices, sales, costs, profits, productivity, inflation, unemployment rate, etc. In each case, objects exist which *differ* on these variables.

We usually say that an object has a *value* on the variable, or that it is at a given *level* of the variable. Also, sometimes we may speak of an object having a *score* on a variable, especially if the variable represents the performance of individuals on something like a paper and pencil test.

A better understanding of the basic nature of variables can be had by linking them to appropriate sets of objects. For example, we have the *scholastic achievement* of *students*, the *body temperature* of *patients*, the *lifetime* of *cells*, the *quality* of *products*, the *price* of *stocks*, the *sales performance* of *retail outlets*, *costs* of *hospital stays*, *profits* of direct-response *advertisements*, *productivity* of *workers*, *inflation rate* of *countries*, *unemployment rate* of *cities*, and so on.

The above examples should make clear how the concepts of objects and variables are closely related to one another. We cannot think of objects without thinking of the variables on which they differ. Similarly, we cannot think of variables without thinking of an accompanying set of objects which differ on the variable. These two concepts, then, objects and variables, form two-thirds of the conceptual bridge that we have set out to construct, creating a framework for the interpretation of *empirical data*—observations made upon our environment.

4. Scales

We can now introduce the final link in the bridge between our empirical observations and the world of numbers. While we have stated that a variable is an object characteristic that can take on two or more different values, we have made no mention of the nature of these values or how they are represented. It is in this respect that we need to introduce the concept of a *scale*; a scheme for the *numerical representation* of the values of a variable.

Although it is a theoretical construct, for our purposes we can think of a scale as a set of numbers. The most familiar example is manifested in the ordinary ruler which can be used to measure—i.e., numerically represent—the variables of an object's height, width, or perhaps the distance between two objects. As we will see, it is the interpretation we place upon the numbers of the scale, rather than the numbers themselves, that make the scale useful.

Since a scale is intended to numerically represent the values of a variable, we should not be surprised to find that there are as many types of scales as there are classes of variable object characteristics found in the real world. The resulting scales are generally of four types—*nominal*, *ordinal*, *interval*, and *ratio*. These scales differ both with respect to the types of variables they represent, and the properties and interpretations of the numbers comprising the scale. We will look at each of these types of scales in turn, in order to learn more about the concept of a scale and the types of variables occurring in our environment.

Nominal scales. Many objects have characteristics that differ in *kind* only. The variables of occupation, blood type, gender, religion, and political affiliation are examples of this type of variable. All we can say is that the values or levels of each of these variables differ. We have no basis on which to arrange them in a meaningful order, or to make conclusions about quantitative differences betwen them. For example, Democrat, Independent, and Republican, as levels of the political affiliation variable, differ in kind only, and not in any quantitative sense.

Consequently, if we attach numbers to the alternative levels of this type of variable, the numbers have no meaning other than as a distinguishing label. Letters of the alphabet could serve just as well. For example, the numbers 1, 2, 3, and 4 could be used to label four different formulations of a soft drink intended for a taste test among consumers. But knowing that these numbers represent a *nominal* scale, there is no implication that formulation 1 is superior to formulation 2, nor any other type of comparative judgment. The one and only interpretation of the numbers comprising a nominal scale is that they represent different values of the given variable. Other examples of variables that could be nominally scaled include brand of toothpaste used, type of transportation (e.g., boat, train, plane, etc.), or type of therapy (e.g., drug vs. radiation), to name a few.

The limited interpretation that we place upon nominal scale values does not mean that such numerical labels have little practical significance. They are convenient for identifying everything from catalog items to taxpayers. They are especially useful for the computer tabulation of a large set of objects that have been nominally coded on such characteristics. Without numerical labels it would be a complicated computer program indeed that could discriminate between objects based on *verbal* descriptions of their characteristics.

Finally, it should be noted that variables of this sort, in which differences between values of the variable cannot be interpreted in a quantitative sense (i.e., *how much* of a difference), are instances of what we refer to as *qualitative* variables, or *categorical* variables.

Ordinal scales. If the values of a variable can be arranged in a meaningful order—i.e., rank ordered—then such a variable can be represented by the numbers on an *ordinal* scale. Examples of the application of such a scale include the quality grades for various foodstuffs, a consumer's ranked liking of alternative products, and bond grades. While the gradings of foodstuffs and bonds may be in the form of letters (e.g., A, B, C, D or AAA, AA, A) they could equivalently be represented by numbers of an ordinal scale; say, 1, 2, 3, 4, etc. Birth order among siblings is another relevant example.

The numerical values of an ordinal scale indicate a hierarchy of the levels of the variable in question. If an object A has an ordinal value above that of object B, and object B in turn has an ordinal value above that of object C, then we can conclude that object A is also above object C in ordinal value. This apparently common-sense relation of an ordered sequence is called the property of *transitivity*. Obvious as it may seem, recall that the nominal scale does not have this property. The numbers in that scale bear no relationship to one another except to designate that they are not identical values.

A limiting characteristic of the ordinal scale is that we cannot make any inference about the *degree* of difference between values on the scale. For example, the difference between the most liked soft drink formulation and the second most liked formulation is not necessarily the same as that between the second and third, or between the third and fourth. Similarly, the differences between three grades of meat are not necessarily equal. All we can infer from the ordinal scale is that one object ranks above another with respect to the given variable. Because of this limitation in interpreting differences between ordinal values in a quantitative sense, such scales, as well as nominal scales, are referred to as *non-metric* scales.

Variables represented on an ordinal scale are classified by some as qualitative variables and by others as quantitative. The qualitative label as it applies to ordinally scaled variables will be recognized as a bit misleading, since the ordering of values on a variable, irrespective of the differences between values, does *suggest* measurement in a quantitative sense. The qualitative label derives from the fact that *equal differences between ordinal values* do not necessarily nor logically *have equal quantitative meaning*. They could, but that would be fortuitous.

Interval scales. A step beyond the ordinal scale is the *interval* scale, in which equal differences between scale values have equal meaning. For example, the difference between scale values of 56 and 59 represents the same quantity as the difference between the values of 114 and 117—the difference of 3 units having the same meaning regardless of where along the scale it occurs. Because of this property it is considered a *metric* scale, and variables measured by it are considered *quantitative* variables.

A limitation of the interval scale, though, is that *ratios* of scale values have no meaning. This is due to the fact that such a scale has an arbitrary zero point, one that does not really represent a zero quantity. Consequently, the values occurring on the scale cannot be interpreted in any *absolute* sense.

These properties of the interval scale can be better understood with a few examples. Perhaps the most familiar interval scales are the Fahrenheit and Centigrade scales used to measure temperature. The difference between 80° *F* and 90° *F* represents the same quantity as the difference between 50° *F* and 60° *F*. However, it is not true that 80° *F* should be interpreted as *twice* the value of 40° *F*. This should be apparent when we convert these two Fahrenheit temperatures to their Centigrade equivalents of 26.7° *C* and 4.4° *C*, respectively. The ratio of 80° *F* to 40° *F* is 2.0; the ratio of 26.7° *C* to 4.4° *C* is 6.1. On one scale the higher temperature is *twice* the lower, while on the other scale it is *six times* as large. One or the other is incorrect—in fact, both are incorrect. The reason for the irrational results is that the zero points on both of these scales have been placed at arbitrary points, and not at the true zero point representing *zero temperature*.

The calendar of years is another example of an interval scale. The year 2000 compared to the year 1000 has meaning only with respect to the arbitrary origin of 0 based on biblical references. With another origin, say the latest ice age, the years would have different values. And, as with the variable of temperature, unless we are able to identify a genuine zero point, the values along the scale have no absolute meaning, only differences between values have significance.

Other examples of interval scales are to be found in fields dealing in human verbal behavior; fields such as education, psychology, sociology, and marketing research. For example, in a market survey aimed at determining the taste appeal of alternative coffee blends, a set of consumers might be asked to rate the products on a scale of *equal appearing intervals* consisting of the values *excellent*, *good*, *fair*, and *poor*; assigning to them the numerical values of 4, 3, 2, and 1, respectively. In such a procedure we are assuming that the difference in taste appeal between an *excellent* and *good* rating is the same as the difference between a *good* and a *fair* rating, which in turn is the same as the difference between the *fair* and *poor* rating. How tenable is this?

In truth, it is virtually impossible to construct such verbal scales with any degree of confidence that we are dealing with equal intervals. The same problem arises in the field of psychological testing in the scaling of intelligence and personality traits. While we know that an intelligence test score of 100 does not represent twice a score of 50, since it is presumably intervally scaled, but we are not even confident that the difference between a score of 90 and 100 is the same as the difference between 100 and 110. It is more realistic when trying to scale human dimensions to recognize that our scales are *approximately* of equal intervals—somewhere between an ordinal scale and a true interval scale—and we might distinguish them as *quasi*-interval scales.

Ratio scales. The fourth type of scale, the *ratio* scale, packs the most information of all. In addition to having the transitive property of the ordinal and interval scales, the ratio scale has the added property that ratios of its

values *do* have meaning, and *equal ratios* have *equal meaning* due to the presence of a genuine zero point on the scale. The ratio of the value 36 to the value of 18 has the same meaning as the ratio of 100 to 50, or 900 to 450, or any other pair of values with a ratio equal to 2. As with the interval scale, the ratio scale is considered a metric scale, and variables that it measures are considered *quantitative* variables.

Examples of ratio scales abound. A scale of inches can be used to measure the variable of length. A scale of seconds can be used to measure the variable of time duration. The monetary scale of dollars can be used to measure the variable of sales performance. The variable of land area can be measured on a scale of acres. All of these scales have one thing in common—a genuine zero point.

It should be noted that a given variable can often be measured on alternative scales. For example, the variable of width can be measured on a scale of inches, centimeters, miles, or any of a number of other scales associated with spatial displacement. The variable of sales performance can be measured on a scale of dollars, marks, yen, or any of a number of other monetary scales.

In the above examples, the alternative scales upon which the variable is measured are simple transformations of one another, much like foreign languages each having a different set of words to denote the same concept. In such situations, where alternative scales can measure the same variable, we need to distinguish one from the other by identifying its *units*. For example, a sales performance of 5,000 means nothing unless we know whether it is 5,000 dollars, marks, yen, or what. Our choice of scale will usually be a question of practicality and convenience. Office floor space would be measured in square feet or square meters, while large parcels of land would be measured in acres or square miles.

A very common ratio scale that does not have units as such, is the scale of *frequency count*. It measures a variable by telling us "how many" rather than "how much". For example, sales performance could be measured in terms of *number of transactions*, rather than in terms of the total dollar value of the transactions. The variable of school absenteeism could be measured in terms of *number* of absentees. Abnormal behavior could be measured in terms of *number* of occurrences of a neurotic pattern. Stock market performance could be measured in terms of *number* of stocks increasing in price. Such a scale, based on a frequency count, has universal meaning and therefore does not require units, other than identifying the numbers as comprising a count or frequency. A value of 25 on a frequency count scale is not ambiguous, as might be a value of 25 on a monetary or distance scale, which require units to further clarify their meaning.

Another ratio scale, which like the frequency count scale can be considered *pure* in the sense that it requires no units, is the *percentage* scale. Again using

the sales performance example, we could not only measure the sales of a given brand in various markets in terms of dollar quantities, or in terms of number of pieces sold, but also in terms of the percentage of the total sales of its product category. Similarly, a laboratory rat's learning performance in a maze could be measured in percentage of correct turns. The units, if any, on such a scale would be percentage points.

From the above discussion it should be clear that a variable is an abstract construct and that it does not take on full meaning until we identify a scale upon which it can be measured. This is especially apparent when we realize that the same variable can be measured on any number of different scales. The importance of distinguishing the various types of scales—nominal, ordinal, interval, and ratio—lies in the validity of the types of interpretations we can place upon the numerical values comprising the scales. Also, the nature of these scales place limitations on the legitimate types of mathematical operations that can be performed on the values of the scales. We will learn more of the respective limitations of these scales as we introduce the various analytical techniques. For now, the most important thing to remember is that numbers arising from different scales cannot be used indiscriminately in data analyses as if it did not matter on what type of scale the numbers originated. Whereas it may be true that a rose is a rose is a rose, it is not true that a number is a number is a number. Figure 1 summarizes the four basic scales we have discussed.

We have now completed our conceptual bridge—a construct system consisting of objects, variables, and scales—that will allow us easy access back

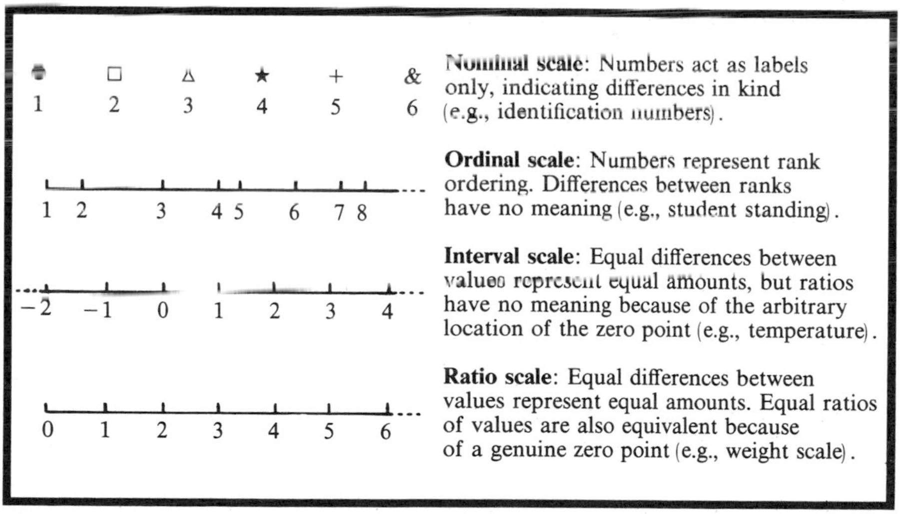

Figure 1 Four basic types of measurement scales.

and forth between the observable world and the world of numbers. Before we begin putting the system to use, we will look at some common types of variables that we will encounter throughout our later discussions.

5. Major Classes of Variables

Earlier we defined a variable as an object characteristic that can take on two or more different values. Within this definition there is room for all types of variables which differ in their special features. For purposes of efficient communication it is useful to give names to these different categories of variables.

Qualitative vs. quantitative. We have already encountered the distinction between qualitative and quantitative variables. More precisely, the distinction is between qualitatively and quantitatively *scaled* variables, with those measured on the non metric nominal (and sometimes ordinal) scales being referred to as *qualitative*, and those measured on the metric interval or ratio scales being referred to as *quantitative*. Student academic achievement measured in terms of grade point average is quantitative in nature, whereas blood type is qualitative in nature since it is based on a nominal scale. Income is quantitative, while religious affiliation is qualitative.

Discrete vs. continuous. Another basis on which we can classify a variable is in terms of the possible values it can assume. An important distinction is between discrete and continuous variables. A *discrete* variable only takes on a finite number of values. Variables measured on the frequency count scale are examples of this type of variable: number of errors on an exam, number of crimes in a city, number of hospital admissions, number of product defects, etc. Although discrete variables will usually take on only integral values (i.e., 1, 2, 3,...) this is not always the case. For example, closing prices of stocks occur only in 1/8 of a dollar increments, rather than assuming intermediate values as well.

In contrast to the discrete variable, the *continuous* variable can potentially take on any numerical value. For any two values that it may assume, there is another value between them that is possible. Examples of continuous variables include height, longevity, grade point average, interest rates, etc. Name any two values of each of these variables and another value between them is possible. For example, interest rates of 8.0% and 8.5% are possible, as well as values of 8.25% or 8.173%, or any of an infinite number of intermediate values. This is not the case with discrete variables.

We often refer to a continuous variable simply as a *continuum*; e.g., the creativity continuum, the intelligence continuum, the quality continuum, the productivity continuum, etc. Also, whether a variable is continuous *or* discrete, so long as it is not nominally scaled, we often refer to it as a *dimension*; e.g.,

the price dimension, the leadership dimension, the taste dimension, etc. While there are slight differences in the connotations of continuum and dimension we will use them more or less interchangeably with the term variable. *Variate* and *factor* will also serve as substitute words for variable.

Even though a variable is scaled as discrete, it is possible that the underlying variable, in a theoretical sense, is continuous in nature. For example, if we measure the sales performance of a group of sales reps in terms of the number of accounts opened—a discrete variable—it is reasonable to believe that the true underlying variable that we are tapping—sales performance—is continuous in nature. Although a fractional account cannot be opened, in theory the sales effort has a degree of success even though it does not end in a sale. So in this sense we should feel comfortable talking about a sales performance continuum even though it is measured as a discrete variable. Similarly, scores on an achievement test are discrete in nature, but the underlying variable being measured is likely to be continuous. In other words, we can distinguish between the *underlying* and the *measured* variable.

Dichotomous variables. In addition to classifying variables in terms of the nature of their values, we can classify them on the basis of *how many* values they can assume. By definition, the minimum number of values that an object characteristic can assume in order to qualify as a variable is *two*. Now a two-valued variable may seem at first glance to be a trivial case, but in fact such variables are both common and of great significance. We refer to these two-valued variables as *dichotomous* variables.

Perhaps the most common dichotomous variable is that of gender, taking on values of male or female. Other general examples include performance on a qualifying test (pass or fail), sales pitch outcome (sale or no sale), product category usage (user or nonuser), and a belief (true or false), to name a few. In the case of dichotomous variables, as with other discrete variables, it is always worthwhile to question whether the underlying variable is continuous in nature or is truly dichotomous. Whereas most would agree that gender is dichotomous, an attitude (agree or disagree) may actually exist as a continuum.

Dichotomous variables are also often referred to as *binary* variables. However, through usage, the expression binary variable has taken on a shade of meaning somewhat more specific than a dichotomous variable in general. Its use suggests a dichotomous variable whose respective values have been assigned the numerical values of 0 and 1. For example, whereas home ownership is a dichotomous variable, once we assign owners a numerical value of 1 and nonowners a value of 0, usually as a preliminary to a quantitative analysis, then we think of it as a binary variable.

Yet another type of dichotomous variable is the *dummy* variable. It is created by converting a given level of a qualitative variable into a binary variable; i.e., the presence (call it 1) or absence (call it 0) of the characteristic.

Consider the familiar qualitative variable of political affiliation, having the levels of Democrat, Independent, and Republican. As we learned in our discussion of nominal scales, attaching numbers to the levels of this type of variable will have little value except as labels. But what we can do is create separate binary variables from the respective levels of the qualitative variable. We refer to these makeshift variables as dummy variables. In our example we have:

First dummy variable: *Democrat*(1) vs. *Non-Democrat*(0)
Second dummy variable: *Independent*(1) vs. *Non-Independent*(0)
Third dummy variable: *Republican*(1) vs. *Non-Republican*(0)

What we have done is convert *one qualitative variable that had three levels, into three separate binary variables with two values each*. Whereas we started with the variable of "political affiliation," we now have the three variables which we might call "Democratness," "Independentness," and "Republicanness." In other words, if someone has a score of 1 on the "Democratness" variable, we can logically interpret this to mean that they have more of this quality than someone who scores 0 on the variable. So too with the "Independentness" and "Republicanness" variables.

The creation of these dummy variables may seem like an idle exercise, but we will see that this approach reaps benefits in allowing us to subject qualitative variables to *quantitative* analyses. We will learn more of their application in later chapters.

Finally, it should be noted that *any* variable can be made into a dichotomous variable simply by dividing its range of values into two categories. For example, the consumer variable of tea consumption measured on a scale of cups consumed per week, could be split down the middle at some convenient intermediate value, and values below the cutoff point would be called *light usage*, and above the cutoff point *heavy usage*. This dichotomizing of the variable would be useful for a rough assessment of whether tea consumption (heavy vs. light usage) was related to variables such as age, gender, education, income, blood pressure, longevity, etc.

There are several other types of variables with special qualities that are worth singling out and identifying with names. However, we will reserve their discussion until that point when their introduction is more appropriate. The classifications of variables introduced here—qualitative vs. quantitative, discrete vs. continuous, and dichotomous variables—are very basic distinctions and will occur repeatedly throughout our subsequent discussions. For this reason it is efficient to communicate their special properties with a single word, rather than spelling them out each time they occur. We are, after all, concerned not only with data reduction, but with word reduction as well.

6. Constants

Since an object characteristic needs to assume only a minimum of two values in order to qualify as a variable, we can seriously wonder if there is any such thing as an object characteristic that is *not* a variable.

In fact, there are properties of objects that do assume *one and only one value*, and we refer to these characteristics as *constants*. They range from the ordinary (the constant velocity of light regardless of its source) to the sublime (the constant ratio of circumference to diameter in all circles, $\pi = 3.1416...$).

When we try to think of examples of constant properties of objects, we arrive at such basic features that we hardly realize that they are the fundamental defining characteristics of the objects: e.g., the number of sides to a cube, the number of hours in a day, the taste of salt, or the color of gold. Constants, then, are the invariable characteristics of objects which differentiate one class of objects from another. Therefore, it is a combination of variables and constants that precisely define any object in our environment. If an object characteristic is not a constant, it is a variable; if it is not a variable, it is a constant.

7. Notation

It is often convenient to use letters as shorthand notation for variables. For example, we can speak of variable x, which can refer to any particular variable we desire, be it age, income, sales, test score, crime rate, etc. By convention, the letters most frequently used for designating a variable come from the latter part of the alphabet; x, y, and z being the ones most commonly used, though u, v, and w are also favored. They are also used in their upper case form; U, V, W, X, Y, Z, depending on one's preference. Other letters may also be used, but generally speaking the choice will be from among r through z.

The observed *values* of a given variable can be designated with the same letter as the variable itself, but we must use subscripts to distinguish between the values obtained from the various *objects* under study. For example, the variable x would have the values x_1, x_2, x_3, and so on: x_1 being the value of Object 1 on the variable, x_2 being the value of Object 2 on the variable, x_3 being the value of Object 3 on the variable, and so on until we reach the final or nth Object in the collection, whose value on the variable would be designated as x_n; where n, the number of observations in the set, would have to be specified for any given set of data. We read x_n, "x sub n."

In general, then, we can specify a value of a variable as x_i, where i can be any particular integer—e.g., $1, 2, 3, \ldots, n$. This is one way of distinguishing between a variable x, and a value of that variable x_i. Table 1 presents the

Table 1 Format of a one-variable data collection.

(a) Specific example		(b) General format	
Individuals	Test scores	Objects	Values on variable x
Anderson, B.	85	O_1	x_1
Andrews, T.	91	O_2	x_2
Barclay, S.	79	O_3	x_3
⋮	⋮	⋮	⋮
Wheeler, E.	86	O_n	x_n

typical data collection format and the accompanying notation for a set of objects measured on a single variable. Notice that we also use subscripted letters ($O_1, O_2, O_3, \ldots$) as abbreviations for the various objects. These subscripts will be recognized as an application of the nominal scale; i.e., numbers used strictly for identification purposes.

Multivariate data collections. In multivariate analysis we are interested in a set of objects which are measured on a number of different variables rather than on just one. For example, a set of employees could be measured with respect to the variables of age, years of experience, amount of education, score on an aptitude test, current salary, etc. The basic format for this type of *multivariate* or many-variable data collection is shown in Table 2, where we have a *matrix*, or tabular arrangement of the data, in which the column headings represent the various *variables*, and the row headings represent the various *objects*. The entries within the body of the matrix represent the *values* of the objects on the various variables.

In such instances, when we are studying more than one variable at a time, we have the option of using different letters to designate the different variables —say u, v, w, x, y, z—but if perchance we have a large number of variables under study, more than the number of letters in the alphabet, which will often be the case, we will have to resort to identifying the variables by numbers. This is most often done by selecting a particular letter and giving it numerical subscripts. For example, in Table 2 we have identified the variables as $x_1, x_2, x_3, \ldots, x_k$, where the value of k, the total number of variables under study, would be designated for a specific set of data. However, we must know the context in which this notation appears, in order to know that x_1 and x_2 are two different variables, as opposed to two different values of a single variable.

Since we have used subscripts to differentiate between the various variables, we must resort to *double subscripts* if we want to designate the values of the variable obtained from each object. For example, x_{32} will represent the value of Object 3 on variable x_2, while x_{95} will represent the value of Object 9 on variable x_5.

Table 2 Format of a multivariate data matrix.

(a) Specific example						(b) General format					
	Variables						Variables				
Individuals	Exam score	Grade point average	Age	...	Family income	Objects	x_1	x_2	x_3	...	x_k
Anderson, B.	85	2.9	21	...	$44,500	O_1	x_{11}	x_{12}	x_{13}	...	x_{1k}
Andrews, T.	91	3.2	19	...	$23,000	O_2	x_{21}	x_{22}	x_{23}	...	x_{2k}
Barclay, S.	79	1.6	20	...	$16,750	O_3	x_{31}	x_{32}	x_{33}	...	x_{3k}
⋮	⋮	⋮	⋮		⋮	⋮	⋮	⋮	⋮		⋮
Wheeler, E.	86	2.7	17	...	$35,000	O_n	x_{n1}	x_{n2}	x_{n3}	...	x_{nk}

In general, we can designate an observation in a multivariate data collection as x_{ij}, where the first subscript identifies the object, and the second subscript identifies the variable being measured, although the order is admittedly arbitrary.

It will be worthwhile to think of the one-variable data collection scheme, as shown in Table 1, as being a special case of the more general multivariate data matrix shown in Table 2.

Constants. While the letters in the latter part of the alphabet are used to designate variables, the letters in the front part are typically used to represent constant quantities. The one which will be encountered most frequently will be n, which will be used primarily to represent the number of observations in a set of data. The letter k or m, on the other hand, is often used to indicate the number of variables involved in a particular analysis. But before we use any letter abbreviations we will be sure to explain what they stand for. Again, their only purpose is for convenience and economy of description.

8. Concluding Comments

Now that we have an overview of the objectives of statistical analysis, and have established a construct system for understanding how our observations of the world can be meaningfully translated into numbers, we can proceed with our study of the principles and techniques available for analyzing these observations. We will discover that all that is to come, without exception, will pivot on the concepts introduced in this chapter—*objects*, *variables*, and *scales*. In fact, whenever we are confronted with a set of data, we should ask ourselves three questions. What *objects* are being measured? On which *variable(s)* are the objects being measured? On what *scale* is the variable being represented?

Part II.
DATA REDUCTION

Chapter 3

Frequency Distributions

1. Introduction

The most fundamental type of data collection problem that lends itself to statistical analysis involves the measurements of each of a set of objects on a single variable; for example, achievement test scores of a set of students. While there are many different types of analysis that such data could be subjected to, in this chapter we will introduce the most basic type of statistical analysis, the *frequency distribution*; a simple tally or count of how frequently each value of the variable occurs among the set of measured objects.

2. Creation of a Frequency Distribution

To better understand the procedure and rationale for creating a frequency distribution, we can consider a typical example of its application. Table 1 presents the test performance of 36 students. The data could just as well represent the number of accounts opened by a set of sales reps, the number of arrests among a set of criminals, the number of absences among a set of employees, the number of operations performed among a set of surgeons, the heights of a set of children; or, if the objects were other than people, the maze-running errors of a set of rats, the crop yields of various plots of land, the prices of stocks, the recall levels for different TV commercials, etc.

While an inspection of the raw data presented in Table 1 provides information about the individual performance of the students, it is difficult to get a concise picture of their overall collective performance. The data would be more comprehensible if it could be summarized into a more compact and interpretable form. This is where the frequency distribution comes into play. In examining the data we notice that many values of the variable occur more than once, so it is reasonable that we can summarize the data by making a *tally* or count of how frequently each value occurs. In constructing such a frequency

Table 1 Test scores of 36 individuals.

Individuals	Test scores	Individuals	Test scores
Anderson, B.	12	Kornfield, L.	11
Andrews, T.	9	Lee, R.	10
Barclay, S.	8	Logan, B.	14
Bishop, C.	10	Marsh, N.	8
Brody, R.	15	Melrose, G.	10
Carlton, M.	11	Moran, C.	9
Clark, D.	7	Noble, V.	10
Cox, S.	14	Parker, L.	12
Dewey, D.	10	Potter, D.	13
Edelman, P.	13	Rhodes, F.	8
Farrell, J.	11	Rubin, B.	10
Frank, R.	7	Schultz, R.	11
Gibbs, J.	9	Silver, W.	9
Gray, W.	11	Stack, E.	13
Harmon, G.	12	Thomas, J.	12
Hodge, N.	14	Vargas, R.	11
Irving, T.	6	Weiss, C.	10
Kent, N.	16	Wheeler, E.	9

distribution, the most logical arrangement of the values of the variable is either from lowest to highest, or from highest to lowest, whichever we prefer.

In the present example, the test scores range from a low of 6 to a high of 16, and are listed in the first column of Table 2. In the second column of Table 2 is a tally of how frequently each value of the variable occurs. It was obtained by working down the list of 36 values appearing in Table 1, and making a tally mark for each one beside its corresponding value in Table 2. Finally, in column (3), the tally counts are shown in their numerical form. Columns (1) and (3), then, comprise a frequency distribution table. The column (2) tallies are typically dropped from the table since they represent an intermediate work step, although they do serve as a rough graphical portrayal of the distribution.

A comparison of Table 1 with columns (1) and (3) of Table 2 reveals the data reduction function of a frequency distribution. We find that our original set of data has been condensed and summarized into a more compact and interpretable form. At a glance we know the range of scores as well as where they cluster most heavily. These features of the data were not readily apparent from a mere inspection of Table 1. With larger collections of data, the benefits of the frequency distribution are even more apparent.

Relative frequencies. It is often helpful to convert the frequencies of a frequency distribution into *relative frequencies*, which are nothing more than

Table 2 Frequency distributions of 36 test scores.

(1) Test scores	(2) Tally of individuals	(3) Frequencies	(4) Relative frequencies	(5) Cumulative frequencies	(6) Cumulative relative frequencies
6	|	1	2.8%	1	2.8%
7	||	2	5.6	3	8.3
8	|||	3	8.3	6	16.7
9	𝍸	5	13.9	11	30.6
10	𝍸 ||	7	19.4	18	50.0
11	𝍸 |	6	16.7	24	66.7
12	||||	4	11.1	28	77.8
13	|||	3	8.3	31	86.1
14	|||	3	8.3	34	94.4
15	|	1	2.8	35	97.2
16	|	1	2.8	36	100.0
		36	100.0%		

the observed frequencies converted into percentages based on the total number of observations.

Such relative frequencies appear in column (4) of Table 2. They were obtained simply by dividing each frequency in column (3) by the total number of observations, 36. The relative frequencies tell us at a glance what percentage of the 36 students had a score of a given value. This information is helpful since it is often easier to interpret a percentage figure than a raw frequency figure. For example, it is somewhat more informative to be told that 19.4% of the students had a score of 10 on the test, than to say that 7 of the 36 students had such a score.

Cumulative frequencies. There are many times when not only do we want to know how many observations in a data collection are of a particular value, but also how many are *above* or *below* a given value. For instance, with respect to the student test scores we might want to know how many students scored, say, 10 or lower. To answer such questions we can create a *cumulative frequency distribution*, a distribution that indicates how many of the observations in the data collection occur *up to and including each particular value*.

Such cumulative frequencies are shown in column (5) of Table 2. They will be seen to represent a running total or "accumulation" of the frequencies appearing in column (3). At a glance, then, we see that 18 of the 36 students had a score of 10 or lower; or that 34 had a score of 14 or lower. Necessarily, the highest score in the distribution, in this case 16, must be associated with a cumulative frequency corresponding to the total number of observations, in this instance 36.

Cumulative relative frequencies. Once we have formed a cumulative frequency distribution, it is a simple matter to convert it into a *cumulative relative frequency distribution*. All we need to do is divide each cumulative frequency by the total frequency of observations. An alternative approach is to accumulate the relative frequencies. Except for differences due to rounding the two methods will yield the same results.

The cumulative relative frequency distribution tells us what percentage of the total observations in our data collection are of a particular value or lower. Column (6) of Table 2 shows the cumulative relative frequencies for the student test score example. We can see at a glance that 66.7% of the students scored 11 or lower, or that 94.4% scored 14 or lower. If we wanted to know what percentage scored above a given value, we merely have to subtract the cumulative relative frequency from 100%. For example, it is easily verifiable that 83.3% of the students scored above 8.

3. Graphical Presentation

It is usually very informative to make a graphical representation of a frequency distribution. Such pictorial presentations can show aspects of the distribution not readily apparent from a tabular presentation. While tally marks, such as those in Table 1, give us a good picture of the frequency distribution, we can create a more formal and aesthetic presentation.

The two most common methods of graphically portraying a frequency distribution are shown in Figure 1. Part *a* of the figure shows a *histogram*, or bar chart, in which the horizontal axis shows the values of the variable in question, and the heights of the bars above each value represent their respective frequencies of occurrence. The data is taken from Table 2, the distribution of student test scores. Notice that Figure 1*a* is simply an alternative representation of the tally marks in Table 2, where the values of the variable are arranged horizontally rather than vertically.

Notice in Figure 1*a* that we have chosen to portray both the raw frequency distribution and the relative frequency distribution with the same histogram. This was accomplished by using the left vertical axis to represent the raw frequencies and the right vertical axis for the relative frequencies; the conversion of the observed frequencies to percentages having no effect on the shape of the distribution.

Figure 1*b* shows an alternative method of graphically portraying a frequency distribution. It is referred to as a *frequency polygon*, and is constructed by connecting the points which have heights corresponding to the

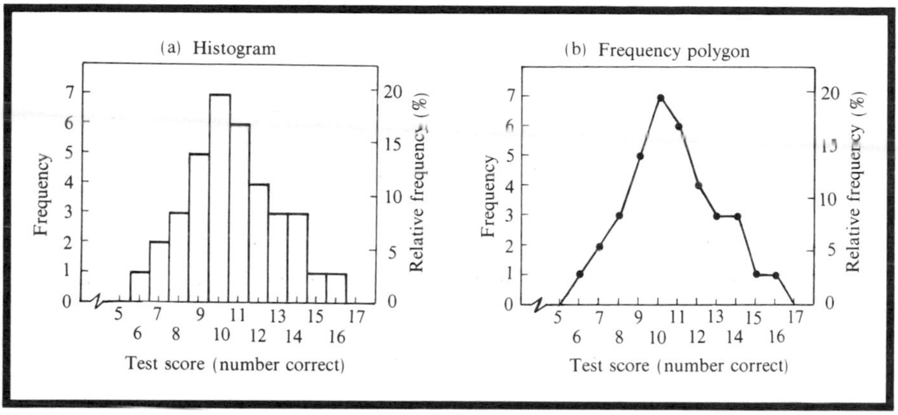

Figure 1 Alternative methods for portraying the frequency distribution of the test scores of 36 individuals as given in Table 2.

frequencies associated with each value along the horizontal axis. Alternatively, it can be thought of as the connection of the midpoints of the tops of the bars forming the histogram in part *a* of the figure. Again, the left vertical axis has been used to scale the observed frequencies, while the right axis shows the relative frequencies.

4. Grouping of Data

In the student test scores example we were fortunate that there were only eleven different values of the variable that occurred, ranging from 6 to 16 in integral values. At other times, we must contend with variables which assume a large number of values. For example, if instead of test scores we measured the students with respect to the variable of weight, we might expect to get values ranging from under 100 to over 200, and if the weight was measured to the nearest tenth of a pound, the number of possible values would be very large. In such situations, it is unlikely that any given value will occur with a frequency greater than one. Consequently, little is to be accomplished by making a frequency distribution of each value. The solution to such a data reduction problem is to create *intervals of values* of the variable in question, and then make a frequency tally of the number of observations falling within each interval.

To demonstrate the grouping procedure we can reanalyze the student test data using score intervals two units wide. The resulting frequency distribution would be as follows:

Score Interval	*Frequency*
5–6	1
7–8	5
9–10	12
11–12	10
13–14	6
15–16	2

We see that such grouping of the data results in an even greater summary of the data, though at the same time sacrificing some detail of the original distribution.

As can be seen from the above example, the key issue is to decide into *how many intervals* the data range should be divided. This in turn will dictate the size of the intervals. Although there are no hard and fast rules as to how many intervals to use, anywhere from ten to twenty intervals, depending upon the nature and extensiveness of the data, will usually result in a satisfactory representation of the data's distribution. To use more than twenty intervals would be defeating the data reduction objective, while using fewer than ten intervals might obscure important features of the distribution.

Another important consideration in creating the intervals is to specify the *interval boundary values* in such a way that there is no ambiguity as to where a particular observation falls. For example, if weights of individuals are recorded to the nearest tenth of a pound, the interval boundaries should be specified to the hundredth of a pound.

A final consideration in developing an interval scheme is to decide where to begin the lowest valued interval; i.e., the one containing the lowest values in the distribution. In the above example, the 5–6 interval was chosen over a 6–7 interval, based on a flip of a coin. This type of random selection of the exact location of the bottom interval is to avoid the possibility that personal judgment affects the shape of the resulting distribution.

In creating a frequency distribution based on grouped data we have lost the identity of the individual values falling within a given interval, but this is a small price to pay for the resultant summary of the data that has been achieved. We will discover this trade-off feature for virtually every statistical technique aimed at data reduction. We lose detail, but we gain an overall look at the data. We may not see the individual trees, but we do see the forest; and that, in most cases, is better than not being able to see the forest because of the trees.

5. Theoretical Frequency Distributions

Until now we have considered only empirical frequency distributions which are based on a finite number of observations. In most analytical situations, however, these distributions serve as estimates of theoretical frequency distributions which are based on an *infinite* number of observations.

To understand the basic nature of such theoretical distributions, consider the successive stages of the frequency distributions shown in Figure 2. Beginning with relatively wide intervals, we make them progressively smaller and smaller, while at the same time increasing the number of measured objects so that each interval still has many observations falling in it. We continue diminishing the size of our intervals until the tiny bars of the histogram blend into a continuous distribution as shown in part *d* of Figure 2. It can be imagined as a frequency polygon connecting an infinite number of very narrow bars of a histogram, bars which approach zero width.

Notice in Figure 2 that once the histogram blends into the continuous distribution based on an infinite number of observations, we no longer label

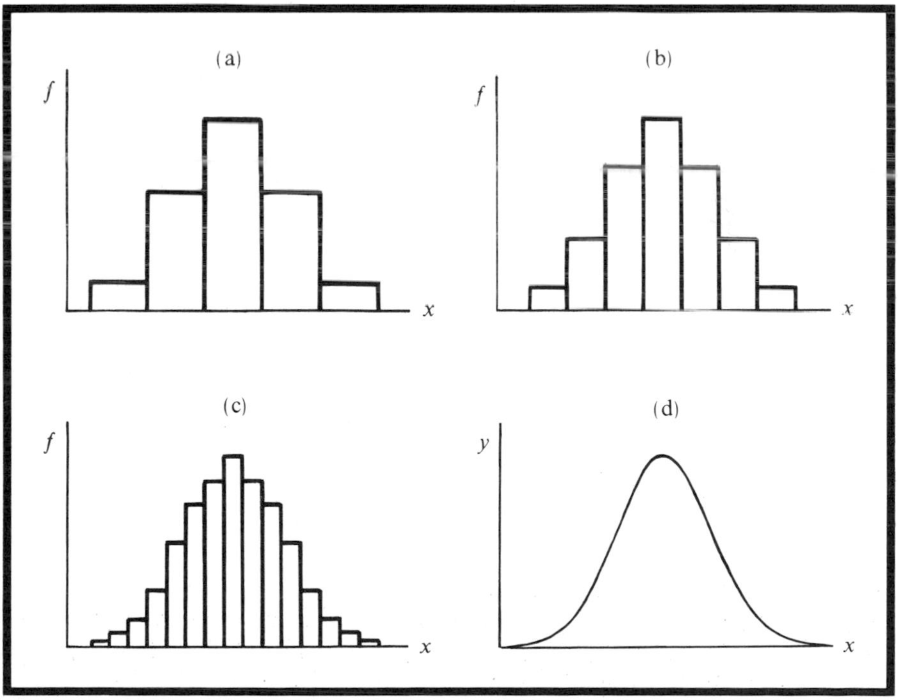

Figure 2 Progression of a histogram to a theoretical continuous distribution.

the vertical axis as a frequency measure. Rather, we identify it simply as a variable y, which now represents the height of the curve and is a function of the value of the variable along the horizontal axis, which we designate simply as variable x, meaning it can be any particular variable.

The height of the continuous distribution shown in Figure 2*d* can no longer signify the percent of observations having a specific value, for if there are an infinite number of values along the horizontal axis, and each of them had a non-zero percentage of occurrence, the sum of those percentages would necessarily be infinite in size. Yet we know the sum of all observations in the distribution must equal 100%. We arrive at the somewhat paradoxical conclusion that we cannot ascertain how often any one value occurs, since there are an infinite number of such values, but yet we know their relative frequency of occurrence must sum to 100%. Can an infinite number of %'s sum to 100%?

We can reconcile this dilemma by studying two pieces of information. For one, let us say that to define a particular value on a continuous variable, we must necessarily create an interval of values. For example, a value of 21 really represents the interval of values from 20.50 to 21.49. A value such as 21.7 actually represents an interval of values from 21.650 to 21.749. Similarly, a value of 21.73 actually represents an interval from 21.7250 to 21.7349. No matter how precisely we try to measure a particular value, it will still be nothing more than the midpoint of an interval of values.

The second piece of information needed to resolve the original dilemma is that the area under the continuous curve, which totals 100%, will now represent the total of our observations.

It is easy to see, then, how we can determine the relative frequency of occurrence of a given value—or actually of a given interval of values—in a distribution based on an infinite number of observations. All we need to do is measure *the area under the curve between any two values* of interest. An example is shown in Figure 3. By measuring what percent of the area under the curve occurs between the two indicated values, we will know the relative frequency of occurrence of that range of scores. And we can make the interval as small or as large as we like. The method for determining the area under the curve between a given pair of values will be presented in detail in Chapter 5.

If the above discussion has seemed a bit abstract, and if it is a little difficult to immediately understand how we can move from a real-world histogram based on a fixed number of observations, to a theoretical distribution based on an infinite number of observations, it should not be the slightest cause of concern, for we have just touched upon the basic concepts of that branch of mathematics known as *Calculus*, the bane of many a student. But for our purposes we need not be concerned about the specific mathematics of the approach. It will be enough for us to understand the *concept* of a theoretical distribution, and be able to imagine how a histogram with progressively

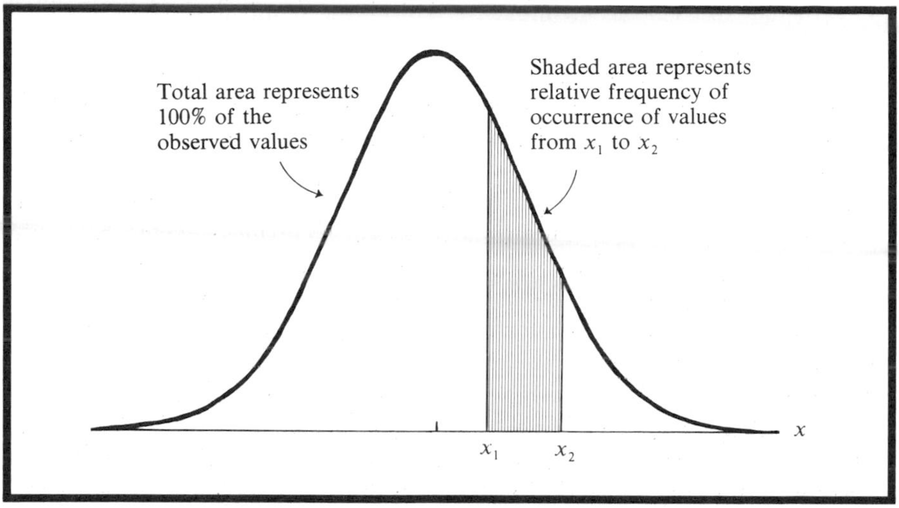

Figure 3 The relative frequency of occurrence of an interval of values.

smaller intervals, and based on an increasingly larger number of observations, eventually evolves into a continuous distribution based on an infinite number of observations. Later we will learn more about such theoretical distributions, and before we are through they will be second nature to us.

6. Shapes of Distributions

To this point we have studied distributions that are more or less alike in shape; namely, with the majority of the observations clustered at the intermediate values of the variable, with progressively fewer observations at the extremes of the distribution. While this type of "bell-shaped" distribution is one of the most common forms that data assumes in the real world, it is not the only type of distribution that we may encounter.

Figure 4 presents five common shapes of data distributions. In each case the distribution is shown in its finite histogram form and in the theoretical continuous form based on an infinite number of observations.

Figure 4*a* shows what is referred to as a *rectangular* or *uniform* distribution, in which each value of the variable occurs equally often. Interestingly, this type of distribution is not as common as one might think. More often data assumes an unequal distribution of occurrence; i.e., with certain values tending to occur more frequently than others. However, one important example of a rectangular distribution is the distribution of ages of the citizens in well-developed nations in which the birth and death rates have stabilized and are more or less equal to each other. In such a nation, the number of five-year-olds is approximately the same as the number of ten-year-olds or the number of twenty-year-olds or the number of any given age; until, or course, we reach the upper age brackets where death due to aging begins to take its toll. But to that point the distribution of ages is more or less rectangular in form.

The *U-shaped* distribution shown in Figure 4*b* reveals a polarization of observed values for a given variable; either they tend to be very high or very low, with relatively few intermediate values occurring among the set of measured objects. An example of this type of data configuration is the distribution of consumer purchase-interest ratings for a particular product. Most people are either very favorably disposed to it, or they are turned off by it, with relatively few having an intermediate degree of interest in the product.

The distribution shown in Figure 4*c*, looking much like half of a *U*-shaped distribution, is often referred to as a *J-shaped* distribution. The one shown actually looks like a backward *J*, but it could also occur in the reverse direction with most of the observations piled up at the right instead of the left. The distribution shown could be characteristic of a variable such as the number of defects found in various batches of a quality controlled product. Most batches would have zero defects, fewer would have one defect, fewer yet would have two defects, even fewer three defects, and so on until the frequency of batches with numerous defects approaches zero. Another example of this type of reverse *J*-shaped distribution is the age distribution in a developing country in which the number of births increases each year and the expected longevity is short, resulting in more one-year-olds than two-year-olds, and more two-year-olds than three-year-olds, etc.

The *bell-shaped* distribution shown in Figure 4*d* is perhaps the most common of all distributions. The examples presented earlier in the chapter were of this type, situations in which observed values of the variable become increasingly more frequent at the intermediate values. Different bell-shaped distributions may vary in their specific profile, but there is one such distribution that has a specific shape, with a precise mathematical definition, that will be of special interest to us, for it is so pervasive in nature, and could well be said to be the most important distribution in the field of statistical analysis, and it is called the *normal distribution*. It is that distribution that we have already seen three times—in Figures 2*d*, 3, and 4*d*—and before the end of our study we should expect to see it or refer to it nearly a hundred times more, and its shape and characteristics will be as familiar to us as the multiplication tables are to a grade schooler, so basic is it to statistical analysis.

Distributions that are not symmetrical in form, those that tail off either to the right or the left, such as the *J*-shaped distribution, are referred to as *skewed* distributions. The distribution shown in Figure 4*e* is an example of a distribution skewed to the *right*, the direction of the skew being the direction of the

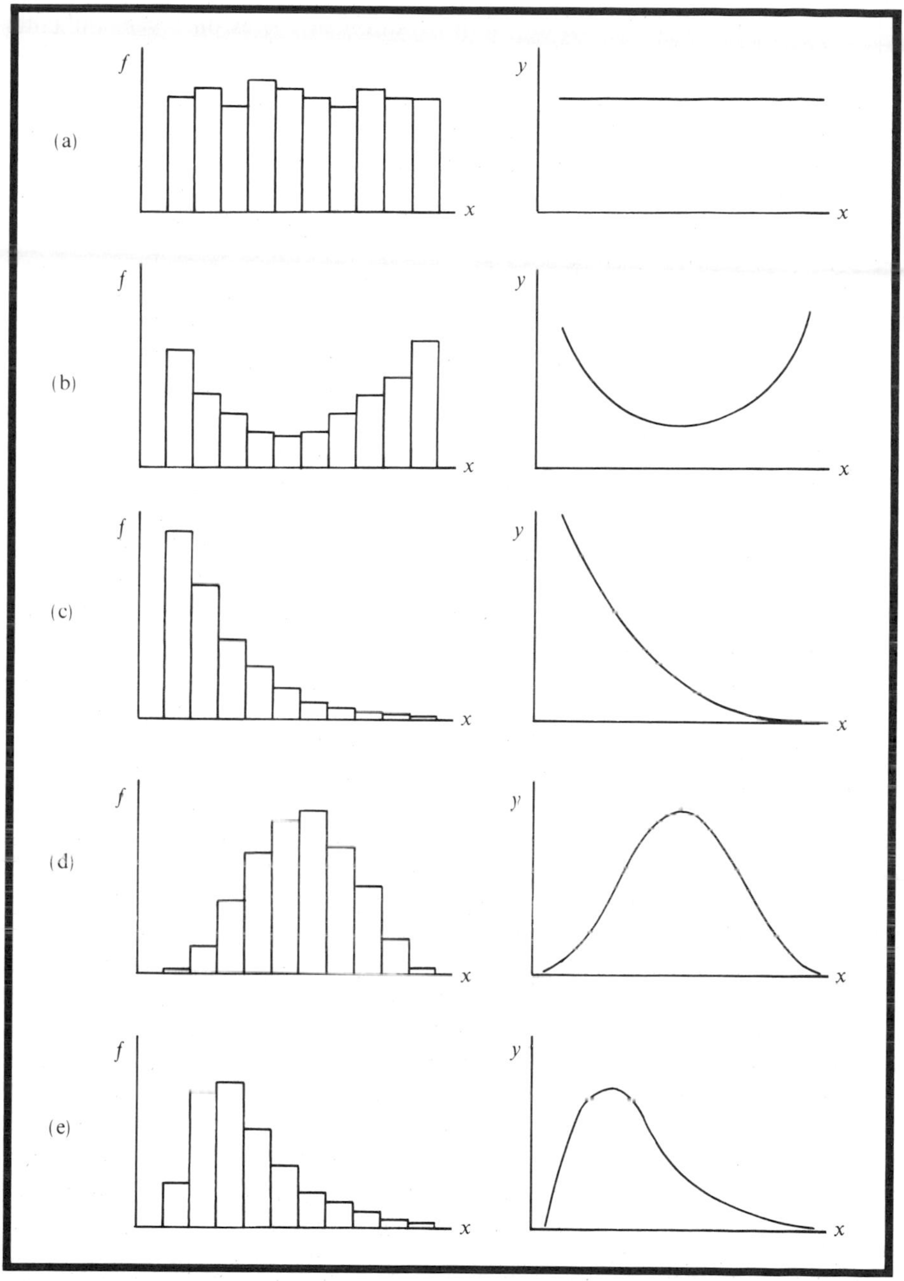

Figure 4 Common shapes of frequency distributions. Histograms and continuous distributions shown side by side: (a) uniform, (b) *U*-shaped, (c) reverse *J*-shaped, (d) bell-shaped, and (e) skewed right.

distribution's tail. It is typical of the distribution of scores on a very difficult exam, or the incomes of a given population. In each case, low values of the variable predominate, with relatively few instances of the higher values. A mirror image of the distribution shown in Figure 4*e*, one that was skewed left, would be typical of the distribution of scores on an easy examination, in which most of the scores pile up at the upper end of the scale. These types of distributions can be viewed as falling between the symmetrical bell-shaped type and the extremely skewed *J*-shaped type.

The shapes introduced above—uniform, *U*-shaped, *J*-shaped, bell-shaped, normal, skewed right, skewed left—are useful for summarizing an entire body of data with a word or two, and extend the data reduction function of creating a tabular frequency distribution.

7. Pie Charts

When we deal with qualitative variables such as political affiliation, religion, race, occupation, etc., frequency distributions in the form of histograms or frequency polygons are not quite appropriate, since the values of such nominally scaled variables represent categories and have no logical order of arrangement, as numerical values do. In such situations it is often convenient to represent the frequency distribution in the form of a *pie chart*, a circular figure with slices of varying size reflecting the relative frequencies of occurrence of the values of the qualitative variable under study.

An example of a pie chart is shown in Figure 5. It is based on a frequency distribution of the causes of 2,523 deaths in a given city in a given year. The sizes of the slices in the pie chart were determined from the frequency distribution shown in Table 3. The frequencies of the various causes of death were first converted into percentages of the total, by dividing each by 2,523 and then these percentages were multiplied times 360°—which corresponds to the total circle—to determine how much of the circle would be occupied by each value of the variable. A "protractor" can then be used to measure and lay down each angle in turn.

As shown in Figure 5, it is customary to arrange the values of the variable from the one most frequently occurring to the one least frequently occurring, except for the "other" category which is placed last and contains a miscellaneous grab-bag of values. Generally, it is advisable to list alternatives until the "other" category is 5% or smaller, so as not to arouse curiosity as to what is included in it.

The actual orientation of the pie chart (i.e., where the first category is placed) is pretty much a matter of judgment and personal preference as to what makes the most dramatic presentation.

Table 3 Frequency distribution of the causes of 2,523 deaths in a given municipality in a given year.

Cause of death	*Frequency*	*Relative frequency*	*Portion of a pie chart's 360°*
1. Heart disease	1,251	49.6%	178.6°
2. Cancer	527	20.9	75.2
3. Stroke	217	8.6	31.0
4. Accidents	134	5.3	19.1
5. Chronic obstructive pulmonary disease	71	2.8	10.1
6. Pneumonia & flu	66	2.6	9.4
7. Diabetes	43	1.7	6.1
8. Liver disease	40	1.6	5.7
9. Arteriosclerosis	38	1.5	5.4
10. Suicide	33	1.3	4.7
11. Other	103	4.1	14.7
Totals:	2,523	100%	360°

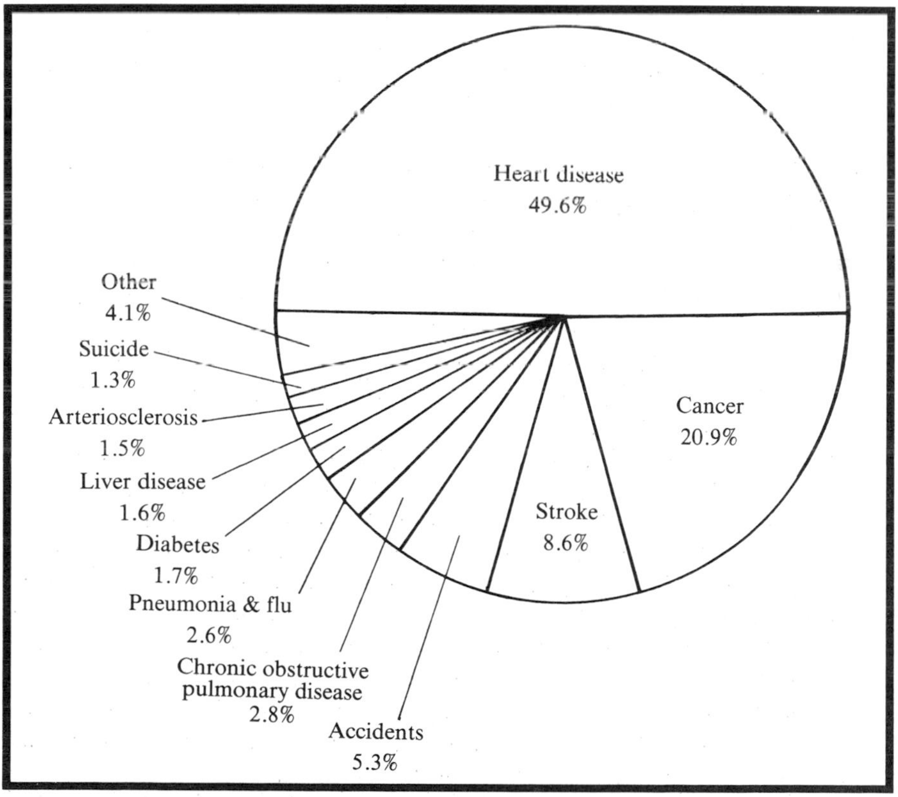

Figure 5 Pie chart of causes of death (see Table 3).

8. Concluding Comments

The creation of a frequency distribution for a set of data is usually the very first analysis that we perform. It gives us a concise overview of the data that is not possible from an examination of the unorganized observations. The shape of the frequency distribution, which we found could vary widely, provides the initial descriptive and interpretive summary of the data. In the following chapters we will discuss further data reduction techniques and interpretations based on the frequency distribution.

Chapter 4

Central Tendency

1. Introduction

Although a frequency distribution reduces a large collection of data into a relatively compact form, it would be highly desirable to further summarize the frequency distribution itself. While we can use verbal expressions such as "bell-shaped" or "skewed to the right" to describe the overall distribution of scores, such descriptions say nothing about the specific numerical values of the variable in question.

What would be beneficial would be a single summary value that would suggest a typical or representative observation, a measure of *central tendency*, or *location* as it is also called; that is, at which value do the observations "tend to center," or equivalently, where along the values of the variable in question are the observations clustered or "located." While such a measure of central tendency, by itself, would necessarily sacrifice much of the information inherent in the frequency distribution, it would nonetheless serve as a very concise description of a body of data; and when used in conjunction with a frequency distribution would provide a richer summary of the observations than either data reduction technique itself.

In the following sections we will discuss the most useful measures of central tendency, including their respective advantages and limitations.

2. The Mode

Since, as we have seen in the preceding sections, the observations in most data distributions tend to cluster heavily about certain values, one logical measure of central tendency would be that value which occurs most frequently; and that value is referred to as the *modal value*, or simply the *mode*.

For example, in the following data collection consisting of nine observations arranged in value from lowest to highest

9 12 15 15 15 16 16 20 26

the modal value is 15, occurring three times, more frequently than any of the other values. Also, it is important to remember that the mode is the value of the variable occurring most often, and not the frequency associated with that value.

When dealing with a distribution based on data grouped into intervals, the mode is often taken as the *midpoint* or *class mark* of that interval containing the highest frequency of observations. Alternatively, the interval itself could be cited as the modal value. For example, if a grouped frequency distribution of the heights of a set of corn stalks revealed that the 60 to 62 inch interval contained the highest frequency of observations, the mode could be reported as either 61 inches, or as the 60–62 inch interval, whichever we prefer.

There are instances in which a data distribution will have two modal values, characterized by two humps like the back of a camel. Such distributions are referred to as *bimodal*, even though one of the two modal values may be associated with a somewhat higher frequency than the other. Whenever we confront a bimodal distribution we should immediately question the composition of the collection of measured objects. Chances are that two distinct populations of objects, differing on the variable in question, are intermingled; e.g., the distribution of heights of a mixture of males and females.

The chief advantage of using the mode as a measure of central tendency is the ease with which it can be obtained, and its common sense interpretation. However, its limited mathematical properties make it less than the ideal measure of central tendency for more advanced analyses.

3. The Median

If we arrange a set of observations from lowest to highest in value, and then single out the middle value, we have identified what is called the *median value*, or simply the *median*, the value above and below which 50% of the observations fall.

For example, in the following simple collection of nine observations

22 24 24 25 27 30 31 35 40

the median value is 27. If we are dealing with an even number of observations instead of an odd number, as in the following collection of ten scores

22 24 24 25 27 30 31 35 40 47

then the median lies between the fifth and sixth values, midway between the values of 27 and 30—namely, 28.5.

In the case of a frequency distribution based on grouped data, the median can be reported roughly as that interval in which the cumulative relative frequency reaches 50%, or as the midpoint of that interval, or, for a more precise measurement, the median value can be determined by interpolation within the said interval.

Aside from its common sense interpretation as a truly central value, the most attractive feature of the median is its insensitivity to the values of the very extreme scores in a distribution, which are atypical and sometimes flukes. For example, if the highest value in the set of ten scores listed in the preceding paragraph were 85 instead of 47, the median would not be affected in the least. The distribution of individual incomes is a good example of how the median, because of its independence from the values of the extreme observations, is a good indicator of central tendency. While most individuals have incomes within a relatively narrow band, there is a minority who have exceptionally high incomes, many times the typical value. Since the definition of the median does not take into account the actual values of the extreme scores, it will not be affected by such deviant data points. For this reason, the median is especially appropriate as a measure of central tendency of a skewed distribution of data. Its main limitation, like that of the mode, is that it does not have mathematical properties that lend itself to more advanced analyses.

4. The Mean

The most important measure of central tendency, and one of the basic building blocks of all statistical analysis, is the *arithmetic mean*, or simply the *mean*. It is nothing more than the sum of a set of values, divided by the number of values involved.

Consider, for example, the following set of seven observations

3 4 4 5 6 8 10

Summing the values and dividing by the number of values, we have

$$\text{Mean} = \frac{3+4+4+5+6+8+10}{7} = \frac{40}{7} = 5.7$$

As in the above example, we typically calculate a mean to one decimal point beyond that occurring in the data itself, for it would be highly misleading as to the precision of our original observations if we reported the mean as, say, 5.714286.

The mean will be recognized as equivalent to the popular concept of an "average" of a set of numbers; more specifically, the arithmetic average. In other uses, the term "average" is used loosely to mean typical or representative, or central tendency in a most general sense.

Since the mean is such a fundamental concept to statistical analysis, it is useful to designate it in a shorthand form. The letter M is sometimes used to denote the mean, or it may be subscripted to identify the variable in question; for example, M_x would indicate the mean of variable x, while M_y would signify the mean of variable y.

An alternative and more common notation for the mean of a variable x, is $\bar{x}$, which is read "x bar" or "the mean value of variable x." By the same token, $\bar{y}$ signifies the mean value of a variable y. With this notation we can define the mean more concisely as

$$\bar{x} = \frac{x_1 + x_2 + \cdots + x_n}{n}$$

where n is understood to be the number of observations in our data collection, and $x_1, x_2, \ldots, x_n$ are the various observations.

Historically, the capital letter S was used to stand for "sum of" which allowed the even more compact definition

$$\bar{x} = \frac{S(x_i)}{n}$$

where $\bar{x}$, the mean of variable x, is shown to be the sum of the individual observations—$S(x_i)$—divided by the number of observations, n. However, since the letter S signifies another important statistical concept, it has been replaced with the capital Greek letter *Sigma*, designated Σ. Thus, the standard mathematical definition of the mean becomes

$$\bar{x} = \frac{\Sigma x_i}{n} \qquad (1)$$

where Σx_i stands for the sum of the individual values of the variable x, n is the number of values in question, and $\bar{x}$ is the resultant mean value of the observations, as before.

The above formula can be made more explicit in the form

$$\boxed{\bar{x} = \frac{\sum_{i=1}^{n} x_i}{n}} \qquad (2)$$

in which the range of the subscripts have been added to the summation sign. We read it, "the mean of variable x is equal to the sum of the values x_i, where i ranges from 1 to n, that sum divided by n." Typically, however, we will just use the summation sign Σ and understand implicitly the range of values being summed. And yet other times, for the sake of simplicity, we will also drop the subscript from x_i and simply write $\bar{x} = \Sigma x / n$.

The $\bar{x}$ notation is reserved for the mean of a *sample* of data. To distinguish the mean of a *population* of observations we use the Greek letter *mu*, designated μ, and pronounced *mew*. Thus, both $\bar{x}$ and μ signify a mean, but $\bar{x}$ refers to a sample, while μ refers to a population of observations. The distinction will take on greater significance when we undertake topics of inference, the drawing of conclusions about populations based on sample observations.

Whether one prefers the verbal or the symbolic definition of the mean makes no difference, so long as the concept is understood. In either case the mean should be made an integral part of one's intellectual hardware; for hardly a topic will go by from now on that does not depend either directly or indirectly on the mean, so important is it to the field of statistical analysis.

The major advantage of the mean over the mode and the median as a measure of central tendency is that it takes into account the *numerical value of every single observation* in the data distribution. It represents a balance point, or center of gravity, in that the sum of the distances to the observations below it, is equal to the sum of the distances to the observations above it. This mathematical characteristic of the mean, as we will see, makes it a cornerstone of statistical analysis.

Ironically, this feature of the mean—its sensitivity to every numerical value—becomes its chief drawback as a measure of central tendency in situations where the data distribution is highly skewed, and when there are one or two freaky "outliers" in the data. For example, the mean of the values 5, 6, 7, 8, and 9 is 7.0; while the mean for the set of values 5, 6, 7, 8, and 30 is 11.2, the dramatic increase in the value of the mean being due to the single exceptional score of 30. The median, on the other hand, uninfluenced by the deviant value, remains at 7.

When a data distribution is basically symmetrical in form, the mode, median, and mean will have very nearly the same value. In a skewed distribution the mean tends to get dragged *toward the tail* of the distribution, toward those few exceptional values, as in the example of the preceding paragraph. The median in such distributions will typically fall between the mode and the mean. This relationship between the values of the mode, median, and mean for symmetrical and skewed distributions is summarized in Figure 1.

Knowing only the values of the mean and the median of a data distribution, we can generally guess the shape of the distribution. If these measures of

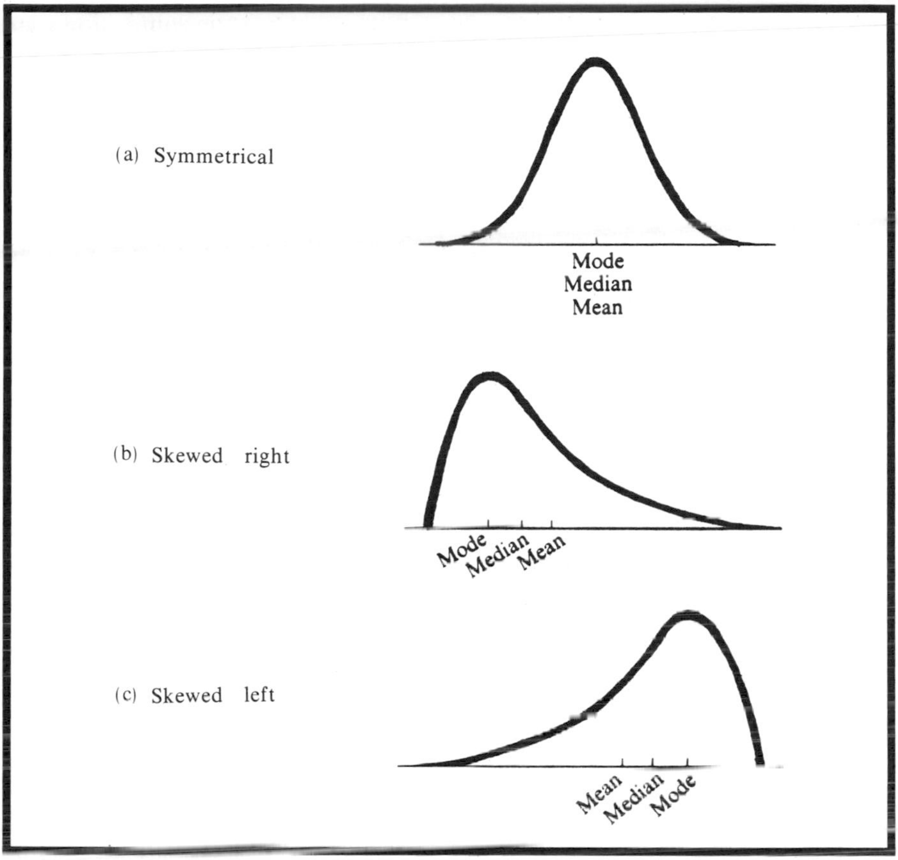

Figure 1 A comparison of the mode, median, and mean for distributions differing in shape.

central tendency are approximately equal, we know that the distribution is probably symmetrical in form. If the mean is *less* than the median, we know that the distribution is skewed to the *left*; i.e., a few low scores have disproportionately affected the mean. When the mean is *higher* than the median, the distribution is skewed to the *right*, the mean being dragged in that direction by a relatively few high scores.

For descriptive purposes, then, we would do well to report the mode, median, and mean when trying to characterize a distribution of data, for each in their own way is a measure of central tendency, though reflecting different features of the data. However, when we are dealing with roughly symmetrical distributions, in which case the mode, median, and mean are more or less

equal, and we are interested in making an inference about the central tendency of the population from which our sample observations have been drawn, we should rely on the mean, for the mean is known to be a more *efficient* estimate, in that for repeated samples of a given size the mean will show less fluctuation in value than either the mode or the median. This should not be surprising since the mean takes into account the information inherent in every single observation, whereas the mode and median do not take advantage of the numerical values of every single data point, wasting information as it were.

5. The Weighted Mean

Typically, every observation in a data collection has equal importance to us. There are situations, however, in which a data collection is based on a set of objects that differ in their *importance* for one reason or another. For example, a student's test scores are not equally important—quiz scores are less important than a midterm score, which in turn is less important than a final exam score. Similarly, a marketer's brand share in a large sales territory is more important than its share in a small territory.

In such situations a straight-forward mean might be misleading. To compensate for the varying importance of the observations, we can calculate a *weighted mean*—more precisely, a *mean based on weighted observations*—in which each observation is multiplied by its importance weight, and then dividing the sum of these weighted observations by the sum of the weights. In symbols, we have

$$\bar{x}_w = \frac{\sum w_i x_i}{\sum w_i} \qquad (3)$$

where $\bar{x}_w$ is the weighted mean, w_i is the importance weight associated with an individual observation x_i, $\Sigma w_i x_i$ is the sum of the products of the observations multiplied by their respective weights, Σw_i is the sum of the weights, and the subscript i is implicitly understood to vary from 1 to n, the number of observations in the data collection.

Consider, for example, the results of a poll that finds a particular referendum is favored by 44%, 56%, 64%, and 51% of the prospective voters in a city's North, South, East, and West voting districts, respectively. The ordinary mean of these values is 53.8%. If, however, we take into account the unequal importance of the four voting districts based on the sizes of their voter bases—2,500, 1,000, 400, and 2,000 in the North, South, East, and West

districts, respectively—the weighted mean becomes

$$\bar{x}_w = \frac{2{,}500(44)+1{,}000(56)+400(64)+2{,}000(51)}{2{,}500+1{,}000+400+2{,}000} = 49.8\%$$

We find that the weighted mean of 49.8% is lower than the unweighted mean of 53.8%, due to the fact that the more important districts—those with a larger voter base—were less in favor of the referendum than were the smaller districts.

The above example is our first encounter with what can be considered a multivariate analysis; namely, a set of objects (the voting districts) were measured on two variables (referendum vote and size of voter base). Had there been no systematic relationship between the two variables, the unweighted and weighted means would have been equal. We will encounter the importance-weighting concept again in later chapters.

6. The Mean of Grouped Data

The above weighting concept can also be used to approximate the mean of a set of observations that have been arranged in a frequency distribution consisting of grouped data. In such situations the midpoint of each interval, x_i', can be weighted by its corresponding frequency of occurrence, f_i, to yield an *estimate* of the mean were it calculated from the original ungrouped data; i.e., $\bar{x} \doteq \Sigma f_i x_i' / \Sigma f_i$, where $\doteq$ stands for "is approximately equal to," and the summation is across $i = 1, 2, \ldots, k$ *intervals*.

The reason the resulting mean will only be approximate in nature is that by taking the midpoint of an interval as representative of the values falling within it, we are assuming the data within the interval are uniformly distributed, which is usually not the case. In any event, with modern computing facilities, there is no reason why the original ungrouped data is not used to obtain an *exact* value of the mean.

7. Proportions

Proportions are of two types—*quantity* proportions and *frequency* proportions. A quantity proportion represents the ratio, usually in decimal form, of a *sub-quantity* to a *total quantity*. For example, the proportion of one's income that goes to taxes represents the ratio of the *amount* of income taken by taxes to the *total amount* of income.

The second type of proportion, the frequency proportion, represents the ratio of a *sub-frequency* to a *total frequency*. For example, the proportion of a

sample of patients with symptom A, represents the ratio of the *number* of patients with the symptom to the *total number* of patients.

While we may not have been aware of it, a frequency proportion represents the *mean* of the 0 and 1 binary values attached to a dichotomous variable. For example, if we let 0 and 1 stand for a "no" and "yes" vote, respectively, on an issue, we might observe the following 15 votes:

1 0 0 1 1 0 1 0 0 1 0 1 0 0 0

By definition, the mean of these values is their sum divided by their number: namely, 6/15 or .40. This, of course, is exactly the same as the proportion of "yes" votes.

Thinking of a proportion as the mean of a binary variable will be helpful in generalizing many of the analyses of later chapters.

8. Arithmetic Operations and the Mean

How would you expect the mean of a set of scores to be affected if we added a constant value to each of the original observations? What if we subtracted a constant value from each of the scores? Or, if we multiplied or divided the scores by a constant?

An illustration of the effects of these arithmetical operations on the mean, for a sample set of five scores, is shown in Figure 2. The mean of the five given scores—2, 3, 4, 5, and 6—is 4.0, and can be seen in Figure 2*a*.

In part *b* of the figure, after adding a constant value of 3 to each of the original values, we see the mean of the new set of scores has changed to 7; namely, the original mean value plus the added constant value (i.e., $4+3=7$).

In Figure 2*c* we see that subtracting a constant value of 5 from each of the original scores lowers the mean by exactly that amount. The new mean is -1.0 (i.e., $4-5=-1$).

Multiplying the original scores by a constant value of 2 results in a new mean of 8, two times as large as the original mean (i.e., $4\times 2=8$), as shown in Figure 2*d*.

Dividing the original scores by the constant value of 4 results in a new mean of 1.0, which is the original mean divided by 4 (i.e, $4\div 4=1$), and is shown in Figure 2*e*.

In summary, bypassing the relatively simple proofs, we can state the following relationships between the mean of a set of scores and the mean of the set after it has been transformed by various arithmetical operations. The letter M stands for either a sample mean $\bar{x}$ or a population mean μ.

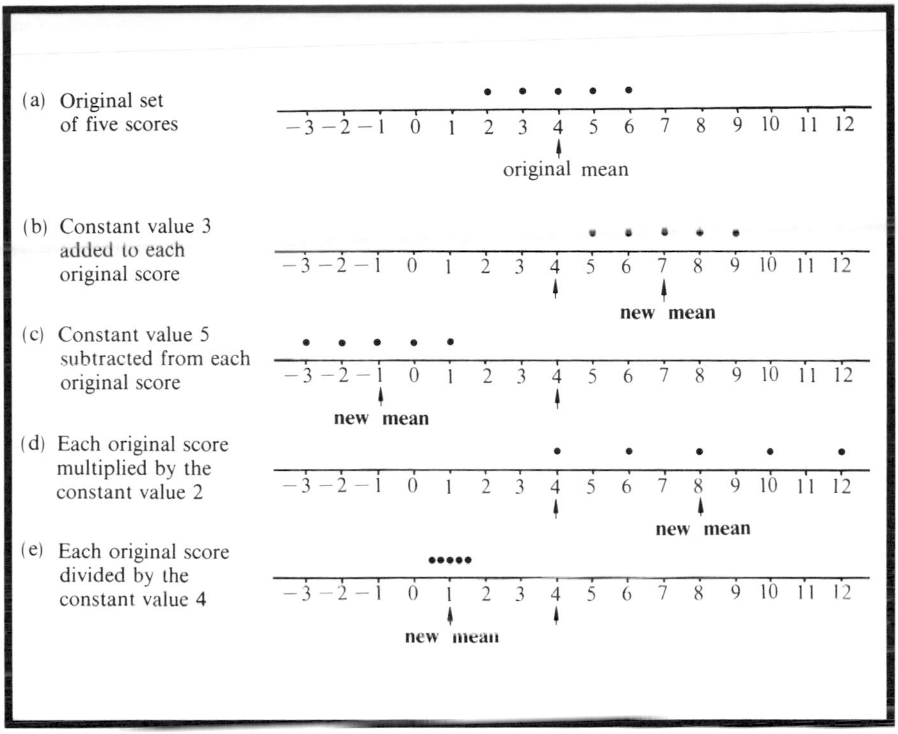

Figure 2 The effects of various arithmetic operations on the mean of a set of data.

- *Adding a constant value c* to each of a set of scores x results in a mean which is equal to the *original mean plus the constant value*. In symbols, $M_{x+c} = M_x + c$

- *Subtracting a constant value c* from each of a set of scores x results in a mean which is equal to the *original mean minus the constant value*. In symbols, $M_{x-c} = M_x - c$

- *Multiplying a constant value c* times each of a set of scores x results in a mean which is equal to the *original mean times the constant value*. In symbols, $M_{cx} = cM_x$

- *Dividing a constant value c* into each of a set of scores x results in a mean which is equal to the *original mean divided by the constant value*. In symbols, $M_{x/c} = M_x/c$

These fundamental relationships, important in themselves, will be very useful in developing subsequent statistical concepts. Learn them well.

9. Concluding Comments

The concept of central tendency, especially as measured by the arithmetic mean, is one of the most fundamental concepts of all statistical analysis. Not only does the mean serve as a data reduction technique, providing a summary value for an entire distribution of data, but it will serve from now on as a building block for the creation of additional concepts and techniques which will help us further in the analysis and interpretation of our data collections. At the end of our course of study we will be amazed at what has evolved out of this simple concept we know as the mean.

Chapter 5

Variation

1. Introduction

A measure of central tendency is not enough to summarize a set of observations. To more fully describe a data distribution we also need a summary measure of the *variation* or *dispersion* of the observed values; i.e., the extent to which the observations differ among themselves in value.

The need for such a measure of variation is evident from the distributions shown in Figure 1. The distributions shown in part *a* of the figure have the same central tendency, or location, but differ markedly in their variation; those in part *b* differ in central tendency, but have the same variation; while those shown in part *c* differ both in variation and central tendency. So just as we have the mean as a measure of central tendency, we need a summary measure of the variation of a distribution of data.

2. The Range

Perhaps the simplest and at the same time the crudest measure of variation is the *range*, the difference between the highest and lowest observed value in a collection of data.

Consider, for example, the following set of ten scores:

7 10 12 15 16 16 19 20 25 29

The range for this set of data is 22, the difference between the extreme values of 29 and 7. We could summarize this collection of observations by stating that it has a mean of 16.9 and a range of 22, extending from 7 to 29; a description which is much more comprehensive than reporting the mean alone.

While the range is an easy measure to determine, easily understood, and a seemingly satisfactory measure of the variation of a set of scores, it suffers from the same weakness as the median; namely, it does not take into account

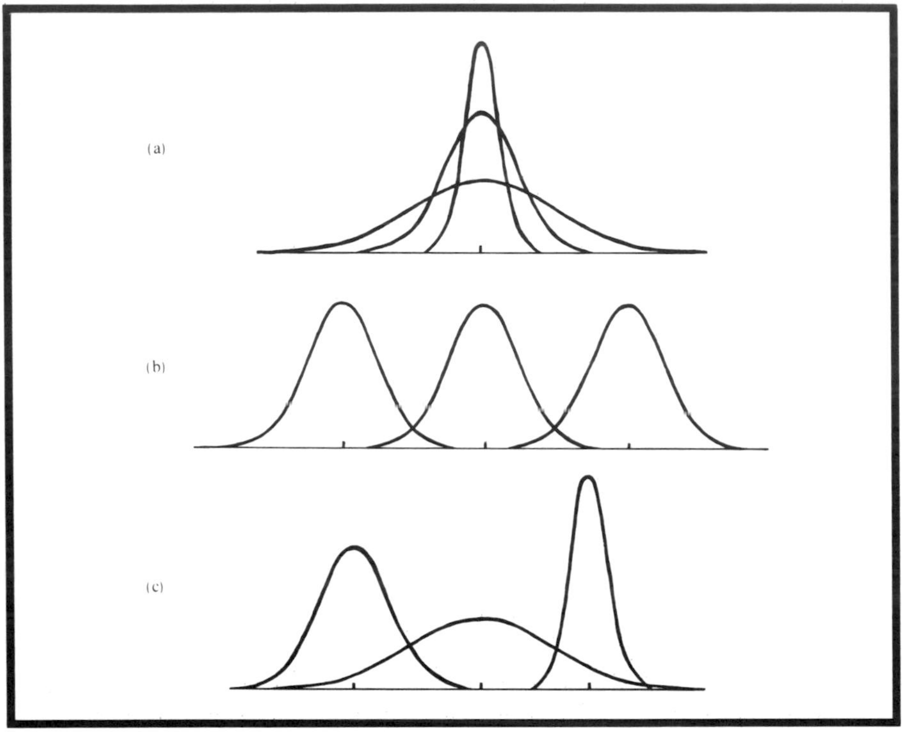

Figure 1 Distributions which have (a) the same central tendency but different variation, (b) different central tendency but the same variation, and (c) different central tendency and variation.

the numerical value of each and every observation. It is based entirely on two values, the highest and the lowest. The shortcoming of the range as a comprehensive measure of variation will be evident from a comparison of the above distribution with the following one:

7 14 15 15 16 16 17 18 22 29

This distribution has exactly the same range as that in the above paragraph, but clearly this set of values, taken as a whole, varies less about the mean than the previous example. Yet, the range, 22 in both instances, does not reflect this facet of the data.

Because of this insensitivity of the range to the "internal" variation of a distribution of data, it should be used only as the roughest measure of variability, primarily to provide information about the values beyond which no observations fall. This, in itself, is very useful information, whatever other limitations the range may have.

3. Mean Absolute Deviation

The greater the variation of a set of scores, the greater will be the deviations of the values from the mean. Consequently, a logical measure of variation would be the average value of those deviations. However, it will be recalled that the mean is the center of gravity, or balance point, of a set of observations, in that the sum of the distances or deviations from the mean of those values falling above it equals the sum of the deviations from the mean of those values falling below it. Since the deviations from the mean, $x_i - \bar{x}$, always sum to zero, their average value will also always equal zero, leaving us no information about the sizes of the individual deviations.

One way to get around the fact that the deviations from the mean sum to zero, is to strip the deviations of their positive and negative signs, and then take the arithmetic mean of these *absolute values* of the deviations. Such a measure of variation is known as the average absolute deviation or *mean absolute deviation* ($M.A.D.$). In symbols, we have

$$M.A.D. = \frac{\sum |x_i - \bar{x}|}{n} \tag{1}$$

where $|x_i - \bar{x}|$ signifies the absolute value of a deviation from the mean, and the summation is across the $i = 1, 2, \ldots, n$ observations in the data set.

As a simple illustration of the calculations of the mean absolute deviation, consider the following set of five scores that have a mean value of 9:

5 6 9 11 14

Their respective deviations from the mean are:

$-4 \quad -3 \quad 0 \quad +2 \quad +5$

As expected, the sum of the positive deviations from the mean $(+7)$ balances the sum of the negative deviations from the mean (-7). However, in calculating the mean absolute deviation we disregard the positive and negative signs of the deviations, averaging instead their absolute values. Summing 4, 3, 0, 2, and 5 yields a total of 14, which when divided by $n = 5$, the number of observations in the data set, results in a mean absolute deviation of 2.8.

The logic of the mean absolute deviation as a measure of variation cannot be faulted, but how do we interpret it? While we can easily visualize the meaning of the range of a given distribution of data, how do we visualize the meaning of an average deviation. There is the further complication that, due to its reliance on absolute values, it cannot be used in certain mathematical operations necessary for the development of more advanced statistical techniques.

Nonetheless, the rationale behind the average deviation as a measure of variation is sound enough, and in the following two sections we will see how a minor variation on its theme will provide us with two much more useful measures of dispersion, measures that we will use throughout our subsequent study of statistical methods.

4. The Variance

An alternative to dealing with the absolute values of the deviations from the mean is to *square* each deviation, thereby yielding all positive quantities. The *mean* of these squared deviations could then be determined. Rather than having an average deviation we would have an *average squared deviation*. This very important measure of variation is called the *variance*, and for a *population* of observations we can define it concisely as follows:

$$\textit{Population Variance} = \frac{\textit{Sum of the squared deviations from the mean}}{\textit{Number of observations}}$$

In symbolic form, a population variation is denoted with the square of the lower case Greek letter *sigma*, σ^2, and can be read "sigma squared" or "the variance of variable x." In notational form, the above definition of a population variance becomes

$$\sigma^2 = \frac{\sum (x_i - \mu)^2}{N} \tag{2}$$

where σ^2 is the variance, $x_i - \mu$ is the deviation of a given observation from the population mean, N is the number of observations in the data set, and the summation Σ is across the $i = 1, 2, \ldots, N$ observations. We could subscript the variance notation as follows, σ_x^2, to signify that we are referring to the variance of variable x, but that is understood in the absence of a subscript. If the variance refers to other than a variable x then it will be subscripted.

The above formula is appropriate for defining the variance of a population of observations. However, if we apply it to a *sample* of observations in an attempt to estimate the variance of the parent population from which the sample was drawn, we will discover that it is a *biased* estimate—it tends to *underestimate* the population variance. That is, if repeated samples were drawn from the population, and the variance calculated for each according to the above formula using the sample mean $\bar{x}$ instead of the population mean μ, the average of these variances would be somewhat lower than the true value of the population variance, were we able to measure every single member of it.

An adequate adjustment to the formula to avoid this bias is to divide the sum of squared deviations not by the number of observations in the sample, but by *one less* than the number of observations. We then have an *unbiased estimate of a population variance based on a sample of observations*, and we define it as follows:

$$\textit{Sample Variance} = \frac{\textit{Sum of the squared deviations from the sample mean}}{\textit{Number of observations less one}}$$

The sample estimate of a population variance is denoted s^2, and is defined symbolically as

$$s^2 = \frac{\sum (x_i - \bar{x})^2}{n-1} \qquad (3)$$

where the summation is across all n observations in the sample. Notice that we have used n to designate the size of a sample, and N for the size of a population. For formula (3) to yield an unbiased estimate of a population variance, we must assume that the population is infinite in size, or for practical purposes that N is much larger than n—say, at least fifty times as large.

So, when we are interested in estimating the variance of a population of values based on a random sample from that population, we will use formula (3). If we are merely interested in describing a body of data that we define as our population of interest, with no intent of generalizing to a larger parent universe, then we will use formula (2).

It should be apparent that as the size of our sample of data increases, the effect of dividing by the number of observations as opposed to one less than the number of observations becomes slight. Still, for theoretical purposes, the distinction between the two variances should be kept in mind.

5. The Standard Deviation

Since the variance is in units of measurement that are squared, as well as for other reasons that will gradually become apparent, it is convenient to take the *square root* of the variance and define a quantity known as the *standard deviation*:

$$\textit{Standard Deviation} = \sqrt{\textit{Variance}}$$

If we are interested in the *standard deviation of a population* of observations, we

will take the square root of the *population variance*. If, on the other hand, we are interested in *estimating* the standard deviation of a population we will take the square root of the *sample variance*, the unbiased estimate of that population parameter.

In symbols, the *standard deviation of a population of values* is given by

$$\sigma = \sqrt{\frac{\sum (x_i - \mu)^2}{N}} \tag{4}$$

In turn, the *unbiased sample estimate of the population standard deviation* is given by

$$s = \sqrt{\frac{\sum (x_i - \bar{x})^2}{n - 1}} \tag{5}$$

The standard deviation formulas (4) and (5) will be recognized as the square roots of the variance formulas (2) and (3), respectively, as they should by definition.

Since the standard deviation equals the "square root" of the "mean" of the "squared deviations" it is also known as the *root mean square* value of a data collection, or simply as the *RMS* value. This terminology is especially common in the engineering fields and physical sciences.

As an example of the calculation of the standard deviation, consider the observations

12.1 15.4 14.1 14.4

which represent the widths in centimeters of four skulls found in an archeological dig. Subtracting their mean value of 14.0 from each, squaring the resultant deviations, summing them, dividing by $n - 1 = 3$, and then taking the square root, we have

$$s = \sqrt{\frac{(12.1 - 14.0)^2 + (15.4 - 14.0)^2 + (14.1 - 14.0)^2 + (14.4 - 14.0)^2}{4 - 1}}$$

$$= \sqrt{\frac{5.74}{3}} = 1.38 \text{ cm}$$

It should be noted that the standard deviation is in units corresponding to those of the variable we are measuring, centimeters in this instance.

Assuming the four studied skulls were a random sample of a larger population (which is probably not tenable), then the standard deviation of $s = 1.38$ would be an unbiased estimate of the variation of the larger population of skulls. If we had wanted to treat the four skulls as a population in itself, then we would divide by $N = 4$ in the above expression, instead of by $n - 1 = 3$. So doing, we obtain a standard deviation of $\sigma = 1.20$ cm for the set of four skulls, when treated as a population.

When data is arranged in a frequency distribution consisting of intervals of values of the variable in question, the variance and standard deviation of the original ungrouped data can be *approximated* by taking the midpoint of each interval as representative of the values falling within them, and then weighting their deviations from the mean by their corresponding frequency of occurrence. However, as pointed out in the discussion of the calculation of an approximate mean from grouped data, the procedure is not recommended, since modern computing facilities make it easy enough to obtain exact values from the original ungrouped data.

6. Variation of the Normal Distribution

While both the variance and the standard deviation have wide applications in statistical analysis, we will concentrate initially on the standard deviation. The applications of the variance will be encountered in later chapters. Though we now understand the simple manner in which the standard deviation is calculated, so far we know little of how to interpret it. What does a standard deviation of 1.2 mean, or one of 23, or 955, or whatever else we might calculate from a set of data. At the very least we can surmise that the greater the value of the standard deviation the greater the variation of scores about the mean. This follows from the fact that the standard deviation is based on the *deviations* of the observations from the mean. For example, if two distributions each have a mean of 100, but one has a standard deviation of 5 while the other has a standard deviation of 12, we can safely conclude that the distribution with the larger standard deviation has a wider dispersion of scores.

We can be more specific in our interpretation of a standard deviation provided our distribution of data is of a well-defined form. For our purposes we will be primarily interested in the *normal distribution*, the characteristic bell-shaped curve that we have been alerted to from the very start. Although the curve representing the normal distribution is defined by a precise mathematical equation, we can do without it for our purposes. Rather, we will recognize the normal distribution by its graphical representation, as shown in Figure 2, and by its special characteristics vis-a-vis its mean and standard deviation.

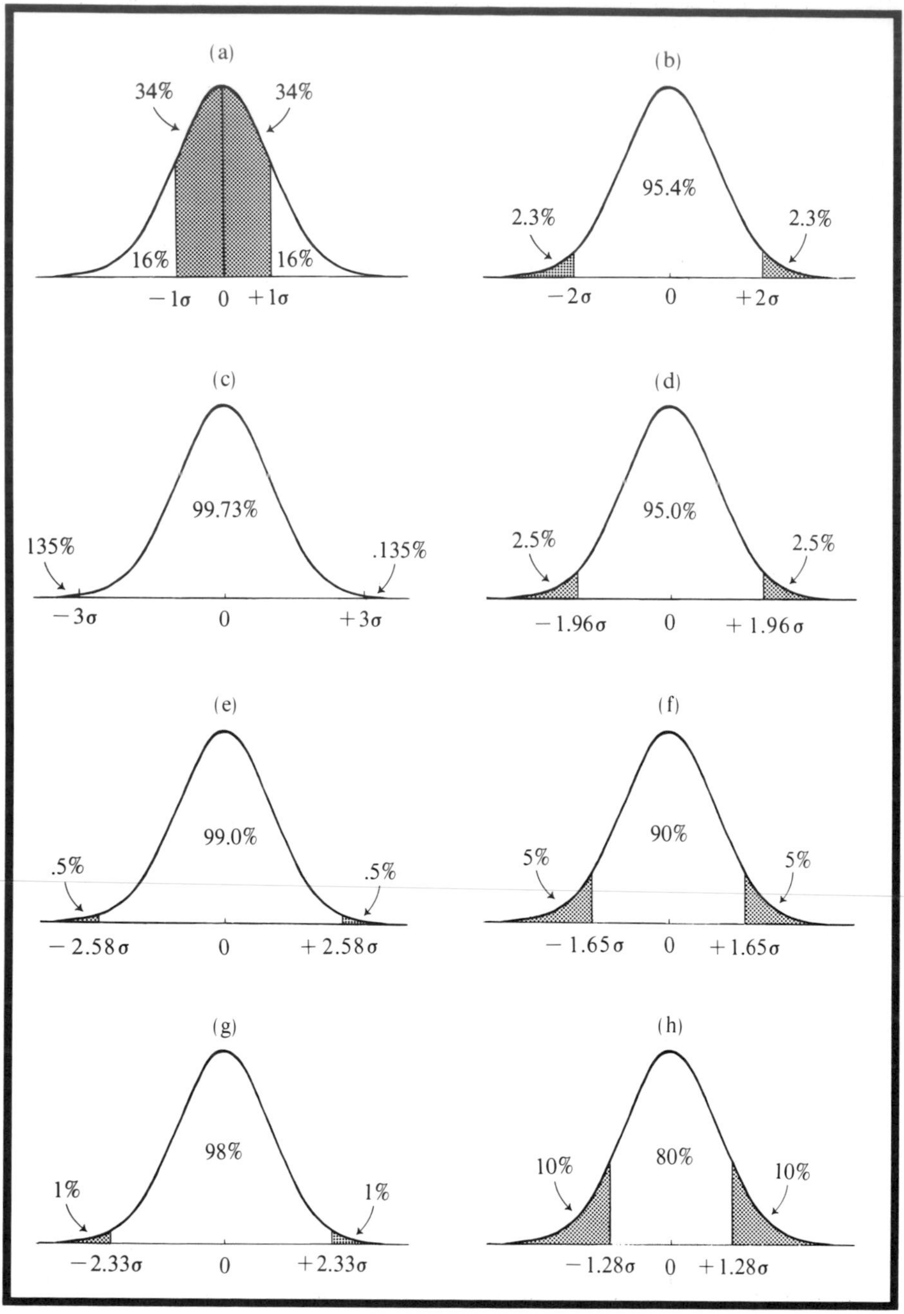

Figure 2 Areas under the normal curve for various standard deviations from the mean.

It will be recalled that the normal distribution is theoretical in nature, based on an infinite number of observations. But we also learned that we can view the area under the curve as representing 100% of the observations, and that to determine the relative frequency of occurrence of values within any given interval we merely need to measure the area under the curve in that interval. It is in this respect that the standard deviation takes on meaning; for we can determine *the percentage of observations which fall between the mean and any other value, when the distance is measured in standard deviations.*

For example, in Figure 2*a* we see that 34% of the total area under the curve occurs between the mean and a value which is *one* standard deviation above the mean. Since the curve is symmetrical it follows that an additional 34% of the area occurs between the mean and a value one standard deviation below the mean; which leads us to a conclusion worth remembering, that 68% of the values occurring in a normal distribution fall within an interval extending from one standard deviation below the mean to one standard deviation above the mean. For example, if we had a distribution with a mean of 25 and a standard deviation of 3, we would expect about 68% of the observations to fall between a value of 22 and 28 (i.e., 25 ± 3). That is, *if* the scores were distributed normally. If the scores were distributed, say, rectangularly, we would not expect these percentages to hold. But it is in this respect that the normal distribution is so important, since so many observations we make upon the world are in fact normally distributed, or very nearly so. And in the next chapters we will see that the normal distribution has theoretical significance even beyond the fact that empirical data often follow a normal distribution.

We see further in Figure 2*b* that 2.3% of the observations in a normal distribution fall beyond a value that is *two* standard deviations above the mean. In total, then, just over 95% of the scores (95.4%) occur within an interval extending from a value two standard deviations below the mean to a value which is two standard deviations above the mean. For example, if a normally distributed set of data has a mean of 75 and a standard deviation of 12, we would expect about 95% of the observations to fall between the values of 51 and 99 (i.e., $75 \pm 2 \times 12$).

While the theoretical normal distribution extends on to infinity both above and below the mean, we can see from Figure 2*c* that virtually 100% of the area under the curve occurs within *three* standard deviations on either side of the mean. Only a small fraction of a percent of the observations occur beyond these extreme values. This too is a worthy piece of information to remember, for it often provides us with a quick method for roughly estimating the size of the standard deviation when all we know is the *range* of the distribution. Since six standard deviations—three above the mean, and three below—encompass virtually 100%of the observations, we could divide the range of a distribution,

which appeared to be roughly normal in shape, by six and get an estimate of the size of the standard deviation. For example, if such a distribution had a mean of 200 and a range of 150, we would estimate the standard deviation as being about 25 (i.e., $150 \div 6$). If our estimate is good we will expect in turn that about 68% of the observations would fall between 175 and 225 (i.e., $200 \pm 1 \times 25$), and about 95% would occur between 150 and 250 (i.e., $200 \pm 2 \times 25$). Parts *d* through *h* of Figure 2 present other important characteristics of the normal distribution that are worth noting.

We see, then, that the standard deviation does take on a tangible meaning when interpreted with respect to a normal distribution. This would not be of great significance if the normal distribution were only a sometime thing, but as we have stressed repeatedly it is everpresent in every field of study. And even when a distribution departs from normality, provided it is not *U*-shaped or extremely skewed, the percentages we have observed above are still fairly accurate descriptions of the data's dispersion.

7. Standardized *z* Scores

By now it is clear that an object's value on a variable, in and of itself, has no meaning as to its relative location in the distribution of observed values. For example, to know that someone scored 115 on a test would be impossible to interpret unless we knew the mean of the distribution of scores. But even if we knew the mean, say it were 100, a score of 115 still could not be interpreted beyond observing that it fell above the mean. To know *how far* above the mean, we would also need to know the standard deviation of the distribution; i.e., to what extent do the scores vary about the mean. If the standard deviation of the distribution were 5 score units, a score of 115 would have quite a different meaning than if the standard deviation were 15 score units. In the first instance, a score of 115 would be *three* standard deviations above the mean, while in the second instance it would be only *one* standard deviation above the mean.

Knowing that the distributions were normal in form, we could further interpret the scores in terms of what percent of the total number of observations fall below or above the given score. Although real-life data distributions, due to their finite size, can never be perfectly normal in form, the approximation is often close enough to allow us to use the theoretical normal distribution as a model for interpreting empirical populations of data.

To identify the relative location of an observed value in a data distribution, then, we need to know not only its deviation from the mean, but that deviation must be *translated into standard deviations*. The manner in which we can transform an observed value into a new value which expresses how many standard deviations it departs from the mean, is relatively simple. We first

determine how far the given score deviates from the mean of its distribution. For example, if we have a score of 450 from a distribution with a mean of 400, the score deviates 50 units from the mean. Now if we know that the standard deviation of this distribution is, say, 25 units, it is our task to transform the observed deviation of 50 units into standard deviation units. By simple division ($50 \div 25$) we determine that a deviation of 50 units from the mean is equivalent to *two* standard deviations, each 25 units in size. Converting a deviation from the mean into standard deviations will be recognized as analogous to the procedure for converting a quantity of eggs into dozens of eggs.

We can summarize the above procedure for converting an observed score into a new score, expressed as standard deviations from the mean, with a simple formula, and we will refer to these transformed values variously as *standard scores*, *standardized values*, or *z scores*:

$$z \text{ score} = \frac{\textit{Observed value minus the mean value}}{\textit{Standard deviation of the values}}$$

In symbols we have

$$\boxed{z = \frac{x_i - \mu}{\sigma}} \tag{6}$$

It is important to note that regardless of the mean and standard deviation of the original variable x, the standardized z variable will have *a mean of 0* and *a standard deviation of 1*. The benefit of this transformation will become apparent in subsequent discussions.

Repeating the earlier example of a score of 450 from a distribution with a mean of 400 and a standard deviation of 25, we have

$$z = \frac{450 - 400}{25} = 2.0$$

Now a z score of 2.0 is far more informative than a raw score of 450, for built into the z score is information about both the *mean* and the *standard deviation* of the distribution. And if the distribution happens to be normal in form, then we also know how many observations in the distribution fall below or above a given score. As we learned in the preceding section based on Figure 2*b*, about 97.7% of the observations in a normal distribution will fall beneath a value corresponding to two standard deviations above the mean; i.e., below a $z = 2$.

We can easily determine what percentage of the observations in a normal distribution fall above or below any given z value by consulting Appendix

Table III. It provides the proportion of area under the normal curve beyond various z scores. For example, we discover that .2843 or 28.43% of the area under the curve falls beyond a $z = .57$. (By symmetry, it also corresponds to the area below a $z = -.57$). By subtracting 28.43% from 100%, we can then determine that 71.57% of the observations fall *below* a $z = .57$. Further familiarity with the table can be obtained by verifying the percentages reported in Figure 2.

It must be stressed again that the interpretations we have placed upon the preceding z scores have assumed that we do in fact have a normal distribution of observations. In real life this will only be approximated by our data collections. Nevertheless, a z score is an extremely powerful descriptive device, for in a single number it communicates an object's relative position in a data distribution. Without the z score we would need to know the object's *raw score* as well as the *mean* and the *standard deviation* of the distribution in order to obtain an equal amount of information. In subsequent chapters we will discover further applications of the z score ratio, as it is used in problems of statistical inference.

8. Arithmetic Operations and the Standard Deviation

In the chapter on central tendency we learned that adding a constant to, or subtracting a constant from, each of a set of data points resulted in a new mean which was increased or decreased, respectively, by the value of the constant that was added or subtracted. Also we found that dividing a set of observations by a constant resulted in a new mean equal to the original mean divided by that constant. And when we multiplied a set of observations by a constant, the result was a new mean equal to the original mean times that constant.

We should now wonder what happens to the *standard deviation* and *variance* of a set of data, when we add or subtract a constant to or from the set of values in our data set, and when we multiply or divide our data by a constant. Figure 3 shows some graphical examples of these operations on a small set of five data points. It is the same data that was presented in Chapter 4. A simple inspection of Figure 3 will show that the *addition and subtraction operations do not affect the variation of the data*; only the mean is influenced. However, when we *multiply* each value by a constant, the standard deviation *increases* by that factor; and similarly, when we *divide* the observations by a constant, the standard deviation is *decreased* by that factor.

Again, bypassing the relatively simple proofs, we can state the following relationships between the standard deviation (and variance) of a set of data before and after various arithmetic operations.

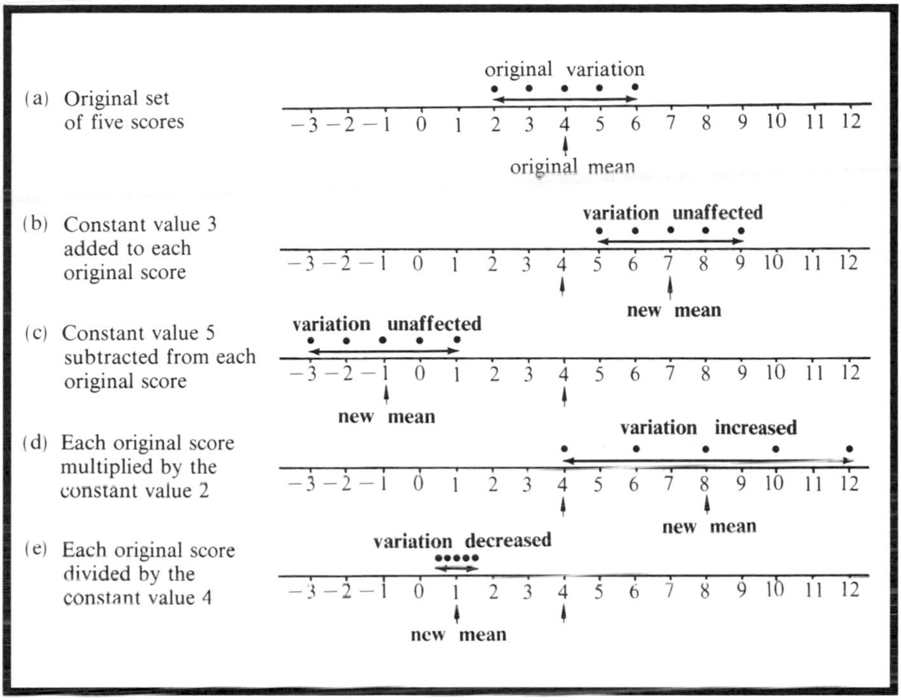

Figure 3 The effects of various arithmetic operations on the variation of a set of data.

- *Adding a constant value c to each of a set of scores x does not affect* the standard deviation or variance of the set of scores. In symbols, $\sigma_{x+c} = \sigma_x$ and $\sigma^2_{x+c} = \sigma^2_x$
- *Subtracting a constant value c* from each of a set of scores *x does not affect* the standard deviation or variance of the set of scores. In symbols, $\sigma_{x-c} = \sigma_x$ and $\sigma^2_{x-c} = \sigma^2_x$
- *Multiplying a constant value c* times each of a set of scores *x* results in a standard deviation *c times as large* as the standard deviation of the original scores; and it results in a variance which is c^2 *times as large* as the variance of the original set of scores. In symbols $\sigma_{cx} = c\sigma_x$ and $\sigma^2_{cx} = c^2\sigma^2_x$
- *Dividing a constant value c* into each of a set of scores *x* results in a standard deviation $1/c$ *times as large* as the standard deviation of the original scores; and it results in a variance which is $1/c^2$ *times as large as* the variance of the original set of scores. In symbols $\sigma_{x/c} = (1/c)\sigma_x$ and $\sigma^2_{x/c} = (1/c)^2\sigma^2_x$

While the above relationships are shown for σ and σ^2, they apply equally to s and s^2.

What we see, then, is that the addition or subtraction of a constant value from a set of observations affects the central tendency but not the variation of the data; whereas the multiplication or division by a constant affects both the central tendency and variation of the original distribution. These fundamental relationships will be very useful in developing and understanding subsequent statistical concepts. They also explain why $\bar{z} = 0$ and $\sigma_z = 1$. The proof is left as an exercise.

The concept of central tendency, along with that of variation, comprise two of the most basic notions of statistical analysis. The measures of variation —most notably the variance and the standard deviation—and the key measure of central tendency, the mean, are not only data reduction techniques, but they will serve as the primary building blocks for the more advanced statistical methods that we will develop. Indeed, we can see the variance and standard deviation as evolving out of the mean itself: The mean was determined by performing simple operations on a set of observations; then, the mean in turn was used to perform further operations on the data, giving rise to deviations from the mean; the concept of the mean was further used to average the squared deviations, the result being a measure of variation. And then, in addition, the mean and standard deviation were used in combination to define standard z scores, an even greater abstraction and summary of the data. So, the *shape* of a data distribution, the *central tendency* of the distribution, and the *variation* of the distribution—these are the primary characteristics of our data to which we must always attend.

9. Covariation

When objects are measured on a single variable x we are primarily concerned with the shape of the data distribution and measures of its central tendency and variation. However, when the objects are measured on two or more variables, say x and y, we are additionally interested in a measure of the *association* or *covariation* between the values of the respective variables—that is, some indication of whether a systematic pairing of the values of the variables exists. If so, we variously say that the variables are *associated*, *related*, *correlated*, *dependent*, *interdependent*, *non-orthogonal*, a *function* of one another, or that they *covary*.

For example, we might be interested in possible relationships between school grade point average and scores on an aptitude test, between employee productivity and scores on a personality scale, between number of criminal arrests and number of childhood beatings, between longevity and blood pressure, between cell lifespan and extent of a trace element, between product

sales and advertising levels, or between any number of other variables originating in any field of study.

As discussed at the very beginning of the text, associations between variables are of two basic types—*correlational* and *experimental*. In the case of *correlational* relationships, a systematic association exists between the values of two *random variables*; that is, situations in which the researcher has no influence over the values of the respective variables. Rather, the values of the variables are a matter of chance, or probabilistic in nature, and all we do is observe their occurrence in their natural setting. For example, the relationships between crop yields and rainfall levels for various plots of land, or between academic achievement and aptitude test scores, or between crime rate and unemployment rate, are instances of correlational relationships between two random variables, in that the variation in their values (as studied) are not under our manipulative control.

The second type of relationship is referred to as an *experimental* relationship in the sense that the values of one of the variables under study is experimentally manipulated by the researcher, and not left to chance. For example, the relationship between crop yield and amount of irrigation would be an experimental relationship, since the amount of irrigation is under our control, as contrasted with the amount of rainfall in the example above. Similarly, if we purposefully, and in a controlled manner, manipulated the total hours of study for different samples of students, the resulting relationship between achievement test scores and hours of study would be experimental in nature; again in contrast to a correlational relationship based on hours of study as determined by each student's personal habits in their natural environment.

There are important differences between correlational and experimental relationships with respect to the manner in which they are measured and interpreted, but since these multivariate issues are among the main objectives of this book their detailed discussion will be reserved for the remaining chapters. For the present, it is most important to recognize the notion of association, or covariation, as a concept as fundamental as central tendency and variation.

10. Concluding Comments

We have now been introduced to the fundamentals of statistical analysis, including its key objectives—*data reduction*, *inference*, and the *identification of relationships*—and some of the basic building block concepts for attaining those objectives. We found that the frequency distribution of a measured variable, along with measures of central tendency (especially the mean) and variation (especially the standard deviation), are the primary tools for achiev-

ing the data reduction function of univariate statistical analysis. In the following set of chapters we will study the *inference* function, beginning with a discussion of basic probability topics. Most of the remainder of the book will then be devoted to multivariate analyses which are aimed at the third function, the study of the relationships that exist among variables and objects. At the same time, we will not be surprised to find that all the concepts of this and the preceding chapters will appear repeatedly throughout our subsequent discussions.

Part III.
INFERENCE

Chapter 6

Basic Probability

1. Introduction

The frequency distribution and the various measures of central tendency and variation introduced in Part II provided us with a look at the first broad function of statistical analysis, that of data reduction—the summary of large sets of observations into more compact and interpretable forms.

We can now undertake the study of the second major function of statistical analysis, its role as an inferential measuring tool; a means of drawing conclusions about population parameters based on sample statistics. This role as a measuring instrument is inherent in the nature of *inferential* statistical analysis, for whenever we make measurements upon a sample of objects our inferences about the corresponding characteristic in the parent population from which the sample was drawn, must necessarily contain an unknown degree of error. This is not a concern of descriptive statistical analysis, for when we measure every single member of a population there is no larger universe to which we want to generalize. When we are dealing with samples, on the other hand, we must make inferences about the parent population, inferences which are subject to uncertainty, and consequently must be couched in the language of *probability*. So before we can address ourselves directly to the issues of statistical inference we must first understand the notion of probability itself.

2. Conceptualizations of Probability

The concept of *probability* is at once intuitively simple, yet impossible to define in a *noncircular* manner; that is, the words we have available to define the concept need defining themselves, and in the end, their definitions are found to depend on the very concept that they are attempting to define.

Consider, for example, the following words and their many variations, which are related in one way or another to the concept of probability:

likelihood, *chance*, *certainty*, *random*, *risk*, *luck*, *gamble*, *maybe*, *speculative*, *haphazard*, *accidental*, *stochastic*, *chaotic*, *entropy*, *perhaps*, *possibility*, *expectation*, *confidence*, *trust*, *hope*, *faith*, *miracle*, etc.

As a *common sense* understanding of the notion of probability, we could say it is that concept represented by the above cluster of words, supposing we have a common intuitive understanding of the meaning of these concepts, even though we may not be able to define any one of the terms independently of the others. This type of interpretation of probability is shown in Figure 1*a*, and its circular nature should be apparent.

A somewhat more objective and quantitative interpretation of probability can be had by thinking of it as the *relative frequency of occurrence* of an event in the "long run;" that is, the ratio of the number of times the event occurs to the total number of opportunities for occurrence of the event:

$$\textit{Probability of an event} = \frac{\textit{Number of occurrences of the event}}{\textit{Number of opportunities of occurrence}}$$

From this relative frequency of occurrence interpretation, it is apparent that the maximum probability of occurrence will be 1.0—when the event occurs at *every* opportunity—and the minimum probability will be 0, when the event *never* occurs. This interpretation is shown graphically in Figure 1*b*.

But is this interpretation of probability anything but a glorified proportion; which, as we have seen, is nothing more than the mean of a binary variable. There is no easy answer to this question, but certainly there are

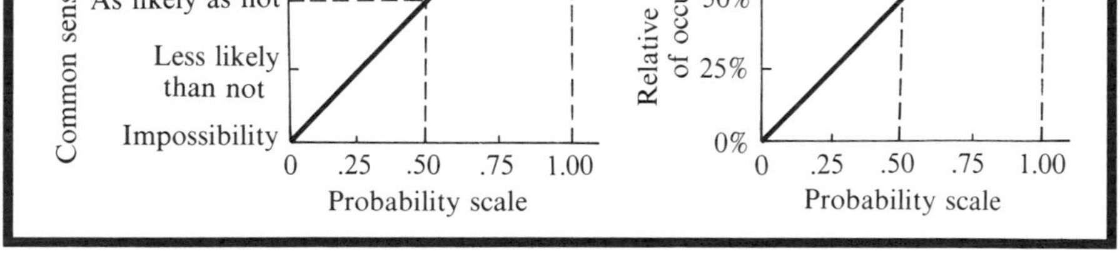

Figure 1 Circular notions of the primitive probability concept.

limitations to this definition, for our notion of probability is something more than just the past proportion of occurrence of an event. For example, we often speak of the probability of events which have occurred only once or twice in the past, or have *never* occurred, according to our knowledge, yet we speak of their probability of occurrence in the future.

For example, what is the probability that a space satellite will fall out of orbit and land in New York City? What is the probability that a nuclear power plant will malfunction and lead to the deaths of more than a thousand people? What is the probability that an earthquake will cause more than $1 million damage in Los Angeles? What is the probability that a direct-response advertisement will draw a profitable response? What is the probability of the existence of extra-terrestrial organic matter?

There are yet other instances in which events have been observed in the past, but it is possible that the circumstances of their occurrence have changed and consequently affected their future probability of occurrence. For example, what is the probability that interest rates will go down in the next year? What is the probability that the stock market will rise tomorrow? What is the probability that the crime rate will increase next year? What is the probability of meeting a sales quota? What is the probability of an economic recession next year? What is the probability of a given disease epidemic?

Theoretical vs. empirical probabilities. From the above examples it is clear that the notion of probability is something more than just a long run average of what has happened in the past. It is useful, therefore, to distinguish between *theoretical* probabilities and *empirical* probabilities, with the latter type further divided into objective and subjective probabilities.

Theoretical probabilities are those which can be determined with a good degree of confidence purely on formal or *logical* grounds, independent of prior experience, most often based on arguments of symmetry. For example, we can state the probability of a tossed coin landing heads as .5 (i.e., 1/2), since it is one of two possibilities that appear equally likely. Similarly, the probability of drawing a spade from a poker deck is .25 (i.e., 1/4), since it is one of four equally likely alternatives. When rolling a single die, the probability of it landing with the three-dot side facing upward is .167 (i.e., 1/6), since the six sides of the cube would seem to have an equal chance of landing face up. In each of these cases we would quote the probability, or guess the long-run relative frequency of occurrence of the event, without actually witnessing their occurrence. We would, of course, be greatly comforted to discover that the events actually did occur in about the same relative frequency as we had anticipated, although there is no guarantee that they would.

In the case of *empirical* probabilities, on the other hand, our basis for estimating the relative frequency of occurrence of an event rests on our past

observational behavior. For example, the probability that a conception will result in twins has no current logical basis for prediction, but yet based on millions of prior births we can state the probability of twins being born; namely, their relative frequency of occurrence in the past. Similarly, on the basis of past empirical observations we can quote the probability that an insurance applicant will file a claim within a given period of time. These types of empirically based probabilities can be considered *objective* in the sense that they are based on observations of past occurrences of the events, under what are hopefully the same conditions that currently prevail.

There is another class of empirical probabilities, however, which can more rightfully be considered *subjective* in nature; empirical in the sense that the assessment of the probability is ultimately based on past observations, but subjective in the sense that the particular observations upon which the probability estimates are based are not well-defined—that is, an independent observer could not be instructed on how to arrive at the same probability. This class of subjective probabilities include those examples mentioned above in which the events of interest have not occurred in the past, or have occurred rarely, or we have not had the opportunity to observe their possible occurrence; for example, the explosion of a nuclear power plant, an economic depression, or the successful job performance of a new employee.

In each of the above types of situation, the notion of probability applies, but we do not have a logical basis, nor a sufficient body of past observations of their occurrence, on which to estimate the probability in an objective sense. Furthermore, for those situations in which we do have past observations to rely upon, as in the past performance of the stock market, the surrounding circumstances may be different, invalidating any conclusions we may draw from the past. So, subjective probabilities may also find application in these types of *dynamic* or changing situations. To the extent that the derivation of these probabilities is not based on well-defined procedures, their full meaning will remain in the head of the person who proclaims them. They remain "hunches" or "educated guesses."

We see, then, that the concept of probability is not so simple as it would seem at first glance. And it should be realized that it is at the center of great controversy. For our purposes, let us think of it as a theoretical concept which we will use to further interpret our empirical observations. And if it succeeds in that task, as it contributes to the success of the gambler in games of chance, then we will have benefited from it. As with all theories, *they are not right or wrong, only useful or not useful.*

Recognizing that probability is an abstract concept, we can treat it much like the notion of a *point* in geometry. Although they are each primitive undefined concepts (but with intuitive meaning), entire logical systems can be built with them at the foundation. In the following sections we will learn the

basics of the theory of probability, which will then allow us to confront the issues of statistical inference which rely heavily upon it. Indeed, probability is the essential thread throughout all statistical inference.

3. Probability Experiments

Whenever we manipulate or make an observation on our environment with an *uncertain outcome*, we have conducted an experiment—or, more specifically, a *probability experiment*. Such experiments include the taking of an exam, the tossing of a coin, the delivering of a sales pitch, the drawing of a card from a poker deck, the asking of a question, the selection of a sample from a population, the hiring of an employee, etc.

In each of the above instances the experiment can be repeated many times, or at least in theory it can, and each such repetition of the experiment is called a *trial* of the experiment. Later we will encounter the expression "experiment" as it applies to an entire collection of probability experiment trials, but in the present context the term experiment will refer to a single trial.

In our analytical framework of objects, variables, and scales presented in Chapter 2, a trial of an experiment corresponds to an object, in the sense that it is the source of our observation; e.g., a coin toss, a sales pitch, an election, the drawing of a sample from a population, etc. The fact that these objects are abstract rather than tangible in nature should not concern us; the important thing is that they yield data.

4. Random Variables

In a probability experiment as defined above, the characteristic that we are measuring, the one with the uncertain value, is called a *random variable*. Consider, for example, the experiment of tossing a coin, for which there are any number of random variables in which we might be interested; the side of the coin that lands upward, the number of times it flips in the air before hitting the ground, the number of times it bounces before coming to rest, the amount of time it takes between landing and coming to rest, etc.

In the experiment consisting of a sales pitch, the uncertain outcomes could be "sale" or "no sale" (a dichotomous variable), or the outcomes could be the various possible monetary returns of the pitch (a continuous variable). Or we could draw a sample of students from a school system and calculate their mean age. In each case the value of the variable is uncertain; that is, it is associated with a probability.

Since a variable, by definition, is a characteristic of an object that can take on two or more values, we may wonder if every variable does not qualify as a random variable. In other words, what is the difference between a random

variable and one that is not. To understand the difference we must study the context in which the variable is being measured. More specifically we need to look at the nature of the object that is being observed, to determine whether or not it is an *experiment*. Consider, for example, a handful of coins. As objects they differ on the variables of weight, diameter, metal composition, and monetary value. Outside of the context of a probability experiment these would not be considered random variables; that is, when the coins themselves are the objects. But once we create an experiment, such as tossing the handful of coins into the air—in which case the coin *toss* is the object—and measure whether they land heads up or heads down, an uncertain outcome, then we are dealing with a random variable. Notice that the variables of weight, diameter, metal composition, or monetary value are not random variables in *this* context since their values are not uncertain; although in times of economic uncertainty, the monetary value of the coins after each toss could well qualify as a random variable.

Another example which will distinguish the nature of a random variable would be the nitrogen content in various sample plots of soil. If we measure the various plots for their uncertain nitrogen content, than it would be a random variable. However, if *we* ourselves manipulate the amount of nitrogen in the various soil samples by adding varying amounts of fertilizer, then it is not an uncertain value and consequently is not a random variable, but rather a "fixed" variable.

While we will be primarily concerned with random variables which are quantitative in nature, we will find that we often need to study qualitative variables (e.g., the "head" or "tail" outcome of a coin toss) as a preliminary to defining the quantitative random variable of interest. The reason for this indirect approach will become clearer in the following sections.

5. Sample Spaces

Before we can even think of the probabilities of the outcomes of an experiment, we must consider the *possibilities*. The set of possible outcomes of a probability experiment is referred to as the *outcome space* or *sample space* of the experiment.

The outcomes forming the sample space must be *mutually exclusive* and *exhaustive*. By mutually exclusive we mean that no two of the outcomes can both occur on a given trial of the experiment; or, stated alternatively, only one of the outcomes can occur on a given trial. By exhaustive we mean that the outcomes defining the sample space include *every possible* outcome of the experiment. Outcomes which fulfill these requirements are called the *simple outcomes* of the experiment. Alternatively, they are called simple events, but we

prefer the use of the term outcomes, since event seems to be more descriptive of the experiment itself rather than the result of the experiment.

Defining the sample space of an experiment is the most crucial and perhaps the most difficult aspect in solving a probability problem. In other areas of statistical analysis, and in other disciplines as well, once we have learned certain basic rules and procedures, the solution to a problem is more or less mechanical in nature. In the definition of a sample space, however, we are left pretty much at the mercy of our own wits and our abilities for logical thought. To understand the fundamental nature of the sample space, we will study a number of different probability experiments, beginning with very simple ones.

Consider the sample space for the experiment of tossing a coin, with the variable of interest being the side of the coin that faces upward upon landing. The two possible outcomes are:

Head Tail

This, then, is the sample space for the experiment as described. The two outcomes are mutually exclusive, since Head and Tail cannot both occur on a given trial of the experiment; and they are also exhaustive, since they represent all possible outcomes of the toss. So, they qualify as the simple outcomes of the experiment.

While the two outcomes of a coin toss may seem like a trivial example, it is a good model for many experiments in real life which have but two outcomes; for example, test performance (pass vs. fail), a sales pitch (sale vs. no sale), a new product introduction (success vs. failure), a medical operation (life vs. death), etc.

Now let us consider a somewhat more elaborate experiment, and try to determine its sample space. Rather than tossing a single coin once, let us define our experiment as consisting of two successive tosses of the coin, with the outcomes of interest being the *ordered sequences* of the coin landings. What are the possible outcomes? We have no rules for determining them; we must rely upon logic. Certainly we could have a Head occur on both tosses. Also we could have a Tail occur on both tosses. And then we could have a Head and a Tail. But wait! We defined our outcomes of interest specifically as being the ordered sequence of landings, therefore a "Head and a Tail" is a *different outcome* than a "Tail and a Head." In all, then, we have four simple outcomes. Letting H stand for Head, and T for Tail, we have as our sample space:

HH HT TH TT

where the first letter of each ordered pair signifies the landing of the first toss, and the second letter the landing of the second coin toss.

To be sure that we have identified the simple outcomes defining the sample space of this experiment, we need to ask if they are mutually exclusive and exhaustive. Only one of the four outcomes could occur on a given trial of the experiment, so they are mutually exclusive. Further, they are exhaustive, since they represent all possible outcomes of the experiment.

Tree diagrams. One useful method of generating the simple outcomes of a sample space, a method that will help ensure that we identify all the possibilities, is the *tree diagram*. An example of a tree diagram as applied to our preceding coin toss experiment is shown in Figure 2. The first two branches of the tree correspond to the *possible results* of the first coin toss; i.e., either Head or Tail. Each of these branches or possibilities, in turn, has two of its own branches, representing the result of the second coin toss; again, either Head or Tail. All told, the diagram terminates with four branches, corresponding to the four simple outcomes which comprise the sample space—HH, HT, TH, and TT. Using this approach we would be less likely to overlook the fact that HT and TH are really two separate outcomes of the experiment.

As another example of the application of the tree diagram, let us use it to define the simple outcomes of an experiment consisting of tossing a coin and rolling a die, with the paired results of these two activities being our variable of interest. Such a tree is shown in Figure 3. The first two branches correspond to the results of the coin toss—Head or Tail—and each of these branches has six branches of its own—the six results of the die roll (1, 2, 3, 4, 5, 6). In all, there are twelve simple outcomes comprising the sample space, which are listed at

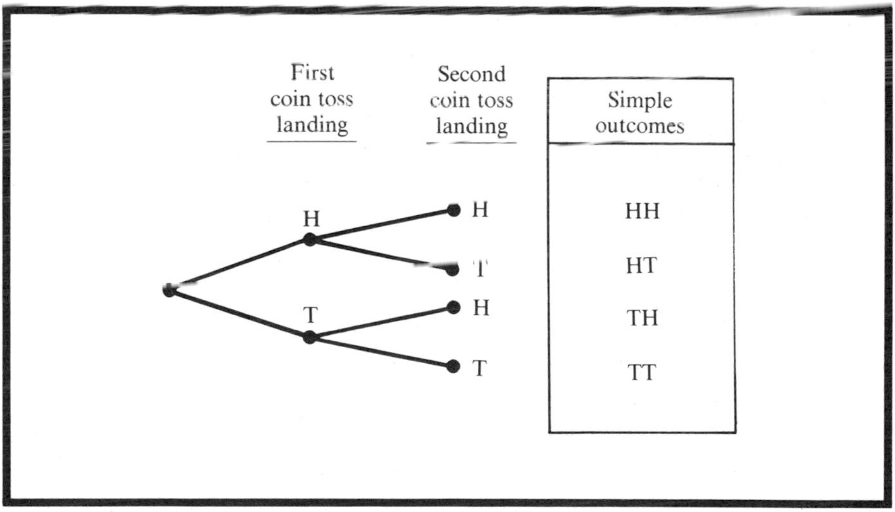

Figure 2 Tree diagram for determining the sample space for a double coin toss experiment.

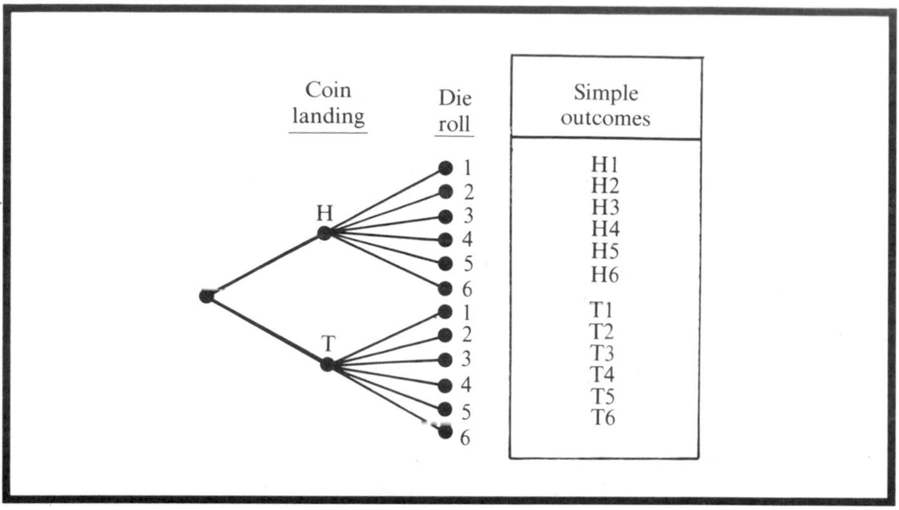

Figure 3 Tree diagram for determining the sample space for a simultaneous coin toss and die roll experiment.

the termination of each branch in the figure. Since the order of the coin toss and die roll was not an issue in this experiment, the outcome H1 (Head-1) is equivalent to the outcome 1H (1-Head). Consequently, an alternative tree diagram could have been used to generate the same outcomes. It would have begun with six branches, corresponding to the result of the die roll, and each of these would have two branches corresponding to the result of the coin toss, again resulting in the same twelve outcomes.

The preceding examples, plus those in the balance of the chapter, will be sufficient for an understanding of the basic nature of a sample space; namely, as a set of mutually exclusive and exhaustive outcomes of a probability experiment. Also, it should be recognized that the use of the term "sample" in "sample space" does not have the same meaning as when it is used to specify a subset of a population. In fact, a sample space can more appropriately be thought of as a population itself, in that it represents an exhaustive set of alternatives. But the confusion of terms can be avoided as long as sample space is always equated with the outcome space of a probability experiment, while a sample is used to denote a subset of a population of objects or observations.

6. Probability Distributions

After having defined the probability experiment, identified the random variable of interest, and listed the possible outcomes, we are in the position to assign *probabilities* to the outcomes. Whether we base these probabilities on

logical considerations, past experience, or sheer guesswork, our probability model will require that they meet the following constraints:

(1) *The probability of a simple outcome will be represented by a number from 0 through 1.*

(2) *The sum of the probabilities of the simple outcomes within a sample space will equal 1.*

The simple outcomes of a sample space and their associated probabilities comprise what is known as a probability function, or a *probability distribution*; in the sense that it tells us how the total probability of 1 is distributed among the alternative outcomes. The similarity to a relative frequency distribution should be apparent. However, instead of each value of the variable being associated with a relative frequency, each value in a probability distribution is associated with a probability. The similarity is greater when we recall that one of the interpretations of probability is as a relative frequency of occurrence of an event in the long run. In this respect, a relative frequency distribution can be viewed as an *estimate* of the underlying theoretical probability distribution.

Table 1*a* shows the assignment of probabilities to the simple outcomes of the experiment of tossing a single coin. The two outcomes—Head and Tail—are each assigned a probability of $\frac{1}{2}$. This was accomplished by piecing together the belief—based on logical considerations—that the two outcomes are equally likely, and the fact that the two probabilities must sum to 1, according to the stipulations of the formal probability model. Hopefully, this assignment of probabilities will conform with reality. But even if we flipped the coin a thousand times, or a million, we might observe a frequency of occurrence of a Head (or Tail) that differs somewhat from the probability value of .5. But does this mean that the assigned probabilities are wrong? No. Again, the validity of a model will be judged on the basis of its usefulness in interpreting empirical observations. If the observed relative frequency of occurrence of a Head

Table 1 Probability distributions for two alternative coin-toss experiments.

(a) Single coin toss		(b) Double coin toss	
Simple outcomes	Probabilities	Simple outcomes	Probabilities
Lands Head up (H)	$\frac{1}{2}$	H on first, H on second	$\frac{1}{4}$
Lands Tail up (T)	$\frac{1}{2}$	H on first, T on second	$\frac{1}{4}$
Sum:	1	T on first, H on second	$\frac{1}{4}$
		T on first, T on second	$\frac{1}{4}$
		Sum:	1

departed substantially from .5, we might begin to doubt the *usefulness* of assuming that a Head and Tail are equally likely, for perhaps this particular coin is biased. Or perhaps the observed deviation was just a matter of chance. These are all questions of statistical inference, the drawing of conclusions about populations based on sample observations, and will be confronted later. For the moment we need to more fully understand the nature of the probability distributions which represent the populations.

Another example of the assignment of probabilities to the simple outcomes of a sample space can be found in part *b* of Table 1, for the experiment of tossing a coin two times in succession. As noted earlier, the four outcomes of this experiment include HH, HT, TH, and TT. Again on the basis of logic and the aid of the tree diagram in Figure 2, we conclude that these four possibilities are equally likely and therefore should be asigned equal probabilities. Since the four probabilities must add to 1, they must each have a value of $\frac{1}{4}$, or .25.

To this point we have been considering sample spaces that are made up of outcomes which are qualitative in nature; e.g., H, T, HH, HT, etc. As a preliminary to some later developments we will now consider an experiment which has outcomes that are quantitative. As a simple example, consider the rolling of a single die in which the outcome of interest is the number of dots on the side facing upwards after the die has come to rest. The six possible outcomes are 1, 2, 3, 4, 5, and 6. Assuming that we have a fair die, we can assign a probability of 1/6 to each of the six simple outcomes of the experiment. This is based, again, on the belief that each of the six sides of the die is equally likely to land facing upward.

As with relative frequency distributions, we can make graphical portrayals of probability distributions. The probability distribution for the outcomes of the die rolling experiment is shown in Figure 4. Notice that we use separate lines rather than adjacent bars to emphasize that this is a *discrete* variable, and that intermediate values are not possible.

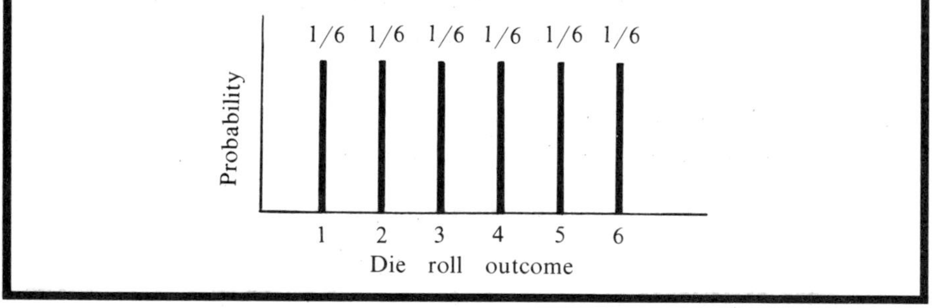

Figure 4 Probability distribution for a die roll experiment.

The preceding examples typify the probability distribution for a discrete random variable; namely, the simple outcomes of a *finite* sample space and their associated probabilities. The normal distribution, on the other hand, is an example of a probability distribution for a *continuous* random variable. However, since the sample space upon which it is based contains an *infinite* number of outcomes, it is impossible to list them as with discrete variables. But, as we have already learned in the preceding chapter, we can determine the relative frequency of occurrence of an *interval* of values—which we will now call the probability of occurrence of those values—by measuring the *proportion of area* under the curve between any two values. That is, since the total area under the curve equals 1—the sum of the probabilities in the infinite sample space—then the proportion of that total area between any two values will represent the probability of occurrence of the values in that interval.

7. Composite Outcomes

Very often our primary concern is not with the simple outcomes themselves, but with a *collection* of simple outcomes. For example, with the roll of the die we might be interested in the outcome corresponding to "at least a 3," which is equivalent to the collection of simple outcomes 3, 4, 5, or 6. Or we might be interested in the outcome of a roll which is "odd" in value, which would be equivalent to the set of simple outcomes 1, 3, or 5.

Such a collection of simple outcomes is referred to as a *composite outcome*. In a given trial of an experiment, the composite outcome is said to occur if any one of the simple outcomes which comprise it occurs. Thus, in the die rolling experiment, the composite outcome "odd value" will have occurred when any of the simple outcomes defining it—1, 3, or 5—occurs.

The probability of a composite outcome is equal to the sum of the probabilities of the simple outcomes of which it is comprised. Therefore, the probability of an "odd value" die roll will be the sum of 1/6 and 1/6 and 1/6, the respective probabilities of the simple outcomes 1, 3, and 5. This yields a probability of 3/6 or .5.

The concept of composite outcomes is also useful for creating new sample spaces. Consider, for example, the experiment of flipping a coin *three* times in succession. A tree diagram will show that there are eight possible outcomes to the experiment:

HHH HHT HTH HTT THH THT TTH TTT

Each of these eight simple outcomes is equally likely so we can assign a

probability of 1/8 to each. But let us suppose that our real interest in the experiment is the *number of Heads* that occurred among the three coin tosses. Toward that end we can translate the simple outcomes above into a new set of outcomes, consisting of the possible number of Heads. This has been done in Table 2. The outcome "3 Heads" corresponds to the original simple outcome HHH, and takes on its probability value of 1/8. But the outcome "2 Heads" is a composite outcome consisting of the simple outcomes HHT, HTH, and THH. The probability of this composite outcome will then be the sum of its component simple outcome probabilities, namely 1/8 plus 1/8 plus 1/8, for a total of 3/8. We also see in the table that the outcome "1 Head" is made up of the simple outcomes HTT, THT, and TTH, and the sum of their probabilities is 3/8. Finally, "0 Heads" corresponds to the outcome TTT, and has a probability of 1/8.

Since the probabilities of the events 3, 2, 1, and 0 Heads sum to 1, they are now *simple* outcomes of a new sample space, even though the outcomes 2 Heads and 1 Head are each *composite* outcomes with respect to the *original* sample space. Why, we might ask, did we go through the bother of constructing the original sample space if we were only going to transform it into another. Why did we not immediately construct the sample space for 3, 2, 1, or 0 Heads. To answer this question, ask yourself how likely we would have been to assign the probabilities 1/8, 3/8, 3/8, and 1/8 to the respective outcomes without the aid of the preliminary sample space.

Table 2 Alternative sample spaces and probability distributions for a triple coin toss experiment where interest is in (a) the *sequence* of Head and Tail landings, vs. (b) the *number* of Heads landing.

(a) Preliminary sample space		(b) Derived sample space	
Simple outcomes	Probabilities	Simple outcomes	Probabilities
HHH	1/8	3 Heads	1/8
HHT	1/8		
HTH	1/8	2 Heads	3/8
THH	1/8		
HTT	1/8		
THT	1/8	1 Head	3/8
TTH	1/8		
TTT	1/8	0 Heads	1/8
Sum:	1	Sum:	1

We see, then, that a sample space of equally likely simple outcomes is a good starting point for the construction of other sample spaces which may include composite outcomes. While the composite outcomes that we created in the preceding example were in turn simple outcomes of a newly defined sample space, this will not always be the case nor our intent. Many times we are interested in composite outcomes that overlap—i.e., are not mutually exclusive—and may not be exhaustive of the sample space from which they are formed, and consequently do not meet the requirements of defining a sample space.

That composite outcomes are not necessarily mutually exclusive nor exhaustive, can be seen in Figure 5. Composite outcomes A and B are alternatively defined from a sample space consisting of ten equally likely simple outcomes $(o_1, o_2, \ldots, o_{10})$. Only in part a of the figure do the composite outcomes A and B qualify as simple outcomes for a newly defined sample space. In each of the other instances, they either are not mutually exclusive or not exhaustive.

We can now summarize the relation between composite and simple outcomes with some convenient shorthand notation. If we let $o_1, o_2, \ldots, o_n$ stand for a general collection of n simple outcomes, and then define a composite outcome A as

$$A = \{o_1, o_2, \ldots, o_i\}$$

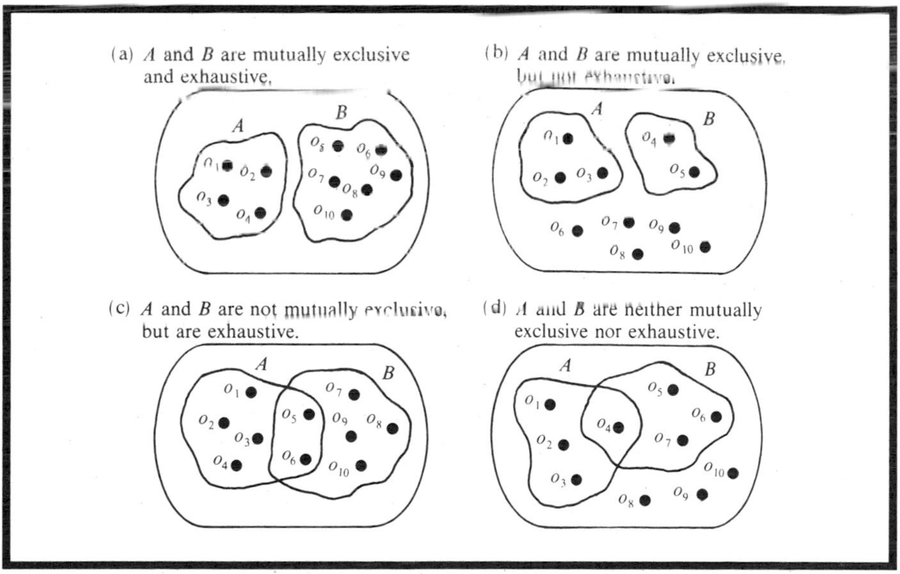

Figure 5 Examples of composite outcomes A and B defined on a sample space consisting of equally likely simple outcomes $o_1, o_2, \ldots, o_{10}$.

then outcome A is said to occur whenever one of the simple outcomes comprising it occurs. Letting $P(A)$ stand for the "probability of outcome A" and $P(o_i)$ stand for the probability of a particular simple outcome o_i, we have:

$$P(A) = P(o_1) + P(o_2) + \cdots + P(o_i)$$

Again, these two statements are nothing more than shorthand notation for the two relationships we have discussed above. The first stating that a composite outcome is a collection of simple outcomes. The second stating that the probability of a composite outcome is equal to the sum of the probabilities of the simple outcomes which define it. By convention we use capital letters to stand for composite outcomes, and lower case letters to stand for simple outcomes. In the following sections we will learn more about the relationships that can exist among composite outcomes.

8. Conditional Probability

When we have composite outcomes which overlap (i.e., are not mutually exclusive), we will often be interested in their *conditional probabilities*; that is, the probability of occurrence of one composite outcome, *given that another composite outcome has occurred*. For example, what is the probability of a die roll being a 6, *given that the roll is even in value*? What is the probability of a consumer being a user of Product A, *given that they are* 35 *years or older*? What is the probability that a person subscribes to Magazine A, *given that they subscribe to Magazine B*? What is the probability of the stock market going up today, *given that it went up yesterday*?

To understand how we determine conditional probabilities such as those above, consider the sample space in Figure 6. In it we have defined two composite outcomes, A and B, in a sample space made up of ten equally likely

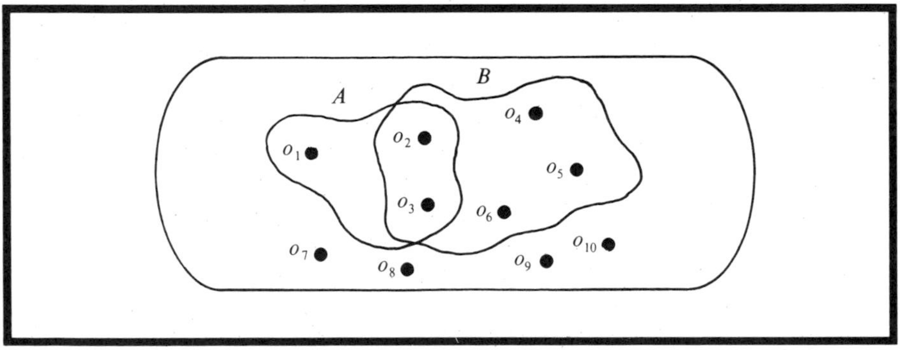

Figure 6 A sample space in which $P(A|B) = 2/5 = .4$.

simple outcomes, $o_1, o_2, \ldots, o_{10}$. Let us say that we are interested in knowing the probability of occurrence of outcome A, *given that outcome B has occurred*. Studying the figure, we see that outcome B is composed of the simple outcomes o_2, o_3, o_4, o_5, and o_6. That is,

$$B = \{o_2, o_3, o_4, o_5, o_6\}$$

Now since we want to know the probability of outcome A given that outcome B occurs, we must determine which of the simple outcomes comprising B *also* correspond to the outcome A. That is, rather than evaluating the probability of outcome A based on the *total* sample space, we want to evaluate it on the *restricted set of outcomes defined as outcome B*.

Inspection of Figure 6 will reveal that simple outcomes o_2 and o_3 in B also correspond to the composite outcome A. In other words, *two of the five* equally likely outcomes in B, are also A. Consequently, the probability of outcome A, *given that outcome B occurs*, is 2/5 or .4. Compare this with the *overall* probability of outcome A; that is, without respect to outcome B. It is 3/10 or .3, consisting of the three simple outcomes o_1, o_2, and o_3, out of the total of ten.

We can now develop a general formula for determining the value of a conditional probability. Let us begin by introducing a shorthand notation for the conditional probability of a composite outcome A, given outcome B:

P(A|B) = The probability of outcome A, given that outcome B occurs.

Based on our preceding example we can further state that the above conditional probability is equal to the ratio of the *number* of simple outcomes which are in *both A and B*, to the *number in B*. That is, when the simple outcomes are equally likely, we have:

$$P(A|B) = \frac{\text{Number of simple outcomes in } both\ A\ and\ B}{\text{Number of simple outcomes in } B}$$

Recalling a general principle of elementary algebra, we can divide the numerator and denominator of the ratio by a constant quantity without affecting the value of the ratio. Let us divide the top and bottom of the ratio by the *total number of simple outcomes in the sample space*.

$$P(A|B) = \frac{\dfrac{\text{Number of simple outcomes in } both\ A\ and\ B}{\text{Total number of simple outcomes}}}{\dfrac{\text{Number of simple outcomes in } B}{\text{Total number of simple outcomes}}}$$

But the top ratio is now nothing more than the definition of the probability of

the outcome corresponding to the occurrence of *A and B*. And the bottom ratio is nothing more than the definition of the probability of B. So we have:

$$P(A|B) = \frac{P(A \text{ and } B)}{P(B)} \tag{1}$$

This definition of conditional probability will be easy to remember if we keep in mind its derivation based on the ratio of the number of simple outcomes in *A and B*, to the number in B (when the simple outcomes are equally likely).

Let us now apply the formula to the example we worked earlier based on Figure 6.

$$P(A|B) = \frac{P(A \text{ and } B)}{P(B)} = \frac{2/10}{5/10} = \frac{.2}{.5} = .4$$

We do get the identical result as before. The reason we need the formula is that we do not always have the sample space of simple outcomes available, since in many situations it may be too large. But as long as we know the joint probability of *A and B*—i.e., $P(A \text{ and } B)$—and the probability of outcome B, we can calculate the conditional probability of outcome A, *given* outcome B. We will find this concept of conditional probability to be very important to the development of other probability concepts.

9. Multiplication Rule

Two simple outcomes, by definition, cannot occur at the same time. Two composite outcomes, however, can. Consequently, not only would we want to know the probabilities of the individual composite outcomes, but also the probability of their *joint* occurrence. For example, what is the probability of a patient having symptom *A and* disease B, what is the probability of a consumer being a subscriber to Magazine *A and* a user of Product B? What is the probability of a job applicant scoring above the mean on Test *A and* above the mean on Test B? What is the probability of a new auto having a defect in the paint job *and* a defect in the upholstery? Or, in general, what is the probability of occurrence of outcome *A and* outcome B; which we designate as $P(A \text{ and } B)$.

Logically, we would expect that the above *joint probabilities*, as they are often called, would be less than the respective probabilities of the individual outcomes. For example, the probability of a randomly selected patient having symptom A and symptom B cannot possibly be greater than the lower of the

two individual symptom probabilities. Also, we might expect that the joint probability of occurrence of two outcomes would somehow be a function of the conditional probabilities of the outcomes, and this is the case.

Using our definition of conditional probability

$$P(A|B) = \frac{P(A \text{ and } B)}{P(B)}$$

we can perform the simple algebra of multiplying both sides of the equation by $P(B)$, exchange the right and left sides, and get

$$\boxed{P(A \text{ and } B) = P(A|B)P(B)} \qquad (2)$$

This is known as the *multiplication rule* of probabilities. In words, it states that the probability of the joint occurrence of outcome A *and* outcome B is equal to the product of the probabilities of (1) the conditional probability of A *given* B, and (2) the probability of B.

As an example of its application, consider the sample space in Figure 7, composed of ten equally likely simple outcomes, and upon which two composite outcomes, A and B, are defined. To determine the probability of the joint occurrence of outcomes A and B, we need to know $P(A|B)$ and $P(B)$. In the absence of the sample space, we would have to be given these values, but since it is available we can calculate their values as

$$P(A|B) = 2/4 = .5 \quad \text{and} \quad P(B) = 4/10 = .4$$

We can now use these values for determining the joint probability of A *and* B:

$$P(A \text{ and } B) = P(A|B)P(B) = (.5)(.4) = .2$$

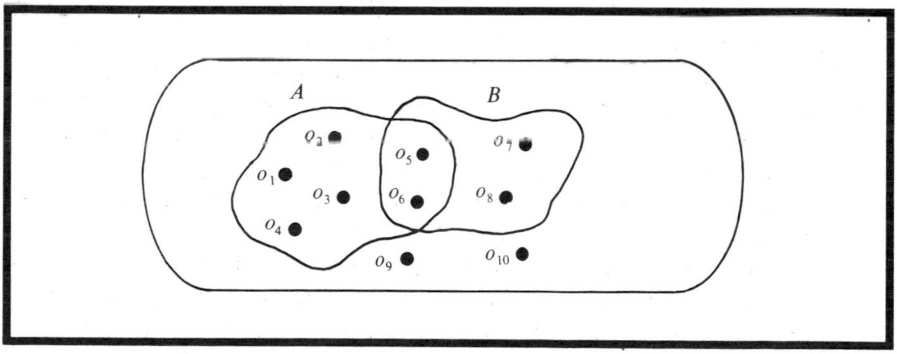

Figure 7 A sample space in which $P(A \text{ and } B) = 2/10 = .2$.

Checking the sample space we see that the result of .2 corresponds with the result we would have obtained directly from the sample space itself. That is, two of the ten simple outcomes—o_5 and o_6—occur in both outcome A and outcome B, independently yielding the probability of 2/10 or .2. So, the multiplication rule does not provide us with any information that is not available directly from the sample space, it is only informing us of the way in which probabilities behave, and therefore investing them with meaning.

10. Addition Rule

If we can determine the probability of outcome A *and* outcome B occurring, we should also be able to determine the probability of outcome A *or* outcome B occurring. For example, what is the probability that a student is above average in mathematics *or* verbal aptitude? What is the probability that a consumer uses product A *or* product B? What is the probability that a patient has high blood pressure *or* high blood sugar? What is the probability that the stock market goes up tomorrow *or* that the bond market goes up? What is the probability that an employee is competent in task A *or* task B? What is the probability that a new auto has a defect in the paint job *or* a defect in the upholstery? In general, what is the probability of occurrence of outcome A *or* outcome B, which we designate $P(A \textit{ or } B)$. It should be noted that in each of these instances the use of the word "or" includes the possibility that both A *and* B occur; i.e., it should be interpreted as meaning *and/or*.

The probability of occurrence of outcome A *or* B is given by the *addition rule*:

$$\boxed{P(A \textit{ or } B) = P(A) + P(B) - P(A \textit{ and } B)} \tag{3}$$

In words, it states that the probability of occurrence of outcome A *or* outcome B is equal to the sum of their respective probabilities, minus the probability of their joint occurrence. The reason we must subtract $P(A \textit{ and } B)$ from the sum of the individual probabilities is that its occurrence is included in both of the individual outcomes, and to not do so would be counting their occurrence twice. This can be seen most clearly in the following example.

The sample space in Figure 8 is made up again of ten equally likely simple outcomes, with the composite outcomes A and B defined upon it. We see that

$$P(A) = 7/10 = .7; \quad P(B) = 5/10 = .5 \quad \text{and} \quad P(A \textit{ and } B) = 4/10 = .4$$

If we merely added the probabilities of outcomes A and B, we would get the absurd probability of 1.2. The reason it exceeds the upper limit of 1, is that the overlapping occurrences of outcomes A and B, represented by simple

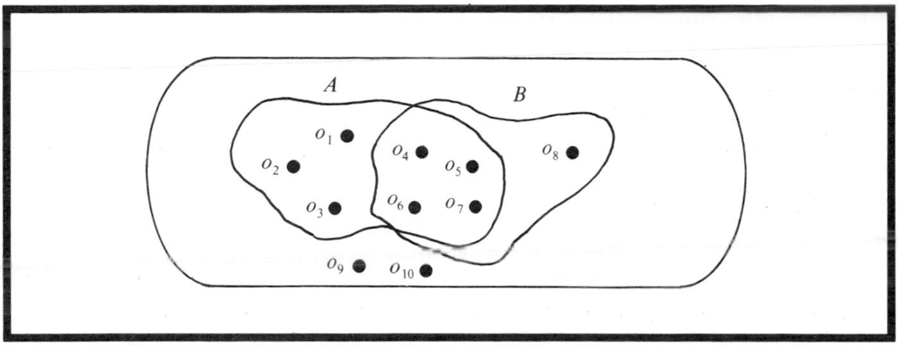

Figure 8 A sample space in which $P(A \text{ or } B) = 8/10 = .8$.

outcomes o_4, o_5, o_6, and o_7, are counted twice. And these simple outcomes happen to define $P(A \text{ and } B)$, the reason it is taken into account in the addition rule as a subtraction term. So, applying the addition rule we have

$$P(A \text{ or } B) = P(A) + P(B) - P(A \text{ and } B) = .7 + .5 - .4 = .8$$

Checking this result against the sample space we find that it does indeed correspond to the probability of outcome *A or B*, representing eight of the ten equally likely simple outcomes.

It should now be clear that if the outcomes *A* and *B* are *mutually exclusive*, the addition rule simplifies to

$$P(A \text{ or } B) = P(A) + P(B)$$

since if there is no overlap between the outcomes, $P(A \text{ and } B) = 0$. If outcomes *A* and *B* are *exhaustive* of a sample space, then $P(A \text{ or } B) = 1$, since one or the other of the exhaustive outcomes *must* occur.

11. Independent Outcomes

If the probability of occurrence of an outcome *A* is the same *regardless* of whether or not an outcome *B* occurs, then outcomes *A* and *B* are said to be *independent* of one another. Symbolically

If $P(A \mid B) = P(A)$, *then A and B are independent outcomes.*	(4*a*)

Recalling that $P(A \text{ and } B) = P(A \mid B)P(B)$, then we can also state the follow-

ing relationship for independent outcomes:

$$P(A \text{ and } B) = P(A)P(B), \text{ if and only if } A \text{ and } B \text{ are independent.} \quad (4b)$$

In other words, the probability of the joint occurrence of *independent* outcomes is equal to the product of their respective probabilities. Conversely, if the probability of the joint occurrence of two outcomes is equal to the product of their respective probabilities, we can conclude that the outcomes are independent.

In Figure 8, outcomes A and B are not independent since the $P(A|B) = 4/5 = .8$ is not the same as the $P(A) = 7/10 = .7$. That is, the probability of occurrence of A *is* dependent on the outcome of B, occurring somewhat more often in its presence than it does overall. Also, verify that (4b) is violated.

In Figure 9, on the other hand, outcomes A and B are independent of one another. The probability of A *given* B is identical to the probability of A in total; i.e., $P(A|B) = 2/5 = .4$, and $P(A) = 4/10 = .4$. In other words, the conditional and unconditional probabilities of outcome A are the same; i.e., the chance of outcome A occurring is not influenced by the occurrence of outcome B. Note, also, that (4b) is satisfied.

While the only sure way to determine whether two outcomes are independent of one another is to test it with formula (4), we can obtain an intuitive idea of whether two outcomes are independent by asking ourselves whether the occurrence of one of the outcomes could in any way affect the chances of the other outcome occurring. For example, in a double toss of a coin it should be clear that the probability of a Head on the second toss is independent of a Head on the first toss. After all, the coin has no memory. On the other hand, the probability of drawing an ace in the card game of Black Jack, is dependent

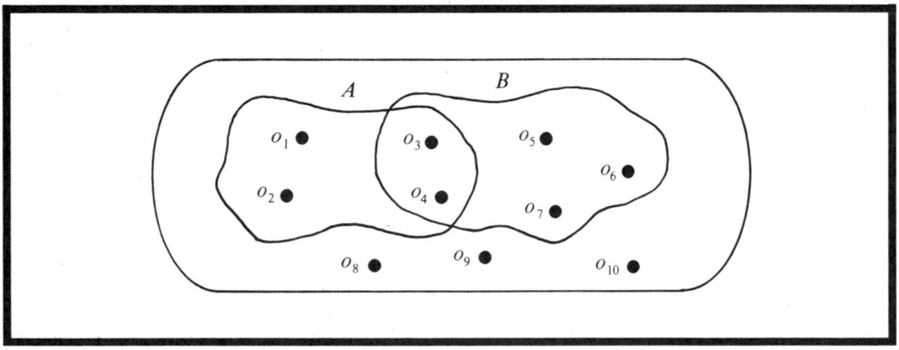

Figure 9 A sample space in which $P(A|B) = P(A) = .4$ (i.e., independent outcomes).

on the drawing of an ace on the previous draw, since the deck (or sample space) has changed by virtue of the preceding outcome.

When we depart from games of chance it is not always so easy to identify clear-cut examples of dependent and independent outcomes. For example, we could ask ourselves whether the probability of a defect in the paint job of a new auto is independent of the occurrence of a defect in the upholstery. In other words, is the probability of a paint job defect any different when we look at all autos as opposed to when we only look at those with an upholstery defect? We might in fact expect such a difference, since the conditions which led to one type of defect (perhaps a "blue Monday" or a gloomy day or a short shift) may also have contributed to the occurrence of the other type of defect.

Let us now consider an example with some actual probability figures. Assume we are interested in the readership of three magazines, A, B, and C, among a large group of consumers. Using the tree diagram in Figure 10 we can generate the sample space for the experiment of determining which combination of the magazines is read by a randomly selected individual. We find that there are eight simple outcomes for this experiment, but unlike our previous examples these outcomes are not equally likely. Their *empirical* probabilities are shown in Table 3. While the probabilities of the eight outcomes are not the same, they do still sum to 1 as required for a sample space.

Based on these simple outcomes, we can create a number of composite outcomes. This is done in the right hand portion of Table 3. We determine the probability of reading Magazine A by summing the probabilities of all those simple outcomes which include it; namely, readership of A, *B and C*, readership of *A and B only*, readership of *A and C only*, and readership of *A only*. We arrive at a probability of readership for Magazine A of $P(A)=.2$. Following a

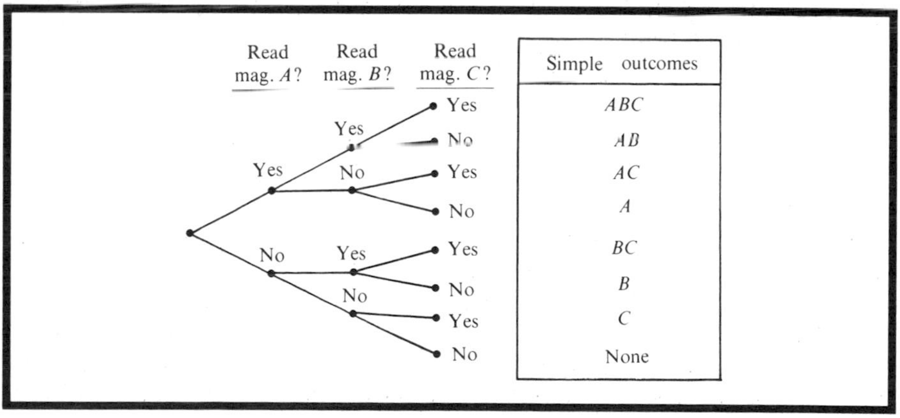

Figure 10 Tree diagram for determining the sample space for the inter-readership of three magazines for a randomly selected individual.

Table 3 Sample space and probability distribution for the inter-readership among three magazines, and selected composite outcomes.

		Examples of composite outcomes				
Simple outcomes (magazines read)	Empirical probabilities	Reads *A*	Reads *B*	Reads *C*	Reads *A* & *B*	Reads *A* & *C*
A, *B*, & *C*	.01	.01	.01	.01	.01	.01
A & *B* only	.05	.05	.05		.05	
A & *C* only	.06	.06		.06		.06
B & *C* only	.09		.09	.09		
A only	.08	.08				
B only	.15		.15			
C only	.09			.09		
None of them	.47					
Sums:	1.00	.20	.30	.25	.06	.07
		Probabilities of the composite outcomes				

similar procedure, we can determine the probability of readership for Magazine *B*, and it is $P(B)=.3$.

Now, to determine whether readership of Magazine *A* is independent of readership for Magazine *B*, we must take the product of their respective probabilities

$$P(A)P(B) = (.2)(.3) = .06$$

and compare it with the probability of readership of both Magazine *A and B*, which is determined in the second last column of the table as

$$P(A \text{ and } B) = .06$$

We see that the product of the individual probabilities does indeed equal the probability of the joint readership of the magazines, and therefore we conclude that the readership of these two magazines is independent of one another.

On the other hand, the probability of readership of Magazines *A and C* are not independent of each other, since

$$P(A)P(C) = (.2)(.25) = .05$$

(a)

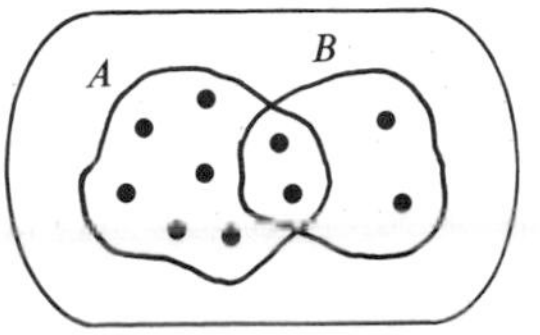

Given:

$P(A) = 8/10 = .8$

$P(B) = 4/10 = .4$

$P(A|B) = 2/4 = .5$

Conclusions:

- $P(A \text{ and } B) = P(A|B)P(B) = (.5)(.4) = .2$
- $P(A \text{ or } B) = P(A) + P(B) - P(A \text{ and } B) = .8 + .4 - .2 = 1.0$
- Since $P(A|B) \neq P(A)$, A and B are not independent

(b)

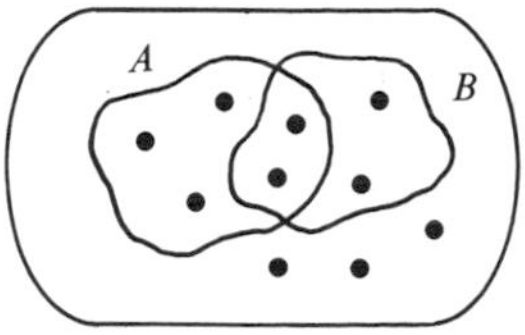

Given:

$P(A) = 5/10 = .5$

$P(B) = 4/10 = .4$

$P(A \text{ and } B) = 2/10 = .2$

Conclusions:

- $P(A|B) = \dfrac{P(A \text{ and } B)}{P(B)} = \dfrac{.2}{.4} = .5$
- $P(A \text{ or } B) = P(A) + P(B) - P(A \text{ and } B) = .5 + .4 - .2 = .7$

(c)

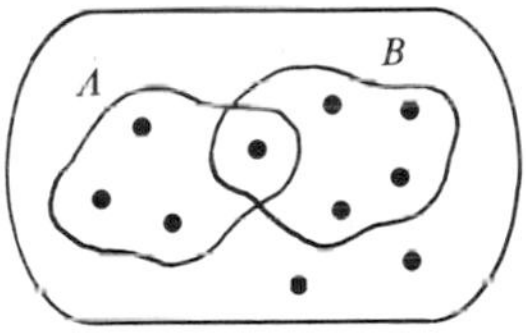

Given:

$P(A) = 4/10 = .4$

$P(B|A) = 1/4 = .25$

$P(A|B) = 1/5 = .2$

Conclusions:

- $P(A \text{ and } B) = P(B \text{ and } A) = P(B|A)P(A) = (.25)(.4) = .1$
- Since $P(A \text{ and } B) = P(A|B)P(B)$,

$$P(B) = \frac{P(A \text{ and } B)}{P(A|B)} = \frac{.1}{.2} = .5$$

(d)

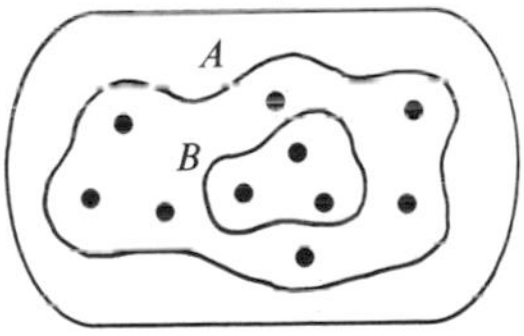

Given:

$P(B) = 3/10 = .3$

$P(B|A) = 3/10 = .3$

$P(A \text{ and } B) = 3/10 = .3$

Conclusions:

- $P(B|A) = P(B)$, therefore A and B are independent
- So, $P(A \text{ and } B) = P(A)P(B)$,

$$\text{or } P(A) = \frac{P(A \text{ and } B)}{P(B)} = \frac{.3}{.3} = 1.0$$

(e)

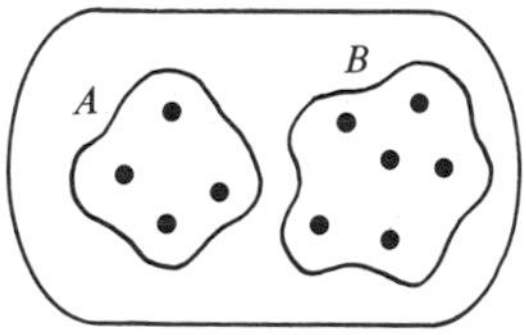

Given:

$P(A \text{ and } B) = 0$

$P(A \text{ or } B) = 1.0$

$P(A) = 4/10 = .4$

$P(A|B) = 0$

Conclusions:

- $P(A|B) \neq P(A)$, therefore A and B are not independent
- Since $P(A \text{ or } B) = P(A) + P(B) - P(A \text{ and } B)$,
  $P(B) = P(A \text{ or } B) - P(A) + P(A \text{ and } B) = 1.0 - .4 + 0 = .6$

Figure 11 Illustrations of the basic rules of probability.

while

$$P(A \text{ and } C) = .07$$

This tells us that a person is more likely to read Magazine *A* given that they read Magazine *C*, suggesting that these two magazines have a more common appeal than *A* and *B*, which were found to have independent probabilities of readership.

Figure 11 shows a number of other examples of independent and dependent outcomes. A full hour's study of the examples will crystallize all of the probability concepts we have studied to this point, since they are all interrelated.

Causality issue. The fact that two outcomes are not independent of one another, does not imply that one *causes* the other. For example, if the probabilities of patients having (a) high blood pressure, and (b) high blood cholesterol, are found to be dependent, it does not necessarily follow that one causes the other. Rather, they may each be dependent on a complex of other intervening variables, with no direct link between the two. We will address this type of confounding influence at length in later chapters when we further study the associations that exist between the values of random variables, such as those illustrated above.

12. Expected Value of a Random Variable

To obtain the mean of a probability distribution we weight each value of the random variable by its associated probability, and then sum these products. Dividing the sum of the products by the sum of the probabilities, as with other weighted means, is academic since the probabilities necessarily sum to 1. The result of this procedure is also known as the *expected value* of the random variable, and is denoted $E(x)$. In symbols, the expected value or mean of a random variable is defined as follows:

$$\boxed{E(x) = \sum x_i P(x_i) = \mu} \qquad (5)$$

In words, the expected value or mean of a random variable is equal to the sum of the products of the variable's values and their respective probabilities.

A simple example of the calculation of the expected value of a discrete random variable is shown in Table 4. In it we have the distribution of the number of Heads resulting from a triple toss of a coin, and their associated probabilities taken from Table 2. We merely multiply each value of the variable by its probability, and sum these products.

$$E(x) = 3(\tfrac{1}{8}) + 2(\tfrac{3}{8}) + 1(\tfrac{3}{8}) + 0(\tfrac{1}{8}) = 1.5 \text{ Heads}$$

Table 4 Calculation of the expected value of the number of Heads in a triple coin toss experiment (See Table 2b).

x_i Number of heads	$P(x_i)$ Probability of x_i	$x_i P(x_i)$
3	1/8	3/8
2	3/8	6/8
1	3/8	3/8
0	1/8	0
		Sum: 12/8

$\mu = E(x) = \Sigma x_i P(x_i) = \frac{12}{8} = 1.5$ heads

The probability distribution and the location of the expected value is shown in Figure 12.

When we are confronted with continuous variables and their *infinite* sample spaces, we can no longer use the above procedure for determining their expected values. Rather, it requires an application of integral calculus. But just as we imagined a histogram with everdiminishing bars blending into a continuous distribution, we can imagine the areas of those diminishing bars as the weights applied to the midpoints of the intervals forming the bases of these bars.

In all subsequent discussions, references to the expected value of a random variable will be equivalent to referring to the mean of its presumed probability distribution. In inferential problems it is usually a hypothesized value which is weighed against empirical evidence.

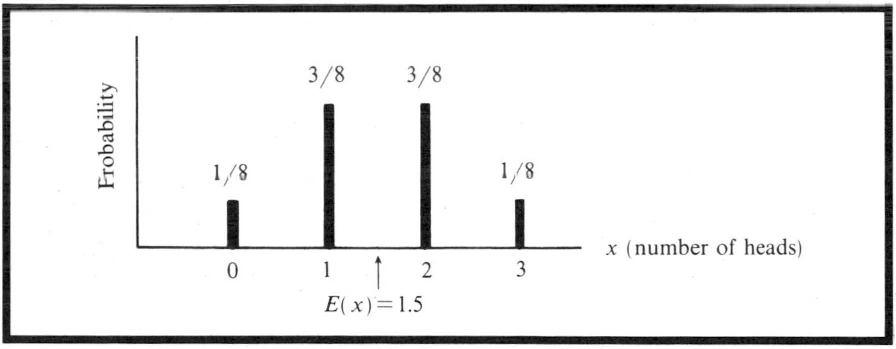

Figure 12 Probability distribution and expected value for the number of Heads in a triple coin toss experiment (See Tables 2 and 4).

13. Odds and the Logit of p

A concept closely related to that of probability is the notion of *odds*. Briefly stated:

$$\text{Odds } \textit{in favor} \text{ of outcome } A = \frac{\text{Expected number of } \textit{occurrences} \text{ of outcome } A}{\text{Expected number of } \textit{non-occurrences} \text{ of outcome } A}$$

This is equivalent to the ratio of the expected number of occurrences of outcome A for *each* non-occurrence. For example, the odds in favor of incorrectly answering a 4-alternative multiple choice question, by guessing blindly, are 3 to 1, usually written 3 : 1. Or, if the odds in favor of avoiding all-out nuclear warfare are 10,000 : 1, then we can expect to avoid such warfare on 10,000 occasions for each occasion that we do not!

The odds *against* an outcome is an alternative way of looking at things, and we have:

$$\text{Odds } \textit{against} \text{ outcome } A = \frac{\text{Expected number of } \textit{non-occurrences} \text{ of outcome } A}{\text{Expected number of } \textit{occurrences} \text{ of outcome } A}$$

For example, the odds against drawing a heart from a poker deck are 3 to 1. The odds against rolling a 2-spot with a single die are 5 : 1. The odds against getting all three heads in a triple coin toss experiment are 7 : 1 (see Table 4). The odds against a head for a single coin toss are 1 : 1, or "even money" in common language.

It is possible, of course, that the odds for or against an outcome could be fractional in value—say, 3.5 : 1—but they are often expressed in their equivalent whole numbers; in this example, 7 : 2.

It is easy enough to translate odds into probabilities and vice versa. Using the following notation

$$\text{Odds } \textit{in favor} \text{ of } A = \frac{N(A)}{N(\sim A)}$$

where $N(A)$ is the expected number of occurrences of outcome A, and $N(\sim A)$ is the expected number of non-occurrences of outcome A, then the probability *in favor* of outcome A is given by

$$p = \frac{N(A)}{N(A) + N(\sim A)}$$

in which the denominator represents the number of all outcomes combined, A and not A.

For example, if the odds in favor of surviving a medical operation are quoted as 5 : 2, the corresponding probability of survival is

$$P(\textit{surviving}) = \frac{5}{5+2} = \frac{5}{7} = .71$$

The probability of *not surviving* will be $q = 1 - p$, or .29.

It is just as easy to convert a probability into odds. Suppose we have a probability of an outcome A given by p. Then, the probability of not getting outcome A will be $q = 1 - p$. The odds are then expressed simply as

$$\boxed{\text{Odds in favor of outcome } A = \frac{p}{q}} \qquad (6)$$

which can also be thought of as the ratio of the relative number of occurrences of the outcome to the relative number of non-occurrences of the outcome.

As an example of translating probabilities into odds, suppose that the probability of an outcome A is given by $p = .75$, then the equivalents odds are

$$\text{Odds in favor of outcome } A = \frac{p}{q} = \frac{.75}{.25} = \frac{3}{1} = 3$$

or 3 : 1 in the more conventional form.

Since the notions of odds and probabilities are conceptually equivalent, it is a matter of personal preference as to which is used in everyday language. However, for problems involving complex sample spaces, it is usually more convenient to work with probabilities since they sum to 1.0 across the alternative outcomes and follow a convenient set of rules. Notice that when $p = 0$, odds will also equal 0. When $p = .5$, odds will equal 1. But when p approaches 1.0 (and consequently q approaches 0), then odds rapidly approach ∞, infinity.

Thus, it is difficult to work with odds mathematically. To get around this problem we typically deal with the natural logarithm of odds, referred to as *log odds*, or

$$\text{log odds} = \log\left(\frac{p}{q}\right)$$

which is also referred to as the *logit of p*. We will see in Chapter 13 a useful application of this concept. As an exercise for the mathematically trained, if $\log(p/q) = y$, solve for p (remember that $q = 1 - p$).

14. Concluding Comments

We now have a firm foundation in the basic theory of probability. Its development was traced through the definition of a probability experiment, the identification of the random variable of interest, the listing in a sample space of the possible outcomes of the experiment, the assigning of probabilities to these simple outcomes, and finally the rules which the probabilities obey, both for simple and composite outcomes. Now that we have this tool available, we can proceed with our study of statistical inference and its application.

Chapter 7

Sampling Distributions

1. Introduction

If we want to make inferences about a population, based on observations made upon a sample, we need to develop a theory which relates our *sample statistics* to the corresponding *population parameters*. We might like to know, for example, with what accuracy a sample mean $\bar{x}$ estimates the population mean μ. Since our sample observations are only a subset of the parent population, any given sample mean must differ from the population mean by an unknown amount. But if we knew how the sample mean behaved—that is, how its value differed upon repeated, say an *infinite* number of samplings—then, using our concept of probability, we could state with a certain degree of confidence that the population mean was within such and such a distance of the observed sample mean.

What is necessary, then, to make inferences about a population, is a *probability distribution for the sample statistic* in question; that is, the various values it can assume, and their associated probabilities. Such a probability distribution is called a *sampling distribution*. Every statistic has its own sampling distribution—the mode, median, variance, standard deviation, range, etc.—not just the mean. However, because of its paramount importance and usefulness, we will be concentrating our attention on the sampling distribution of the mean, and differences between means.

While, like every other probability distribution, the sampling distribution of the mean is strictly theoretical in nature, we can imagine it to be the result of taking an infinite number of samples from a population, calculating the mean of each, and then making a frequency distribution of the resulting means. But before we can consider the characteristics of such a theoretical sampling distribution, we need to specify how these samples must be drawn from the population in order that valid inferences can be made.

2. Random Sampling

From a probability point of view, the simplest and most useful sampling distributions will result from *random samples*, samples in which each member of the population has an equal chance of being included; or, more precisely, *every possible sample of the given size* has an equal chance of being drawn.

Since, in the case of random sampling, every possible sample of a given size that *can* be drawn from the population has an equal chance of being our *selected* sample, we will be able to make inferences about the population based on such samples that would not be possible with other types of sampling schemes. Examples of non-random sampling are rife. Most common are those surveys conducted by magazines among their readers. Readers are invited to mail in questionnaires and the results are then reported as though they apply to the population at large. Not only are they not representative of the general population, inasmuch as the readers of the magazine are not a random sample of the overall populace, but they are not even representative of the *magazine's* population of readers, since those who acquiesce to mailing in the questionnaire are not a random sample of the entire readership. We must ask ourselves whether the views and characteristics of the mailers differ from those of the non-mailers. Most likely they do. Similarly, many studies of the effectiveness of medical or psychological therapies are based on individuals who do not represent random samples of the populations to which the results are generalized. Volunteers recruited through newspaper ads, college classrooms, or hospital clinics, are not likely to be random samples of the populations to which generalizations wish to be made. Also, members of consumer panels who agree to keep diaries of their purchases, or TV viewing habits, do not represent random samples of the nation's population, and therefore cannot be viewed as representative of consumers at large.

In Chapter 1 it was stated that a random sample was equivalent to drawing slips of paper out of a hat in which each member of the population was represented by a separate slip. This is the common practice in sweepstakes and lotteries, presumably assuring every entrant an equal chance of being drawn. But even in this situation, if the entries are not well mixed, both at the start and after each draw, it is conceivable that the early or late entrants have a greater or lesser chance than others. Or if the slips of paper are of different size or texture, their chances of being drawn may not be the same as the others. Consequently, this procedure for drawing a random sample is not ideal. In addition to these drawbacks, it is often impractical to write down the names of every population member on a separate slip of paper, especially when the population is very large—which will usually be the case, otherwise why bother with just a sample when a little extra effort could result in the actual population parameter.

A solution to the practical problem of random sampling lies in the concept of *random numbers*. They can be thought of as the result of experiments such as drawing numbers from a hat. More often, though, they are generated by computers which attempt to *simulate* drawing-from-a-hat. Appendix Table I presents a set of random numbers consisting of the digits 0, 1, 2, 3, 4, 5, 6, 7, 8, and 9, which represent the values of a discrete random variable, each with the same probability of occurrence of 1/10. The probability distribution from which these digits were drawn is shown in Figure 1, which will be recognized as a rectangular or *uniform* probability distribution.

Each of the ten digits has the same probability of occurrence at any given point in the table. Not only that, each digit is equally likely to be followed or preceded by each of the digits, *including itself*. Consequently, every *pair* of digits—00, 01, 02, 03, ..., 99—is equally likely. Also, every *triplet* of digits—000, 001, 002, ..., 999—is equally likely. So too for sets of digits of any size.

The above property of the random number table is convenient for the drawing of random samples from populations. Suppose we have a population of 1,000 members, from which we want to draw a sample of 50 members. All we need to do is label the population members with identification numbers from 000 through 999. Entering the random number table at a random location, we can then select the first fifty 3-digit numbers that we encounter, and these will be the identification numbers of the members which will form our sample. The point of entry into the table can be determined with successive flips of a coin; e.g., an initial toss to determine whether to start in the first two pages or the second two pages; and a second toss to determine which of the two chosen pages to use as a point of entry. The same type of coin-flipping principle can be used to determine where on the page to enter, or that can be done less objectively by pointing into the table with one's eyes closed. Also, the

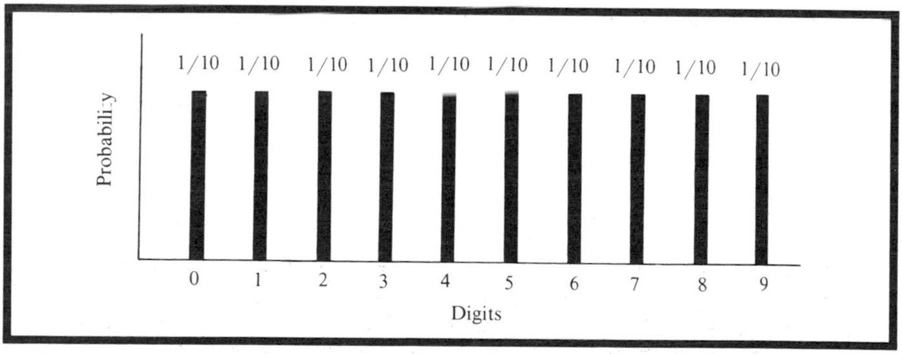

Figure 1 Probability distribution used to generate the random digits in appendix Table I.

decision as to read upward, downward, or sideways, should have been made prior to locating the entry point.

If, in the above example, the population consisted of 1,500 members instead of 1,000, we would have had to assign 4-digit identification numbers, from 0000 to 1499. Now, in drawing a sample of fifty members, we would have to select the first fifty 4-digit numbers encountered. But the task would be somewhat more cumbersome since we would have to skip over all those 4-digit numbers between 1500 and 9999, since no members in the population correspond to those nominal values. If perchance we come upon the same number twice, we would simply skip over it the second time, though this is fairly unlikely to occur.

It should be clear that since the assignment of identification numbers to the members of the population was done *independently* of the creation and contents of the random number table, each sample of identification numbers has an equal chance of being our chosen sample, and therefore the sample qualifies as being a random selection.

It is not always possible or practical to draw a strictly random sample. This occurs when the population is very large or when we cannot enumerate its every member. For example, imagine the difficult task of selecting a random sample of 100 college students from the nation's total college student population. Or selecting a random sample of 75 households from among the countries of Europe. So while practical and cost considerations often obviate the drawing of a completely random sample, we should always strive for a sampling scheme which is in the *spirit* of random sampling; i.e., drawing a sample that is believed on logical grounds to be representative of the parent population to which we want to generalize, a sample that does not blatantly overlook segments of the population that can be regarded as being different than the sampled sections.

When we want to sample a geographic region, or more generally any *spatial area* (such as a rangeland, skin, or organ surface), we need to divide it into well-defined subunits. A straightforward approach is to superimpose a cross-ruled grid, like a checkerboard pattern, over the area to be sampled, number the subunits and then draw a sample of random numbers to decide which will be included in the sample. As an exercise, design a method for sampling a *volume*, such as a kidney, a lake, a suspected ore deposit, or an archeological dig site.

While there are alternatives to random sampling, they do not always allow us to apply the rules of probability in making inferences from sample to population. In other instances, random sampling plays a role within a more complicated sampling scheme, with a resulting complication in the relevant sampling distribution. The many variations and unique properties of these alternative sampling schemes are subjects for advanced study and therefore

will not be introduced here. Our concern will be strictly with random sampling, since it is the most basic type of sampling scheme which allows us to make probability statements which relate sample statistics to population parameters, and it is also the basis for the more complicated techniques.

3. Sampling Distribution of the Mean

At the beginning of the chapter we defined a sampling distribution as the probability distribution of a sample statistic; that is, the values it can assume and their associated probabilities. Consequently, the *sampling distribution of the mean* will be that probability distribution comprised of the alternative values that the sample mean $\bar{x}$ can assume and their associated probabilities.

Although the sampling distribution of the mean has a purely mathematical derivation, based on characteristics of the parent population and the *size* of the sample, we can think of it as the result of taking repeated random samples from a population and forming a frequency distribution of the sample means. As such, the sampling distribution of the mean (or any statistic) will have a mean and standard deviation just like any other probability distribution.

One of our primary tasks is to determine how the mean and standard deviation of a sampling distribution compares with the mean and standard deviation of the parent population from which it was derived. Intuitively, we would think that the distribution of sample means would have a mean value equal to that of the parent population. But what about the *variation* of the sampling distribution? Will it be the same as the parent population, or will it be greater or less? And if it is different, how will it depend on the size of our samples? The answers to these questions will provide us with an understanding of the basic nature of a sampling distribution.

A simple example of a sampling distribution of the mean will provide preliminary insight into the above issues. Let us consider a simple population consisting of a random variable with the possible values of 1, 2, 3, and 4. Further, let us assume that it is a uniform distribution, the four values having an equal probability of occurrence; namely, 1/4. Now let us imagine drawing a sample of two observations ($n = 2$) from this population. The outcome space of this experiment will consist of all possible pairings of values, and is shown in Table 1*a*. We see that there is a total of *sixteen* different possible samples of size $n = 2$.

In Table 1*b* the *means* of these sixteen samples have been calculated. Notice that different samples—e.g., (3,1), (2,2) and (1,3)—result in the identical mean of $\bar{x} = 2.0$. Since each of the sixteen sample pairs in Table 1*a* is equally likely, by virtue of the equal likelihood of the individual observations comprising the population, the sixteen means listed in Table 1*b* are also equally likely. But since some of the means are exactly the same value, we can add their respective probabilities to obtain the overall probability of obtaining

Table 1 Possible samples and sample means based on two observations from a uniform population distribution of the values 1, 2, 3, and 4.

(a) Possible samples of size $n = 2$					(b) Resultant sample means				
	Second observation					Second observation			
First Observation	1	2	3	4	First Observation	1	2	3	4
1	1,1	1,2	1,3	1,4	1	1.0	1.5	2.0	2.5
2	2,1	2,2	2,3	2,4	2	1.5	2.0	2.5	3.0
3	3,1	3,2	3,3	3,4	3	2.0	2.5	3.0	3.5
4	4,1	4,2	4,3	4,4	4	2.5	3.0	3.5	4.0

a particular mean value. The composite outcomes in the table are highlighted with bold underlines.

Having identified a set of mutually exclusive and exhaustive outcomes of our sampling experiment, we can now create the sampling distribution of the mean $\bar{x}$ for a sample of size $n = 2$. The sampling distribution of the mean, and the parent population from which the samples were drawn, are shown in Figure 2. The probabilities of the various means were determined by adding together the probabilities of the simple outcomes in Table 1*b* which yielded a particular mean value. For example, a mean of $\bar{x} = 2.5$ has a probability of 4/16, which is the sum of the probabilities of the means arising from the samples (4,1), (3,2), (2,3), and (1,4), each with a mean $\bar{x} = 2.5$ and a probability of occurrence of 1/16.

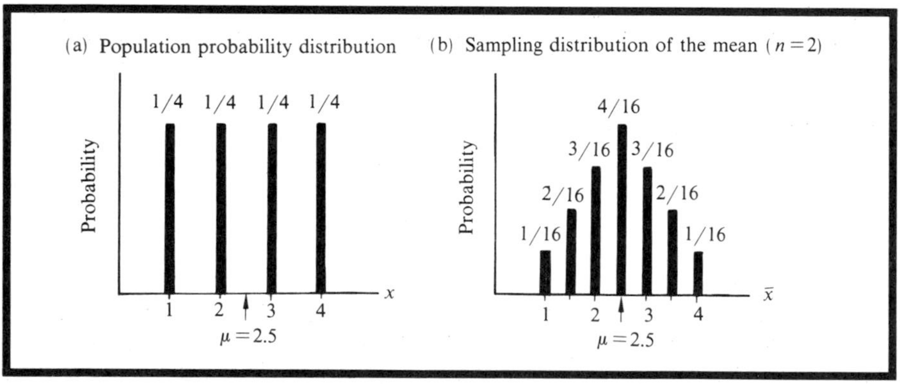

Figure 2 The (a) parent population, and (b) sampling distribution of the mean based on a sample size of $n = 2$ (See Table 1).

A comparison of the sampling distribution of the mean shown in Figure 2*b* with the parent population shown in Figure 2*a*, reveals a number of very interesting things. The first striking observation is that the sampling distribution is not rectangular in shape as is the parent population. Next, without need for calculating its expected value, we see by inspection that the *mean of the sampling distribution is equal to the mean of the population*, namely 2.5. However, we see that the *variation* of the sampling distribution is *less* than that of the parent population. Whereas the probability of an observation with an extreme value of $x_i = 1$ in the population is 1/4, the probability of a *mean* of $\bar{x} = 1.0$ is only 1/16. And this will make sense after a certain amount of reflection, for to obtain a mean of $\bar{x} = 1$, based on a sample of two observations, would require that both observations be a 1. Remembering that the probability of the joint occurrence of independent outcomes is the product of their respective probabilities, we find that the probability of a mean of 1 is indeed equal to 1/4 times 1/4, the respective probabilities of drawing a sample consisting of a 1 on the first draw and a 1 on the second draw.

4. Sampling from a Normal Distribution

With the preceding discrete variable example providing us with the flavor of a sampling distribution, we can now state, without benefit of proof, an important theorem with regard to sampling from a *normal* distribution.

Normal distribution sampling theorem:
If a variable x is normally distributed with a mean μ and a standard deviation σ, then the sampling distribution of the mean $\bar{x}$, based on random samples of size n, will also be normally distributed and have a mean μ and a standard deviation $\sigma_{\bar{x}}$ given by:

$$\sigma_{\bar{x}} = \frac{\sigma}{\sqrt{n}}$$

The various points of the above theorem are portrayed graphically in Figure 3. Beginning with a *normal* population distribution having a mean μ and a standard deviation σ, we create various sampling distributions of the mean, differing in the size n of the sample drawn. There are three points that the theorem makes. Firstly, the sampling distribution of the mean is *normal* in shape when the parent population is *normal*. Secondly, the sampling distribution has the *identical mean* as the population from which it was formed. Thirdly, and perhaps the most interesting relationship, is that the standard deviation of the sampling distribution of $\bar{x}$ is equal to *the standard deviation of*

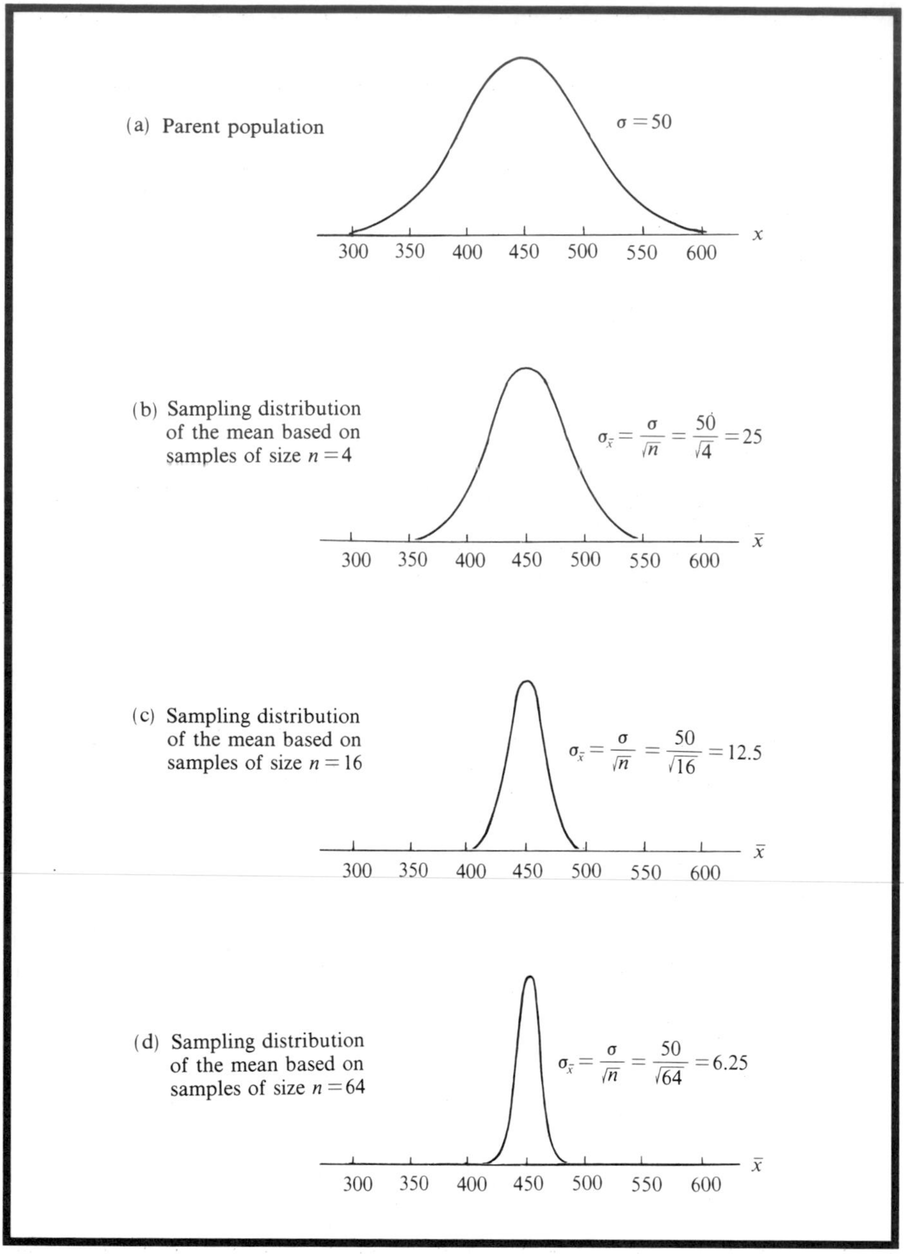

Figure 3 Parent population and sampling distributions of the mean based on different sample sizes. (The implicit vertical axes in parts, *b*, *c*, and *d* differ in order to accommodate the increasing height of the distribution; the areas under all curves being equal to 1).

the parent population divided by the square root of the sample size:

$$\boxed{\sigma_{\bar{x}} = \frac{\sigma}{\sqrt{n}}} \tag{1}$$

We refer to this standard deviation as the *standard error of the mean* and subscript it as $\sigma_{\bar{x}}$ to distinguish it from the standard deviation σ of a population of individual x_i's. By convention we use the terminology *standard error* rather than *standard deviation* when referring to the variation of a sampling distribution.

That the standard error of the sampling distribution of the mean is a function of *sample size* should not be too surprising. In Figure 2 we saw how, with a sample size of only $n = 2$, the extreme values of the mean were greatly curtailed in terms of probability. Based on that relatively simple example, it is not hard to imagine that as the sample size increases it will become more and more improbable to get a sample composed preponderantly of extreme values resulting in an extreme sample mean. As to why the standard error of the mean, $\sigma_{\bar{x}}$, should turn out to be equal to $\sigma/\sqrt{n}$ rather than any other function of the population standard deviation σ and the sample size n requires a mathematical proof beyond the scope of our presentation. However, even without a mathematical proof, we could observe the relationship $\sigma_{\bar{x}} = \sigma/\sqrt{n}$ empirically. That is, if we drew a large number of independent samples from a population (replacing the sample members after each draw), and made a frequency distribution of the resulting sample means, we would find that the standard deviation of this distribution would, within the limits of variation due to our finite number of samples, have a value close to the population standard deviation divided by the square root of the particular sample size that we employed.

In practice, though, we usually obtain only one sample mean, so it is necessary to know the exact nature of the sampling distribution of the mean in order to make inferences about the population. The above theorem, whether proven mathematically or empirically, provides us with the desired information.

5. Sampling from a Non-Normal Distribution

We may now wonder about the nature of the sampling distribution of the mean when the parent population from which the samples are drawn is *not normal* in form. In this context we have another theorem which is perhaps even more remarkable than the preceding one, and it is known as the *Central Limit Theorem*, presented at the top of page 111.

The key difference between the Central Limit Theorem and the earlier one is that in this instance we are drawing samples from a *non-normal* population,

> ***Central Limit Theorem:***
> *If a variable x has a distribution with a mean μ and a standard deviation σ, then the sampling distribution of the mean $\bar{x}$, based on random samples of size n, will have a mean equal to μ and a standard deviation*
>
> $$\sigma_{\bar{x}} = \frac{\sigma}{\sqrt{n}}$$
>
> *and will tend to be normal in form as the sample size becomes large.*

with the difference in result that the sampling distribution will only *approximate* a normal distribution, and then *only when the sample size is large*. But how large is large? It is impossible to generalize on this point, for much depends on the actual form of the non-normal population, for there are any number of possible variations from a normal distribution; e.g., *J*-shaped, *U*-shaped, rectangular, moderately skewed bell-shaped, etc. However, based on various empirical sampling studies, it has been found somewhat surprisingly that a sample size as small as $n = 30$ will often result in a sampling distribution that is very nearly normal in form, even when the original population deviates quite markedly from a normal distribution. A graphical illustration of this facet of the Central Limit Theorem is shown in Figure 4.

That the sampling distribution of the mean based on a non-normal population should eventually approach normality as the sample size increases, should be easy to accept if we refer back to the simple sampling distribution shown in Figure 2. In that example, a sample size of only $n = 2$ drawn from a *rectangular* population distribution resulted in a sampling distribution of the mean that already was taking on the appearance of a normal distribution, at least in the respect that the majority of sample means clustered in the center of the distribution. It is left as an exercise to see what happens to that sampling distribution when the sample size is increased to $n = 3$.

The importance of the normal distribution, and our reasons for placing such emphasis upon it earlier, should now be apparent; for not only is it a common form of *population* distribution, but even in those instances in which the population is not normal, the *sampling distribution of the mean* is normal, or nearly so. As we will see, this property makes the task of drawing inferences about a population based on sample observations so much easier.

6. Sampling Distribution of the Difference Between Two Independent Means

Let us now consider the situation in which we draw samples from *two* populations for comparison. This is a very common procedure. We might

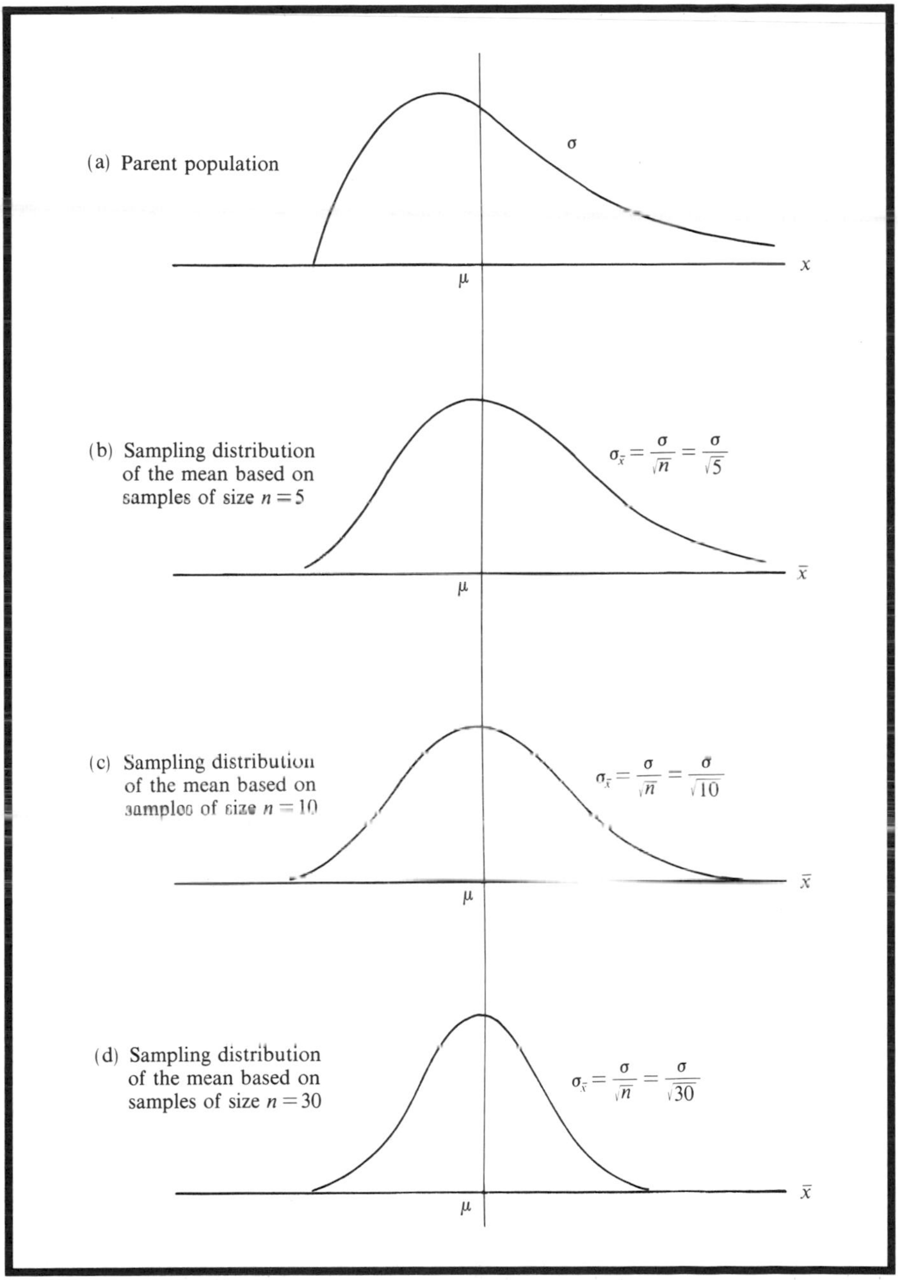

Figure 4 A non-normal parent population and sampling distributions of the mean based upon it. As sample size increases, the sampling distribution approaches normality. (The horizontal and vertical scales differ from distribution to distribution.)

compare students receiving alternative types of instruction, patients treated with alternative drugs, land plots receiving alternative fertilizers, cities with alternative criminal codes, etc. In each case, samples from the two populations are measured on some variable of interest to us. Ultimately, we want to decide whether the two populations differ with respect to the variable in question. It will take the next two chapters to fully address the question, but we begin here by laying down the theoretical underpinnings of the analytical approach.

The means of the two populations in question can be designated μ_1 and μ_2, respectively. Similarly, we can designate their respective standard deviations as σ_1 and σ_2. If we draw samples of size n_1 and n_2 from the two populations, respectively, then we know from (1) that the standard errors for the sampling distributions of the mean for the two populations will be

$$\sigma_{\bar{x}_1} = \frac{\sigma_1}{\sqrt{n_1}}$$

and

$$\sigma_{\bar{x}_2} = \frac{\sigma_2}{\sqrt{n_2}}$$

respectively. Keep these two expressions in mind, for as we might expect, they should logically be used as components in defining the standard error of the *difference* between the two sample means, $\bar{x}_1 - \bar{x}_2$.

Since our main interest is not so much in the parameters of the two individual populations, but rather in the difference between the means of the populations, $\mu_1 - \mu_2$, our focus will center on the sampling distribution of the statistic $\bar{x}_1 - \bar{x}_2$, the difference between the pair of sample means drawn from the respective population distributions. In other words, we want to ask what the sampling distribution of the difference between two sample means would look like. *What would be its mean value*? *What would be its standard error*? *And what shape would it assume*? If we set out to answer these questions empirically, we would take pairs of samples from the respective populations, calculate their means, and then take the difference between the two means. Repeating this experiment a large number of times, we could then make a frequency distribution of the resulting differences between the sample means. An examination of this empirical sampling distribution would then give us a clue as to the characteristics of the theoretical sampling distribution of the difference between two means based on an infinite number of pairs of sample means randomly drawn from the populations in question.

Fortunately, as before, there is a purely mathematical solution to the problem. The relevant theorem is as follows:

Theorem for sampling two independent populations:
If two independent variables x_1 and x_2 are normally distributed with the respective means μ_1 and μ_2 and standard deviations σ_1 and σ_2, then the sampling distribution of $\bar{x}_1 - \bar{x}_2$ will also be normally distributed, with a mean $\mu_1 - \mu_2$, and a standard error equal to

$$\sigma_{\bar{x}_1 - \bar{x}_2} = \sqrt{\sigma_{\bar{x}_1}^2 + \sigma_{\bar{x}_2}^2} = \sqrt{\frac{\sigma_1^2}{n_1} + \frac{\sigma_2^2}{n_2}}$$

where n_1 and n_2 are the respective random sample sizes.

Let us take a closer look at this apparently cumbersome theorem. But first inspect Figure 5 which presents the concepts of the theorem in graphical form, and show it to be less complicated than it would appear at first glance.

First of all, we assume that the two population variables are *independent* of one another. Using our concept of independent outcomes introduced in the preceding chapter, we can interpret this to mean that the value obtained for one sample mean in no way influences or is related to the value obtained for the other sample mean. This is likely to be the case when the variables are measured on separate groups of objects; e.g., the income of high school graduates vs. the income of college graduates, or, say, a student universe *randomly* divided into two groups. In such a situation the mean income calculated for a sample of high school graduates in no way will affect, or be related to, the mean income calculated for a sample of college graduates, even though these means may be different. In other words, repeated sampling of high school graduates will yield different means. So too for college graduates. But for any given sample drawing, and across all drawings, the high school outcome is not going to have a bearing on the college student outcome. Examples of variables which are *not* independent of one another arise when the *same* or *related* objects are measured on both variables. For example, if we measured the incomes of each of a number of pairs of twins, we would expect the value obtained for the first twin to be related to the income of the second twin. This would not be the case if people were paired at random. We will learn about the sampling distribution for this type of situation in the subsequent section.

Next, our theorem states that if the distributions of the two variables are *normal* in form, the sampling distribution of the difference between the sample means of the variables will also be *normal* in form.

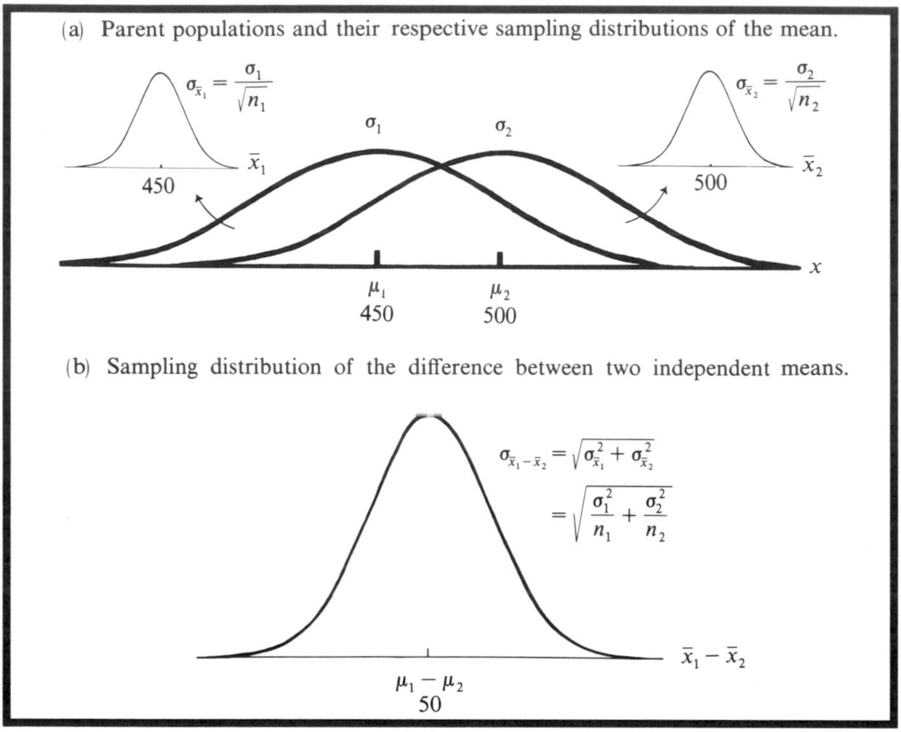

Figure 5 Independent parent populations with unequal means, and the sampling distribution of the difference between two sample means drawn from them.

Furthermore, the *mean* of the sampling distribution of $\bar{x}_1 - \bar{x}_2$ will be equal to $\mu_1 - \mu_2$, the difference between the means of the respective populations from which the samples were drawn.

In addition, the *standard error* of the sampling distribution of the difference between the means will be equal to the *square root of the sum of the variances of the sampling distributions of the mean for the respective populations*; *i.e.*,

$$\sigma_{\bar{x}_1 - \bar{x}_2} = \sqrt{\sigma_{\bar{x}_1}^2 + \sigma_{\bar{x}_2}^2}$$

This relationship is modelled geometrically in Figure 6 as an application of the Pythagorean Theorem for a right triangle, in which the two legs of the triangle correspond to the respective standard errors of the sampling distributions of the two means, and the hypotenuse corresponds to the standard error of the difference between the means. The right angle in Figure 6 gives meaning to the

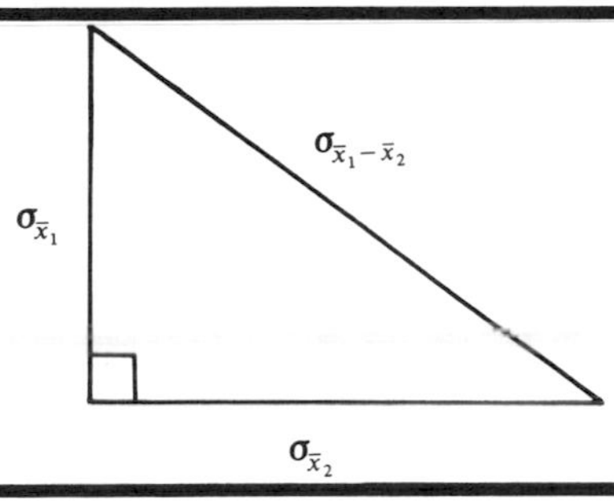

$$\sigma^2_{\bar{x}_1 - \bar{x}_2} = \sigma^2_{\bar{x}_1} + \sigma^2_{\bar{x}_2}$$

$$\sigma_{\bar{x}_1 - \bar{x}_2} = \sqrt{\sigma^2_{\bar{x}_1} + \sigma^2_{\bar{x}_2}} = \sqrt{\frac{\sigma_1^2}{n_1} + \frac{\sigma_2^2}{n_2}}$$

Figure 6 Geometric model of the standard error of the difference between two independent sample means.

term *orthogonal* as a synonym for *independent* variables, the requirement for the relationship to be true.

Because we know from formula (1) that the standard error for the first population's sampling distribution is

$$\sigma^2_{\bar{x}_1} = \frac{\sigma_1^2}{n_1}$$

and that the standard error for the second population's sampling distribution is

$$\sigma^2_{\bar{x}_2} = \frac{\sigma_2^2}{n_2}$$

we can define the standard error of the sampling distribution of the *difference between two independent sample means* more generally as:

$$\sigma_{\bar{x}_1 - \bar{x}_2} = \sqrt{\sigma^2_{\bar{x}_1} + \sigma^2_{\bar{x}_2}} = \sqrt{\frac{\sigma_1^2}{n_1} + \frac{\sigma_2^2}{n_2}} \qquad (2)$$

While the standard deviations of the two parent populations, σ_1 and σ_2, shown in Figure 5 appear to be equal to one another, they need not be. Nor is the standard error of the difference between the means dependent upon the magnitude of the difference between the population *means*. These two facts are illustrated in Figure 7, where we have two parent populations which differ in variation and have different mean values. The sampling distribution of the difference between the means based on these populations has a standard error

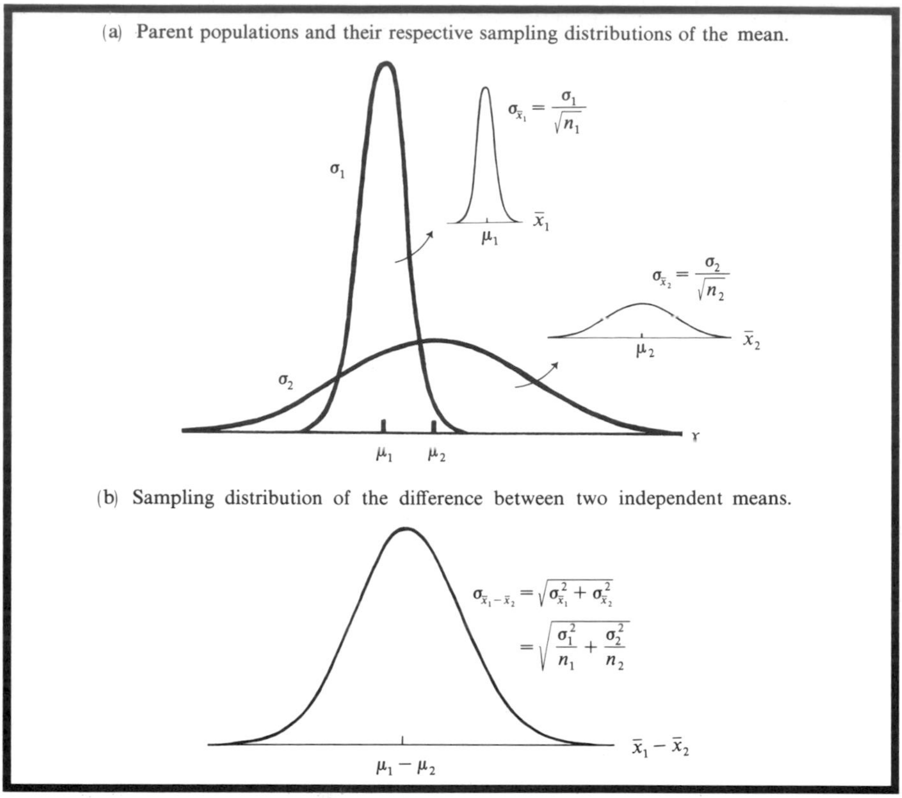

Figure 7 Independent parent populations with unequal means and standard deviations, and the sampling distribution of the difference between two sample means drawn from them.

that only reflects the variation of the parent populations. The mean of the sampling distribution, in turn, depends only on the means of the two parent populations.

We may now wonder about the sampling distribution of the difference between two means when the two parent populations are *not normal* in form. Depending upon the nature and degree of departure from normality of the populations, the sampling distribution will only be *approximately* normal in form. Since, because of the Central Limit Theorem, the sampling distributions of the mean for the separate populations will tend to be normal in form when the sample sizes are large (say, over thirty), then we can also expect that the sampling distribution of the *difference between the means* will also approach normality when the respective sample sizes *each* exceed thirty or so. This fact results in an even wider application of the normal distribution to problems of

making inferences about population parameters based on sample statistics drawn from populations that may or may not be normal in form.

7. Sampling Distribution of the Mean Difference Between Non-Independent Paired Measures

Very often we measure a set of objects on two separate variables, x and y. For example, we might measure a group of consumers' purchase interest in a product before and after viewing a commercial for the product. Or a sample of students might rate two alternative cafeteria menus.

Other times, we might measure *matched pairs* of objects on separate variables, the objects being paired in some meaningful way. For example, achievement test scores of dormitory room-mates, assigned alternative textbooks, might be measured. Or twins could be assigned alternative vitamin diets and then compared in frequency of colds.

On still other occasions, we measure pairs of objects on the *same* variable. For example, we might compare husbands and wives with respect to their scores on a particular social attitude scale. Or first-born children could be compared with their second-born sibling on some variable of interest.

In all of the above types of situations we are usually interested in comparing the *means* from the two sets of observations. But in each of the above examples, the two variables are not likely to be *independent* of each other; for in each case, knowledge of the score of one member of a pair, say in z score form, would suggest the value of the other member, at least better than a blind guess. That is, there are factors operating which influence both variables simultaneously. A consumer's purchase interest in a product before and after exposure to a commercial will be more similar than a pre-exposure and post-exposure score selected at random from *different* individuals. Similarly, room-mates will have more similar aptitudes than *randomly selected* pairs of students, due perhaps to a similarity in study habits or to the underlying basis of their room assignment. Also, husbands and wives are more likely to be similar in their attitude test scores than randomly paired males and females. The same is true for twins, litter mates, adjacent plots of land, etc.

Since none of the above types of situations represent pairs of independently distributed variables, the theorem of the preceding section cannot be used to describe the sampling distribution of the difference between the means. But we do have an alternative. Let us, instead of thinking of the problem as consisting of two non-independent variables, x and y, think of it as a single variable D defined as the *difference* between the paired values of the two measured variables. That is,

$$D = x - y$$

Any particular value of D will then represent the difference between a *pair of*

observations. For example, the difference between the pre-exposure and post-exposure purchase interest score; the difference between the achievement scores of the room-mates; the difference in attitude test score between husband and wife; or the difference in colds between each of the pair of twins. Since the value of one of the pair will have been arbitrarily predesignated as x and the other as y, the individual $D_i = x_i - y_i$ values that we obtain will sometimes be positive and sometimes be negative. Similarly, the resulting mean difference $\overline{D}$ may be positive or negative. It is also true that the *mean difference* of the paired measures will equal the *difference of the means* of the respective measures—i.e., $\overline{D} = \bar{x} - \bar{y}$.

The mean difference $\overline{D}$ has a sampling distribution just like any other mean. The *standard error of the sampling distribution of the mean difference between non-independent paired measures* is given by

$$\sigma_{\overline{D}} = \frac{\sigma_D}{\sqrt{n}} \tag{3}$$

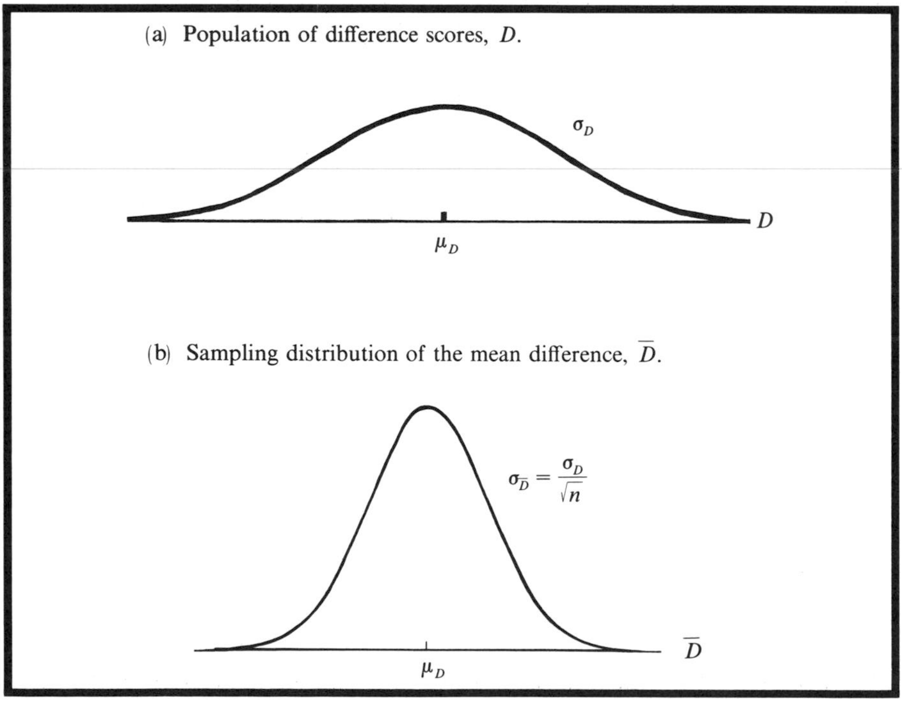

Figure 8 Population of difference scores, and the sampling distribution of the mean difference.

where σ_D is the standard deviation of the population of difference scores D_i, and n is the number of *pairs* of observations in the sample. This relationship can be seen graphically in Figure 8. It should be recognized as being no different in principle than the situation portrayed in Figure 3; the only difference being in the nature of the population variable. In Figure 8 we are merely dealing with a *derived* variable composed of *difference scores* between two other variables. But once the new variable has been created, we can treat it like any other variable, having its own mean μ_D, standard deviation σ_D, and standard error of the mean $\sigma_D/\sqrt{n}$.

As with the sampling distribution of the mean of any variable, the sampling distribution of the mean difference $\bar{D}$ will be normal in form if the parent population of difference scores is normally distributed. And if it is not, the sampling distribution will approach normality if the number of *pairs* of observations upon which the mean difference $\bar{D}$ is based exceeds thirty or so. We will return to this sampling distribution, as well as the preceding ones, in the following chapter, and discover their applications to solving problems of inference from sample to population.

8. Sampling Distribution of a Proportion

In Chapter 4 we learned that a proportion is essentially the mean of a binary variable for which we have assigned the numerical values 0 and 1 to its respective levels. As such, we would expect that a proportion should have a sampling distribution too; and so it does, just as every statistic has its sampling distribution.

To best understand the nature of the sampling distribution of a proportion, we should begin with the parent population which generates the proportions. The prototype of such a two-valued population is shown in Figure 9, alternatively portrayed as a line graph and a histogram. The line graph is a more accurate depiction of the discrete nature of the population, but the histogram has advantages for the following exposition.

By convention we let p stand for the probability of one of the outcomes, and q for the probability of the other outcome. Since the two outcomes are mutually exclusive and exhaustive of the possibilities it follows that

$$p + q = 1$$

Again, by convention, we refer to the outcome associated with probability p as a "success" while the outcome associated with q is referred to as a "failure." But these are only labels of convenience, for if we speak of the probability of an insurance policy holder dying in the next twelve months, this can hardly be thought of as a success in the popular sense of the word.

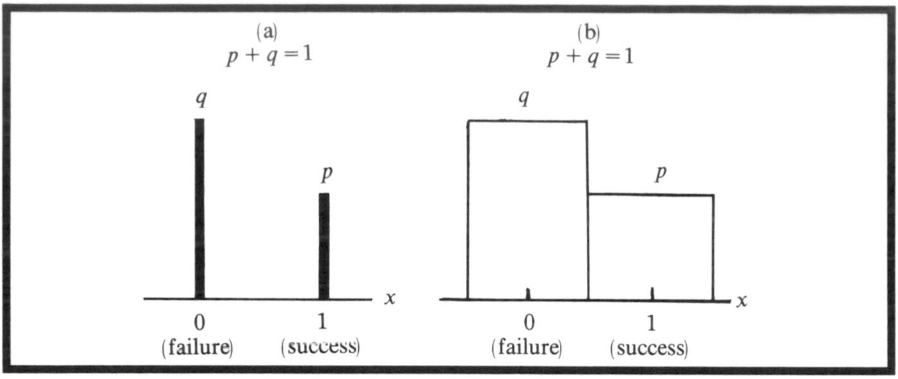

Figure 9 Probability distribution for a binary variable: (a) line graph vs. (b) histogram.

Since $p+q=1$, all we need to know is one of the two probabilities in order to determine the other. That is, if we know the value of p, we can always determine the value of q as being equal to $1-p$. For example, if the probability of surviving an operation is $p=.7$, then the probability of not surviving it is $q=1-p=.3$.

Given the probability distribution for the parent population, we can now consider the sampling distributions that it can give rise to. Let us begin with a sample of only two observations; i.e., $n=2$. We must then ask what are the possible sample proportions that could result from a sample of this size, and we would find that they are:

$$\frac{0}{2}, \frac{1}{2}, \frac{2}{2}$$

That is, we could obtain 0 successes out of 2 tries, 1 success out of 2 tries, or 2 successes out of 2 tries, which correspond to *proportions* of 0, .5, and 1.0, respectively. The total number of "tries" that we have referred to are the *trials* of the probability experiment. If instead of $n=2$ trials of the experiment we conducted $n=5$ trials, then the possible proportions that could result would include:

$$\frac{0}{5}, \frac{1}{5}, \frac{2}{5}, \frac{3}{5}, \frac{4}{5}, \frac{5}{5}$$

It is easy to see, then, that the *variety* of proportions that can result from sampling a given number of observations from a binary population is not difficult to determine. The more difficult problem is to determine the

probabilities of the alternative proportions that are possible. For this, we can use the *binomial distribution*, introduced in the following section.

Binomial distribution. If we make some key assumptions about our sampling procedure and the population probabilities, we can then obtain the probabilities of the alternative sample proportions relatively easily using the *binomial probability distribution*, a set of probabilities given by the successive terms of the expansion of the expression

$$(p+q)^n$$

where n refers to the number of independent observations made upon the binary population—i.e., the number of *binomial trials*—and p and q refer to the respective probabilities of the two values of the binary random variable.

For example, if we conducted three observational trials upon the population we would have

$$(p+q)^3 = p^3 + 3p^2q + 3pq^2 + q^3$$

The expansion is carried out by ordinary algebra, but for convenience other expansions of $(p+q)^n$ are provided in Table 2 for values up to $n = 7$. The first term in the above expression, p^3, equals the probability of *all three* trials

Table 2 The binomial distribution for selected numbers of trials.

Binomial expression	*Binomial expansion*
$(p+q)^1 =$	$p+q$
$(p+q)^2 =$	$p^2 + 2pq + q^2$
$(p+q)^3 =$	$p^3 + 3p^2q + 3pq^2 + q^3$
$(p+q)^4 =$	$p^4 + 4p^3q + 6p^2q^2 + 4pq^3 + q^4$
$(p+q)^5 =$	$p^5 + 5p^4q + 10p^3q^2 + 10p^2q^3 + 5pq^4 + q^5$
$(p+q)^6 =$	$p^6 + 6p^5q + 15p^4q^2 + 20p^3q^3 + 15p^2q^4 + 6pq^5 + q^6$
$(p+q)^7 =$	$p^7 + 7p^6q + \boxed{21p^5q^2} + 35p^4q^3 + 35p^3q^4 + \boxed{21p^2q^5 + 7pq^6 + q^7}$
↑ p: Probability of a success on a given trial. ↑ q: Probability of a failure on a given trial. ↑ 7: Number of trials.	↑ $21p^5q^2$: This term is the probability of *exactly* 5 successes and 2 failures out of 7 trials. ↑ $21p^2q^5 + 7pq^6 + q^7$: The sum of these terms is the probability of getting 2 *or fewer* successes out of the 7 trials.

ending in a "success"; the second term, $3p^2q$, is the probability of 2 successes and 1 failure; the third term, $3pq^2$ is the probability of 1 success and 2 failures; and the final term, q^3, is the probability of a failure on all three trials. Note that the *exponents* of p and q correspond to the number of successes and failures, respectively. (Remember, that in the absence of an exponent for p or q, it is understood to be 1; e.g., $3p^2q$ is understood to be $3p^2q^1$.)

As an example of the application of the binomial distribution, imagine the three trials of the sampling experiment to consist of three sales pitches made by a given individual whose past performance indicates a probability of success (i.e., sale) of $p = 1/4$ for any single sales presentation. Consequently, the probability of failure (i.e., no sale) is $q = 1 - p = 3/4$. We can now determine the probability of this salesperson making 3, 2, 1, or 0 sales out of three independent attempts, by evaluating the successive terms of the binomial expansion:

$$\left(\frac{1}{4}+\frac{3}{4}\right)^3=\left(\frac{1}{4}\right)^3+3\left(\frac{1}{4}\right)^2\left(\frac{3}{4}\right)+3\left(\frac{1}{4}\right)\left(\frac{3}{4}\right)^2+\left(\frac{3}{4}\right)^3$$

The first term, $(\frac{1}{4})^3$, is the probability of making a sale on all 3 tries (trials); i.e., $\frac{1}{4} \times \frac{1}{4} \times \frac{1}{4} = 1/64$. We see that it is relatively unlikely that a sale will result in three out of three attempts. In the long run, only one time out of 64 sets of three sales efforts, will a sale be made three out of three, or 100%, of the tries. The second term $3(\frac{1}{4})^2(\frac{3}{4})$, is the probability of exactly two sales out of three attempts, for a success proportion of .667. Its *probability* of occurrence is equal to $3 \times \frac{1}{4} \times \frac{1}{4} \times \frac{3}{4}$ or 9/64, or .14, still a relatively unlikely outcome. Evaluating the third term, $3(\frac{1}{4})(\frac{3}{4})^2$, will give us the probability of getting exactly one sale out of three. The final term, $(\frac{3}{4})^3$, gives us the probability of getting absolutely no sales out of three tries. Since these probabilities must sum to 1, representing mutually exclusive and exhaustive outcomes, any given combination of the probabilities may be added together to obtain the probability of a desired *composite* outcome. For example, if we wanted to know the probability of getting *at least two sales* (as opposed to getting *exactly* two sales) we would sum the probabilities corresponding to *three* sales and *two* sales; namely, the first two terms of the binomial expansion ($\frac{1}{64} + \frac{9}{64} = \frac{10}{64}$).

We can now address ourselves to the assumptions that must be met before we can apply the binomial probability distribution to the type of sampling distribution we have been discussing. The first key assumption is that the n observational trials are *independent* of one another; that is, the outcome on any given trial is not influenced by the outcome on any other trial. This is very likely to be true for successive tosses of a coin, but is it tenable in the case of a salesperson's performance. Probably not, since the psychological effect of a

Table 3 Pascal's Triangle: Coefficients of the terms of the binomial expansion (see Table 2).

Binomial expression	*Coefficients of successive terms of the expansion*
$(p+q)^1$	1 1
$(p+q)^2$	1 2 1
$(p+q)^3$	1 3 3 1
$(p+q)^4$	1 4 6 4 1
$(p+q)^5$	1 5 10 10 5 1
$(p+q)^6$	1 6 15 20 15 6 1
$(p+q)^7$	1 7 21 35 35 21 7 1
$(p+q)^8$	1 8 28 56 70 56 28 8 1
$(p+q)^9$	1 9 36 84 126 126 84 36 9 1
$(p+q)^{10}$	1 10 45 120 210 252 210 120 45 10 1

given sales effort outcome will most likely influence the next performance. As such, the binomial distribution is only an approximation of the real sampling distribution. The other key assumption is that the population probabilities are *constant* from trial to trial. Again, with the coin toss, it is safe to assume that the probability of a Head remains the same throughout the successive trials, but with the sales example, there may be daily, weekly, or other periodic variations in the salesperson's probability of making a sale—due either to market conditions or personal variables. Therefore, the validity of applying the binomial distribution to sampling problems such as those above will depend on the extent to which these assumptions are met.

While we will postpone the derivation of the binomial distribution until Chapter 20, a careful study of Table 2 will show the systematic nature of the terms comprising it. In each case it will be noticed that the exponents of p decrease, while the exponents of q increase, and for any given term in the expansion the sum of the exponents is always equal to n, the number of trials. As for the coefficients of these terms, they too follow a systematic pattern, and can be found in Table 3. Known as *Pascal's Triangle*, it shows how each coefficient is the sum of the two coefficients above it when they are arranged in the manner of Table 3.

Since the binomial terms are difficult to evaluate when n is large, their values are presented in Appendix Table X for values of n up to 20, and for various values of p.

Standard error of a proportion. Now that we have a means of determining the probabilities of alternative proportions that can result from sampling values of a binary random variable, we can now portray the sampling distribution of a proportion. This is done in Figure 10. Beginning with a parent population having a value of $p=.5$ and $q=.5$, we have created the sampling

distribution of the proportion based on sample sizes of $n=2$, $n=5$, $n=10$, and $n=20$. We designate a *sample proportion* as

$$\hat{p}$$

so as not to confuse it with the probability of a success p in the parent population. Thus, the sample proportion $\hat{p}$ is an estimate of the population probability p. We can read it as "p hat," "estimate of p," or "p estimate."

We can proceed to determine the standard error of the sampling distribution of a proportion just as for the sampling distribution of any other mean. Since the standard error of the mean is defined by $\sigma/\sqrt{n}$, we need to determine the value of σ, the *standard deviation* of the binary population in question. This could be done by subjecting the 0 and 1 binary data to the definitional formula for a population standard deviation, but there is an easier approach, by virtue of the special characteristic of a binary population that its mean happens to equal p. As a result, the standard deviation of a binary population simplifies to the expression

$$\boxed{\sigma = \sqrt{pq}} \tag{4}$$

The derivation of this fundamental formula is shown in Table 4.

It is a simple step now to define the standard error of a sample proportion, based on a sample of size n from the binary population. We have already learned that the standard error of a mean is given by $\sigma_{\bar{x}} = \sigma/\sqrt{n}$, so in the case of the standard error of a proportion (which is a mean) we have, using (4) above,

$$\sigma_{\hat{p}} = \frac{\sigma}{\sqrt{n}} = \frac{\sqrt{pq}}{\sqrt{n}}$$

or, putting both denominator and numerator under the radical sign, we have

$$\boxed{\sigma_{\hat{p}} = \sqrt{\frac{pq}{n}}} \tag{5}$$

as the *standard error of a sample proportion*.

The similarity of Figure 10 to Figure 3 should be apparent in a number of respects. Firstly, the variation of the sampling distribution decreases as the sample size increases. Secondly, we see that the distribution approaches normality as the sample size increases. And finally, by inspection, and which is

Table 4 Derivation of the standard deviation of a binary population, $\sigma = \sqrt{pq}$.

Given: $\mu = p$ *for a binary population*

x_i	f_i	$x_i - \mu$	$(x_i - \mu)^2$	$f_i(x_i - \mu)^2$
0	n_0	$0 - p$	p^2	$n_0 p^2$
1	n_1	$1 - p$	q^2	$n_1 q^2$
SUMS:	N			$n_0 p^2 + n_1 q^2$

	Comments
$\sigma^2 = \dfrac{\sum f_i (x_i - \mu)^2}{N}$	Definition of population variance
$= \dfrac{n_0 p^2 + n_1 q^2}{N}$	From summations above
$= \left(\dfrac{n_0}{N}\right)p^2 + \left(\dfrac{n_1}{N}\right)q^2$	Algebraic manipulation
$= qp^2 + pq^2$	Since $n_0/N = q$ and $n_1/N = p$
$= pq(p + q)$	Factoring out pq, but $p + q = 1$, so
$\boxed{\sigma^2 = pq}$	Variance of a binary population
$\boxed{\sigma = \sqrt{pq}}$	Standard deviation of a binary population

easily supported by mathematical proof, the mean of the sampling distribution is equal to p, the population mean, i.e.,

$$\boxed{\mu_{\hat{p}} = p} \tag{6}$$

Figure 10 shows sampling distributions generated from a population that is symmetrical in form. However, we see in Figure 11 that even when the population is skewed in form, the sampling distribution rapidly approaches normality as the sample size increases. As a rule, the sampling distribution for a proportion can be considered to be very nearly normal in form as long as the product of np (or nq if it is smaller) exceeds a value of 5. In Figure 11*e*, $np = (20)(.3) = 6$, and we see that the sampling distribution does indeed appear normal in form.

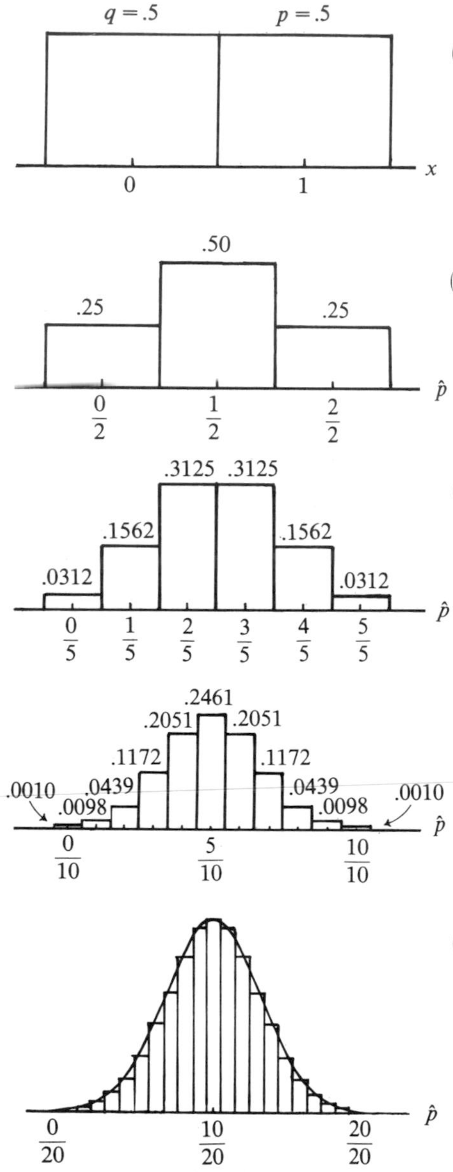

(a) Binary population with $p = .5$.

$$\sigma = \sqrt{pq} = \sqrt{(.5)(.5)} = .5$$

(b) Sampling distribution based on $n = 2$.

$$\sigma_{\hat{p}} = \sqrt{\frac{pq}{n}} = \sqrt{\frac{(.5)(.5)}{2}} = .354$$

(c) Sampling distribution based on $n = 5$.

$$\sigma_{\hat{p}} = \sqrt{\frac{pq}{n}} = \sqrt{\frac{(.5)(.5)}{5}} = .224$$

(d) Sampling distribution based on $n = 10$.

$$\sigma_{\hat{p}} = \sqrt{\frac{pq}{n}} = \sqrt{\frac{(.5)(.5)}{10}} = .158$$

(e) Sampling distribution based on $n = 20$.

$$\sigma_{\hat{p}} = \sqrt{\frac{pq}{n}} = \sqrt{\frac{(.5)(.5)}{20}} = .112$$

Figure 10 Binary population with $p = .5$, and sampling distributions of a sample proportion $\hat{p}$ based on different sample sizes n.

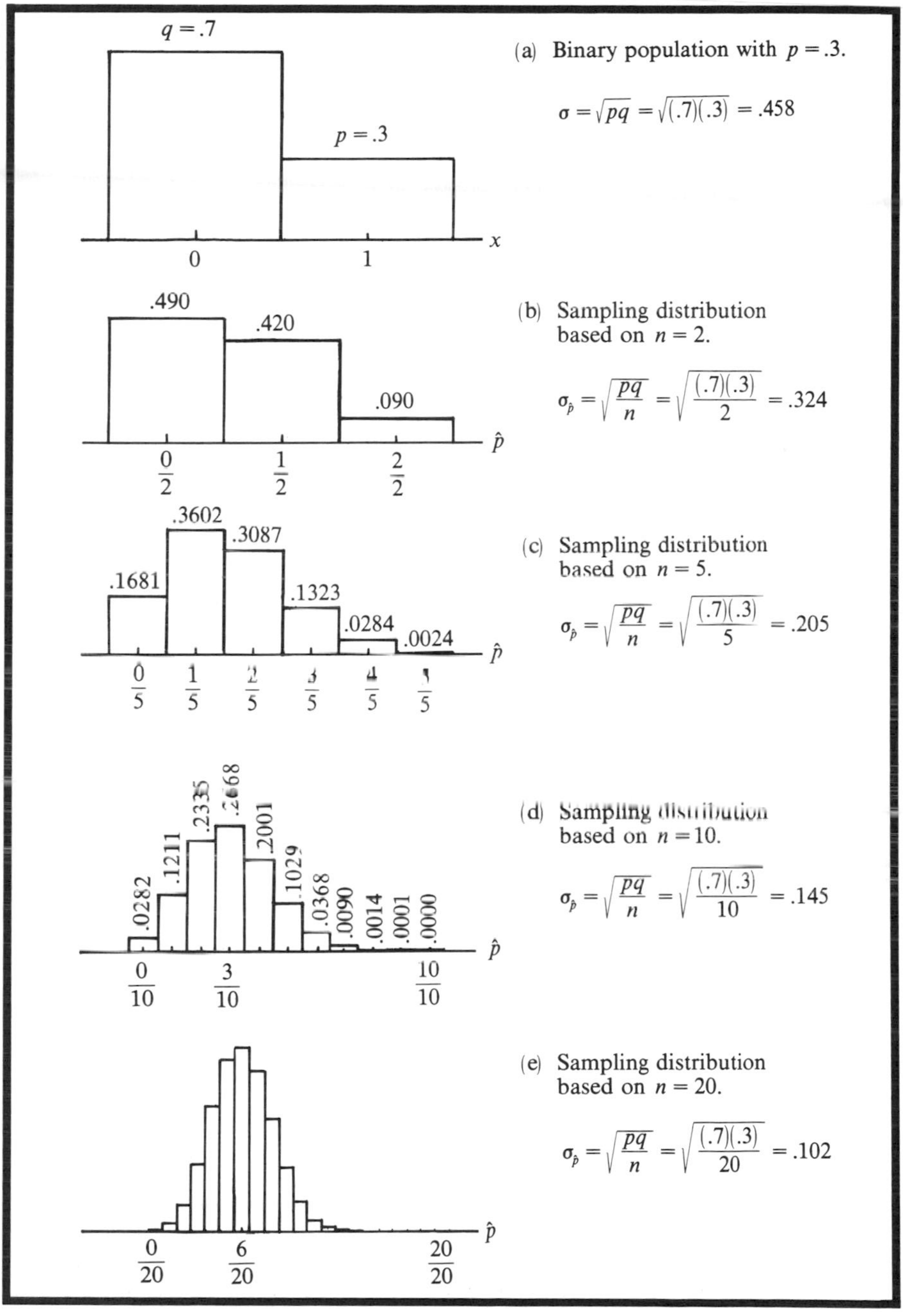

Figure 11 Binary population with $p = .3$, and sampling distributions of a sample proportion $\hat{p}$ based on different sample sizes n.

A comparison of Figures 10 and 11 will show that each supports the Central Limit Theorem, that the sampling distribution of a mean (in this case a mean of a binary variable) will approach a normal distribution regardless of the shape of the parent population, provided the sample size is large. Further, we see that the sampling distribution based on the symmetrical population approaches normality faster than that based on a highly skewed population. That the normal distribution should find application in a situation based on a discrete binary variable further attests to its pervasive importance.

Standard error of the number of successes. In the preceding discussion we developed the sampling distribution of a proportion and its mean and standard error. Instead of dealing with the proportion of successes, we could alternatively be interested in the *number* of successes among a given number of sampling trials.

To obtain the number of successes we merely have to multiply the sample proportion times the number of trials; i.e., $n\hat{p}$. This can be seen most clearly in the sampling distributions in Figure 12, based on four binomial trials drawn from a binary population with $p = .5$. It can be likened to the outcomes that would be expected from tossing a coin four times. In Figure 12a the *proportions* of success are shown as

$$\frac{0}{4}, \frac{1}{4}, \frac{2}{4}, \frac{3}{4}, \frac{4}{4}$$

In terms of *number* of successes, those correspond to the values

$$0, 1, 2, 3, 4$$

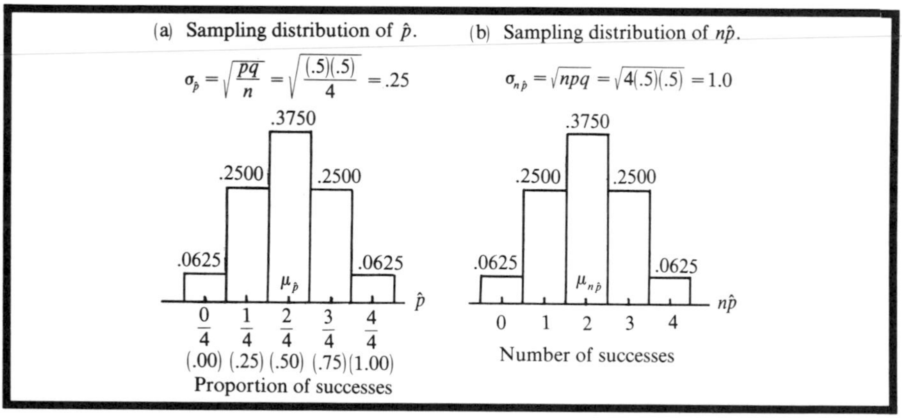

Figure 12 A comparison of the sampling distribution of (a) a sample proportion $\hat{p}$, and (b) the sample number of successes $n\hat{p}$, for random samples of size $n = 4$ drawn from a binary population with $p = .5$.

All we have done to obtain the *number* of successes is multiply the *proportion* of successes $\hat{p}$ by the number of trials n, in this case 4.

As we know, multiplying a variable by a constant results in a mean which is equal to that constant times the original mean. Since the mean of the sampling distribution of the proportion is p, the mean of the sampling distribution of the *number of successes* will be

$$\boxed{\mu_{n\hat{p}} = np} \qquad (7)$$

We also know that multiplying a variable by a constant results in a standard deviation which is equal to that constant times the original standard deviation. Consequently, since the standard error of the *proportion* of successes is

$$\sigma_{\hat{p}} = \sqrt{\frac{pq}{n}}$$

then the standard error of the *number* of successes will be

$$\sigma_{n\hat{p}} = n\sigma_{\hat{p}} = n\sqrt{\frac{pq}{n}} = \sqrt{n^2}\sqrt{\frac{pq}{n}} = \sqrt{\frac{n^2pq}{n}}$$

which simplifies to

$$\boxed{\sigma_{n\hat{p}} = \sqrt{npq}} \qquad (8)$$

A comparison of the two parts of Figure 12 will help crystalize the distinction between these alternative approaches to the sampling distribution based on a binary population; i.e., the sample *proportion* vs. the sample *number* of successes.

Table 5 presents a summary of the developments of this section, including the characteristics of a binary population and the resulting sampling distributions of the *proportion* and *number* of successes for a random sample of n observations, or trials. These relationships should be thoroughly understood, for they are the foundation of the statistical tests involving proportions, which are so common in all areas of scientific study. If we want to deal with *percentages* instead of proportions, then it is easy to verify that the corresponding mean and standard error will be increased by a factor of 100. These concepts will be developed further in the next two chapters.

Table 5 Characteristics of a binary population and its key sampling distributions.

Characteristic	**Binary Population** ($x_i = 0$ or 1, proportion p of which are 1's— "successes")	**Sampling Distributions** $\hat{p}$ Sample *proportion* of "successes"	$n\hat{p}$ Sample *number* of "successes"
Mean	$\mu = p$	$\mu_{\hat{p}} = p$	$\mu_{n\hat{p}} = np$
Standard Deviation/ Standard Error	$\sigma = \sqrt{pq}$	$\sigma_{\hat{p}} = \sqrt{\dfrac{pq}{n}}$	$\sigma_{n\hat{p}} = \sqrt{npq}$

9. Sampling Distribution of the Difference Between Two Independent Proportions

Just as we created a sampling distribution for the difference between two sample means, $\bar{x}_1 - \bar{x}_2$; we can develop a sampling distribution for the difference between two sample proportions, $\hat{p}_1 - \hat{p}_2$. The approach is exactly the same. Satisfying the assumptions of the binomial distribution, we sample n_1 observations from population 1, n_2 observations from population 2, and then determine the difference between the respective proportion of successes, $\hat{p}_1 - \hat{p}_2$. When this is done a theoretical infinite number of times, and the probabilities determined for each possible difference, we have the *sampling distribution of the difference between two independent proportions.*

The mean of the sampling distribution will be

$$\boxed{\mu_{\hat{p}_1 - \hat{p}_2} = p_1 - p_2} \tag{9}$$

equivalent to the difference between the population proportions.

The standard error of the sampling distribution, when the proportions are independent of one another, will be

$$\sigma_{\hat{p}_1 - \hat{p}_2} = \sqrt{\sigma^2_{\hat{p}_1} + \sigma^2_{\hat{p}_2}}$$

which is perfectly analogous to the standard error of the difference between independent means. And since the standard error of each proportion, as learned in the preceding section, is $\sqrt{pq/n}$, we have as the *standard error of the*

difference between independent proportions:

$$\sigma_{\hat{p}_1-\hat{p}_2}=\sqrt{\frac{p_1q_1}{n_1}+\frac{p_2q_2}{n_2}} \tag{10}$$

Again, it must be stressed that this standard error applies to a sampling distribution for which the proportions are independent of one another. For example, the proportion of a random sample of females who smoke, and the proportion of a random sample of males who smoke, are independent proportions; since the value obtained for one in no way can influence the value of the other. On the other hand, if the samples of males and females represented husbands and wives, the proportions would not be independent of one another, for the institution of marriage could have a common influence. We will deal with these types of non-independent proportions in Chapter 13.

At this point it will be worthwhile to reread Section 6 on the Sampling Distribution of the Difference Between Two Independent Means and note the parallelism with this section. They key difference between the sampling distribution of a difference between two independent means based on multi-valued variables, and the sampling distribution of the difference between two independent proportions, is the size of the sample required before the distribution approaches normality. When dealing with means based on continuous variables, or at least multi-valued discrete variables, we need sample sizes of at least thirty or so before the sampling distributions approach a normal distribution. When dealing with proportions, the product of np or nq, whichever is smaller, must exceed 5 or so before the sampling distribution of the proportion approaches normality. If $n=30$ and $p=.5$, we have $np=(30)(.5)=15$, so we could expect its sampling distribution to be very close to normal in form. If $n=30$ and $p=.1$, we have $np=(30)(.1)=3$, and consequently the sampling distribution of this proportion would not be likely to be normal. We would need a sample size of $n=50$ (in which case $np=5$) before we could be confident that its distribution was normal. When dealing with the difference between two proportions, n_1p_1, n_1q_1, n_2p_2, and n_2q_2 should *all* exceed 5, if we expect the sampling distribution to be nearly normal in form. The importance of the normality approximation will become apparent in the following two chapters.

10. Concluding Comments

In earlier chapters we progressed from the *empirical frequency distribution* to the *theoretical probability distribution*, and now we have introduced the concept of a *sampling distribution*; i.e., the probability distribution of a sample

statistic—the values that the statistic can assume, and their associated probabilities.

We studied the sampling distributions of the mean, the difference between two means (independent vs. non-independent), a proportion, and the difference between two independent proportions. We also compared the means and standard deviations (standard errors) of these sampling distributions with the means and standard deviations of the parent populations upon which they were based. In addition, and perhaps most importantly, we discovered that, by virtue of the Central Limit Theorem, these sampling distributions approach a normal shape when the sample size becomes large, regardless of the shape of the parent populations.

In the following chapter we will learn the practical applications of these sampling distributions, and discover that the developments of this chapter, despite what they may have seemed, were more than idle intellectual exercises.

Chapter 8

Parameter Estimation

1. Introduction

The problem of statistical inference—the drawing of conclusions about populations based on sample observations—can be approached from two different directions. On the one hand, the population parameters can be estimated directly, with the sample observations serving as approximations to the true population characteristic. On the other hand, the sample observations can serve to support or discredit (i.e., test) *a priori* hypotheses about the population. In this chapter we will address ourselves to the first of these approaches, *direct* parameter estimation. The indirect hypothesis testing approach will be the subject of the following chapter.

2. Point Estimation

For a given population there may be any number of different parameters in which we might be interested—e.g., its mean, median, mode, variance, standard deviation, range, etc. One approach to this estimation problem is to obtain a *single value* based upon a sample of observations, a value which we feel is the best possible approximation of the true value of the population parameter. This type of *point estimate* does not provide us with information as to how close we might expect it to be to the population parameter, but it does at least suggest its value.

The mean $\bar{x}$ of a sample is an example of a point estimate of the population mean μ. Similarly, the sample standard deviation s and variance s^2 are respective estimates of the population standard deviation σ and variance σ^2.

One of the properties that we would want a point estimate to have, is that it be an *unbiased* estimate of the population parameter in question. By unbiased we mean that the long run average or *expected value* of the estimate

will equal the population parameter; or, equivalently, that the mean of its sampling distribution equals the population parameter. This is certainly true for the sample mean $\bar{x}$ as we learned in the preceding chapter. It is also true for the sample variance s^2, defined as the sum of the squared deviations from the mean, divided by the number of observations *less one*. However, if we had divided by the number of observations n rather than by $n-1$, we would have had a biased estimate of σ^2, tending to underestimate it. A corresponding argument holds for s and σ.

We see, then, that the definition of a population parameter, when applied to a sample of observations, does not necessarily yield an unbiased estimate. Perhaps a more striking example than the population variance or standard deviation is the population range. It should be clear that the range of a small sample of observations would tend to underestimate the range of the parent population from which the sample was drawn, for how likely is it that our sample includes the two extreme scores of the population? Very unlikely, unless the population is *U*-shaped; and even then, a small sample would be unlikely to contain those very two scores that lie at the extremes of the population and define its range. So, we cannot take for granted that a sample statistic will be an unbiased estimate of the corresponding population parameter. Rather, we must be sure that the mean of the statistic's sampling distribution equals the population parameter in question.

A second property that we want a point estimate to have, is that it be relatively *efficient* compared to alternative point estimates. By efficient we mean the speed with which the estimate becomes a more accurate estimate of the population parameter as the sample size increases. For example, while both the mean and the median are unbiased estimates of central tendency, the mean is a relatively more efficient estimate than the median. More technically, for a given sample size n, the sampling distribution of the mean has a smaller standard error than that of the median. Similarly, the sample variance s^2 and standard deviation s are more efficient estimates of variation than other statistics, such as the range or average deviation.

In summary, the sampling distribution of a statistic can inform us whether a particular statistic has the desired properties of being an unbiased and efficient estimate of a population parameter. It is because the mean $\bar{x}$, the variance s^2, and standard deviation s have these properties that they play such an important role in statistical inference.

3. Interval Estimation

While a point estimate of a population parameter is better than no estimate at all, it would certainly be desirable to have some idea of *how close* the true parameter might be to the estimate. We cannot expect to know exactly

how close our sample statistic is to the population parameter, for that would be equivalent to knowing the value of the parameter itself; but we could state with a certain *probability* that it is *within a particular distance* of the parameter. Such an estimate is referred to as an *interval estimate*, an interval of values within which we can state with a certain degree of confidence (i.e., probability) that the parameter falls.

For example, rather than using a sample mean $\bar{x}$ as a point estimate of the population mean μ, we could state an interval of values within which we strongly believe the true mean to fall, such as

$$a < \mu < b$$

where the symbol $<$ stands for "is less than" and a and b are two numerical values. The above expression can then be read "a is less than μ which in turn is less than b," or equivalently, "μ lies between a and b." For instance, we might state

$$60 < \mu < 80$$

which suggests that the true mean μ probably lies somewhere between a value of 60 and 80. But why 60 and 80, rather than between 65 and 75, or between 68 and 72, or any other pair of values for that matter which form an interval either narrower or wider. The key phrase is "probably lies between," for in actuality any interval can be used; but for each different interval we will have a different degree of confidence that it contains the population parameter. For example, if we are 95% confident that a population mean μ falls between 60 and 80, we would logically have to have less confidence that it falls within a narrower band, say 65 to 75. Similarly, we would have greater confidence that it falls within a wider band, say 55 to 85. Or to go to an absurd extreme, we could say that we are 100% confident that the population mean is between $-\infty$ and $+\infty$ (negative and positive infinity). The point is, that for any given interval of values, there is an associated probability which expresses the likelihood that the population parameter is contained within it.

Confidence intervals and limits. The two key elements of an interval estimate, then, include the specification of the limits or boundary values of the interval, and the associated probability that the population parameter is contained in that interval of values. The interval of values is referred to as a *confidence interval*, and the boundary values which define it are referred to as the *confidence limits* of the interval.

For each confidence interval there is an associated probability indicating how certain we are that the population parameter falls within the interval. By convention, the probability associated with a particular confidence interval is usually expressed as a *percentage* statement rather than as a decimal probabil-

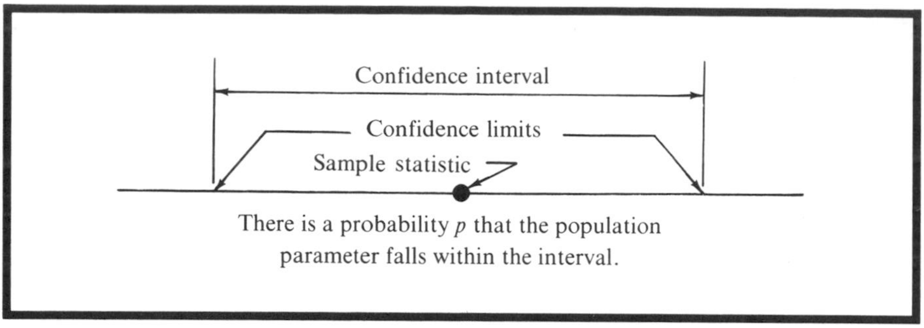

Figure 1 Schematic diagram of an interval estimate of a parameter.

ity. For example, if the probability of a parameter being included in a particular interval is $p = .95$, we state instead that we are 95% confident that the interval includes the parameter value; and the interval in turn is referred to as the 95% confidence interval for the parameter.

Depending on the limits of the particular interval, we can have a 99% confidence interval, a 95% confidence interval, a 90% confidence interval, or any specific level we desire. Figure 1 shows the three key features involved in an interval estimate—the confidence limits, the confidence interval, and the probability that the parameter is contained within the interval, all of which we will see are inter-related with one another.

4. Interval Estimate of a Mean

Since we rarely have the luxury of being able to measure every single member of a population in order to determine a parameter value, we are surrounded by situations in which we must resort to interval estimates of those parameters.

For example, based on sample observations, we might be interested in interval estimates of the mean for math aptitude scores of ten-year-olds, the diameters of a manufacturer's ball bearings, the lifetimes of a certain brand battery, the ages of tea drinkers, the fat contents of a fast-food chain's hamburgers, the tensile strengths of steel rods, the crime rates of cities, the weights of golf balls, the cholestoral contents of blood specimens, the hours of teen TV viewing, the masses of atomic particles, or any of a number of other variables.

The question naturally arises as to how we go about creating an interval estimate of a population parameter; i.e., how do we determine the confidence interval and its associated probability of containing the parameter in question. To gain an insight into the necessary procedure, we can begin by analyzing the

relationship between a sample statistic and the corresponding population parameter. In the most general of terms we can state that the values of a statistic and the relevant parameter are related as follows:

$$\textit{Statistic} = \text{Parameter} \pm \textit{error}$$

where $\pm$ stands for "plus or minus," and italics are used to identify the variable quantities. In other words, our observed sample *statistic* deviates from the fixed parameter value by some *unknown and variable amount of error.*

It should also be apparent, on the basis of logic or a bit of algebra, that the above expression can be written equivalently as

$$\text{Parameter} = \textit{Statistic} \pm \textit{error}$$

Again, we interpret this to mean that the unknown parameter value is equal to the observed statistic plus or minus some unknown amount of error. If we are dealing with a mean, we can state

$$\mu = \bar{x} \pm \textit{error}$$

which is to say that the true population mean μ is equal to the observed sample mean x plus or minus some unknown amount of error.

The key to creating a confidence interval for the mean μ, then, rests on a knowledge of the "unknown error." While we will never know the size of the error for any single sample mean, we could certainly attach a *probability* to errors of a given size, provided we knew the probability distribution of those errors; namely, what is the *mean* and *standard deviation* and *shape* of the *distribution* of errors.

To see how we might understand the source of the needed information, let us rewrite the preceding expression in a slightly different, but the algebraically equivalent form

$$\bar{x} = \mu \pm \textit{error}$$

which simply states that the sample mean $\bar{x}$ is equal to the population mean μ plus or minus some unknown error. Now let us imagine that we take repeated samples from the population, each time calculating a sample mean $\bar{x}$. When performed an infinite number of times we have nothing less than the sampling distribution of the mean. The variability of this distribution, as we learned in the preceding chapter, is nothing other than $\sigma_{\bar{x}}$, the standard error of the mean. But $\sigma_{\bar{x}}$ must also be the standard error of the right side of the above equation, namely the standard deviation of the distribution of *errors*, since we know from Chapter 5 that adding a constant μ to each of the errors will not affect

the standard deviation of the errors. That is,

$$\sigma_{\bar{x}} = \sigma_{(\mu \pm error)} = \sigma_{error}$$

Therefore, we arrive at the very important conclusion that *the variation of the unknown estimation errors is exactly the same as the variation of the sample means*; or, more specifically, the standard error of the mean is also the standard deviation of the distribution of *sampling errors*—the discrepancies between the statistic and parameter. Perhaps now it is clear why we call it the standard *error* of the mean. Figure 2 shows the distribution of these errors.

Now that we know the standard deviation of sampling errors equals the standard error of the mean, we are over the biggest hurdle in arriving at an interval estimate for the population mean μ. All we need to know now is the *shape* of the distribution of sampling errors. But again, we know that its shape will be identical to the shape of the sampling distribution of the mean, for adding the constant value μ to each error affects neither its standard deviation, nor the shape of the distribution. So, we can conclude that the shape of the distribution of sampling errors will be identical to the shape of the sampling distribution of the mean $\bar{x}$. In the preceding chapter we learned that the sampling distribution of the mean will be *normal* in form if the parent population from which the sample is drawn is normal, or, nearly so, if the sample size is large—say $n = 30$ or greater—regardless of the shape of the

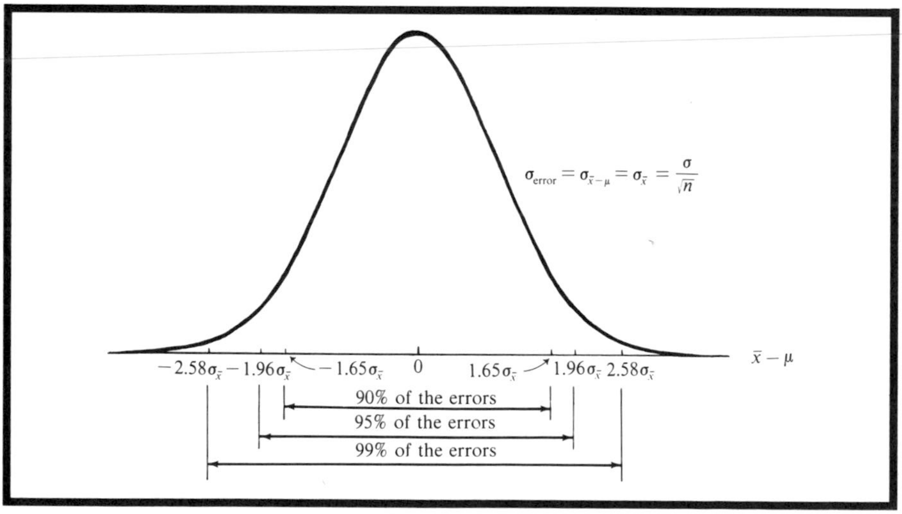

Figure 2 Distribution of sampling errors, $\bar{x} - \mu$.

parent population. It follows, then, that the distribution of sampling errors will be normal in shape under exactly the same conditions.

95% confidence intervals. Now that we know the shape and standard deviation of the distribution of sampling errors is the same as that of the sampling distribution of the mean, it is a simple step to creating an interval estimate for the population mean μ. We know from Chapter 5 that in a normal distribution, 95% of the observations fall within ± 1.96 standard deviations from the mean. Since the standard deviation of the distribution of sampling errors is equal to the standard error of the mean $\sigma_{\bar{x}}$ (as shown in Figure 2) and since $\mu = \bar{x} \pm \textit{error}$, it follows that

$$\mu = \bar{x} \pm 1.96\sigma_{\bar{x}} \qquad \text{(with 95\% certainty)}$$

Alternatively, we can write

$$\boxed{(\bar{x} - 1.96\sigma_{\bar{x}}) < \mu < (\bar{x} + 1.96\sigma_{\bar{x}})} \tag{1}$$

as the 95% confidence interval for the population mean μ, when $\bar{x}$ is based on a random sample from a normal population.

In other words, we are 95% certain that the true mean μ will be less than 1.96 standard errors above the observed sample mean $\bar{x}$, but greater than 1.96 standard errors below the sample mean. We do not know exactly where in the interval it may fall, only that we are 95% sure that it will. On the other hand, there is a 5% chance that it will fall *outside* the stated interval. What this means is that *in the long run*, 95% of the confidence intervals so constructed will contain the true population mean μ. Figure 3 shows a number of hypothetical confidence intervals based on successively drawn samples, and we see that in most cases the confidence intervals do indeed encompass the population mean μ, as they should.

To demonstrate the procedure for creating a confidence interval, let us imagine that we randomly sample 100 ball bearings from a manufacturer's daily production, with the purpose of estimating the mean weight of all ball bearings produced that day. (Of course, we could just as well be estimating the mean test scores of individuals, the mean sugar content of beets, the mean blood pressure of newborns, the mean sales of retail outlets, etc.) Let us say we obtain a mean weight for the sample of 100 ball bearings of $\bar{x} = 149.8$ grams, and that we know from historical records that the standard deviation of the population of ball bearing weights is $\sigma = 3$ grams. From this information we can calculate the standard error of the mean as

$$\sigma_{\bar{x}} = \frac{\sigma}{\sqrt{n}} = \frac{3}{\sqrt{100}} = .3 \text{ grams}$$

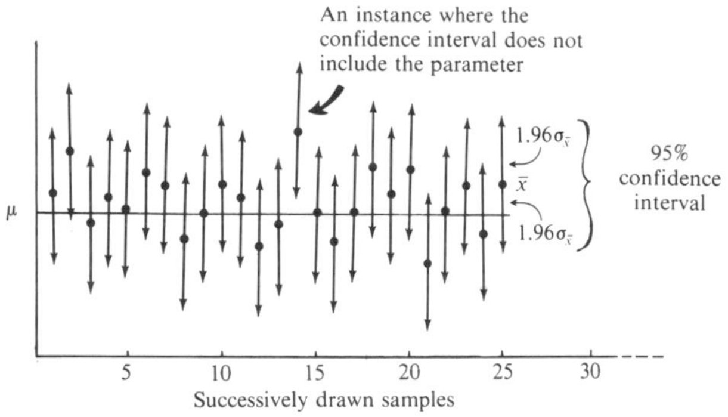

Figure 3 Confidence intervals created from hypothetical successively drawn samples.

With our 95% confidence interval defined as

$$(\bar{x} - 1.96\sigma_{\bar{x}}) < \mu < (\bar{x} + 1.96\sigma_{\bar{x}})$$

we can substitute our values for $\bar{x}$ and $\sigma_{\bar{x}}$, and we have

$$(149.8 - 1.96 \times .3) < \mu < (149.8 + 1.96 \times .3)$$

or

$$149.21 < \mu < 150.39 \text{ grams.}$$

In other words, we are 95% sure that the true mean weight of the ball-bearings, had we measured every single one of the day's production, would be somewhere between 149.21 and 150.39 grams. More technically, we are 95% sure that the *procedure* for creating the obtained confidence interval would produce an interval encompassing the true population mean. For the *actual* interval produced, the true mean either *is* or *is not* in it, so strictly speaking we cannot say that there is a .95 probability that the mean falls in the interval. This is a moot point from a practical standpoint, but has importance in a theoretical context in which no quantitative probability-type statement is allowable for a specific interval. We will adhere to the more practical view that "95% sure" or "95% confident" are meaningful common sense statements with respect to specific intervals, and are as legitimate as statements which invoke "odds" or phrases such as "substantial assurance" to circumvent the probability issue.

99% confidence intervals. If, instead of a 95% confidence interval for the mean, we wanted to create a more stringent 99% confidence interval, we would

rely on the fact that 99% of the observations in a normal distribution lie within ± 2.58 standard deviations of the mean. Then, it follows that the 99% confidence interval for a population mean μ will be given by

$$(\bar{x} - 2.58\sigma_{\bar{x}}) < \mu < (\bar{x} + 2.58\sigma_{\bar{x}}) \tag{2}$$

provided, again, that the sample mean $\bar{x}$ is based on a random sample from a normal population.

For the preceding ball bearing example we have as the 99% confidence interval

$$(149.8 - 2.58 \times .3) < \mu < (149.8 + 2.58 \times .3)$$

or

$$149.03 < \mu < 150.57 \text{ grams}$$

Note that the 99% confidence interval is wider than the 95% confidence interval, and well it should be if we are to be more certain that the population mean is contained within it.

Effect of sample size. Since the standard error of the mean is given by $\sigma_{\bar{x}} = \sigma/\sqrt{n}$, it follows that the width of our interval estimate will decrease as our sample size n increases; more specifically, it will vary inversely as the square root of the sample size. For example, we can decrease the width of our interval estimate by half, by quadrupling the sample size.

However, sheer sample size alone is not sufficient to assure accurate estimates of a mean, which seems to be the belief behind much non-scientific research which boasts *huge* samples, while neglecting to draw the sample in a *random* fashion. In this respect it is important to remember the statistical adage that *it is better to be approximately correct, than precisely wrong.*

General confidence interval. Having learned how to create a confidence interval for a mean, it is a relatively simple matter to generalize the procedure to other types of parameters with normal sampling distributions. For instance, in its most general form, the confidence interval for any given parameter θ (Greek letter *theta*) can be stated

$$(\hat{\theta} - z_c\sigma_{\hat{\theta}}) < \theta < (\hat{\theta} + z_c\sigma_{\hat{\theta}}) \tag{3}$$

where $\hat{\theta}$ ("theta hat") is the sample estimate of θ, and z_c is the critical z value associated with a given confidence interval; $z_c = 1.96$ for a 95% confidence interval, and $z_c = 2.58$ for a 99% confidence interval. We assume, of course, for strict accuracy, that the sampling distribution of $\hat{\theta}$ is normally distributed,

which is not the case for all statistics. The sampling distribution of each statistic must be studied in its own right to verify its shape and standard error. Such investigations are subjects of advanced study, since there are so many statistics other than the mean, to which we have confined our analysis.

Approximate confidence intervals. In the preceding discussion we created confidence intervals based on the knowledge of the population standard deviation σ and the understanding that the population was normally distributed. However, we do not usually know the true value of the population standard deviation σ, so we must settle for the sample standard deviation s as an estimate of σ. If we then use $s_{\bar{x}} = s/\sqrt{n}$ instead of $\sigma_{\bar{x}} = \sigma/\sqrt{n}$, we have

$$(x - 1.96 s_{\bar{x}}) < \mu < (\bar{x} + 1.96 s_{\bar{x}})$$

as the *approximate* 95% confidence interval for the mean μ, and

$$(\bar{x} - 2.58 s_{\bar{x}}) < \mu < (\bar{x} + 2.58 s_{\bar{x}})$$

as the *approximate* 99% confidence interval for the mean μ.

The above are only approximately correct intervals, because $s_{\bar{x}}$ is only an estimate of $\sigma_{\bar{x}}$, the actual standard error of the sampling distribution of the mean. The assumption of the normality of the sampled population is less of a cause for concern, for as we learned in the preceding chapter, the Central Limit Theorem assures us that the sampling distribution will be very nearly normal in form even when the sampled population is not, provided our sample size is at least thirty or so.

In the following section we will learn a technique for creating exact confidence intervals when our sample sizes are small, say less than thirty, and drawn from a normal population with an unknown standard deviation σ. For now, we must remember that the confidence intervals studied so far are strictly accurate only when we know the population standard deviation σ, and are approximately correct when we use the sample standard deviation s as an estimate of σ, and in turn $s_{\bar{x}} = s/\sqrt{n}$ as an estimate of $\sigma_{\bar{x}} = \sigma/\sqrt{n}$. Also, for full accuracy, we must be assured that the sampling distribution of the mean is normally distributed, which will be the case if the sampled population is normal, and will very nearly be the case whenever our sample size is large, say thirty or greater, regardless of the shape of the sampled population.

5. Student's *t* Distribution

In the preceding section we had to resort to approximate confidence intervals when we did not know the value of the population standard deviation σ. Instead, we used the sample standard deviation s as an estimate of σ, and in

turn $s_{\bar{x}} = s/\sqrt{n}$ as an estimate of $\sigma_{\bar{x}} = \sigma/\sqrt{n}$, the standard error of the sampling distribution of the mean. This approximation procedure is satisfactory provided we are dealing with large samples, for then $s_{\bar{x}}$ is likely to be a close estimate of $\sigma_{\bar{x}}$. However, when the sample sizes are small, say less than thirty or so, the resulting confidence intervals may be substantially in error.

Fortunately, we can obtain *exact* confidence intervals in situations involving small samples from normal populations with unknown standard deviations, by using *Student's t distribution*, a probability distribution similar to the normal distribution but with important differences. To better understand the nature of Student's t distribution we should first review the normal distribution. Let us say we are interested in some population parameter θ. If its sample estimate $\hat{\theta}$ has a normal sampling distribution, we also know that the quantity $\hat{\theta} - \theta$ will also be normally distributed, for subtracting a constant θ from a variable $\hat{\theta}$ does not alter the shape nor the standard deviation of the variable's distribution. Furthermore, if we divide the quantity $\hat{\theta} - \theta$ by a constant quantity, say $\sigma_{\hat{\theta}}$, the resulting distribution is *still normal* in form, only its standard deviation has been altered. In fact, the resulting quantity is nothing other than a standardized normal deviate

$$z = \frac{\hat{\theta} - \theta}{\sigma_{\hat{\theta}}}$$

in which z expresses the sampling error $\hat{\theta} - \theta$ in terms of standard error units. The distribution of z is normally distributed just as the sample statistic $\hat{\theta}$, for all we have done is subtract a constant θ from each value of $\hat{\theta}$, and divided the result with another constant, $\sigma_{\hat{\theta}}$. These arithmetic operations have changed the mean and standard deviation of the distribution but have not affected its shape —it is still normal.

But what happens when we do not know the value of the standard error $\sigma_{\hat{\theta}}$ of the sampling distribution and must resort to estimating it with the *sample* standard deviation, $s_{\hat{\theta}}$. Since $s_{\hat{\theta}}$ is a statistic it will exhibit variability. Therefore, in the case of Student's t variable, defined as

$$\boxed{t = \frac{\hat{\theta} - \theta}{s_{\hat{\theta}}}} \tag{4}$$

the value of t will not have the same distribution as $\hat{\theta}$, as in the case of z when we divided by a constant $\sigma_{\hat{\theta}}$. Although the t distribution is similar in shape to the normal distribution, it tends to be flatter with more pronounced tails. In fact, there is not just one t distribution, but many, a few of which are shown in Figure 4, superimposed upon a normal distribution.

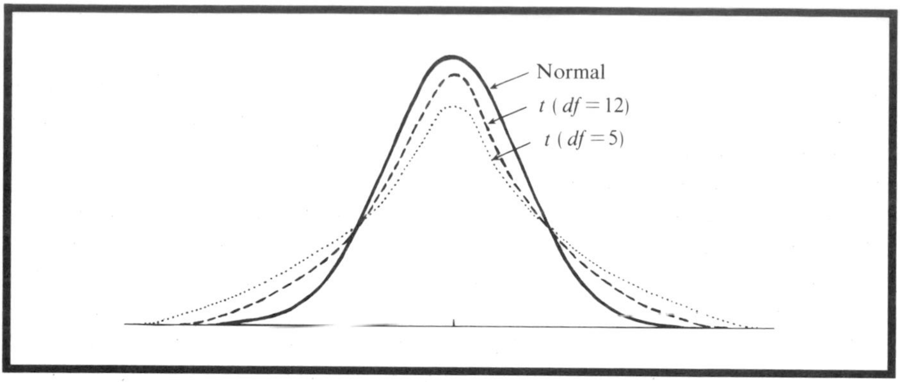

Figure 4 Comparison of selected t distributions with the normal distribution.

Degrees of freedom. Associated with each t distribution is a defining characteristic referred to as *degrees of freedom*, often abbreviated *df*. Notice in Figure 4 that the t distributions with the smaller degrees of freedom tend to be flatter in form. As the degrees of freedom increase, the t distribution approaches the shape of a normal distribution. In theory, the t distribution approaches the normal distribution when the degrees of freedom are infinite in size. For all practical purposes, it is very nearly the shape of the normal distribution once the degrees of freedom exceed thirty or so.

The number of degrees of freedom associated with a t distribution is dependent upon the number of observations that went into calculating the sample standard error estimate in the denominator of the t ratio. If we are dealing with the standard error of a *mean*, then $df = n - 1$, where n is the sample size. Other statistics will be associated with other degrees of freedom.

The term *degrees of freedom* is quite descriptive, in the sense that it is the number of observations in the data collection that are *free* to vary after the sample statistics have been calculated. For example, in calculating a sample standard deviation, we must subtract the sample mean from each of the n data observations in order to get the deviations from the mean. But once we have completed the second last subtraction, the final deviation is automatically determined, since the deviations prior to squaring *must* sum to zero. So the last deviation is not free to vary, only $n - 1$ are free to vary.

Since the denominator of the t ratio is itself a statistic, subject to sampling fluctuations, we should not be surprised that when it is based on a relatively small number of observations, the resulting t distribution will tend to be more spread out than when the sample size is large. Whereas in the normal distribution, 95% of the area lies within ± 1.96 standard deviations of the mean, and 99% within ± 2.58 standard deviations of the mean, this is not the

case with the t distribution. Since the t distribution is flatter with more pronounced tails, *more* than 5% of the area will be beyond ± 1.96 standard deviations, and more than 1% of the area will be beyond ± 2.58 standard deviations. How much more will depend on the particular t distribution and its degrees of freedom: The smaller the degrees of freedom, the flatter the central part of the distribution and the more pronounced the tails. Therefore, we will have to travel more than ± 1.96 standard deviations from the mean before we encompass 95% of the area under the curve. And we will have to travel more than ± 2.58 standard deviations from the mean to encompass 99% of the area under the curve. And as the degrees of freedom get smaller, the further from the mean we will have to go before encompassing these areas.

Appendix Table IV provides us with the necessary values of t to encompass various amounts of the area under the curve, for various degrees of freedom. It will be seen, for example, that for $df = 11$, a t of ± 2.20 is necessary to encompass 95% of the area; and $t = \pm 3.11$ is needed to embrace 99% of the area. When $df = 30$, the corresponding values of t are ± 2.04 and ± 2.75, respectively; not that much different than the values of $z = \pm 1.96$ and $z = \pm 2.58$, needed to embrace 95% and 99%, respectively, of the area under the normal curve.

Interval estimates using *t*. Now that we are familiar with the characteristics of the t distribution, we can use it to create interval estimates of population parameters based on statistics from small samples. It will be recalled from earlier in the chapter that when we know the standard error of the sampling distribution of our statistic, $\sigma_{\hat{\theta}}$, we determine the 95% confidence interval for the parameter θ as follows

$$(\hat{\theta} - 1.96\sigma_{\hat{\theta}}) < \theta < (\hat{\theta} + 1.96\sigma_{\hat{\theta}})$$

However, when we are dealing with small samples and an unknown value of $\sigma_{\hat{\theta}}$, we must replace 1.96 with the critical value of t that embraces 95% of the area under the curve, which we will designate t_c, and we must further replace $\sigma_{\hat{\theta}}$ with its sample estimate $s_{\hat{\theta}}$. The general form of an interval estimate of a parameter θ based on a small sample from a normal population is then

$$\boxed{(\hat{\theta} - t_c s_{\hat{\theta}}) < \theta < (\hat{\theta} + t_c s_{\hat{\theta}})} \qquad (5)$$

where t_c is the critical t value for a given confidence interval. The value of t_c will depend upon the number of degrees of freedom associated with $s_{\hat{\theta}}$. Correspondingly, t_c will assume different values depending on whether we are interested in, say, a 90%, 95% or 99% confidence interval.

As an example of the application of the above procedure, consider a governmental agency report that a particular model automobile gets 24 miles per gallon of gasoline. Such a blanket statement should immediately raise a number of questions in our minds. Even granting that the figure was based on standardized driving tests, we would like to know (1) how many autos were tested, and (2) what was the standard deviation s of the resulting *mpg* figures. Knowing the sample size n and the standard deviation s, we could then calculate the standard error of the mean $s_{\bar{x}} = s/\sqrt{n}$. Let us suppose we are told that the $\bar{x} = 24$ *mpg* figure was based on tests with $n = 9$ cars, and that the standard deviation was $s = 4$ *mpg*. We can then calculate the standard error of the mean as $s_{\bar{x}} = s/\sqrt{n} = 4/\sqrt{9} = 1.33$ *mpg*. Now we look up the value of t for 8 degrees of freedom ($df = 9-1$) that embraces 95% of the area under the t distribution and find it to be ± 2.31. We now have all the necessary information to substitute into the general form of the confidence interval for the mean

$$(\bar{x} - t_c s_{\bar{x}}) < \mu < (\bar{x} + t_c s_{\bar{x}})$$

and we have

$$(24 - 2.31 \times 1.33) < \mu < (24 + 2.31 \times 1.33)$$

or

$$20.9 < \mu < 27.1 \; mpg$$

Thus, we are 95% confident that the true mileage figure for the tested car model is anywhere between 20.9 and 27.1 *mpg*. This, of course, is a much more meaningful statement than the point estimate of 24 *mpg*.

It must be reiterated that the use of the t distribution assumes that the population from which the sample statistic is drawn is *normally* distributed. Since we are dealing with small samples we can no longer invoke the Central Limit Theorem to assure us that the sampling distribution of the statistic will be very nearly normally distributed, even when the sampled population is not. However, even the assumption of a normal population is not really that crucial, since several computer simulation studies have found that the t ratio is relatively insensitive to departures from normality in the sampled population, provided those departures are not too severe.

We have now seen how the notion of a sampling distribution aids us in estimating the values of population parameters based on sample statistics. Most importantly, we have discovered that interval estimates of a parameter are necessarily probabilistic in nature, and that we cannot be absolutely certain of the accuracy of our estimates. In the following sections we will look at other parameter estimation problems.

6. Interval Estimates of Other Parameters

In the preceding sections we looked at the alternative procedures for constructing an interval estimate for a single mean μ. But in everyday research there are a number of other parameters which we often wish to estimate. In the following four sections we will consider four such parameters which are commonly encountered in basic research designs, including (1) the mean difference between paired measures, μ_D, (2) the difference between two independent means, $\mu_1 - \mu_2$, (3) a proportion, p, and (4) the difference between two independent proportions, $p_1 - p_2$.

Upon the first reading of the four sections, it may seem that we are being onslaughted with a morass of new concepts and formulas, but this is not the case. Nothing really new is introduced in the presentations, only variations on the theme of the preceding sections. Namely, we have a *parameter* θ that we wish to estimate with a *statistic* $\hat{\theta}$, which has a sampling distribution with a *standard error* $\sigma_{\hat{\theta}}$. The interval estimate is of the general form

$$\theta = \hat{\theta} \pm z_c \sigma_{\hat{\theta}}$$

or equivalently

$$(\hat{\theta} - z_c \sigma_{\hat{\theta}}) < \theta < (\hat{\theta} + z_c \sigma_{\hat{\theta}})$$

where z_c is selected depending upon whether we want a 95%, 99% or other confidence level for our interval estimate.

Our only real task, then, is to identify the value of $\sigma_{\hat{\theta}}$, the standard error of the statistic in question. They have all been introduced in Chapter 7, so they will not be entirely new to us. They are also summarized in Table 1, which should be studied as a preview of what is to come.

It will be recognized that the standard error of a statistic's sampling distribution is always some function of the sample size n and of the standard deviation σ of the population of measures upon which the statistic is based. Since we rarely know this population standard deviation, in practice we resort to the sample estimate of it, s, when constructing our interval estimates, realizing that they are now approximate in nature by virtue of the substitution of s for σ. Consequently, we should think of a parallel set of standard errors in Table 1, in which s replaces σ wherever it occurs. In most cases the resulting approximation will be good, provided the sample size(s) exceeds thirty or so. When samples are smaller, then t_c can be used instead of z_c in constructing the interval estimates. If the sampled populations are normally distributed, the use of t will give us accurate interval estimates, but even if not normal, the estimates with t will be better than if z is used—this for the small sample situation. When samples are large the value of t is very nearly equal to z, so it

Table 1 Summary of terms for various parameter estimation problems of the form $(\hat{\theta} - z_c\sigma_{\theta}) < \theta < (\hat{\theta} + z_c\sigma_{\hat{\theta}})$.

θ Parameter to be estimated: Name	Symbol	$\hat{\theta}$ Statistic used to estimate θ	$\sigma_{\hat{\theta}}$ Standard error of the statistic's sampling distribution
A mean	μ	$\bar{x}$	$\sigma_{\bar{x}} = \dfrac{\sigma}{\sqrt{n}}$
A mean difference of paired measures	μ_D	$\bar{D}$	$\sigma_{\bar{D}} = \dfrac{\sigma_D}{\sqrt{n}}$
A difference between two independent means	$\mu_1 - \mu_2$	$\bar{x}_1 - \bar{x}_2$	$\sigma_{\bar{x}_1 - \bar{x}_2} = \sqrt{\sigma_{\bar{x}_1}^2 + \sigma_{\bar{x}_2}^2} = \sqrt{\dfrac{\sigma_1^2}{n_1} + \dfrac{\sigma_2^2}{n_2}}$
A proportion	p	$\hat{p}$	$\sigma_{\hat{p}} = \dfrac{\sigma}{\sqrt{n}} = \sqrt{\dfrac{pq}{n}}$
A difference between two independent proportions	$p_1 - p_2$	$\hat{p}_1 - \hat{p}_2$	$\sigma_{\hat{p}_1 - \hat{p}_2} = \sqrt{\sigma_{\hat{p}_1}^2 + \sigma_{\hat{p}_2}^2} = \sqrt{\dfrac{p_1q_1}{n_1} + \dfrac{p_2q_2}{n_2}}$
			Note: *When the population* σ *is unknown, and sample size is large, the sample* s *is used instead of* σ, *for constructing an approximate interval (see text for details).*

makes little difference which we use. See, for example, Appendix Table IV for the similarity of t to z when samples are large.

7. Interval Estimate of the Mean Difference Between Non-Independent Paired Measures

Let us now consider the situation in which we have paired measures on *two* variables, say x and y. As we learned in the preceding chapter, the paired measures might arise from naturally occuring couplets such as twins, spouses, or before and after comparisons on the same individuals; or they might be created for analytical purposes, such as by pairing students on similarity of academic aptitude and then assigning one member of each pair to one teaching method and the other member to another teaching method.

For each pair of measures we can obtain a difference score, defined as $D_i = x_i - y_i$. Next, we can calculate the mean and standard deviation of the difference scores, which we denote $\bar{D}$ and s_D, respectively. Knowing that we have n pairs of measures (and consequently n difference scores), we can

calculate the standard error of the mean difference between the paired measures with $s_{\bar{D}} = s_D/\sqrt{n}$, which is our estimate of the true standard error $\sigma_{\bar{D}} = \sigma_D/\sqrt{n}$. This provides us with all the information we will need to create an interval estimate for μ_D, the true mean difference between the population of paired measures under study. The following example will demonstrate the procedure.

Suppose a developer of educational materials is interested in evaluating the difference in effectiveness of two alternative home-study courses, *A* vs. *B*. A sample of 200 students is rank-ordered in terms of grade point average, and 100 pairs are then formed by coupling each two students in turn, working down the list. One member of each pair is randomly chosen for course *A*, and the other member receives course *B*. All students are then administered an achievement test at the end of the prespecified time period for the course. Always subtracting in the same order—course *A* score minus course *B* score—a set of 100 difference scores D_i are obtained, a mixture of positive and negative values depending upon whether the *A* or *B* member of the pair had a higher achievement score. The mean difference score is found to be $\bar{D} = 2.7$, and the standard deviation of the difference scores is $s_D = 15$ points. From this, we can calculate the estimated standard error of the mean difference as

$$s_{\bar{D}} = \frac{s_D}{\sqrt{n}} = \frac{15}{\sqrt{100}} = 1.5$$

Then for the approximate 95% confidence interval we merely have to substitute into

$$(\bar{D} - z_c s_{\bar{D}}) < \mu_D < (\bar{D} + z_c s_{\bar{D}})$$

and we have

$$[2.7 - 1.96(1.5)] < \mu_D < [2.7 + 1.96(1.5)]$$

or

$$-.24 < \mu_D < 5.64$$

as our approximate 95% confidence interval for the true population mean difference score between the two home-study courses. In other words, if this procedure were repeated an infinite number of times, we would expect that the resulting confidence interval would encompass the true mean difference about 95% of the time.

Notice that the lower limit of the interval is $-.24$, an outcome in favor of course *B*. Thus, while the sample results suggest that course *A* is superior to

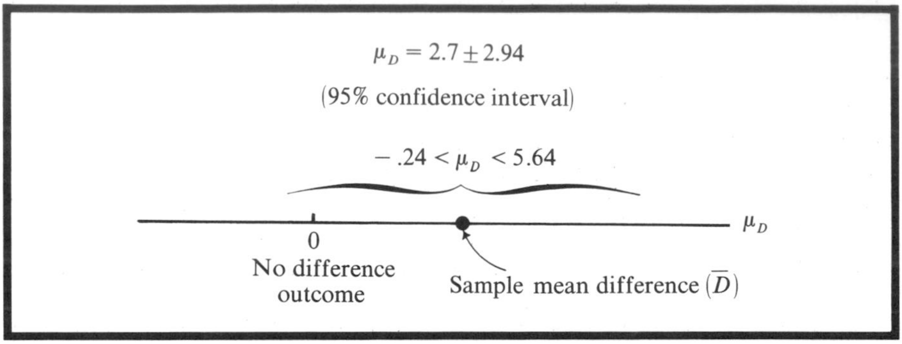

Figure 5 The 95% confidence interval for the mean difference between achievement scores for two home-study courses, *A* and *B*.

course *B*, we cannot reject the possibility that course *B* will outperform course *A* in the population at large. That the true mean difference could be in favor of course *B* is easily seen in Figure 5, where the confidence interval is shown to extend slightly into the negative mean difference values, those signifying a superiority of course *B*. Thus, the results are not entirely conclusive. However, since the data came very close to suggesting the better performance of one method over the other, the researcher may be encouraged to repeat the study with a larger sample of students. This would narrow the interval estimate, perhaps enough to exclude the negative values which support the *B* course, providing the assurance that one course was in fact better than the other.

8. Interval Estimate of the Difference Between Two Independent Means

Very often we want to estimate the difference between the means of two distinct populations, populations which can be thought of as being *independent* of one another, in the sense that there is no naturally occurring or contrived relationship between members of the two populations, as was the case in the preceding section when we compared populations which were comprised of systematically paired individuals.

For example, we might want to estimate the difference between males and females, or schizophrenics and non-schizophrenics, or felons and non-felons, or endorsers and non-endorsers of an attitude, or productive and non-productive well sites, or diagnostic group *A* and *B*, etc., on any particular variable of interest. Other times, we can form two independent *experimental* groups by randomly partioning a sample into two treatment groups, and administer one treatment to one group and another to the other group, with an aim toward assessing the difference between the two treatments with respect to some

variable of interest; e.g., the difference in effect between two drugs on blood pressure, the difference in persuasiveness of two communications messages, the difference in student achievement for two different class sizes, the difference in crop yield for two different fertilizers, the difference in blood cholesterol resulting from two different diets, etc.

In order to estimate the difference between the means of the two populations in question on a given variable of interest, designated $\mu_1 - \mu_2$, we proceed in the traditional manner outlined throughout the chapter. Since the difference in the population means, $\mu_1 - \mu_2$, is our parameter of interest, we need to obtain a sample estimate of that difference, namely $\bar{x}_1 - \bar{x}_2$, and an estimate of the standard error of the sampling distribution of the statistic, namely $s_{\bar{x}_1 - \bar{x}_2}$. This is all the information we need to substitute into our familiar formula for the approximate interval estimate of a parameter θ; namely, $\theta = \hat{\theta} \pm z_c s_{\hat{\theta}}$ (or, for small samples, $\theta = \hat{\theta} \pm t_c s_{\hat{\theta}}$). The following example will show the similarity of the approach to the problems discussed in earlier sections. The game is the same, only the players differ. Whereas before, θ was a single mean μ, or the mean of difference scores μ_D, it will now be the difference between two independent means, $\mu_1 - \mu_2$.

The price elasticity of goods and services—the differential response to different price levels—is a constant issue with government officials, economists, and marketers. Suppose a company wants to assess the difference in sales response to its product at two different price levels, *A* and *B*. A total of 100 retail outlets are chosen at random from across the country to participate in the price test. Fifty of the stores are randomly chosen to carry the product at price level *A*, and the other fifty at price level *B* (although there is no requirement that the sample sizes be equal). Sales of the product over a one-month period are then recorded for each of the stores. Average store sales of the product at price *A* is $\bar{x} = 174$ units, and at price *B* it is $\bar{x} = 163$ units. The respective standard deviations for the *A* and *B* samples are $s_1 = 15$ and $s_2 = 17$. With this information in hand we estimate the standard error of the sampling distribution of the difference between the means, with

$$s_{\bar{x}_1 - \bar{x}_2} = \sqrt{s_{\bar{x}_1}^2 + s_{\bar{x}_2}^2} = \sqrt{\frac{s_1^2}{n_1} + \frac{s_2^2}{n_2}} = \sqrt{\frac{15^2}{50} + \frac{17^2}{50}} = 3.2$$

We now have all the information needed to substitute into the now familiar formula for the approximate 95% confidence interval estimate, which in this instance only *looks* more complicated than earlier ones:

$$\left[(\bar{x}_1 - \bar{x}_2) - 1.96 s_{\bar{x}_1 - \bar{x}_2}\right] < (\mu_1 - \mu_2) < \left[(\bar{x}_1 - \bar{x}_2) + 1.96 s_{\bar{x}_1 - \bar{x}_2}\right]$$

Substituting our sample statistics, we have

$$[(174-163)-1.96(3.2)] < (\mu_1 - \mu_2) < [(174-163)+1.96(3.2)]$$

which, after a little arithmetic, yields

$$4.73 < (\mu_1 - \mu_2) < 17.27$$

In other words, we are approximately 95% confident, in a betting sense, that the difference in average sales for the product under the two price levels will be between 4.73 and 17.27 units per store per month. This information, in addition to product cost factors, could then be used to help reach a decision as to whether to go with the lower or higher price.

9. Interval Estimate of a Proportion

Estimating the population value of a proportion p is one of the most common of all research problems. For example, we are often interested in the population proportion, or *incidence*, or a particular disease, of usage of a certain product or drug, of student truancy, of a given crime, of infected plants, of death by a specific cause, of endorsement of a certain belief or attitude, of disease cure by a specific treatment, etc. In all instances we are interested in the proportion (or percentage, if you prefer) of a stipulated population that possesses a particular characteristic, where each member of the population either has or does not have the characteristic, referred to as a binary or dichotomous characteristic in that only two values are possible.

Earlier, in Chapter 4, we learned that a proportion was nothing more than the mean of the 0 and 1 values of a binary variable. And in Chapter 7 we learned that the sampling distribution of a sample proportion $\hat{p}$, based on a random sample of size n, was very nearly normally distributed provided the values of np and nq exceeded 5 or so. Consequently, we should expect that the interval estimate of a proportion based on a large sample should be very close in form to the interval estimate for any other mean. Similar to the interval estimate for a mean μ, the interval estimate for a proportion p will be of the form

$$(\hat{p} - z_c\sigma_{\hat{p}}) < p < (\hat{p} + z_c\sigma_{\hat{p}})$$

provided the sample size is large, np and nq greater than 5. We also know that $\sigma_{\hat{p}}$, the standard error of the sampling distribution of a proportion, is given by

$$\sigma_{\hat{p}} = \frac{\sigma}{\sqrt{n}} = \sqrt{\frac{pq}{n}}$$

since $\sigma = \sqrt{pq}$ for a binary population. But we do not know the value of p in the above definition, otherwise why would we be trying to estimate it. Instead, we can use the sample estimates $\hat{p}$ and $\hat{q}$ in place of the population values p and q, respectively, and realize that our subsequent interval estimate of p will be approximate in nature. The estimate of the standard error of the proportion then becomes

$$s_{\hat{p}} = \sqrt{\frac{\hat{p}\hat{q}}{n}}$$

which will be used in the approximate interval estimate

$$(\hat{p} - z_c s_{\hat{p}}) < p < (\hat{p} + z_c s_{\hat{p}})$$

where the value of z_c will depend upon whether we want a 95%, 99%, or other confidence level.

As an example, suppose that a random poll of 1,000 registered voters results in candidate A receiving 270 votes, for a sample proportion of $\hat{p} = .27$. To obtain the 95% confidence interval for the true but unknown population proportion p, we begin by estimating the standard error of the sampling distribution of the sample proportion with

$$s_{\hat{p}} = \sqrt{\frac{\hat{p}\hat{q}}{n}} = \sqrt{\frac{(.27)(.73)}{1{,}000}} = .014$$

Then, substituting this result and our sample proportion $\hat{p} = .27$, and $z_c = 1.96$ for a 95% confidence level, into the above formula, we have

$$[.27 - 1.96(.014)] < p < [.27 + 1.96(.014)]$$

which simplifies to

$$.243 < p < .297$$

as our approximate 95% confidence interval for the proportion of voters expected to favor candidate A. That is, we are 95% sure that the candidate will poll between 24.3% and 29.7% of the vote, assuming the voter preferences hold up over time between when the sample was taken and when the voting booth is reached. It also assumes that the sample is representative of those who actually do vote.

Again, the above 95% confidence interval is only approximate in nature, inasmuch as we estimated the standard error of the proportion, and further-

more because we invoked the characteristics of the continuous normal distribution to substitute for the discrete binomial distribution, which is the precise representation of the sampling distribution of a proportion based on independent observations from a binary population. Nonetheless, the approximation is quite good when the values of np and nq are each greater than five or so, as we have seen in the preceding chapter.

10. Interval Estimate of the Difference Between Two Independent Proportions

When two independent populations are measured on some binary characteristic, we will be interested in the difference between the two populations in their proportion or incidence of that characteristic. For example, how do males and females differ in their incidence of a particular disease, how do liberals and conservatives differ in the proportion of their votes for a particular referendum, how do two different therapy populations differ in cure rate, how do two fertilizer treatments affect seed germination rates, how do two different instructional populations differ in their proportion of correctly answering a test item, etc.

Our interest is in forming a confidence interval for the true difference between the two given populations with respect to the proportion in question. Again, our familiar confidence interval will be of the form $\theta = \hat{\theta} \pm z_c \sigma_{\hat{\theta}}$ where θ now represents the true but unknown population difference, $p_1 - p_2$, and $\hat{\theta}$ corresponds to the observed difference in the sample proportions, $\hat{p}_1 - \hat{p}_2$. The standard error $\sigma_{\hat{\theta}}$ in this instance is the standard error of the difference between the sample proportions, given by

$$\sigma_{\hat{p}_1 - \hat{p}_2} = \sqrt{\sigma^2_{\hat{p}_1} + \sigma^2_{\hat{p}_2}} = \sqrt{\frac{p_1 q_1}{n_1} + \frac{p_2 q_2}{n_2}}$$

where n_1 and n_2 are the respective sample sizes drawn from the two populations. Again, since the true values of p_1 and p_2 are unknown to us, we resort to their sample estimates $\hat{p}_1$ and $\hat{p}_2$ to obtain an estimate of the standard error of the sampling distribution of the difference between the proportions, or

$$s_{\hat{p}_1 - \hat{p}_2} = \sqrt{\frac{\hat{p}_1 \hat{q}_1}{n_1} + \frac{\hat{p}_2 \hat{q}_2}{n_2}}$$

We can be relatively confident that the sampling distribution will be nearly normal if form provided that $n_1 p_1$, $n_1 q_1$, $n_2 p_2$, and $n_2 q_2$ all exceed a value of 5 (which can be estimated using $\hat{p}$'s and $\hat{q}$'s). In the next chapter we will see

an alternative approach to estimating this standard error when p_1 and p_2 are hypothesized as being equal.

All we need, then, to form an approximate confidence interval are values of $\hat{p}_1$ and $\hat{p}_2$, which are sample estimates of the corresponding population proportions p_1 and p_2. An example will demonstrate the procedure.

A survey which included $n_1 = 500$ males and $n_2 = 700$ females showed that $\hat{p}_1 = .264$ males vs. $\hat{p}_2 = .323$ females agreed with a particular attitudinal statement. Our estimate of the standard error of the difference between the two independent proportions is as follows:

$$s_{\hat{p}_1 - \hat{p}_2} = \sqrt{\frac{\hat{p}_1\hat{q}_1}{n_1} + \frac{\hat{p}_2\hat{q}_2}{n_2}} = \sqrt{\frac{(.264)(.736)}{500} + \frac{(.323)(.677)}{700}} = .0265$$

Substituting, then, into the appropriate formula for estimating the confidence interval for the difference between two independent proportions

$$\left[(\hat{p}_1 - \hat{p}_2) - z_c s_{\hat{p}_1 - \hat{p}_2}\right] < (p_1 - p_2) < \left[(\hat{p}_1 - \hat{p}_2) + z_c s_{\hat{p}_1 - \hat{p}_2}\right],$$

we have, for the 95% confidence level,

$$[(.264 - .323) - 1.96(.0265)] < (p_1 - p_2) < [(.264 - .323) + 1.96(.0265)]$$

which simplifies to

$$-.111 < (p_1 - p_2) < -.007$$

Thus, we are about 95% sure that the proportion of females endorsing the attitudinal statement in question is anywhere from .111 to .007 (or 11.1% to .7%) higher than it is for males in the sampled population at large. That is, with the procedure we have followed, we will, in the long run, have encompassed the true population parameter within our interval estimate about 95% of the time.

By now, the creation of confidence intervals should be close to second nature to us. We have constructed interval estimates for five different parameters, summarized earlier in Table 1. The approach was always the same. We obtained a sample estimate $\hat{\theta}$ of a parameter θ. From the sample estimate we added and subtracted an allowable amount of error, namely $z_c\sigma_{\hat{\theta}}$, where $\sigma_{\hat{\theta}}$ is the standard error of the sampling distribution of the statistic in question, and z_c is the critical standardized normal deviate associated with a given confidence level, most commonly 1.96 or 2.58, for a 95% or 99% confidence level, respectively. In most cases, we had to estimate the standard error $\sigma_{\hat{\theta}}$ with sample values $s_{\hat{\theta}}$, resulting in approximate interval estimates. When sample sizes are small, we can use Student's t instead of z for more accurate estimates.

When we are dealing with proportions, we must be sure that np and nq (estimated with $n\hat{p}$ and $n\hat{q}$) are always greater than 5 or so if we are to treat their sampling distributions as normal in form.

At this point it will be worthwhile to review again in Table 1 the five parameters for which we have constructed interval estimates, and then reread these past five sections to appreciate that they are simply variations on a single theme.

11. Sample Size Selection for Limiting Error

In the planning stages of our research we always have to decide upon what size sample to use. While time, cost, and other practical considerations must be taken into account, we can also make our decision based on how large of an *error* we are willing to tolerate in our estimate of the population parameter in question.

Since our interval estimate of the parameter θ is given by $\theta = \hat{\theta} \pm z_c\sigma_{\hat{\theta}}$, it follows that $\theta - \hat{\theta} = \pm z_c\sigma_{\hat{\theta}}$, where $\theta - \hat{\theta}$ is the maximum estimation error. Consequently, the quantity $z_c\sigma_{\hat{\theta}}$ represents the maximum error for a given confidence level, which we can designate E, and write

$$\boxed{E = \theta - \hat{\theta} = z_c\sigma_{\hat{\theta}}}$$

Since the standard error of a statistic $\sigma_{\hat{\theta}}$ is always a function of the sample size upon which it is based, we can solve for the value of n for any specified value of the maximum error E.

For instance, when we are interested in estimating a single mean μ we deal with the standard error of the mean and we have

$$E = z_c\sigma_{\bar{x}} = z_c\left(\frac{\sigma}{\sqrt{n}}\right)$$

Solving for the value of the sample size n, we have

$$\boxed{n = \left(\frac{z_c\sigma}{E}\right)^2}$$

as the sample size required in order that the maximum error in our estimate of the population mean μ is no larger than E, with the level of confidence associated with the value of z_c that we choose.

Suppose, for example, that we are interested in estimating the mean diameter of the trees in a large stand of timber. Further assume that we do not

want to be in error by more than .5 inches, with 95% assurance. How many trees should we sample and measure for this degree of precision? Since we do not know the value of σ, the standard deviation of the tree diameters, we can consult past records or perform a little pilot study to obtain a sample estimate of it, s. Let us say we obtain an estimate of $s = 4.7$ inches. So, with $E = .5$, $s = 4.7$, and $z_c = 1.96$ for a 95% confidence level, we have

$$n = \left(\frac{z_c s}{E}\right)^2 = \left[\frac{(1.96)(4.7)}{.5}\right]^2 = 339.4$$

We conclude, then, that we should measure a random sample of 340 trees in order to be 95% sure that our estimate $\bar{x}$ of the true mean tree diameter μ is not in error by more than .5 inches. As an exercise, verify that $1.96 s_{\bar{x}}$ does not exceed .5 inches.

Let us now consider another example, this time determining the sample size required to estimate a population proportion p with a specified degree of precision. Since the standard deviation of the parent binary population is $\sigma = \sqrt{pq}$, the expression for the needed sample size becomes

$$n = \left(\frac{z_c \sigma}{E}\right)^2 = pq\left(\frac{z_c}{E}\right)^2$$

But we do not know the value of p, for that is what we are trying to estimate with a certain degree of precision. There are basically two approaches to the problem. One possibility is to use an estimate of p (and consequently q) based on available information. We would then have $n = \hat{p}\hat{q}(z_c/E)^2$ for determining the sample size. An alternative, if we do not have the faintest idea of the value of p, is to let $p = .5$. This choice of p will result in the largest value of pq, namely $pq = (.5)(.5) = .25$ (e.g., compare: when $p = .4$, $pq = (.4)(.6) = .24$; when $p = .3$, $pq = (.3)(.7) = .21$; etc.). This "worse case" scenario of assuming $p = .5$, which results in the largest possible standard error, will lead to a sample size n which is larger than is really needed for the desired precision, and so it is a prudent course of action.

As an application of the approach, suppose we want to determine the size of the sample required to estimate the proportion of the nation's high school students that consumed any alchoholic beverage during the preceding 7-day period, and that we do not want to be in error in our estimate of the proportion by more than .03, with a 95% confidence level. Substituting into our conservative formula for the sample size n, we have

$$n = pq\left(\frac{z_c}{E}\right)^2 = (.5)(.5)\left(\frac{1.96}{.03}\right)^2 = 1067.1$$

We conclude, then, that if we poll a random sample of 1,067 students our obtained proportion will be within .03 of the true population proportion value, with 95% assurance. Verify that $1.96s_{\hat{p}}$ does not exceed .03.

If we want to estimate the sample sizes required to estimate the *difference* between *two means*, or between *two proportions*, with a specified degree of precision at a given confidence level, we can follow the same approach as above, with the exception that we must use their respective standard errors. Since two sample sizes, n_1 and n_2, are involved in their standard errors, it is easiest to assume they will be equal in size, and then the formula for n can be solved with a little bit of arithmetic. This is left as an exercise.

12. Concluding Comments

We have seen how the notion of a sampling distribution aids us in estimating the values of population parameters based on sample statistics. Most importantly, we have discovered that interval estimates of a parameter are necessarily probabilistic in nature, and that we cannot be absolutely certain of the precision of our estimates. In the following chapter we will learn another approach to the inference problem, which also relies on the concept of a sampling distribution, and which will be seen as a mirror image of the interval estimation approach.

Chapter 9

Hypothesis Testing

1. Introduction

We have seen that one approach to the problem of statistical inference is the *direct* estimation of population parameters from sample observations—either as point estimates or interval estimates. In this section we will consider an alternative approach, one that is *indirect* in nature; i.e., *hypothesis testing*. Rather than using our sample observations to derive statistics which are approximations of the population parameter in question, we will use our sample statistics to support or discredit *a priori* hypotheses, or speculations, about the true value of the population parameter.

The hypotheses about the population parameters that we wish to test can be based either on prior observations or on theoretical grounds. Whatever the basis for the hypothesis, our sample observations will be used to test the likelihood or tenability of its being true. If it is found to be untenable, from a probability point of view, then we are forced to believe in an alternative hypothesis. But just as we cannot be 100 percent sure with regard to our interval estimates of a parameter, neither can we be absolutely certain of our conclusions with regard to the truth or falsity of our tested hypotheses. We can be 90 percent sure, 95 percent sure, 99 percent sure, and so on, but never 100 percent sure. With this in mind let us begin our discussion of hypothesis testing by considering the various types of hypotheses in which we might be interested.

2. Types of Hypotheses Sets

The most common approach to hypothesis testing is to establish a set of two mutually exclusive and exhaustive hypotheses about the true value of the parameter in question. Then, our sample statistics will be used to support one or the other of the hypothesized alternatives.

Exhaustive hypotheses sets. A researcher might be interested in testing the hypothesis that the mean weight loss from a particular three-week diet is 10 pounds. We designate this *working or "null" hypothesis as* H_0, and write

$$H_0: \mu = 10 \text{ pounds}$$

The alternative to this hypothesis, which we designate H_1, is that the mean weight loss is *not* equal to 10 pounds, and we write

$$H_1: \mu \neq 10 \text{ pounds}$$

where $\neq$ is the symbol for "is not equal to." These two hypotheses, then, cover all possibilities, for either the true weight loss due to the diet *is* equal to 10 pounds (H_0) or it *is not* (H_1), and therefore can be referred to as an *exhaustive* hypotheses set.

To be perfectly general, we can again let θ stand for any particular parameter we wish—whether it be a mean, a difference between two means, a mean difference of paired measures, a proportion, a difference between two proportions, etc.—and then our hypotheses set can be expressed

$$H_0: \theta = a$$

$$H_1: \theta \neq a$$

where a is some specified value.

Truncated hypotheses sets. The hypotheses set outlined above is not the only kind in which we might be interested. There are some situations in which the hypotheses set

$$H_0: \theta = a$$

$$H_1: \theta < a$$

is more meaningful, where $<$ is the symbol for "is less than." That is, if the hypothesis $\theta = a$ is not true then the only alternative is that $\theta < a$. But here we might object that while these two alternatives are mutually exclusive, they are not exhaustive, since we have not admitted the possibility $\theta > a$, that θ is greater than a. And this is a crucial point, for when using such a *truncated* or *curtailed* or *one-sided* hypotheses set we must be absolutely certain, usually on logical grounds, that the third and omitted possibility has a zero probability of occurrence. We cannot use an exhaustive hypotheses set such as

$$H_0: \theta \geq a$$

$$H_1: \theta < a$$

since *we have no idea what the sampling distribution for* H_0: $\theta \geq a$ *looks like if we admit the possibility that* $\theta > a$. And it is the H_0 sampling distribution upon which the theory of our test rests.

We might use a truncated or one-sided hypotheses set, for example, when evaluating the breakage level of a product shipped with additional packing material. If the breakage rate is not the same as that which occurred with the old packing technique (H_0: $\mu = a$), then it must be lower (H_1: $\mu < a$). On purely logical grounds we can rule out the possibility that the breakage rate could be higher.

Of course there is another truncated hypotheses set possible, which is simply a variation of the preceding one, namely

$$H_0: \theta = a$$

$$H_1: \theta > a$$

In this type of situation, the alternative to $\theta = a$ is $\theta > a$, and we rule out the possibility that $\theta < a$. For example, if we revise a billboard advertisement by increasing the size of the print with which the brand name appears, we can safely assume that the percentage of consumers recalling the brand name will either be unchanged (H_0: $\theta = a$) or it will be higher (H_1: $\theta > a$), where a is a level of recall for the original version of the billboard.

Whether we use an exhaustive or truncated hypotheses set will be dictated by the nature of the particular problem, that is, which alternatives are possible. The importance of the distinction between these types of hypotheses sets will become more apparent in the following sections where we will consider the techniques for choosing among the alternative hypotheses. Which *type of hypotheses set we choose* is a logical decision, which *hypothesis in the set* we eventually believe is a statistical question.

3. Test of a Mean

Suppose a municipality wants to test the hypothesis that a particular model school bus averages 20 miles per gallon of gasoline. The hypotheses set is

$$H_0: \mu = 20\ mpg$$

$$H_1: \mu \neq 20\ mpg$$

Our task is to test the credibility of H_0 vs. H_1.

Since we cannot realistically test every single bus that came off the assembly line, we must test a sample of buses of the particular model of

interest. The mean gas mileage $\bar{x}$ of this sample would then be used to decide the credibility of the hypothesis that $\mu = 20$. What if we obtained a mean value of $\bar{x} = 21$, would the hypothesis of $\mu = 20$ be credible? What if we obtained a sample mean of $\bar{x} = 22$? Or $\bar{x} = 16$?

By now we realize that to answer these questions we need additional information—namely, characteristics of the sampling distribution of the mean $\bar{x}$; which, as we have learned, is dependent on the size of the sample n, and the shape and standard deviation σ of the population of values which was sampled. Knowing the population standard deviation σ and the size of the sample n, we can easily determine the standard error of the sampling distribution of the mean using

$$\sigma_{\bar{x}} = \frac{\sigma}{\sqrt{n}}$$

Now if we further know that the sampled population is normally distributed, we will also know that the sampling distribution of $\bar{x}$ is normally distributed. Therefore, we can express our observed sample mean $\bar{x}$ as a deviation from the hypothesized mean μ in standard error units; namely

$$\boxed{z = \frac{\bar{x} - \mu}{\sigma_{\bar{x}}} = \frac{\bar{x} - \mu}{\dfrac{\sigma}{\sqrt{n}}}} \qquad (1)$$

However, if we do not know the value of the population standard deviation σ, which we need to calculate the standard error of the mean $\sigma_{\bar{x}}$, then we must use the sample standard deviation s as a substitute estimate. Our observed sample mean $\bar{x}$ can then be expressed as a deviation from the hypothesized population mean μ in estimated standard error units; namely

$$z \doteq \frac{\bar{x} - \mu}{s_{\bar{x}}} = \frac{\bar{x} - \mu}{\dfrac{s}{\sqrt{n}}}$$

where $\doteq$ is the symbol for "is approximately equal to." So long as our sample size n is sufficiently large, say greater than thirty, we can be confident that the above z will be very nearly normally distributed, and approximately equal to the value which would be obtained if we knew the value of $\sigma_{\bar{x}}$.

Returning, then, to our example, let us assume that we test $n = 100$ buses and obtain a mean mileage figure of $\bar{x} = 19.1$ *mpg*. Let us also assume, for the time being, that we know the population standard deviation to be $\sigma = 3$ *mpg*.

We can then calculate the standard error of the mean as

$$\sigma_{\bar{x}} = \frac{\sigma}{\sqrt{n}} = \frac{3}{\sqrt{100}} = .3 \ mpg$$

Next, we can express the observed mean $\bar{x} = 19.1$ as a deviation from the hypothesized value $\mu = 20.0$, expressing it as a normal deviate

$$z = \frac{\bar{x} - \mu}{\sigma_{\bar{x}}} = \frac{19.1 - 20.0}{.3} = -3.00$$

In other words, our sample mean of $\bar{x} = 19.1$ deviates from the hypothesized value of $\mu = 20.0$ by three standard errors of the mean. This deviation is shown graphically in Figure 1.

The question we must ask is *whether it is likely that we would get such a deviant sample mean $\bar{x}$ if in fact the true population value were $\mu = 20.0$*. If it is too unlikely, then we will tend to disbelieve the hypothesis. But what is our

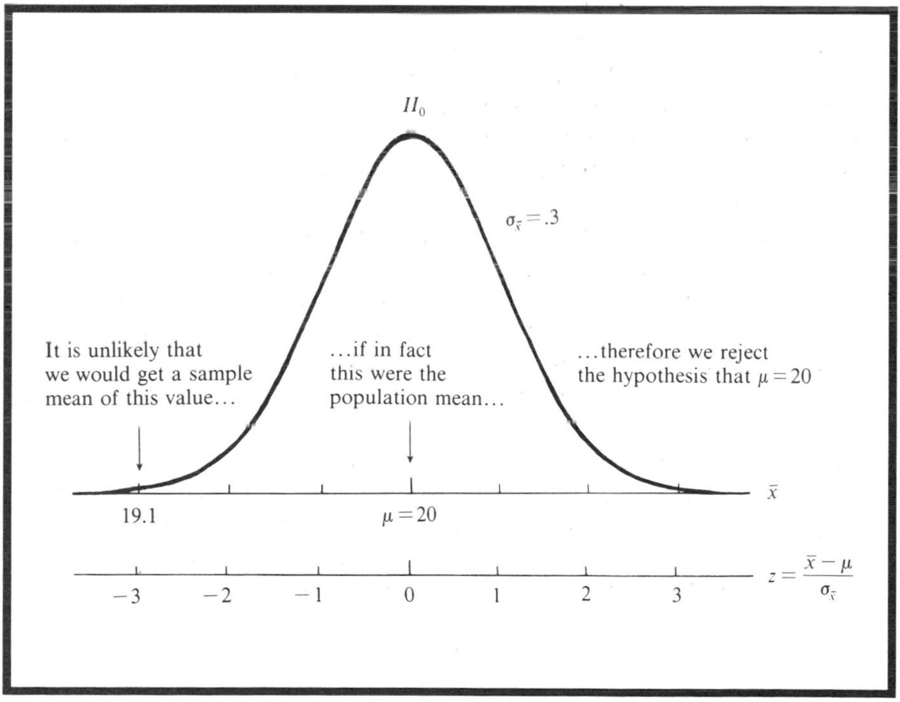

Figure 1 The deviation of a sample mean $\bar{x} = 19.1$ from the hypothesized mean $\mu = 20$.

standard for "unlikely." By convention, a probability less than .05 (1 chance in 20), or less than .01 (1 chance in 100), is usually considered an unlikely event. In other situations we may want to establish either a more stringent or a less stringent criterion for what we consider an unlikely event under the condition that H_0 is true, but in most instances .05 and .01 are accepted standards.

Significance levels. The particular probability that we use as our criterion for an unlikely outcome, supposing that our working hypothesis H_0 is true, is referred to as the *significance level* of our statistical test and is designated with the Greek letter *alpha*, α. It is essential that we stipulate α *prior to conducting our statistical test*, for to choose one after the data has been analyzed would be less than objective. In many research studies "p values" are reported, signifying the after-the-fact probability of obtaining the test statistic in question, given that the null hypothesis were true. While such information is useful, it is not a substitute for prespecifying α, our criterion for rejecting the null hypothesis. For if we do not specify beforehand our standard for an unlikely event, then we will be tempted to decide after the fact, which is not a "well-defined" procedure. It would be like bragging about the number of fish we caught at a particular fishing hole, without first having specified our criterion for what would be a good catch.

Three common significance levels ($\alpha = .01$, $\alpha = .05$, and $\alpha = .10$) are shown in Figure 2. The extreme values of the normal distribution corresponding to these probabilities are designated by the shaded areas, and are referred to as the *critical regions*, or *rejection regions*, for our test of the hypothesis H_0.

It will be further noted in Figure 2 that the probability corresponding to the various significance levels can either be split between the two tails of the distribution, or placed all in one tail. If the probability of an unlikely outcome is divided between the two tails of the distribution, as in the left parts of Figure 2, we refer to our statistical test as a *two-tailed* or *two-sided* test. If, however, we are dealing with a truncated hypotheses set in which we assume that any departure from the H_0 hypothesis must be in a particular direction, then we will choose to put the entire probability of an unlikely outcome in one tail or the other of the distribution, as shown in parts b, d, and f of Figure 2, and we call this type of test a *one-tailed* or *one-sided* test.

Compare, for example, the value of z needed to reach the critical region for a significance level of $\alpha = .05$ when we are dealing with a two-tailed vs. a one-tailed test. For a two-tailed test, a deviation from the mean beyond $z = 1.96$ standard errors will occur by chance 5% of the time. For a one-tailed test, a deviation of only $z = 1.65$ standard errors from the mean, in a given direction, will occur by chance 5% of the time. For a significance level $\alpha = .01$, a value of z greater than 2.58 vs. 2.33 is needed to reach the rejection region for a two-tailed vs. one-tailed test, respectively, as shown in parts e and f of Figure 2. Of course, the z's reported above are their absolute values.

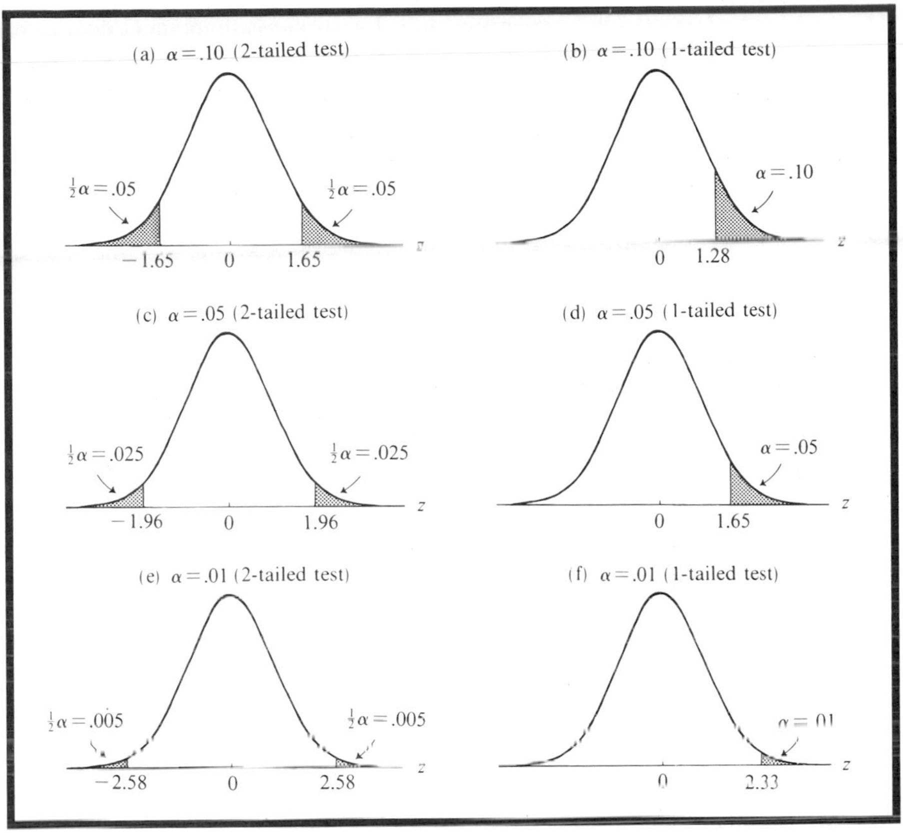

Figure 2 The most commonly used significance levels α, and their corresponding critical regions (when σ is known or sample size is large).

In our bus gas mileage example, assuming we had specified beforehand an $\alpha = .05$ significance level for judging the hypotheses H_0: $\mu = 20$, our observed sample mean $\bar{x} = 19.1$, which deviates from $\mu = 20$ by 3 standard errors, can be considered an unlikely outcome if indeed the hypothesis $\mu = 20$ were true. Rather than believe that we have simply witnessed a *chance* deviation from $\mu = 20$ due to sampling error, we prefer to disbelieve the hypothesis that $\mu = 20$. In "rejecting" the hypothesis H_0: $\mu = 20.0$, we are "accepting" the alternative hypothesis H_1: $\mu \neq 20.0$. We use the terminology "reject" and "accept" very guardedly, for we recognize that our conclusions are based on probabilities, therefore "believe" and "disbelieve" might be more appropriate terminology, suggesting that we might be wrong.

If we had not been provided with the population standard deviation σ, from which we calculated $\sigma_{\bar{x}}$, we would have had to use the sample standard

deviation s in its place, and estimate the standard error of the mean as $s_{\bar{x}} = s/\sqrt{n}$. Suppose, for example, that the sample standard deviation were $s = 3.2$. The estimated standard error of the mean would then be $s_{\bar{x}} = s/\sqrt{n} = 3.2/\sqrt{100} = .32$, and we would have

$$z = \frac{\bar{x} - \mu}{s_{\bar{x}}} = \frac{19.1 - 20.0}{.32} = -2.81$$

However, since we have estimated $\sigma_{\bar{x}}$ with $s_{\bar{x}}$, this test and the corresponding significance levels will only be *approximate* in nature. Recognizing this, our conclusion is the same as above. Since $z = -2.81$ is beyond the value $z = -1.96$ needed to enter the critical region for $\alpha = .05$, we must reject the hypothesis that $\mu = 20$. That is, rather than believe our sample mean $\bar{x} = 19.1$ was a "fluke" from a population with $\mu = 20$, we prefer, rather, to disbelieve that $\mu - 20$. Consequently, we place more credence in the alternative hypothesis that $\mu \neq 20$. We do not know the exact value of μ but we are fairly confident that it is not 20.

Small sample technique. In situations involving a small sample from a *normal population* with an unknown standard deviation σ, we can use Student's t variable

$$\boxed{t = \frac{\bar{x} - \mu}{s_{\bar{x}}} = \frac{\bar{x} - \mu}{\frac{s}{\sqrt{n}}}} \qquad (2)$$

to *exactly* test hypotheses about population means. Associated with the t ratio are degrees of freedom $df = n - 1$, as discussed in the Parameter Estimation chapter.

Suppose, in the preceding example, we had tested only $n = 10$ buses and obtained a mean mileage figure of $\bar{x} = 21.2$ with a standard deviation of $s = 3.4$. The estimated standard error of the mean is then $s_{\bar{x}} = 3.4/\sqrt{10} = 1.08$. We then calculate t as

$$t = \frac{\bar{x} - \mu}{s_{\bar{x}}} = \frac{21.2 - 20.0}{1.08} = 1.11$$

with degrees of freedom $df = 9$. We look in Appendix Table IV and find that for $df = 9$, a t of 2.26 is required before we enter the area of an unlikely event for a significance level of $\alpha = .05$. In other words, our value of $t = 1.11$ could well have happened *by chance alone* if in fact the hypothesis $\mu = 20.0$ were true. Figure 3 shows the relative location of $t = 1.11$. Consequently, we *will not*

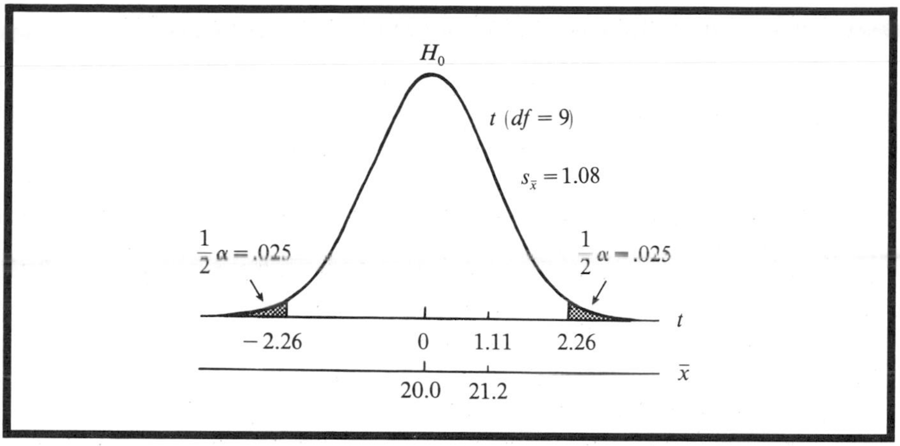

Figure 3 The relative location of $t = 1.11$ $(df = 9)$.

reject that hypothesis; we do not have enough evidence to discredit it. We never "accept" H_0, only refuse to reject it.

The preceding examples demonstrate the probabilistic foundations of our conclusions with regard to the believability of the alternative hypotheses that we are testing. As such, we cannot be 100 percent certain that we have reached the correct conclusion. In the following section we will consider the types of errors we are liable to make.

4. Type I and Type II Errors

In the American system of justice an accused person is assumed to be innocent until proven guilty. We could express these alternatives as hypotheses:

H_0: Innocent

H_1: Not innocent (i.e., guilty)

Now just as with the earlier statistical hypotheses, we as a jury might come to believe either of these alternatives. But since we base our conclusions only on sample evidence, we are liable to make mistakes in our judgments. For instance, we could conclude that the accused person is guilty when in fact they are innocent; or we could conclude that the person is innocent when in fact they are guilty. These two types of errors are shown in Table 1a, along with the two possible correct conclusions—acquitting an innocent person and convicting a guilty one.

In the case of deciding between statistical hypotheses, analogous types of errors can arise. If, for example, we have the hypotheses

$$H_0\colon \mu = a$$

$$H_1\colon \mu \neq a$$

where a is some specified value, we could *conclude that* H_0 *is false when it is in fact true*; or, on the other hand, we could *fail to reject* H_0 *when in fact it is false*. These two types of errors are referred to as *type I* and *type II errors*, respectively, and are shown in Table 1 along with the two correct conclusions —not rejecting H_0 when it is true, and concluding that it is false when it is false.

The probability of committing a type I error—rejecting H_0 when it is true —is easy enough to determine. It is simply equal to the significance level α which we use as our criterion for judging whether our sample statistic deviates an unlikely amount from the hypothesized value. For example, if our hypothesis H_0: $\mu = a$ is actually true, there will be a certain number of instances in which our sample mean $\bar{x}$ deviates more than 1.96 standard errors from it; to be more specific, the probability of occurrence of such a happening is $\alpha = .05$. We will be less likely to commit a type I error if we have a more stringent significance level, say $\alpha = .01$, for then we are even more unlikely to obtain such a deviant result when in fact H_0 is true.

It is not such an easy matter to determine the probability of committing a type II error—i.e., concluding that H_0 is true when in fact it is false. The probability of making a type II error, which we designate with the Greek letter *beta*, β, depends on a number of factors, including (1) the *true* value of the parameter in question; (2) the significance level α we use to evaluate our working hypothesis H_0 and whether we use a one-tailed or two-tailed test; (3) the standard deviation σ of the sampled population; and (4) the size of our

Table 1 Outcomes of (a) a jury trial, and (b) a statistical test of an hypothesis H_0.

(a) Jury Trial			(b) Test of an hypothesis H_0		
	True state of affairs			True state of affairs	
Jury verdict	Innocent	Guilty	Conclusion of statistical test	H_0 is true	H_0 is false
Innocent	Correct judgment	Error	H_0 is not rejected	Correct conclusion	Type II error
Guilty	Error	Correct judgment	H_0 is rejected	Type I error	Correct conclusion

sample n, where the latter two factors combine to determine the standard error of the sampling distribution of the statistic in question.

We can generalize that the probability of a type II error, β, will *decrease* as the difference between the hypothesized and true value of the parameter *increases*, will *decrease* when the chosen significance level α is *increased* in size, will *increase* for populations with *larger* standard deviations, and will *decrease* when the sample size is *increased*. Thus, the professional analyst must take all these considerations into account, as well as the comparative costs of committing a type I vs. type II error, when designing a particular piece of research. That the probability of a type II error should depend on the true value of the parameter in question will be apparent from an inspection of Figure 4. The overlapping sampling distributions represent H_0: $\mu = \mu_0$ (solid curve) and the *true but unknown* H_1: $\mu = \mu_1$ (dashed curve), where $\mu_0 \neq \mu_1$. Naturally, the smaller the difference between the hypothesized value of μ and

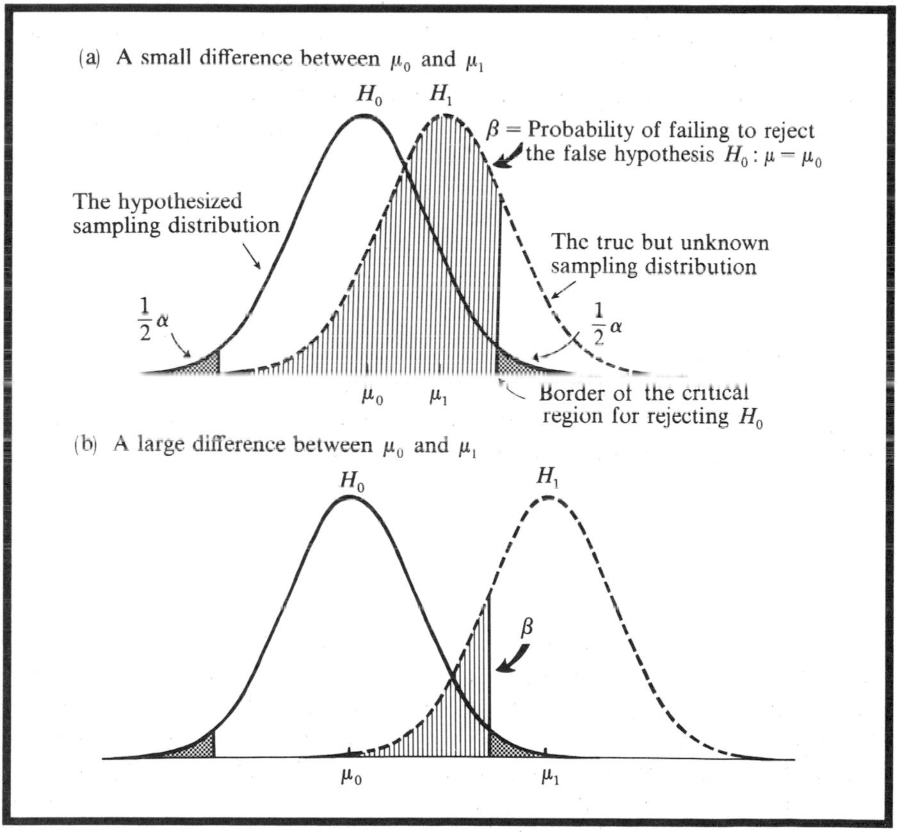

Figure 4 How the probability of a type II error, β, depends on the deviation of the true parameter value from the hypothesized value.

its true value, the less likely we are to detect it; which is to say, we are more likely to commit a type II error, accepting H_0 when it is false. Compare, for example, the striped areas in parts *a* and *b* of Figure 4. They represent the probabilities of obtaining a sample mean (or whatever the statistic may be) within the "acceptance" region for H_0. We see that the greater the difference between the *hypothesized* value of μ and its *true* value, the smaller the probability β of a type II error.

That the probability of a type II error also depends on the standard deviation σ of the population and the sample size n, which together determine the standard error of the sampling distribution, is apparent in Figure 5. The

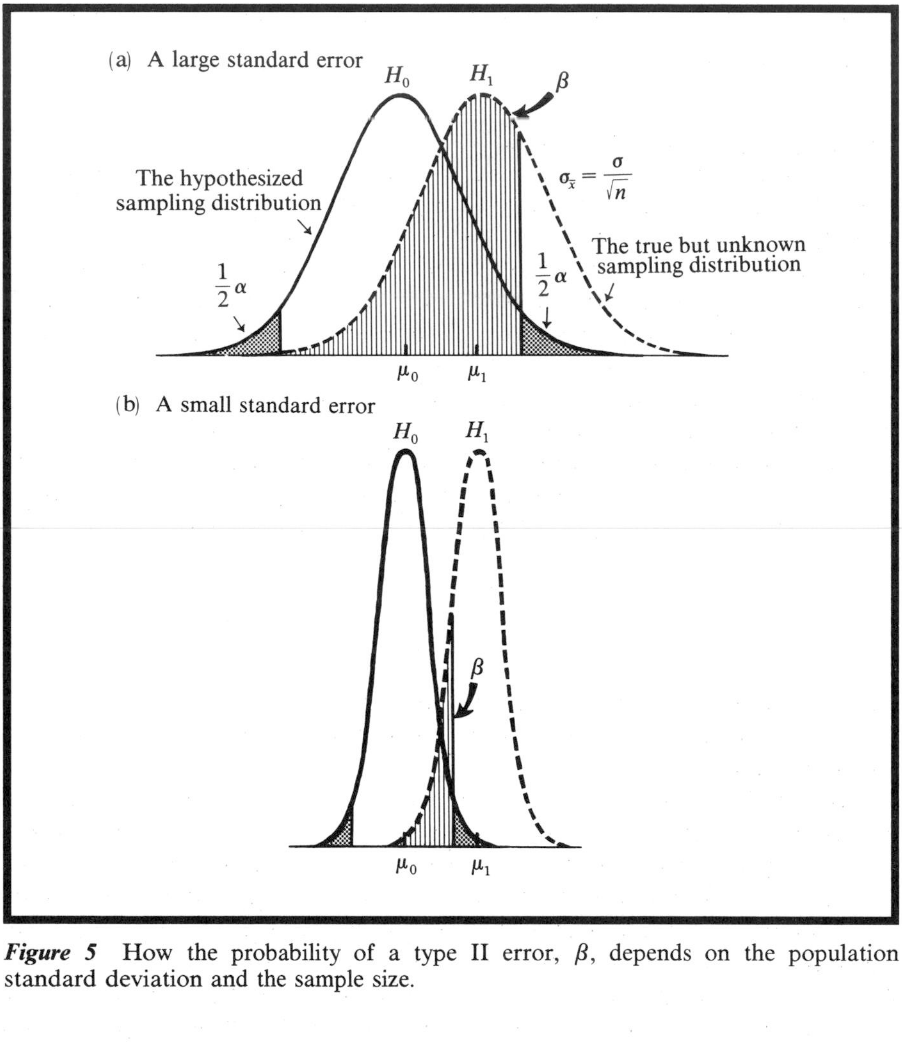

Figure 5 How the probability of a type II error, β, depends on the population standard deviation and the sample size.

smaller the standard error of the sampling distribution, the less the likelihood that the two sampling distributions—the hypothesized one and the true one—will overlap. And the less the overlap, the less likely that a sample mean will fall within the "acceptance" region of the H_0 sampling distribution; and consequently, the lower the probability β of a type II error.

Finally, the probability β of a type II error depends on the significance level α we choose to evaluate the truth of H_0. Notice in Figure 6, that for a given difference between the hypothesized mean and its true value, that the *smaller* the size of α the more likely that a sample mean will fall within the "acceptance" region for H_0, and wrongly so. We see, then, that as we strive to reduce the chance of making a type I error by making α small, we risk increasing the probability β of making a type II error, should H_0 be false.

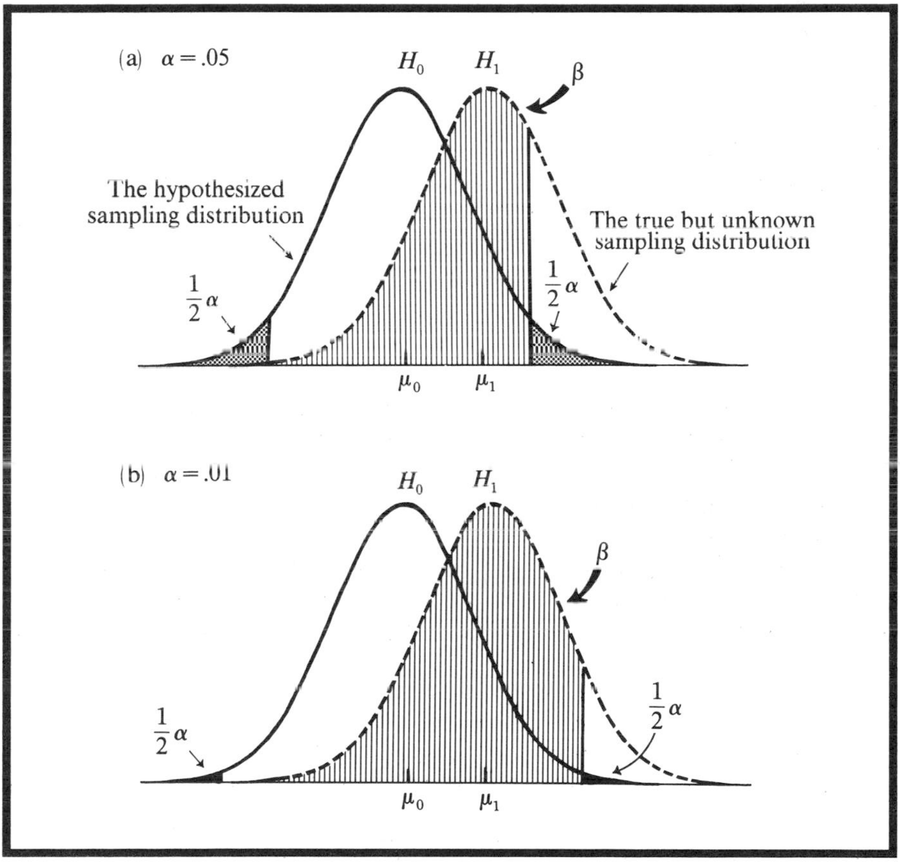

Figure 6 How the probability of a type II error, β, depends on the significance level α used.

Later in the chapter, when we discuss the "power" of a statistical test, we will return to type II errors and learn how to calculate the probability β of such an error when we suppose the true value of the population parameter, the population standard deviation, the sample size, and the significance level α. For now, let us just keep in mind why the probability of a type II error increases or decreases as the above factors are varied. A review of Figures 4, 5, and 6 will help fix the logic of these relationships.

5. Test of the Mean Difference Between Non-Independent Paired Measures

Very often we measure a set of objects on two different variables, say x and y. For example, we might measure the purchase interest aroused in a sample of consumers by advertisements A and B, where each consumer rates *both* advertisements. Or we might measure the effectiveness of razor A vs. B, used on opposite sides of the *same* face among a sample of men. Yet other times we measure *matched pairs* of objects on separate variables. For example, twins could be assigned alternative tooth paste formulations and then compared with respect to the incidence of cavities. Or we might compare the effectiveness of two alternative teaching methods by first pairing students in terms of similarity of aptitude test scores, and then randomly assigning the members of the pairs to the alternative methods. Similarly, treatment effects could be assessed among animal litter pairs, contiguous tissue samples, pairs of fruits drawn from the same branch, adjacent strips of land, etc.

In each of the above examples, the resulting data is in the form of *n pairs* of observations, one of each pair being a measurement on variable x, and the other a measurement on variable y. Although this is a multivariate analysis problem, it can be reduced to a univariate analysis by creating a single *derived* variable D defined as the difference between the paired values on variables x and y; i.e., $D = x - y$. Consequently, a difference score $D_i = x_i - y_i$ can be determined for each individual object, or pair of objects, whichever the case may be. Furthermore, we can determine the mean of the n difference scores, $\overline{D}$, as well as their standard deviation s_D. All this noted in Chapters 7 and 8.

We can now establish the hypotheses set

$$H_0\colon \mu_D = 0$$

$$H_1\colon \mu_D \neq 0$$

in which the hypothesis $H_0\colon \mu_D = 0$ corresponds to the working assumption that the mean of the population of difference scores is zero. This is tantamount to stating that there is no difference in the effects of variables x and y, inasmuch as $D = x - y$.

After choosing a significance level α we can then evaluate the observed sample mean difference $\overline{D}$ as a deviation from the hypothesized value μ_D using

$$z = \frac{\overline{D} - \mu_D}{\sigma_{\overline{D}}} = \frac{\overline{D} - \mu_D}{\dfrac{\sigma_D}{\sqrt{n}}}$$

where $\sigma_{\overline{D}}$ is the standard error of the mean difference, σ_D is the standard deviation of the population of difference scores, and n is the *number of pairs* of observations upon which the sample difference $\overline{D}$ is based. The above ratio will be recognized as equivalent to the one used to test the hypothesis that a sample mean $\bar{x}$ differed from the hypothesized mean μ. The only difference is in the nature of the empirical variable with which we are dealing; difference scores, D_i, versus single variable scores, x_i.

If σ_D is unknown, which is usually the case, we can use the sample value s_D as its estimate, and in turn estimate the standard error $\sigma_{\overline{D}}$ with $s_{\overline{D}} = s_D/\sqrt{n}$. The value of z, so defined, will then be approximately normally distributed, provided our sample size is sufficiently large, say thirty or more pairs of observations. If not, we can use the exact Student t distribution with $df = n - 1$ to evaluate the stated hypothesis, provided that the population of difference scores can be assumed to be normally distributed.

As a concrete example of testing a mean difference between paired measures, consider a study of the effectiveness of two alternative tooth paste formulations. Forty sets of twins are used, one member of each pair assigned at random to formulation A and the other to formulation B. After a period of two years, the number of cavities obtained by the formulation B individuals is subtracted from the number of cavities obtained by their respective twin mates using formulation A, resulting in a distribution of D_i scores that could be a mixture of positive and negative values. It is found that the mean difference score is $\overline{D} = -2.3$, which tells us that on average the B formulation resulted in 2.3 more cavities per individual than the A formulation. The standard deviation of the difference scores was found to be $s_D = 5.1$, which can be used to obtain an estimate of the standard error of the mean difference; $s_{\overline{D}} = s_D/\sqrt{n} = 5.1/\sqrt{40} = .81$. We then have as our approximate test

$$z \doteq \frac{\overline{D} - \mu_D}{s_{\overline{D}}} = \frac{-2.3 - 0}{.81} = -2.84$$

Having preselected a significance level of $\alpha = .01$, we reject the hypothesis that the population mean difference is zero, since the obtained $z = -2.84$ is beyond the required value of $z = \pm 2.58$ forming the borders of the rejection areas. We

conclude, therefore, that there is a real difference in the efficacies of the two toothpaste formulations.

The above type of experimental design in which paired objects (or the same object) receive two treatments must be distinguished from the design in which *independent samples* of objects receive the alternative treatments; e.g., when a sample of individuals is randomly divided into two groups, one receiving treatment A and the other treatment B. The analytical model for such an experiment is quite different from the one above for paired measures. For now, it is important only to know that the "paired-measures" design, in which alternative treatments are administered to the same or meaningfully paired objects is to be preferred to the design where independent samples of objects receive the respective treatments, since in the former design less inter-object variation is included in the variation between the observed treatment differences. This makes sense since paired individuals, such as twins, will display less variation between themselves than two individuals selected at random.

However, the paired-measures design is not always feasible, for the administration of one treatment might obviate the administration of the other treatment to the same individual. For example, we could not test two alternative fever reducing drugs on the same patients, and would have no recourse but to test the drugs using independent samples of patients. The methods for analyzing the difference between the means of two independent groups is presented in the following section. The approach for testing the differences among more than two groups is presented in the Analysis of Variance chapter.

6. Test of the Difference Between Two Independent Means

In the preceding chapter we learned that we could create an interval estimate not only for a mean μ, but for other parameters as well—e.g., the difference between two independent means, $\mu_1 - \mu_2$; the mean difference of paired measures, μ_D; a proportion, p; or a difference between two independent proportions, $p_1 - p_2$; to name the parameters that we have studied. Similarly, we can test *hypotheses* about these parameters.

If we again let θ stand for any particular parameter, we can express its sample estimate $\hat{\theta}$ as a normal deviate

$$z = \frac{\hat{\theta} - \theta}{\sigma_{\hat{\theta}}}$$

where $\sigma_{\hat{\theta}}$ is the standard error of the sampling distribution of the statistic $\hat{\theta}$, which is assumed to be normal in form. If, for instance, our parameter of

interest is the difference between two population means, $\mu_1 - \mu_2$, the difference between our observed sample means, $\bar{x}_1 - \bar{x}_2$, can be expressed as a normal deviate

$$z = \frac{(\bar{x}_1 - \bar{x}_2) - (\mu_1 - \mu_2)}{\sigma_{\bar{x}_1 - \bar{x}_2}} = \frac{(\bar{x}_1 - \bar{x}_2) - (\mu_1 - \mu_2)}{\sqrt{\dfrac{\sigma_1^2}{n_1} + \dfrac{\sigma_2^2}{n_2}}} \tag{3}$$

where $\sigma_{\bar{x}_1 - \bar{x}_2}$ is the standard error of the sampling distribution of the difference between two independent means $\bar{x}_1 - \bar{x}_2$, and σ_1^2 and σ_2^2 are the respective variances of the two populations. The resulting value of z can then be evaluated for statistical significance using a suitable significance level α.

To more clearly see the perfect analogy between the above test of a difference between two means and the test of an individual mean, as presented in Section 3, let us begin by stating our alternative hypotheses

$$H_0\colon \mu_1 - \mu_2 = a$$

$$H_1\colon \mu_1 - \mu_2 \neq a$$

What we have done is hypothesize that the difference between the two population means, $\mu_1 - \mu_2$, is equal to some value a. The alternative is that the difference is something other than the quantity a.

While the preceding hypothesis set is perfectly general, allowing us to hypothesize any particular value of a as being the true difference between our population means, we typically set a to zero, which is to say that we are hypothesizing that there is *no* difference between our population means. The hypotheses then become

$$H_0\colon \mu_1 - \mu_2 = 0$$

$$H_1\colon \mu_1 - \mu_2 \neq 0$$

The hypothesis $H_0\colon \mu_1 - \mu_2 = 0$ is generally referred to as the *null hypothesis* in the sense that it supposes *no difference* between the population means. In fact, the null hypothesis is often set up as a "straw-figure" which we really want to knock over or discredit, and as a consequence put greater credence in the alternative hypothesis that there really is a difference between the population means.

For instance, we might be interested in determining whether there is a difference in the effectiveness of two alternative teaching methods, or between the taste appeal of two soft drink formulations, or between the lifetimes of two different types of lightbulb, or between the job performances of men and women, or between the crop yields for two fertilizers, or between the efficacies of two medical treatments, or between the appeal of two car designs, or between the incomes of two ethnic groups, or between any number of other alternatively defined populations of observations. In each case, what we usually want to do is to detect a difference between the two sets of observations, often for practical or theoretical reasons, but we begin by hypothesizing that there is *no* difference and proceed to determine whether this hypothesis is likely to be true.

Consider, for example, an oil company that wishes to determine whether or not a motor oil it has developed improves gas mileage. A random sample of 100 automobiles is chosen for the test; 50 are randomly assigned to use the test motor oil, and 50 to a control standard motor oil. None of the 100 drivers know which type of oil they are using. This "blind" experimental procedure precludes the possibility that the drivers in either the "test" or "control" group consciously or unconsciously alter their driving habits so as to spuriously affect their gas mileage figures.

The hypothesis to be tested is that there is no difference in gas mileage between cars using the test motor oil and those using the control oil. Symbolically,

$$H_0: \mu_1 - \mu_2 = 0$$

The alternative hypothesis is that there is a difference in gas mileage resulting from the two types of oil

$$H_1: \mu_1 - \mu_2 \neq 0$$

Our task is to decide which of the hypotheses is more credible based on our sample observations. We will use an $\alpha = .01$ significance level to evaluate the null hypothesis in order to minimize the chance of a type I error; i.e., wrongly concluding there is a difference when there is not, for to do so might lead us to spend further developmental funds unduly.

After two months of typical driving, the cars' gas mileage figures are calculated, with the following results: Mean mileage with the new motor oil is $\bar{x}_1 = 22.4$ *mpg*, with a standard deviation $s_1 = 2.1$ *mpg*; and the mean mileage for cars with the standard oil is $\bar{x}_2 = 18.2$ *mpg*, with a standard deviation $s_2 = 2.0$ *mpg*. Using s_1 and s_2 as estimates of σ_1 and σ_2, respectively, we can substitute our statistics into formula (3) at the beginning of the section to yield

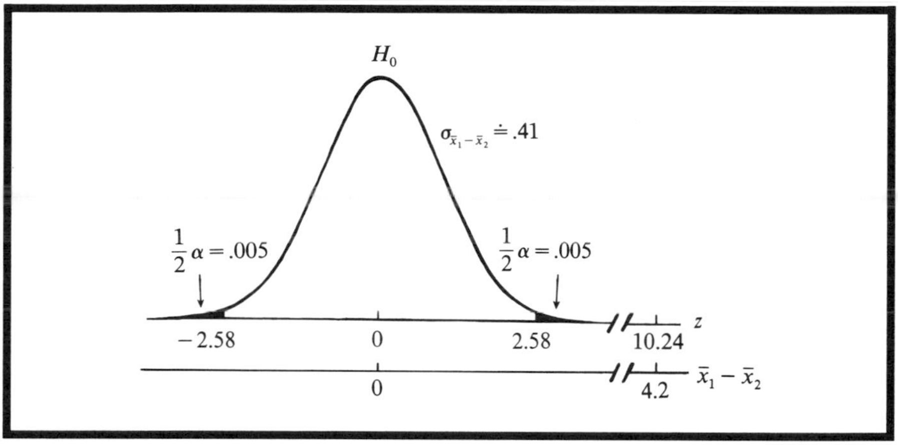

Figure 7 The relative location of $z = 10.24$.

a z which will very nearly be normally distributed by virtue of our large sample sizes:

$$z = \frac{(\bar{x}_1 - \bar{x}_2) - (\mu_1 - \mu_2)}{\sqrt{\dfrac{s_1^2}{n_1} + \dfrac{s_2^2}{n_2}}} = \frac{(22.4 - 18.2) - (0)}{\sqrt{\dfrac{(2.1)^2}{50} + \dfrac{(2.0)^2}{50}}} = \frac{4.2}{.41} = 10.24$$

The deviate $z = 10.24$ is well beyond the value of $z = 2.58$ needed to reach the critical region for our $\alpha = .01$ significance level, so we reject the null hypothesis that there is no difference in the gas mileage resulting from the two types of motor oil. The result is shown graphically in Figure 7.

The standard error of the difference between two means is shown in Figure 7 as $\sigma_{\bar{x}_1-\bar{x}_2} \doteq .41$ to indicate that it is an approximate value based on the sample estimates of σ_1 and σ_2. Alternatively, we can write $s_{\bar{x}_1-\bar{x}_2} = .41$.

Small sample technique. In the preceding example, the use of large sample sizes assured us that the use of s_1 and s_2 in place of the unknown values of σ_1 and σ_2 would provide us with a fairly accurate estimate of the standard error of the sampling distribution of the difference between the two means. If, however, we are dealing with means based on small sample sizes (say, less than 30) then we can resort to the use of Student's t distribution if we want to make strictly accurate probability statements about our data. We must assume, however, that the two sampled populations are each *normally distributed* with the *same standard deviation* σ.

The relevant t ratio for testing the difference between two independent means meeting the above conditions is given by

$$t = \frac{(\bar{x}_1 - \bar{x}_2) - (\mu_1 - \mu_2)}{\sqrt{\dfrac{(n_1 - 1)s_1^2 + (n_2 - 1)s_2^2}{n_1 + n_2 - 2}\left(\dfrac{1}{n_1} + \dfrac{1}{n_2}\right)}} \tag{4}$$

with degrees of freedom, $df = n_1 + n_2 - 2$. The above formula is not as forbidding as it may seem, for it will be recognized that the first expression under the square root sign is merely a kind of weighted average of the variances of the two samples. Alternatively, it can be seen as a pooling of the deviations around the means of the respective samples, since $(n_1 - 1)s_1^2 = \Sigma(x_i - x_1)^2$ and $(n_2 - 1)s_2^2 = \Sigma(x_i - x_2)^2$. We divide these pooled deviations by the pooled degrees of freedom for the two samples, namely $(n_1 - 1) + (n_2 - 1)$ or $n_1 + n_2 - 2$, where we have lost one degree of freedom for each sample variance. The entire denominator of (4) will be further recognized as an estimate of the denominator of (3) under the condition that the populations have equal variances, the requirement for the t variable.

As an application of the t formula above consider the researcher who wishes to determine if two alternative drugs have a differential effect on the pain symptoms of a specific rare disease. From hospitals across the country a random sample of ten patients with the disease is drawn. Five are randomly assigned to drug A and the other five to drug B. The individual drug administrations are labelled with secret code numbers so that neither the patients nor the drug administrators know which is being used. This "double blind" procedure precludes the introduction of extraneous biases into the data from either the patients or the administrators, either consciously or unconsciously. Three hours after the drug administration, each patient is asked to rate the amount of pain reduction as a percentage figure.

Having selected an $\alpha = .05$ significance level to evaluate the null hypothesis of no difference between the drugs, the data is then analyzed. For the patients receiving drug A, there is a mean pain reduction rating of 27 percentage points; for those receiving drug B, the mean pain reduction rating is 31 percentage points. The respective standard deviations for the two samples were $s_1 = 9$ and $s_2 = 12$. Substituting these statistics into (4) yields

$$t = \frac{(27 - 31) - (0)}{\sqrt{\dfrac{4(9)^2 + 4(12)^2}{5 + 5 - 2}\left(\dfrac{1}{5} + \dfrac{1}{5}\right)}} = \frac{-4}{6.7} = -.60$$

with degrees of freedom $df = n_1 + n_2 - 2 = 8$. Appendix Table IV shows that for $df = 8$, a value of t beyond 2.31 is needed to enter the rejection region for a significance level $\alpha = .05$. Since our result $t = -.60$ does not fall in the critical region, we *cannot reject* the null hypothesis that there is no difference in the efficacies of the two test drugs. Figure 8 shows that $t = -.60$ is not that an unlikely an outcome if indeed the null hypothesis were true, and it is for that reason that we do not reject the hypothesis. In other words, we could well have gotten the observed difference between our sample means if both groups of patients had received the identical drug.

The preceding example should demonstrate the difficulty of obtaining statistically significant differences when we are dealing with small sample sizes, for the denominator of the t ratio is largely a function of n_1 and n_2. Although we did not reject the null hypothesis we must be aware that we could be committing a type II error, failing to reject the hypothesis when it is in fact false. The procedure for determining the probability of such type II errors will be outlined later in section 9.

In the above example, it should also be noted that although drugs A and B were each associated with sizeable pain reduction ratings, albeit not significantly different, it does not follow that the drugs *per se* caused the pain reduction. To draw such a conclusion would require a third and fourth experimental group; one group which receives a *placebo*, a chemically inert substance, to control for pain reduction effects due solely to the administration of a "treatment" and accompanying patient expectation effects. The other necessary group would receive no treatment at all, in order to control for reductions in pain due solely to the passage of time. The analysis of such

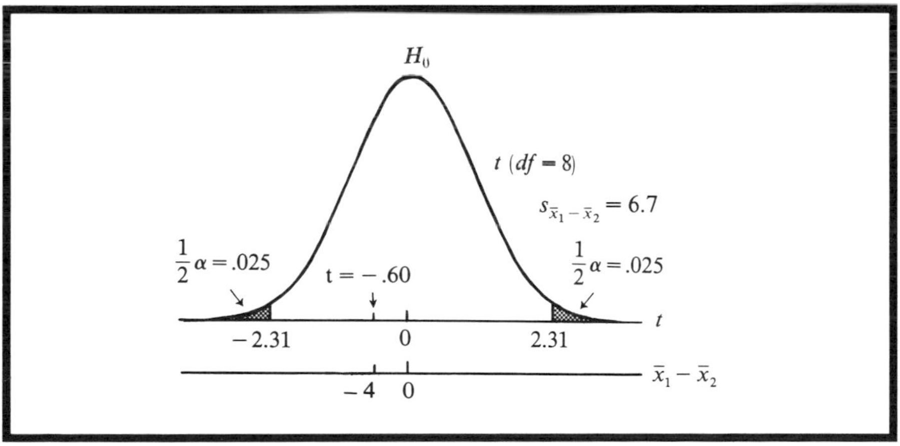

Figure 8 The relative location of $t = -.60$ ($df = 8$).

experiments involving more than two groups can be accomplished with Analysis of Variance techniques introduced in Chapter 12.

7. Test of a Proportion

There are many situations in which we wish to determine whether a sample proportion $\hat{p}$ differs from some hypothesized value p. We establish the hypotheses

$$H_0: p = a$$

$$H_1: p \neq a$$

just as we do when we wish to test a mean μ. Since the sampling distribution of a proportion based on n independent observations is very nearly normal in form when np and nq are both greater than five, we can express a sample proportion $\hat{p}$ as an approximate normal deviate

$$\boxed{z = \frac{\hat{p} - p}{\sigma_{\hat{p}}} = \frac{\hat{p} - p}{\sqrt{\dfrac{pq}{n}}}} \tag{5}$$

where $\sigma_{\hat{p}}$ is the standard error of the sampling distribution of a proportion, and n is the size of the sample upon which the sample proportion $\hat{p}$ is based.

Suppose, for example, we measure the preference for political candidates A and B among a random sample of $n = 50$ prospective voters. We establish as our hypothesis that candidate A (or B) is favored by 50 percent of the total voting population

$$H_0: p = .50$$

which is equivalent to hypothesizing that it is an even race. The alternative is that the preference for the candidate is not 50 percent

$$H_1: p \neq .50$$

Suppose, then, that after stipulating an $\alpha = .05$ significance level for evaluating H_0, we find that 56% of our sample of 50 voters prefers candidate A. The question we are asking with our statistical test is whether it is likely that we would get a sample result of 56% if in fact the population value were 50%. We begin our test by first calculating the standard error of the proportion

$$\sigma_{\hat{p}} = \sqrt{\frac{pq}{n}} = \sqrt{\frac{(.50)(.50)}{50}} = .071$$

and then substituting into (5) we have

$$z = \frac{\hat{p} - p}{\sigma_{\hat{p}}} = \frac{.56 - .50}{.071} = .85$$

The $z = .85$ value is well below the $z = 1.96$ that is necessary to reject the null hypothesis H_0 at the $\alpha = .05$ level of significance. Therefore, we have no good reason to disbelieve the hypothesis that the preference for candidate A is 50%, which is the same to say that the two candidates are equally preferred. As always, we must keep in mind that we might be committing a type II error, failing to reject the hypothesis H_0 when it is in fact false. This is especially a danger with our small sample size, as pointed out earlier.

Notice that in the preceding example we compared a sample proportion $\hat{p} = .56$ against the hypothesized proportion $p = .50$, and not against $\hat{q} = .44$ the proportion favoring the other candidate. The latter comparison is a common mistake made by beginning students, and can be avoided once it is understood that we are testing a variable sample quantity $\hat{p}$ against a theoretical and constant quantity p, and not against another variable sample quantity $\hat{q}$.

The above procedure, then, is appropriate only for comparing a sample proportion $\hat{p}$ against some *hypothetical* value p. It need not be .50 as in our example, but any particular value which we would have reason to hypothesize. It is just that very often occasions arise where we wish to determine if a sample proportion deviates from an even split; i.e., $p = .50$. Also, in the outlined procedure, we are justified in using the normal curve as an approximation to the more exact *binomial* distribution because both np and nq exceed five by quite some margin, assuring us that the approximation will be quite accurate.

8. Test of the Difference Between Two Independent Proportions

Very often we want to determine whether two populations differ with respect to the proportion of some characteristic; for example, whether men differ from women in the incidence of a certain disease, whether two ethnic groups differ in unemployment rate, whether two student populations differ in the failure rate on an achievement test, whether two drug treatments differ in cure rate, whether products manufactured under quality control procedures A and B differ in proportion of defects, etc.

The hypotheses set for such inference problems can be stated

$$H_0\colon\ p_1 - p_2 = a$$

$$H_1\colon\ p_1 - p_2 \neq a$$

where a is the hypothesized difference between the population proportions. Most often, we let $a = 0$, which is to hypothesize that there is *no* difference between the populations with respect to the incidence of the particular characteristic in question. This will be recognized as the same null hypothesis used to test the difference between the means for two independent populations. The proportions are independent in the sense that the sample proportion $\hat{p}_1$ drawn from one population is in no way related to the value of the sample proportion $\hat{p}_2$ drawn from the other population.

To determine whether the difference between two sample proportions, $\hat{p}_1 - \hat{p}_2$, differs more than a chance amount from the hypothesized difference, $p_1 \quad p_2$, we must first know the standard error of the sampling distribution of the difference between two independent proportions, given by

$$\sigma_{\hat{p}_1 - \hat{p}_2} = \sqrt{\sigma_{\hat{p}_1}^2 + \sigma_{\hat{p}_2}^2} = \sqrt{\frac{p_1 q_1}{n_1} + \frac{p_2 q_2}{n_2}}$$

where n_1 and n_2 are the sample sizes drawn from the respective populations.

Since we do not know the population proportions p_1 and p_2, we could consider using the sample estimates $\hat{p}_1$ and $\hat{p}_2$ in place of p_1 and p_2 in the above formula, or we can create a *weighted average* of the two sample proportions as a more accurate estimate of p_1 and p_2, which under the null hypothesis *are assumed to be equal* to one another. The weighted average of the sample proportions is given by

$$\hat{p}_w = \frac{n_1 \hat{p}_1 + n_2 \hat{p}_2}{n_1 + n_2}$$

which we can use to estimate the standard error of the difference

$$s_{\hat{p}_1 - \hat{p}_2} = \sqrt{\frac{\hat{p}_w \hat{q}_w}{n_1} + \frac{\hat{p}_w \hat{q}_w}{n_2}} = \sqrt{\hat{p}_w \hat{q}_w \left(\frac{1}{n_1} + \frac{1}{n_2}\right)}$$

This then becomes the denominator of the z ratio needed to evaluate H_0

$$z = \frac{(\hat{p}_1 - \hat{p}_2) - (p_1 - p_2)}{s_{\hat{p}_1 - \hat{p}_2}} = \frac{(\hat{p}_1 - \hat{p}_2) - (p_1 - p_2)}{\sqrt{\hat{p}_w \hat{q}_w \left(\frac{1}{n_1} + \frac{1}{n_2}\right)}} \qquad (6)$$

This z will be *approximately* normally distributed if the values of $n_1 \hat{p}_1$, $n_1 \hat{q}_1$, $n_2 \hat{p}_2$, and $n_2 \hat{q}_2$ all exceed 5.

To understand the application of the procedure more clearly, let us consider a specific example. Suppose an advertiser wants to determine the difference in the consumer recall for two alternative commercials. Each commercial is shown in a wide variety of local markets with non-overlapping audiences—the commercial shown in any given area is determined at random. After two weeks of advertising, 100 consumers are phoned at random in each of the two advertising areas. Each of the 200 consumers is asked if they recalled seeing a commercial for the advertised brand. Of the 100 consumers polled in the commercial A territory, 17% claimed to recall it. Of the 100 in the commercial B territory, 23% claimed to recall it. If there were in fact no difference in the memorability of the two commercials, is it likely that we would get a sample difference of six percentage points?

The problem is approached by establishing the hypothesis set

$$H_0\colon p_1 - p_2 = 0$$

$$H_1\colon p_1 - p_2 \neq 0$$

and choosing a significance level α for evaluating the tenability of H_0 vs. H_1, *prior* to the collection and analysis of the data.

We then begin our data analysis by obtaining the weighted average of the two sample proportions

$$p_w = \frac{n_1\hat{p}_1 + n_2\hat{p}_2}{n_1 + n_2} = \frac{100(.17) + 100(.23)}{100 + 100} = .20$$

This weighted average will serve as an estimate of the population proportions p_1 and p_2, which under the null hypothesis are assumed to be equal. We use it to estimate the standard error of the difference between two independent proportions drawn from identical populations:

$$s_{\hat{p}_1 - \hat{p}_2} = \sqrt{\hat{p}_w\hat{q}_w\left(\frac{1}{n_1} + \frac{1}{n_2}\right)} = \sqrt{(.20)(.80)\left(\frac{1}{100} + \frac{1}{100}\right)} = .057$$

We now have the necessary information with which to evaluate the difference between the sample proportions as a deviation from the hypothesized difference of zero:

$$z = \frac{(\hat{p}_1 - \hat{p}_2) - (p_1 - p_2)}{s_{\hat{p}_1 - \hat{p}_2}} = \frac{(.17 - .23) - (0)}{.057} = -1.05$$

This value does not approach the value of $z = 1.96$ needed to discredit the null

hypothesis at the $\alpha = .05$ level of significance. Consequently, we cannot confidently reject the hypothesis that there is no difference in the memorability of the two commercials. While our failure to reject the null hypothesis does not prove that the two commercials are equivalent, we must realize that the possibility is real. On the other hand, we could be committing a type II error, concluding that there is no difference between the commercials when in fact there is. In the following sections we will delve deeper into the factors which influence our probabilities of committing type I and type II errors.

9. The Power of a Test

From the many preceding examples it should be clear that whenever we test a statistical hypothesis we are prone to making mistakes in our conclusions; namely, *rejecting* our working hypothesis H_0 when it is in fact *true* (type I error), or *failing to reject* the hypothesis H_0 when it is in fact *false* (type II error). We have designated the probabilities of these two types of errors as α and β, respectively.

The value of α, the probability of rejecting our working hypothesis H_0 when it is in fact true, corresponds to the significance level which we choose as our criterion for evaluating an "unlikely" event under the condition that H_0 is true. Typically, we set $\alpha = .05$ or $\alpha = .01$, so that if the hypothesis H_0 is in fact true we will erroneously reject it only 1 time in 20, or 1 time in 100, respectively.

Determining the probability β of committing a type II error—i.e., failing to reject a false hypothesis H_0—is a somewhat more complicated matter, as alluded to earlier, for it depends on four separate factors: (1) the size of the difference between the true value of the parameter in question and the hypothesized value; (2) the significance level α we use to test the hypothesis and whether we use a one-tailed or two-tailed test, (3) the standard deviation of each sampled population; and (4) the sample size drawn from each population in question. These four factors are identified graphically in Figure 9. The value of β is represented by the area under the normal curve with vertical stripes, instances in which the false hypothesis H_0 is not rejected.

Now if β is the probability of making a type II error—i.e., failing to reject a false hypothesis H_0—then it follows that $1-\beta$ must be the probability of *correctly rejecting the false hypothesis*. For example, if the probability of failing to reject a false hypothesis is $\beta = .25$, then the probability of correctly rejecting the false hypothesis is .75 (i.e., $1-.25$). We refer to this probability $1-\beta$ as the *power* of a test, the *probability of correctly rejecting a false hypothesis*.

To determine the power of a particular statistical test, then, we will need to know the values of the four factors influencing the value of β, which are listed above and itemized in Figure 9. Our task is to calculate the $1-\beta$ area under

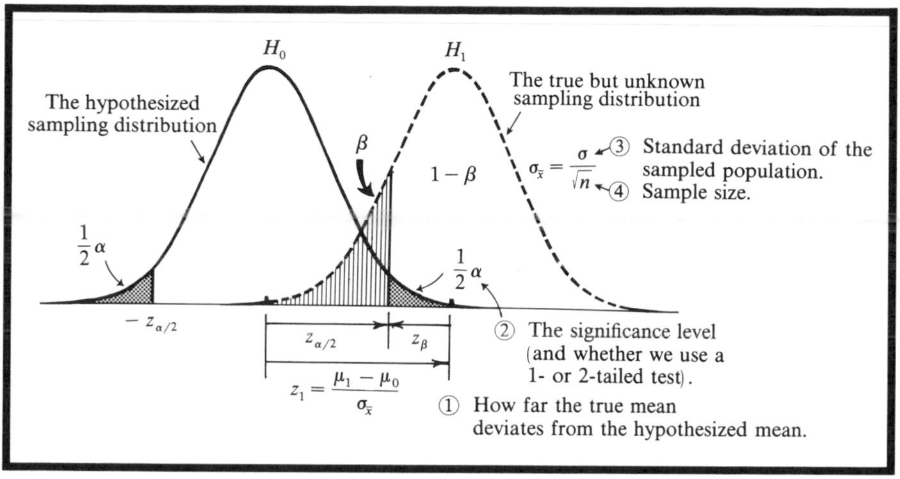

Figure 9 Summary of factors affecting the probability of a type II error, β (and the power of the test, $1-\beta$).

the normal curve at the right in Figure 9, for that area represents the probability of obtaining a sample result which falls in the critical rejection region defined with respect to the normal curve at the left. That is, *the true but unknown distribution* (H_1) *at the right in Figure 9 determines the sample results that we observe, but we evaluate those sample results as though they originated from the hypothesized distribution at the left* (H_0).

It is not really that difficult to determine the $1-\beta$ area in Figure 9, which represents the power of the test. First, we must express the difference between the true parameter θ_1 and the hypothesized value θ_0 in standard error units

$$z_1 = \frac{\theta_1 - \theta_0}{\sigma_{\hat{\theta}}}$$

where $\sigma_{\hat{\theta}}$ is the standard error of the sampling distribution of the statistic $\hat{\theta}$, which is the sample estimate of the parameter θ. If we are dealing with a two-tailed test and have chosen $\alpha = .05$ as our criterion for statistical significance, then

$$z_{\alpha/2} = 1.96$$

represents the value beyond which a sample statistic $\hat{\theta}$ will render the hypothesis H_0 untenable. It follows, then, that

$$z_\beta = z_{\alpha/2} - z_1$$

will represent the number of standard errors *from the mean of the true sampling distribution to the border of the rejection area defined with respect to the hypothesized sampling distribution*. This relationship is more easily seen in Figure 9. To determine the power of our statistical test, we have only to determine the area of the true sampling distribution beyond the critical value $z_{\alpha/2}$ for rejecting H_0, the $1-\beta$ area in Figure 9. The following example will make this procedure perfectly clear.

Suppose we hypothesize that the mean of a particular population of values is $\mu = 19$, and we proceed to test the hypothesis by drawing a random sample of $n = 50$ observations and calculate the sample mean $\bar{x}$. Furthermore, we decide to test our hypothesis

$$H_0\colon \mu = 19$$

and its alternative

$$H_1\colon \mu \neq 19$$

at the significance level of $\alpha = .05$. That is, the sample mean $\bar{x}$ must deviate from the hypothesized mean $\mu = 19$ by 1.96 standard errors before we reject the H_0 hypothesis. But now let us say that the true but unknown value of the population mean is really $\mu = 21$. What, then, is the probability of correctly rejecting H_0—i.e., the power of our statistical test? We proceed to solve the problem by sketching the situation as in Figure 10. The sampling distributions

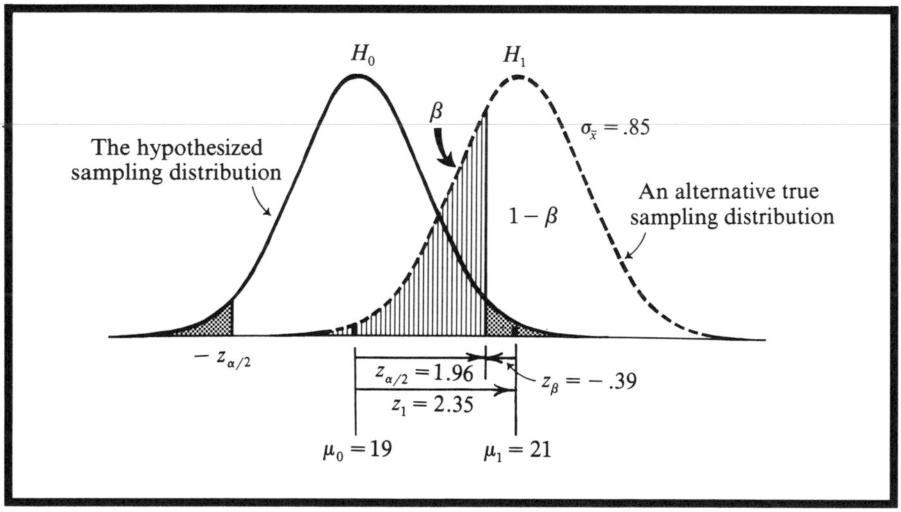

Figure 10 The determination of the power, $1-\beta$, of a test for given values of $\sigma_{\bar{x}}$, α, and the true mean μ_1.

of the mean are shown for the hypothesis $\mu_0 = 19$ and for the true value $\mu_1 = 21$. We must know the value of the sampled population's standard deviation, so that we can calculate the standard error of the sampling distributions $\sigma_{\bar{x}} = \sigma/\sqrt{n}$. Let us assume that we know the population standard deviation to be $\sigma = 6$, so we can calculate the standard error of the mean as

$$\sigma_{\bar{x}} = \frac{\sigma}{\sqrt{n}} = \frac{6}{\sqrt{50}} = .85$$

Now it is a simple matter to determine how far the actual mean $\mu_1 = 21$ deviates from the hypothesized mean $\mu_0 = 19$ *in standard error units.*

$$z_1 = \frac{\mu_1 - \mu_0}{\sigma_{\bar{x}}} = \frac{21 - 19}{.85} = 2.35$$

But the cut-off level of z for declaring the hypothesis H_0 untenable is $z_{\alpha/2} = 1.96$, which corresponds to a $z_\beta = -.39$ *as measured from the mean of the true sampling distribution*; i.e.,

$$z_\beta = z_{\alpha/2} - z_1 = 1.96 - 2.35 = -.39$$

This calculation can be seen graphically in Figure 10. And, of course, we must keep in mind that it is .39 standard errors *below* the mean when we look up the area under the normal curve for values which exceed it. Referring to Appendix Table III we find that 65 percent of the area under the normal curve exceeds $z_\beta = -.39$, so the power of our test, $1 - \beta$, corresponding to the probability of correctly rejecting the hypothesis that $\mu = 19$, is equal to .65. Since the power of our test is given by $1 - \beta = .65$, it follows that the probability of a type II error is given by $\beta = 1 - .65 = .35$, shown as the striped area in Figure 10.

In the preceding example we assumed we knew the value of σ, the population standard deviation. Since it is not usually available, we can use the sample estimate s in its place when calculating the standard error of the sampling distribution. In such a situation we must realize that our determination of the power of our statistical test will only be approximately correct.

Power curves. In the above example we determined the power of the statistical test for only one alternative ($\mu_1 = 21$) to the hypothesis $\mu_0 = 19$. In fact, we usually calculate the power of the test for many alternative values of μ.

Following the identical procedure outlined above, we have calculated the power of the statistical test for values of μ equal to 19.5, 20.0, 20.5, 21.5, 22.0, 22.5 and 23.0, and plotted the resulting probabilities in Figure 11. While not shown, the corresponding "power curve" for values of μ below 19 is a mirror-image of the one shown. As we would expect, the power of our

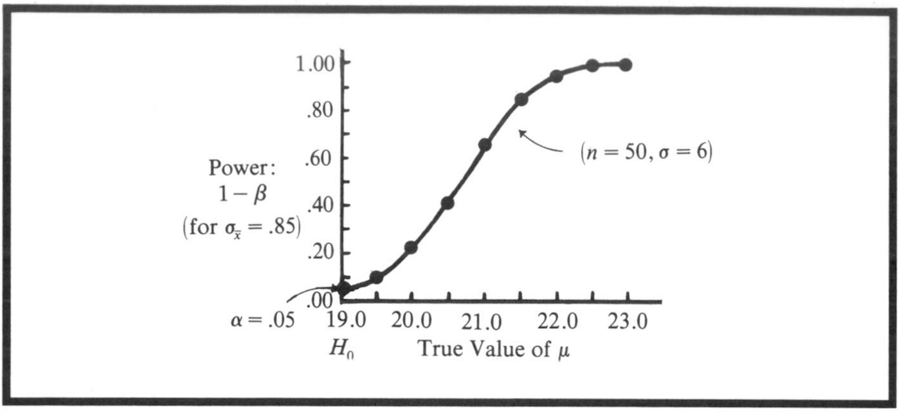

Figure 11 Power curve for given values of α, σ, and n.

statistical test decreases as the difference between the true population parameter value and the hypothesized value decreases. See Figure 11, for example, for the low value of $1-\beta$ when the true mean is 19.5.

The power curve shown in Figure 11 is for the particular values of α, σ, and n studied in the above example. We could, of course, create other power curves by varying the values of α, σ and n, using the same procedure outlined above. Would the curve in Figure 11 be higher or lower if we increased our sample size n? What if we used a significance level of $\alpha = .01$ instead of $\alpha = .05$? What if a one-tailed test were used instead of a two-tailed test? What if the population standard deviation were larger? All of these questions can be easily answered by sketching the sampling distributions as shown in Figures 9 and 10 and determining the size of the $1-\beta$ area under the true sampling distribution which corresponds to the rejection of the incorrect hypothesis H_0.

10. Sample Size Selection for Desired Power

Whenever we draw a sample from a population in order to test a hypothesis about a particular population parameter, we want to have a powerful test. But of the factors that influence the power of a test, only two are under our control—namely, the sample size n and the significance level α we choose to evaluate the hypothesis H_0. Since our concurrent concern about committing a type I error rather restricts our variation of α, we are left with the sample size n as our main tool for increasing the power of our test.

Just as we were able to calculate the power of a test when given (1) the true value of the parameter, (2) the population standard deviation σ, (3) the significance level α, and (4) the sample size n, we can easily turn the problem

around and ask ourselves what sample size n would be needed in order to have a test of (a) a desired power when (b) the true population parameter value is a given value, for (c) a given value of the population standard deviation σ and for (d) the chosen significance level α.

For example, we could ask what sample size n would be needed in order to be 75% sure—i.e., to have a power $1-\beta=.75$—that we will reject the hypothesis H_0: $\mu=19$ if in fact the true value is $\mu=22$, and given that the population standard deviation is $\sigma=6$, and we choose the $\alpha=.05$ level of significance as our criterion for rejecting H_0.

We begin again by sketching the problem in the form of two overlapping normal distributions, as in Figure 12, the one on the left being the hypothesized sampling distribution and the one on the right being the actual sampling distribution under the condition that $\mu=22$. The distributions have been arranged so that 75% of the area of the true sampling distribution falls in the critical rejection region with reference to the hypothesized sampling distribution—that is, so the probability of a sample mean $\bar{x}$ falling in this region is $1-\beta=.75$. Appendix Table III tells us that a $z=-.67$ is the point above which 75% of the area in the normal curve falls. This very same point corresponds to a $z_{\alpha/2}=1.96$ *with reference to the hypothesized H_0 distribution on the left*. Therefore, the difference between the hypothesized value of μ and

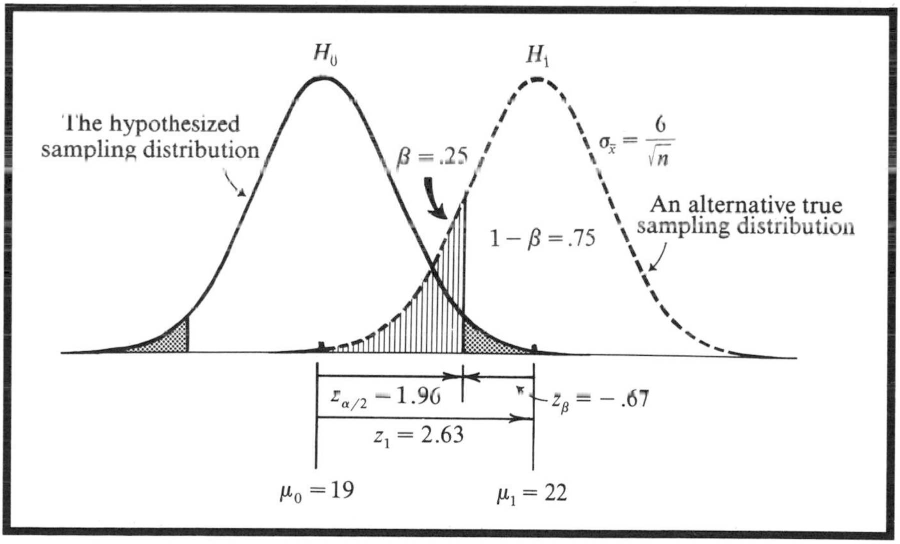

Figure 12 The determination of the required sample size n, for given values of σ, α, and power $1-\beta$ (based on a given departure from the hypothesized value of the mean).

its actual value, expressed in standard error units is

$$z_1 = z_{\alpha/2} - z_\beta$$

where $z_{\alpha/2}$ and z_β are the standard error distances to the borderline of the rejection area, measured from the *hypothesized* and *actual* sampling distribution means, respectively. Expanding on this relationship we can write

$$z_1 = z_{\alpha/2} - z_\beta = \frac{\mu_1 - \mu_0}{\sigma_{\bar{x}}}$$

where μ_0 is the hypothesized mean, μ_1 is the actual mean, and $\sigma_{\bar{x}}$ is the standard error of the sampling distribution. This relationship can be seen graphically in Figure 12.

Now since $\sigma_{\bar{x}} = \sigma/\sqrt{n}$, we can rewrite the above formula as

$$z_{\alpha/2} - z_\beta = \frac{\mu_1 - \mu_0}{\dfrac{\sigma}{\sqrt{n}}}$$

and then solving for the value of n we have

$$n = \frac{(z_{\alpha/2} - z_\beta)^2 \sigma^2}{(\mu_1 - \mu_0)^2}$$

For the present example we have

$$n = \frac{[1.96 - (-.67)]^2 6^2}{(22-19)^2} = 27.7$$

In other words, if we draw a random sample of size $n = 28$ we will be 75% certain of rejecting the hypothesis that $\mu = 19$ when in fact it is $\mu = 22$. Stated alternatively, if we drew an infinite number of samples of size $n = 28$ from our true population with $\mu = 22$ and $\sigma = 6$, 75% of the resulting sample means would deviate a sufficient amount from the hypothesized value $\mu = 19$ to reject that hypothesis. On the other hand, we still have a probability of $\beta = .25$ of making a type II error; i.e., obtaining a sample mean $\bar{x}$ which does not deviate far enough from the wrongly hypothesized value of $\mu = 19$ in order to reject that erroneous hypothesis.

The above examples involved parameters which were means. However, the same procedures apply to any other population parameter which has a normal

sampling distribution, whether it be a mean difference of paired measures, a difference between independent means, proportions, or differences between proportions, assuming the sample sizes turn out to be large enough to assume roughly normal sampling distributions.

The importance of choosing appropriate sample sizes is based on practical considerations. That is, what are the consequences of our reaching an incorrect conclusion with a statistical test. Is a type I or type II error more crucial? Will time be lost? Will money be lost? Will lives be lost? These are the types of things that must be taken into account when deciding what power we will demand from our sample design and statistical test. If we are comparing the efficacies of two alternatives vaccines, that is one thing; if we are comparing the appeal of two alternative soft drink formulations, that is another. What size difference is relevant to us, and what are the consequences of our failing to detect that difference? What are the consequences of concluding there is a difference when there is not? And what is the cost of obtaining our sample measurements? These are the questions we must ask ourselves when deciding upon the desired power of our statistical tests, and consequently the appropriate sample sizes and significance levels that we use.

11. Concluding Comments

We now have a firm foundation in the basic principles of statistical inference, having begun with the fundamentals of the concept of probability, then moving on to the notion of a sampling distribution, which in turn was used to address the practical problems of parameter estimation and hypothesis testing. Now that we have touched upon two of the major functions of statistical analysis—data reduction and inference—we will build upon these concepts in the remaining chapters as we introduce the third major function of statistical analysis—the study of the relationships among variables and objects.

Part IV.

ASSOCIATION
(Multivariate Analysis)

Chapter 10

Correlation Analysis

1. Introduction

As discussed in the early chapters, whenever we measure objects on *two* variables, not only are we interested in measures of the central tendency and variation of the individual variables, but also in an assessment of the *association*, if any, which exists between the two variables. Generally, we are interested in whether above average values on one variable tend to be associated with above average values on the other variable; or in other instances whether above average values on one tend to be associated with below average values on the other.

For example, we might be interested in the association between the lengths and weights of newborn babies, between crop yields and rainfall levels for various plots of land, between crime rates and unemployment rates of cities, between sales and advertising levels for brands within a given product category, between salt intakes and blood pressures of individuals, between school grades and aptitude test scores of students, between hours of psychotherapy and behavior measures of mental patients, etc.

Table 1 shows the general format for the input data for such problems; namely, each of a number of objects possess values on two variables. It is also possible that the objects represent meaningful pairs of other entities measured on the same variable; e.g., twins measured on an intelligence test. Later in the chapter we will see how the two-variable situation can be generalized to a many-variable analysis.

We also learned in earlier chapters that associations between variables can be of two basic types: *correlational* and *experimental*. In experimentally based relationships, *we* control the values of one of the variables by assigning them at random to the objects under study, and observe accompanying changes in another variable. For example, we randomly assign alternative drug dosages to patients and observe changes in bodily functions; we randomly assign fertilizer

Table 1 Input data format for studying the relationship between two variables.

Objects	Variables	
	x	y
Object 1	x_1	y_1
Object 2	x_2	y_2
Object 3	x_3	y_3
Object 4	x_4	y_4
⋮	⋮	⋮
Object i	x_i	y_i
⋮	⋮	⋮
Object n	x_n	y_n

levels to various land plots and observe differences in crop yield; we randomly assign different levels of classroom instruction to various students and observe differences in achievement test scores; we randomly assign alternative advertising levels to various geographic regions and observe differences in sales; etc. In each case, each object under study has an equal chance of possessing each value of the experimental variable.

In *correlational* relationships, on the other hand, we have no control over the values of the variables possessed by the objects under study. Instead we merely observe how the two variables of interest covary in the natural environment. They are *random variables* in that any given object has a probability of possessing a given value of those variables which is not under our control. Examples of such correlational relationships include the association between intelligence test scores and school grades, between salt intake and blood pressure, between crime rate and unemployment rate, between historical advertising and product sales levels, etc. In each instance the values of the variables were not systematically controlled by us but by unknown factors.

This is not to say that a given relationship cannot be studied both in a correlational and experimental context. For example, an association between salt intake and blood pressure would be experimentally based if we as investigators randomly assigned alternative levels of dietary salt to the sample of subjects, while it would be correlational in nature if we simply observed how the two variables were associated without external control. Similarly, crop yield as related to the nitrogen content of the soil plots would be an experimentally based association if the nitrogen were controlled by us through fertilizer application, whereas it would be a correlational relationship if the association were based purely on observations as they occurred in the natural environ-

ment. Thus, the distinction rests on whether *we* assigned values of the variable to the objects in an unbiased random fashion, or whether the "assignment" was historical in nature by unknown means, as exemplified in the association between intelligence test scores and academic achievement.

This chapter will be devoted to the basics of *correlation analysis*, the study of the relationships that exist among random variables, including the identification and summary of such relationships. In the following two chapters techniques will be introduced that can be used for dealing with experimental relationships. As we will see, the importance of the distinction between experimental and correlational associations lies in the interpretations we place upon them. In properly executed experimental studies, *causal* interpretations can be made about the observed relationships. In correlational investigations, on the other hand, no such causal interpretations can be safely made. Nonetheless, we will discover that there are many practical applications of correlation analyses, especially when it is realized that correlation studies are the most common and inexpensive investigations being conducted.

2. Patterns of Association

Whether relationships are correlational or experimental in nature, they can be labelled with various shorthand expressions which communicate the form or pattern of the existing associations between the values of the involved variables.

The most commonly encountered relationships are shown in Figure 1. In each instance we observe a systematic, easily describable, pairing of the values on the two variables. When *high* values on one variable are associated with *high* values on the other variable, and *low* values on one of the variables are associated with *low* values on the other, we say that the variables are *positively* related, as shown in Figures 1*a*, *b*, and *c*. For example, we would expect positive relationships to exist between college performance and aptitude test scores, between sales and advertising levels, and between crop yields and rainfall.

A second possibility is that *high* values on one variable are associated with *low* values on the other variable, and vice versa. In this type of relationship we say that the variables are *negatively* related, or *inversely* related, as shown in Figures 1*d*, *e*, and *f*. We might expect a negative relationship, for example, between interest rates and housing starts, or between student academic achievement and hours of TV viewing of the entertainment type.

A third possibility occurs when the values of the two variables are positively related within a certain range of values, and negatively related in another range, as shown in Figures 1*g*, *h*, and *i*. These types of relationships are described as *non-monotonic*, in the sense that they are not ever-increasing

or ever-decreasing. For example, the association between student test performance and anxiety level has been found to follow a non-monotonic relationship. Up to a certain point, increases in anxiety result in increases in test score, but beyond that point further increases in anxiety lead to a decrease in test performance. A similar relationship is observed between profits and advertising levels. Increased advertising spending leads to increased profits, but beyond a certain point increased advertising expenditures eat into profits.

In addition to the above descriptions, relationships can be described as *linear*, following a straight line, as in Figures 1*a* and *d*, or as *nonlinear*, or *curvilinear*, as in the other parts of Figure 1. When a relationship turns from positive to negative more than once, as in Figure 1*i*, it is often called a *cyclical* relationship. Such relationships most often occur when the x variable represents consecutive time periods, while the y variable could be unemployment rate, fertility rate, product sales levels, sunspot activity, etc.

Finally, when we discover *no* systematic association between the values of two variables—when high or low values on one variable are just as likely to be paired with high or low values on the other, throughout their entire range of values—then we variously say that the variables are *unrelated*, *uncorrelated*, *orthogonal*, or *independent*. For example, we might expect that physical height is independent of grade point average.

It should be noted that the relationships shown in Figure 1 are theoretical in nature. In real life, since we must depend upon limited samples of observations which are subject to sampling error, we rarely encounter such smooth and regular relationships. Rather, our task is to determine whether or not such relationships do in fact underlie our irregular data patterns, and to measure and describe them efficiently as well as to find practical applications for them, our ultimate goal.

3. Gross Indicators of Correlation

If we are faced with the task of determining the presence or absence of a relationship between two random variables, how would we go about it? Suppose, for example, we wanted to determine whether the job performance of 100 employees is related to their scores on a personnel test.

The 2×2 contingency table. Since a correlation between two random variables reflects the extent to which high scores on one variable go with high scores on the other, in the case of positive relationships, and high scores on one variable are associated with low scores on the other, and vice versa, in the case of negative relationships, we might start our analysis by identifying each individual as being high or low on the respective variables.

In our employee evaluation example, each individual's job performance, as measured by supervisor ratings, can be keyed as being above or below the

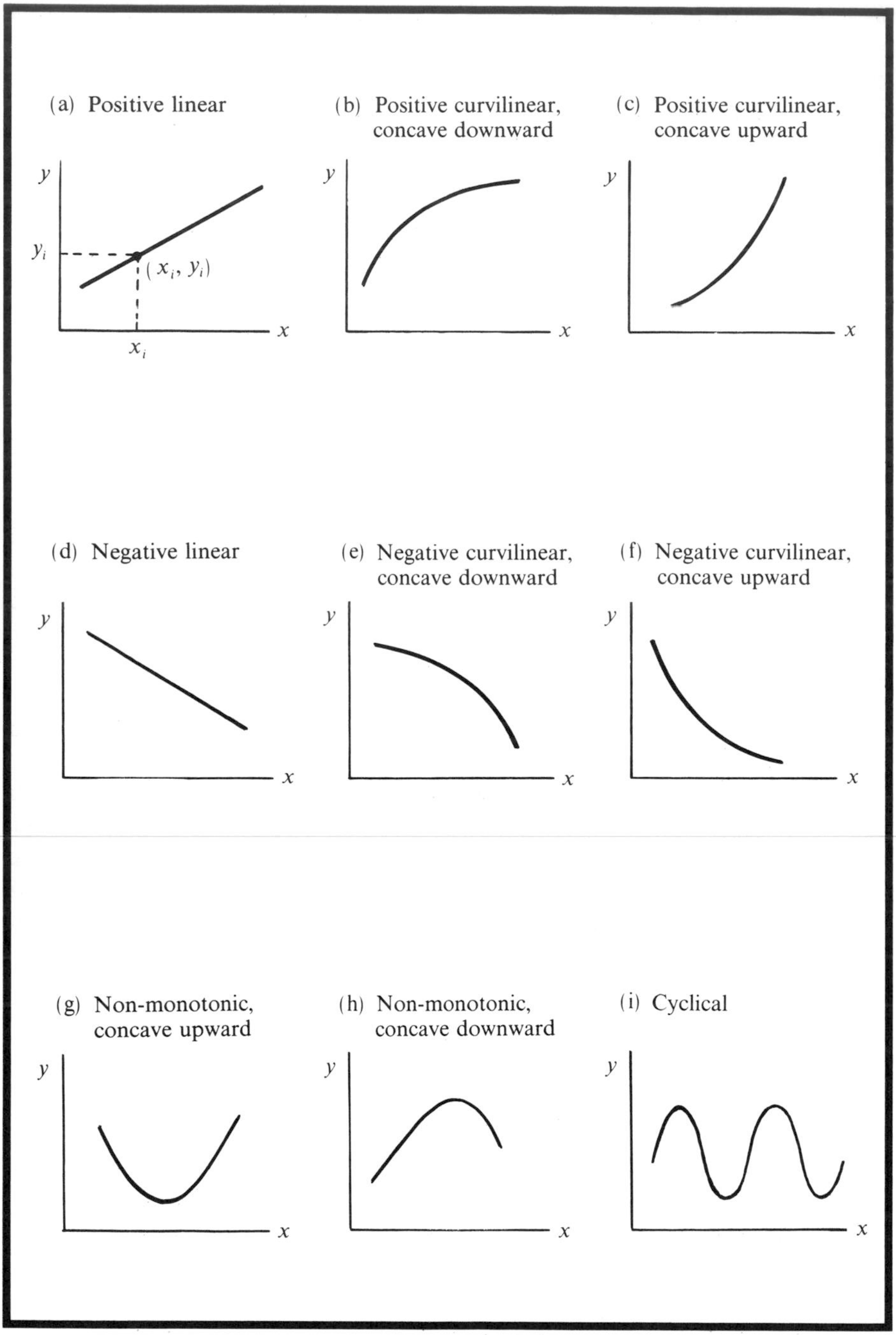

Figure 1 Various forms of relationship between two variables, x and y.

median supervisor rating. Similarly, each person's personnel test score can be keyed as being above or below the median for that variable. Four possibilities result. An employee could be:

- above median both on the test and in job performance (+ +)
- above median on the test, but below median in job performance (+ −)
- below median both on the test and in job performance (− −)
- below median on the test, but above median in job performance (− +)

The above four possibilities are shown in Table 2 in the form of a 2×2 tabular arrangement, often referred to as a *contingency table* since it indicates in a broad way whether values on two dichotomous variables are contingent or dependent on each other. Beginning with the upper right hand cell and moving in a clockwise direction, the four cells of the table correspond to the (+ +), (+ −), (− −), and (− +) outcomes outlined above.

Now if there were no overall relationship between job performance and the test scores we would expect an equal number of individuals to fall into each of the four classification cells. That is, whether a person scored above average on the test had nothing to do with whether they performed above or below average on the job. On the other hand, if above average test scores tended to be associated with above average job performance, and low test scores were associated with low job performance, we would have evidence that the two variables were positively related.

As can be seen in Table 2 there is a strong suggestion that the two variables are indeed correlated with one another. Rather than an equal number of individuals falling into each of the four cells, there is a disproportionate number scoring *above* average on *both* variables or *below* average on *both* variables. Of the 50 individuals scoring above average on the test, 40 of them

Table 2 A 2×2 contingency table of job performance and personnel test scores for 100 employees.

	Score on personnel test	
Job performance	Below median	Above median
Above median	(−, +) 10 employees	(+, +) 40 employees
Below median	(−, −) 40 employees	(+, −) 10 employees

(or 80%) scored above average in job performance. Similarly, of the 50 scoring below average on the test, 40 scored below average in job performance. If there were no correlation between the variables we would expect the 50 high test scorers to split about evenly on job performance—25 performing above average and 25 below average.

Table 3 shows other possible outcomes of this type of cross-classification scheme. Again, 100 objects, chosen for convenience, are classified into one of four cells depending on whether they are above or below the median on each of two hypothetical variables, *x* and *y*. Part *a* of Table 3 shows a positive relationship, in which high scores go with high, and low scores go with low. Part *b* shows a relationship of comparable strength but in the opposite direction—that is, high scores on one variable are associated with low scores on the other, and vice versa. Part *c* shows the situation that would be expected if no correlation existed between the two variables; an equal number of objects falling into each of the four cells. In this case, whether an object scores above or below the median on one variable, it is equally likely to score above or below the median on the second variable. If we know only that an object scored above average on variable *x*, we could do no better than chance in predicting its relative position on variable *y*. Half the time we would be correct, and half the time we would be wrong. Compare this with the situation in which we know that, say, a positive relationship exists between the two variables. Then, if we knew that an object was above the median on variable *x*, we would predict that it was also above the median on variable *y*, and we would be correct more than half of the time.

If the data in these contingency tables represented samples from a larger population, the observed discrepancies from a chance distribution could then be tested for statistical significance by techniques which will be introduced in Chapter 13, Analysis of Category Data.

Table 3 Gross indications of the presence of a relationship between two variables using a 2×2 contingency table.

(a) Positive relationship			(b) Negative relationship			(c) No relationship		
	Value on variable *y*			Value on variable *y*			Value on variable *y*	
Value on variable *x*	Below median	Above median	Value on variable *x*	Below median	Above median	Value on variable *x*	Below median	Above median
Above median	15 objects	35 objects	Above median	35 objects	15 objects	Above median	25 objects	25 objects
Below median	35 objects	15 objects	Below median	15 objects	35 objects	Below median	25 objects	25 objects

The scatter diagram. The 2×2 contingency table provides only a very gross indication of the correlation between two random variables. Aside from their being above or below the median, the *actual magnitudes* of the scores are lost. All the objects falling in a particular classification cell may just as well have identical scores, for all we know, but which is probably not the case. Ideally we would want to use all the detail of our original paired observations in the determination of the relationship between our two variables. Such a technique is available, and to understand it we will consider another example involving measurements on two continuous variables.

A marketer of a line of spices, curious about the wide variation in sales from store to store, developed the hypothesis that the sales of the product was related to the amount of shelf space allocated to the line by each grocer. Ten stores are selected at random from across the country and observations are made on the following variables: (1) total shelf-facing allotted to the product line—width times height, measured in square inches, and (2) total dollar sales of the product over the past week. The data is presented in Table 4 in the data format introduced in Table 1.

It is difficult to tell from a mere visual inspection of the raw data whether a relationship exists between the two variables. To get a better idea of how, if at all, the variables are related we can make a pictorial representation of the data in what is called a *scatter diagram*, or scatter plot. In the scatter diagram shown in Figure 2, the horizontal axis represents one of the variables (Shelf Space), and the vertical axis represents the other variable (Sales). Each point on the graph represents a store and is numerically labelled for reference purposes.

The location of a store on the graph is determined by its scores on the respective variables. Consider store 2, for example. It has 230 sq. in. of shelf-facing devoted to the product (its score on the horizontal axis), and

Table 4 Spice sales vs. shelf space in ten stores.

Store	Shelf space (sq. in.)	Weekly sales (dollars)
1	340	71
2	230	65
3	405	83
4	325	74
5	280	67
6	195	56
7	265	57
8	300	78
9	350	84
10	310	65

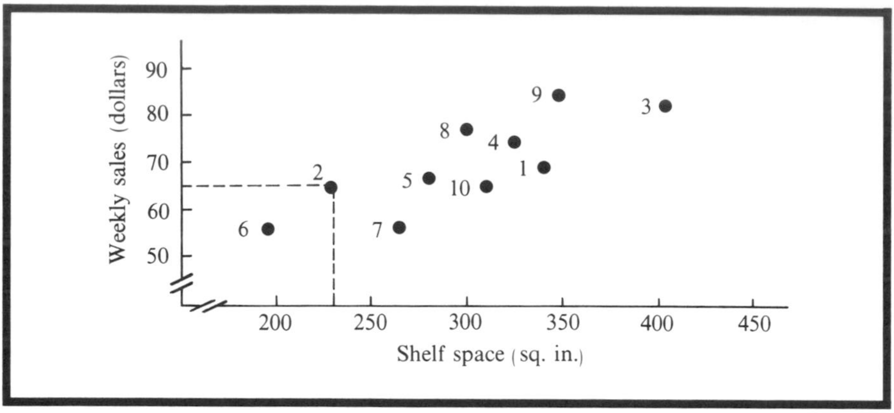

Figure 2 Scatter diagram of spice sales vs. shelf space for ten randomly selected stores (Data from Table 4).

weekly sales of \$65 for the product (its score on the vertical axis). The dashed lines show how the store's data point was located on the basis of these two scores. Similarly, each of the other nine stores was positioned based on its scores on the two variables of interest.

A simple glance at the scatter diagram indicates that there *is* a relationship between these two variables. High values on one variable are clearly associated with high values on the other variable, and low values on each variable are associated with each other; i.e., stores with above average shelf space allocated to the spice line are associated with above average sales for the products. If there had been no relationship between these two variables, the stores would have been distributed on the graph in a random blob with no discernable pattern.

Later in the chapter we will return to this example and discuss possible interpretations of the findings. At this point we want only to see how a scatter diagram can graphically portray a correlation between two variables, and provide more detail of the relationship than does the 2×2 contingency table.

4. The Correlation Coefficient *r*

While the scatter diagram provides us with some idea of the extent of the relationship between two random variables, it would be highly desirable to have a well-defined numerical index of the relationship; a summary measure that would communicate the extent of correlation between two variables in a single number, just as the mean $\bar{x}$ serves as a summary measure of central tendency and the standard deviation s serves as a summary measure of variation.

Although there are a number of such measures of the degree of association between two variables, the *product moment correlation coefficient*, or simply the *correlation coefficient*, finds application in the widest range of data analysis problems. When based on a sample of data, the correlation coefficient is designated with the letter r, and it in turn is an estimate of the population correlation coefficient designated with the Greek letter ρ, *rho*.

The correlation coefficient r can range in value from -1.00 to $+1.00$, as shown in Figure 3. A correlation coefficient of $r = +1.00$ signifies a *perfect* positive linear relationship; the paired values on the respective variables being exactly equal in terms of standardized z scores—i.e., their deviation from the mean in standard deviation units. For example, an object scoring 1.5 standard deviations above the mean on one variable, also scores 1.5 standard deviations above the mean on the other variable. With such a perfect correlation, to know an object's value on one variable is to know its exact value on the other.

A correlation coefficient of $r = -1.00$ indicates a perfect negative or inverse linear relationship between the two variables. In this case an object's standardized score on each variable would be identical in absolute value and differ in *sign* only. An object with a z score of 1.5 on one variable would have a z score of -1.5 on the other. As in the case of the perfect positive correlation, a perfect negative correlation allows us to predict exactly an object's score on one of the variables if we know its score on the other.

Rarely, if ever, though, will two variables have perfect correlations of $+1.00$ or -1.00. Chance variation alone would tend to preclude this possibility. More often, variables are either uncorrelated or have intermediate degrees of correlation. A correlation coefficient of $r = 0$, for example, suggests that there is no relationship between the respective values of the two variables: High values on one are just as likely to be paired with high or low values on the other variable; similarly with low values.

Figure 4 shows some sample scatter diagrams representing differing degrees of correlation between two variables. Notice that the higher the value of the correlation coefficient, the more closely bunched are the data points representing each object's score on the respective variables. At one extreme is the case of the perfect correlation in which the points lie in a straight line (Figure

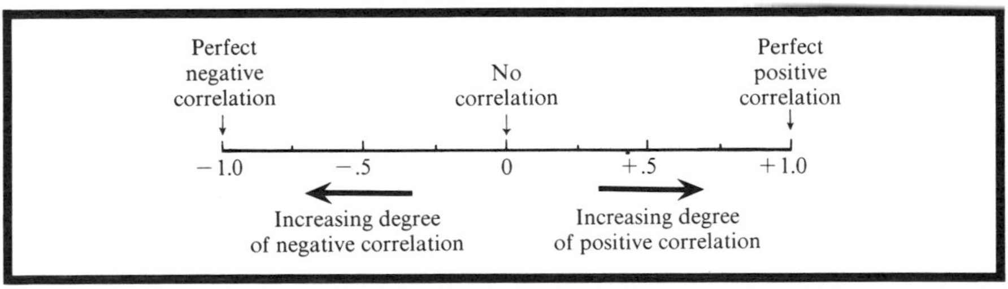

Figure 3 Range of values of the correlation coefficient r.

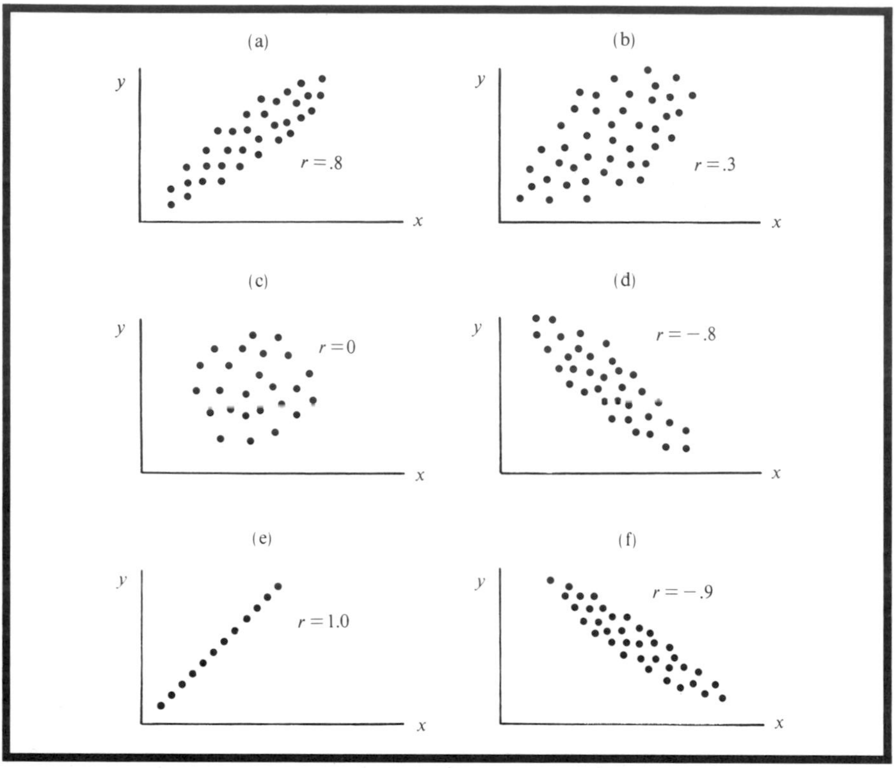

Figure 4 Sample scatter diagrams for varying degrees of correlation between two variables.

4*e*); and at the opposite extreme is the case of a zero correlation in which the points fall in a widely dispersed random pattern (Figure 4*c*). The intermediate degrees of correlation are represented by scatter diagrams in which the data points are clustered together in varying degrees of compactness. The narrower the band of data points, the stronger the relationship between the two variables.

Key assumptions. In the scatter diagrams of Figure 4 the data points are found to lie in groupings that are *linear* in form. That is, the points tend to fall along and about an imaginary straight line that passes through the cluster. The one exception is the case of the zero correlation in which the points fall in no discernable pattern and an infinite number of lines could be passed through the set of points equally well.

Figure 5 provides a comparison of a linear and non-linear relationship. In the case of the linear relationship, a straight line can easily be passed through the body of points. In the case of the non-linear relationship, a curved line is necessary to represent the distribution of points. The importance of this

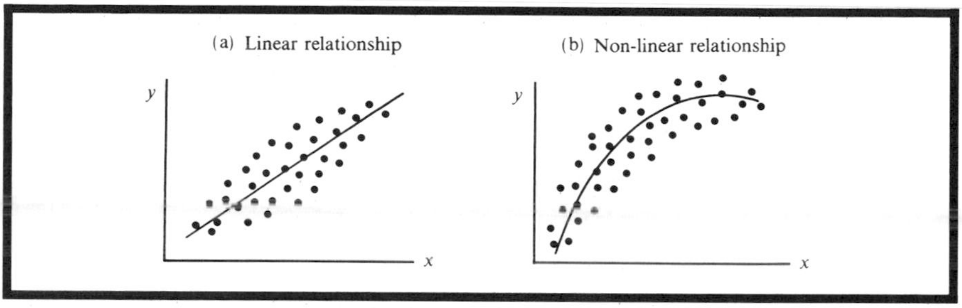

Figure 5 Comparison of a linear and non-linear relationship.

distinction between linear and non-linear relationships is that the correlation coefficient r is only appropriate for measuring the degree of relationship between variables which are linearly related. It is possible, for example, for two variables to be highly related in a non-linear fashion, and yet result in a correlation coefficient of $r = 0$, were it to be inappropriately used in the situation. Such an instance is shown in Figure 6.

A second key assumption for the use of the correlation coefficient is that the variables are *random* variables and are measured on either an *interval* or *ratio* scale, as defined in Chapter 2. It is not appropriate for nominally or ordinally scaled variables, nor when one of the variables is experimentally manipulated, for then *our choice of the values* of the experimental variable would influence the value of r obtained.

A third major assumption for the use of the correlation coefficient is that the two variables have a *bivariate normal distribution* as in Figure 7. It can be imagined as a hill or mount of data points, each representing a pair of values on the two variables in question. A cross-section of the mound, parallel to either the x or y axis, will result in the typical normal distribution for a single

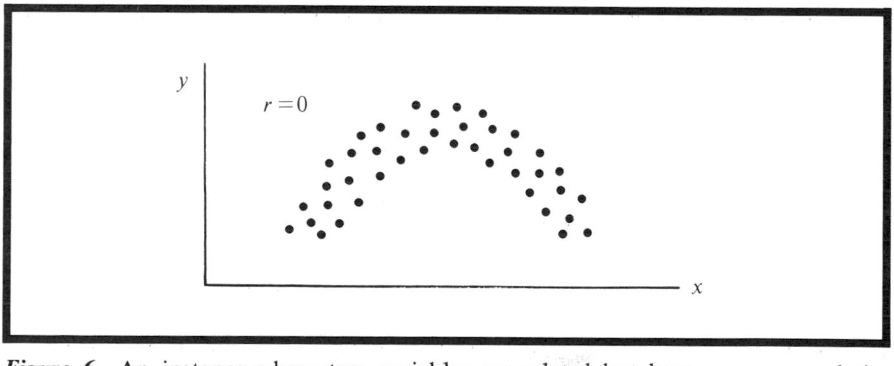

Figure 6 An instance where two variables are related but have a zero correlation coefficient.

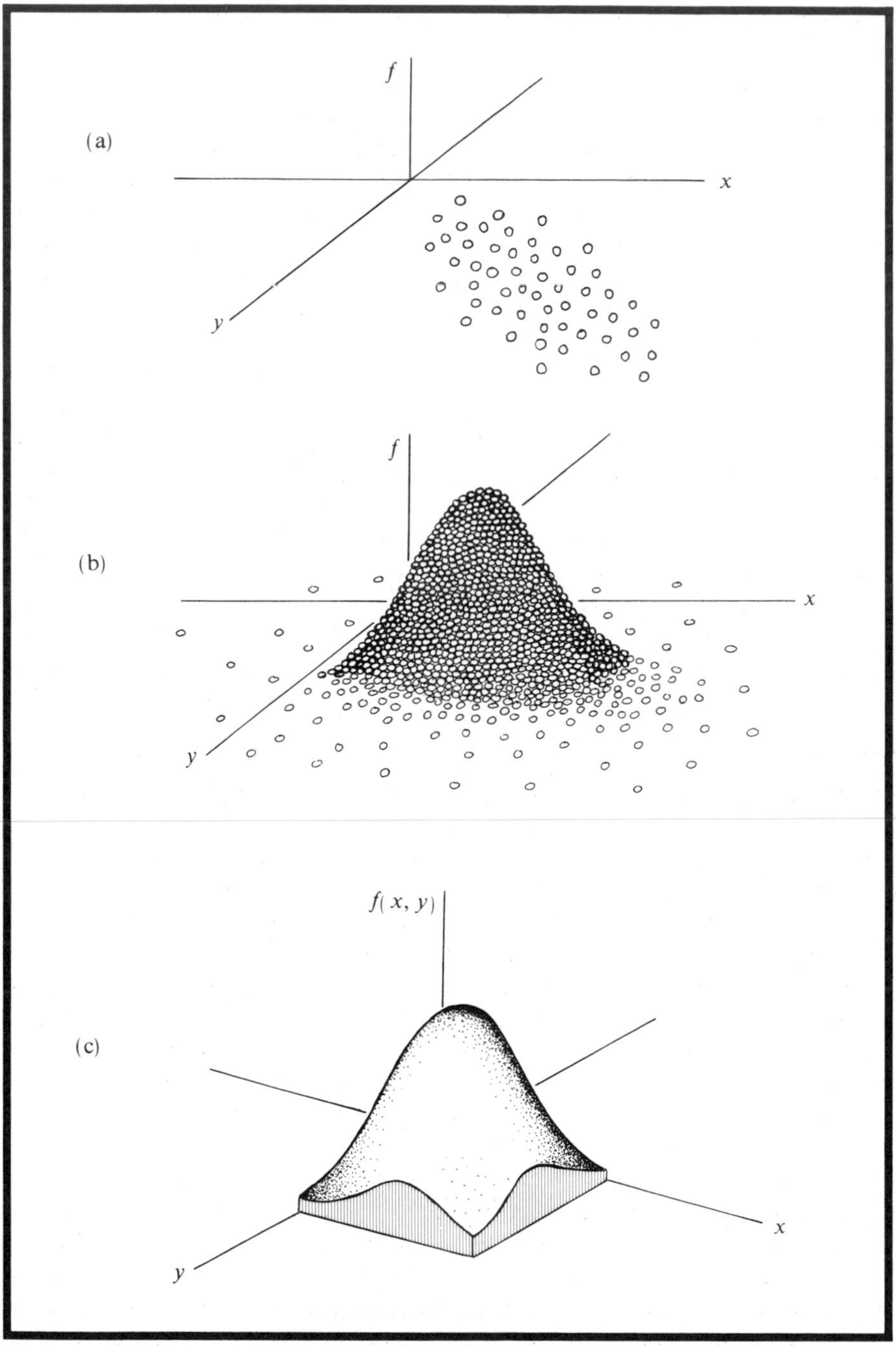

Figure 7 Sample scatter diagrams approaching a bivariate normal distribution.

variable with which we are already quite familiar. This is equivalent to saying that for any given value of the x variable, the y variable is normally distributed. Likewise, for any given value of the y variable, the x variable must be normally distributed. A study of Figure 7 will reveal that as the correlation between variables x and y increases, the "mound" will become narrower when viewed from above.

5. Calculation of *r*

Since the correlation coefficient r is a summary measure of the linear relationship between the paired values of two random variables, we should not be surprised to find that its mathematical definition takes into account each pair of values. The definition is given by

$$r = \frac{\Sigma(x_i - \bar{x})(y_i - \bar{y})}{(n-1)s_x s_y} \tag{1}$$

where $(x_i - \bar{x})$ is the deviation of an individual object's value on the x variable from the mean of that variable, $(y_i - \bar{y})$ is the deviation of that object's value on the y variable from the mean of that variable, s_x and s_y are the sample standard deviations of the x and y variables, respectively, n is the number of pairs of observations, and the summation Σ is across the $i = 1, 2, \ldots, n$ pairs.

If it is recognized that the quantities $(x_i - \bar{x})/s_x$ and $(y_i - y)/s_y$ are standard scores, call them z_x and z_y, representing the deviations of each pair of scores (x_i, y_i) from their respective means in terms of standard deviation units, then formula (1) can be rewritten

$$\boxed{r = \frac{\Sigma z_x z_y}{n-1}} \tag{2}$$

This is a conceptually more meaningful definition, for we see r as a kind of average value of the products of paired z scores.

Using formula (2), Table 5 demonstrates the calculation of the correlation coefficient for our earlier example involving the relationship between spice sales and shelf space for a set of $n = 10$ stores. This worked example points out a number of features of the correlation coefficient that may not have been obvious until now. Notice that we are correlating variables measured on completely different *scales* with completely different *units*—shelf space as measured in square inches and sales as measured in dollars. We could quite literally correlate sales of apples and oranges in these very same stores.

Table 5 Calculation of the correlation coefficient r for the spice sales vs. shelf space example.

i Store	x_i Shelf space (sq. in.)	y_i Spice sales (dollars)	$z_{x_i} = \frac{x_i - \bar{x}}{s_x}$	$z_{y_i} = \frac{y_i - \bar{y}}{s_y}$	$z_{x_i} z_{y_i}$
1	340	71	+.657	+.102	+.0670
2	230	65	−1.149	−.509	+.5848
3	405	83	+1.724	+1.322	+2.2791
4	325	74	+.410	+.407	+.1669
5	280	67	+.328	+.305	+.1000
6	195	56	−1.724	−1.424	+2.4550
7	265	57	−.575	−1.322	+.7602
8	300	78	.000	+.814	.0000
9	350	84	+.821	+1.424	+1.1691
10	310	65	+.164	−.509	−.0835
	$\Sigma x = 3000$	$\Sigma y = 700$	$\Sigma z_x = 0$	$\Sigma z_y = 0$	$\Sigma z_x z_y = 7.4986$
	$\bar{x} = 300.0$	$\bar{y} = 70.0$	$\bar{z}_x = 0$	$\bar{z}_y = 0$	
	$s_x = 60.92$	$s_y = 9.83$	$s_{z_x} = 1$	$s_{z_y} = 1$	

$$r = \frac{\Sigma z_x z_y}{n-1} = \frac{7.4986}{9} = .833$$

Another aspect of the data to notice is that the *means* and *standard deviations* of the *raw* scores are quite different from one another. Once they are standardized, though, each variable has a mean of 0 and a standard deviation of 1. In other words, the value of the correlation coefficient is *independent* of the values of the *means* and *standard deviations* of the individual variables. What is important is the manner in which values on the separate variables are *paired* with one another. To understand this concept more fully, imagine that the x_i and y_i numbers in Table 5 were scrambled in order, resulting in different pairings. Neither the means nor the standard deviations of the variables would change by virtue of the scrambling procedure, but an entirely different correlation coefficient would result, since it is dependent on the products of the *paired scores*.

Based on the example in Table 5 it is also easy to see why a coefficient of $r = 0$ will result when there is no relationship between two sets of scores. Positive products between the standardized scores (resulting from + + pairings and − − pairings) would occur equally often as negative products (resulting from + − and − + pairings), and consequently would cancel each

other out resulting in a sum of zero. On the other hand, the largest sums of the z score products, and consequently the largest correlations, will result when the products are all of the *same* sign, either all +, or all −. And we know that positive products will occur only when *positive* z_x scores are paired with *positive* z_y scores, and *negative* ones with *negative* ones; the case of a positive relationship. All negative products on the other hand will result when *positive* z_x scores are associated with *negative* z_y scores, and vice versa; the case of a strong negative relationship.

The result of our calculations for the spice sales example in Table 5 shows a correlation of $r = .833$ between shelf space and sales for the spice line. A correlation coefficient is subject to sampling variation just like the mean, standard deviation, or any other statistic. To determine if our correlation of $r = .833$ can be considered to be more than a chance deviation from a hypothesized population correlation of $\rho = 0$, we can consult Appendix Table VII for critical values of r for selected significance levels. We enter the table for degrees of freedom $df = n - 2$, where n is the number of pairs in our analysis. Thus, for the present example $df = 10 - 2 = 8$. Having chosen a significance level $\alpha = .01$ ahead of time, and having justified the use of a one-tailed test, we find that an r of .715 must be exceeded in order to reject the hypothesis of a zero correlation between the variables under study. Since our data yielded an $r = .833$ we can reject the null hypothesis that spice sales and the shelf space occupied by the spices are uncorrelated.

For the sake of comparison, the correlation coefficient for the above example is calculated in Table 6*a* using formula (1). Computationally, it is somewhat more convenient since calculation of z scores are not necessary. If the means and standard deviations of the variables are not available, Table 6*b* presents a means of calculating the correlation coefficient from the set of raw scores. Algebraically, the formulas used in Tables 5, 6*a*, and 6*b* are equivalent, and will result in the same value of r, within rounding limitations. Which procedure is used will be a matter of personal preference, the nature of the data, and the availability of computing facilities.

Finally, it is interesting to note that the standard deviation of the differences between the paired values, $D_i = x_i - y_i$, is given by

$$s_D = \sqrt{s_x^2 + s_y^2 - 2rs_xs_y} \tag{3}$$

Similarly, the standard error of the mean difference is given by

$$s_{\bar{D}} = \sqrt{s_{\bar{x}}^2 + s_{\bar{y}}^2 - 2rs_{\bar{x}}s_{\bar{y}}} \tag{4}$$

It follows that the standard error of the mean difference will be small when

r is positive, and will be large when r is negative, vis-a-vis the situation when $r = 0$, the condition of *independent* variables. Mathematically, formula (4) is equivalent to determining the standard error from the standard deviation of the difference scores—i.e., $s_{\bar{D}} = s_D/\sqrt{n}$, where s_D is based directly on the difference scores D_i rather than via formula (3). In either case, the standard error of the mean difference can be used as the denominator of a z or t ratio for testing the difference between the means of the two variables in question, which, in addition to their inter-correlation, is often of interest to us; for example, when "before" and "after," or "test" and "control" measures are made on a set of objects.

6. Other Bivariate Correlation Coefficients

Since the *product moment* correlation coefficient is limited to continuous random variables that are linearly related and that have a joint bivariate normal distribution, there are many instances in which it is not appropriate. But to the extent that necessity is the mother of invention, measures of correlation have been developed for virtually every type of relationship not within the domain of the product moment correlation coefficient.

Since these supplementary techniques add little to our basic understanding of the notion of correlation, they will not be discussed in detail. For our purposes it is enough to know of their existence, should the occasion for their use arise. Among the various bivariate correlation coefficients is the *rank* correlation coefficient, for relating two ordinally scaled variables; the *biserial* correlation coefficient, for relating a continuous normally distributed variable with a dichotomous one which has an underlying normal distribution; the *point biserial* correlation coefficient, for relating a continuous normally distributed variable with a truly dichotomous variable; the *tetrachoric* correlation coefficient, for relating two dichotomous variables, each with underlying normal distributions; the *four-fold point* correlation coefficient, for relating two truly dichotomous variables; the *contingency* coefficient, for relating two nominally scaled variables; and the *correlation ratio* or *eta coefficient*, for relating two variables which are curvilinearly related. They are mentioned here solely to promote familiarity with their names and purposes. Details of their definition and mathematical properties can be found in advanced reference books on correlational methods.

Despite the many different indices of correlation mentioned above, the product moment correlation coefficient remains the most basic. In fact, many of the coefficients named in the preceding paragraph are calculated in the identical manner as the product moment coefficient, differing only in the statistical properties of the resulting coefficient; i.e., in their sampling distributions.

Table 6 Alternative methods of computing r.

(a) Calculation of r using the definitional formula.

x_i	y_i	$x_i - \bar{x}$	$y_i - \bar{y}$	$(x_i - \bar{x})(y_i - \bar{y})$
340	71	40	1	40
230	65	−70	−5	350
405	83	105	13	1365
325	74	25	4	100
280	67	−20	−3	60
195	56	−105	−14	1470
265	57	−35	−13	455
300	78	0	8	0
350	84	50	14	700
310	65	10	−5	−50
$\bar{x} = 300$	$\bar{y} = 70$			Sum: 4490
$s_x = 60.92$	$s_y = 9.83$			

$$r = \frac{\Sigma(x_i - \bar{x})(y_i - \bar{y})}{(n-1)s_x s_y} = \frac{4490}{9(60.92)(9.83)} = .833$$

(b) Calculation of r using a raw score formula.

x_i	y_i	x_i^2	y_i^2	$x_i y_i$
340	71	115,600	5,041	24,140
230	65	52,900	4,225	14,950
405	83	164,025	6,889	33,615
325	74	105,625	5,476	24,050
280	67	78,400	4,489	18,760
195	56	38,025	3,136	10,920
265	57	70,225	3,249	15,105
300	78	90,000	6,084	23,400
350	84	122,500	7,056	29,400
310	65	96,100	4,225	20,150
Sums: 3,000	700	933,400	49,870	214,490

$$r = \frac{n\Sigma x_i y_i - \Sigma x_i \Sigma y_i}{\sqrt{\left[n\Sigma x_i^2 - (\Sigma x_i)^2\right]\left[n\Sigma y_i^2 - (\Sigma y_i)^2\right]}}$$

$$r = \frac{10(214{,}490) - (3000)(700)}{\sqrt{\left[10(933{,}400) - (3000)^2\right]\left[10(49{,}870) - (700)^2\right]}} = .833$$

In the balance of the text we will continue to use the term *correlation* in a most general sense, meaning the degree of relationship between two random variables as measured by an appropriate index. Correlation is the concept; the various coefficients are measures of that concept.

7. The Interpretation of Correlation

Perhaps the best way to begin to understand the meaning of correlation is to understand what it does not mean.

The issue of causality. First of all, the existence of a correlation between two variables *does not imply causality*. Suppose, for example, that we came upon a collection of objects that varied in *shape* (spherical, cubical, or pyramidal) and in *color* (red, blue, or yellow), and that an analysis of these objects reveals that the variables of shape and color are correlated. *Red* objects are more likely than others to be *spherical*; *blue* objects are more likely than others to be *cubical*; and *yellow* objects are more likely than others to be *pyramidal* in shape. That is, levels of one variable are systematically paired with levels of the other variable. How are we to interpret these findings? Can we conclude from this evidence that the *color* of the object *caused* its *shape*, or that its *shape caused* its *color*? While that conclusion *may* be true, there is only one way to really find out, and that is to bring the variables under our control. If we ourselves experimentally manipulate the values on one variable and observe an accompanying change in the paired value on the other variable, then we have evidence of causality. If we color the spheres yellow and they turn into pyramids, or if we mold a cube into a sphere and it turns red, then we have evidence of a cause and effect relationship between our two variables. In the absence of such data we can only interpret the observed correlation as "that's the way things are." *God knows* why the pyramids tended to be yellow and the cubes blue.

The preceding example involved qualitative variables. Let us now consider quantitative variables such as in the earlier spice sales example. In that analysis a correlation was found between the amount of shelf space occupied by the spices and its sales. On face value we might interpret this result to mean that increasing the amount of shelf space for the product will increase its sales. But this is not a valid conclusion on the basis of the correlational information alone. We ourselves did not manipulate shelf space and observe accompanying changes in sales. All we did was observe that larger shelf space tended to be *associated* with larger sales, just as spherical objects tended to be red in our preceding example.

It is quite possible that there were other *confounding* variables responsible for the observed correlation, either in whole or in part. In other words, it is possible that levels of other unidentified intervening variables were intertwined or confused with the variable of shelf space. One obvious possibility is that

larger stores, those with greater customer traffic, allocated more space to the spice line. In this case, size of store was *confounded* with amount of shelf space. This type of confounding relation is portrayed in Figure 8. The real cause of increased spice sales was the size of the store, and shelf space for the line was statistically related to that variable making it appear that it was responsible for the relationship. While it may be true that amount of shelf space would lead to greater sales, we would not know this for sure unless we conducted a controlled experiment in which we systematically varied the degree of shelf space, assigning stores at random to the various levels, thus making sure that no other variables such as store size were associated with different degrees of shelf space.

A second common misinterpretation of the correlation between two random variables concerns the *direction* of causality, in the event that a causal interpretation seems plausible. In the spice example, for instance, ruling out the possibility that size of store or some other confounding variable accounted for the space-sales correlation, it is possible that the *sales* of the spice line itself led to the extent of shelf space. In those stores where the product line did especially well against the competitor, the amount of shelf space allocated to it could have been increased. Or, as another example, consider the observed correlation between sales and advertising expenditure across a set of brands in a particular product category. Were the high levels of advertising responsible for the high level of sales, or did the high level of sales lead to increased advertising budgets?

These problems of cause and effect interpretation are inherent in all correlation analyses, since the observations are made on values of random variables not of our making but as found in the world at large. To appreciate the scope of the interpretation problem consider that the world is filled not only with objects which tend to be red spheres, blue cubes, or yellow pyramids,

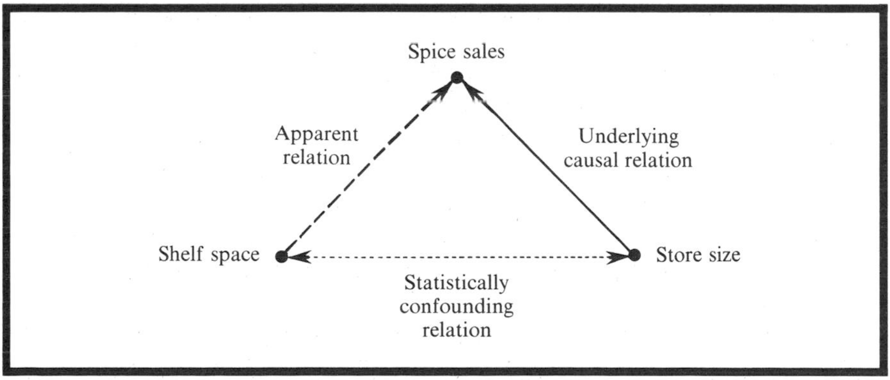

Figure 8 Schematic diagram of a confounding variable situation.

but these objects also vary in size (large or small), vary in texture (rough or smooth), vary in consistency (soft or hard), etc., and all simultaneously. When the objects we study are people, families, societies, cultures, races, plant and animal species, cells, organs, schools, hospitals, companies, products, cities, countries, etc., the number of variables that can become confounded with one another is enormous.

Perhaps the one type of situation in which we can infer causality without the benefit of a well-designed experiment in which objects are assigned at random to the alternative levels of the variable of interest, is when natural *catastrophes* occur. For example, if there is a sudden increase in certain symptoms in a vicinity where a nuclear power plant has contaminated the air, food, or water supply, or if there has been an outbreak of a disease among employees of a given factory, then we can be pretty sure that the "associated" conditions were causal in nature. Rather than wait for such catastrophic events, though, it is usually more prudent to conduct controlled experiments with lower animal forms, for although the full generalization from animal experiments to human beings may be questionable, the issue of causality will not be.

Description. If the correlation between two variables does not tell us that variation in one variable causes variation in the other variable, despite an observed association between values on the two variables, then what exactly does a correlation mean? While it may not be as much as we would hope for, a correlation does serve a data reduction *descriptive* function. It tells us how things are in the world, which is certainly something worth knowing, even though it does not put the world under our control.

Prediction. The descriptive power of correlation analysis is most evident in its potential for *predicting information* about the values on one variable given information on another variable. In all of the previous examples, as shown in Tables 2, 3, and 5, and in Figures 1, 2, 4, and 5 the common theme was that knowledge of an object's score on one variable helped us to do better than chance in predicting its score on the other variable. This is not a trivial feat, despite the limitations on its theoretical interpretation, since it has practical applications as we will see later.

Common variance. Another interpretation of the correlation between two variables is concerned with the degree to which they *covary*. That is, how much of the variation in one of the variables can be attributed to variation in the other, or vice versa. This aspect of correlation analysis is really no different than the prediction feature, just another way of looking at it. Figure 9 presents a schematic diagram of the concept. Two circles, representing two variables, are shown to overlap to varying degrees. The greater the overlap, the greater the degree to which values on the two variables covary. The non-overlapping circles represent uncorrelated variables, since none of the variation in the

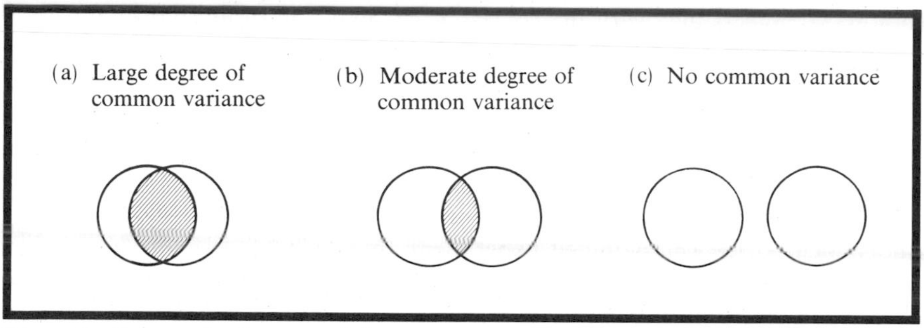

Figure 9 Schematic diagram of the common variance between two variables.

values of one variable is systematically associated with variation in the values of the other variable.

It so happens that a quantitative measure of this common variance is available. The square of the correlation coefficient, r^2, indicates the proportion of variance in one of the variables accounted for, "explained," or predictable from the variance of scores of the other variable. For example, a correlation of $r = .8$ between two variables means that .64 (.8 squared) or 64% of the variance in the scores of one variable is associated with the variance of the scores on the other variable. Loosely speaking, it is a measure of the amount of *information* in one variable accounted for by information in the other variable. We will get a better idea of this concept in the following chapter on Regression Analysis. It is closely related to the accuracy of prediction.

So whether we speak about predictive ability, or reduction in prediction errors, or common variance, it all comes down to the fact that the correlation coefficient is a *summary description* of the extent of systematic linear association between values on two random variables. The practical applications of this concept will be discussed in the following section.

8. Applications of Correlation Analysis

Now that we have an idea of the definition and interpretation of the concept of correlation, we can consider some of its most important practical applications.

Understanding key variables. Every field of study has its own set of important variables. Education is concerned with student achievement. The fields of business, economics, and government are interested in product quality, employee productivity, profits, inflation rates, etc. Psychology, sociology, and anthropology are interested in behavioral and belief characteristics of individuals, families, societies, cultures, and races. Biology and medicine are

interested in blood, urine, tissue, bone, cell, organ, and disease variables. Similarly, a set of variables can be identified for other disciplines.

Whatever the pivotal variable of a field of study may be, we are on a constant search for other variables that are in turn related to them. It is in this respect that correlation analysis can be helpful. We might want to correlate employee performance with scores on various tests or on school performance. We might want to correlate crime rate with unemployment rate. We might want to correlate disease characteristics with diet variables or genetic characteristics.

In each case an attempt is made to determine whether variation in some key variable of interest is related to variation in some other variable. A variety of benefits result from such an analysis. If *no* relationship is found, whatever attention or expenses had been attached to that variable in the past can be abandoned. Or, on the other hand, if a relationship is identified, a number of options are available as described in the following paragraphs.

Experimental input. Variables identified in a correlation analysis can be used in follow-up research investigations conducted under controlled conditions. Such experimental studies can then refute or confirm the efficacy of a variable in contributing to a criterion variable of interest. For example, a food packer could determine if changing the nitrogen content of the soil with the use of chemicals would result in greater crop yields. Or it may be concluded that the original results, based on correlational evidence, were due to some soil variable confounded with nitrogen.

Hypothesis generation. The identification of a correlation between variables may confirm a hypothesis or provide input for additional hypotheses. It may suggest new marketing techniques, teaching methods, or medical treatments, or the correlation may cast new light on old practices. If a number of relationships are studied it may serve as the basis of a broader theory. In short, it will get the investigator thinking about the problem. We refer to this goading and guiding feature of an analysis as its *heuristic* function.

Prediction. The ability to predict values on one variable from a knowledge of values on another variable has wide-reaching applications. Whether for personnel selection, sales forecasting, graduate school admissions, or disease control, the identification of correlated variables is essential for prediction. While the actual prediction procedure will be discussed in the following chapter on Regression Analysis, it can be said now that the *accuracy* of prediction depends on the extent of correlation between the variables in question.

Reliability assessment. Another major application of correlation analysis is in the determination of the *reliability* of measurements made on a variable. By reliability we mean reproducibility. If measurements on a set of objects

cannot be replicated, we must conclude that the scores are extremely unstable or that the score obtained by each object was a matter of chance. In either case, the measurement is no good if it does not reflect a characteristic of the object itself.

The notion of reliability is basic to every measurement situation. It is fundamental to any interpretation we place on our measurements and to any subsequent analysis to which the numbers may be subjected. The question of reliability might be phrased as follows: When we observe the scores of a set of objects on a variable, is this distribution of scores the result of *chance*, or is it due to some *intrinsic characteristic* of the object, such that on remeasurement each object will retain approximately its same score. If the objects do retain their scores, we say the measurements are reliable. If the objects have drastically different z scores upon remeasurement, we say the measurements are unreliable.

For example, if an identical test is administered to a set of job applicants on two successive days, and if these two sets of scores do not correlate with one another, something is wrong with either our test or our applicants. If the dimension we were measuring was stable, we would expect a high scorer on the first administration to be a high scorer on the second administration of the test. Similarly, a low scorer on the first administration should be a low scorer the next time too. If not, we may as well have assigned scores to the applicants using random numbers.

The correlation between successive measurements on a set of objects with respect to a variable is referred to as *test retest* reliability. Depending on the particular variable, and depending on the amount of time passing between the two sets of measurements, we would hope to get a reliability coefficient (test-retest correlation) of at least .90, and hopefully higher. It should be noted that an overall increase in the mean performance on the second test administration, perhaps due to memory effects, does not rule out the reliability of the test. Nor does an equal mean performance on both administrations imply reliability. It is the manner in which the scores on the test and retest are paired that determines the reliability, not the means of the two administrations.

An alternative method of assessing the reliability of the test scores in the preceding example would be to obtain two scores for each individual, perhaps one score on the odd items and one score on the even items; or alternatively, one score on the front half of the test and one score on the second half. The correlation between the two sets of scores, referred to as the *split-half* reliability, would indicate the extent to which the test was measuring a coherent trait. Consider the alternative finding: If there was no correlation between the scores obtained on the odd set of items and even set, then those two sets of items evidently must have been measuring *unrelated* traits, an unlikely possibility if the order of the items in the test was determined completely at random.

Often the reliability of measurements is not obtained since it can be a costly process. But with a little foresight most data collection schemes can be designed in order to permit the determination of the reliability of the measurements. A prudent course is to set aside a random subsample for remeasurement, and then the reliability of the full sample can be inferred from the smaller sample.

In conclusion, the reliability of our measurements should be the first question asked of any data analysis, for if the raw data have no meaning, what possible meaning could the summary statistics have.

Validity assessment. In addition to determining the reliability of our measurements, correlation analysis can be used to determine the *validity* of our measurements. By validity we mean the extent to which our measurements reflect *what we intend them to*, or what we claim they do. For example, if a test administered to job trainees is intended to measure future job success, it has validity only if it is demonstrated to do just that. If scores on the test are found to be uncorrelated with job success, as measured by, say, supervisor ratings or by some productivity criterion, then the test has no validity: It does not do what we intended it to do. To the extent that the correlation exists, the test would have validity.

It is important to note at this point that a set of measurements *can be reliable without being valid*, but *cannot be valid without being reliable*. The reproducibility of scores in itself is no guarantee that the scores will correlate with another set of scores, despite our sincerest hopes. Consider the test administered to the job trainees. It could be highly reliable, with a high test-retest correlation, but the scores could turn out to have no relationship to job success as measured by some objective criterion. While such a test proved to be invalid as a measure of job success, it is possible that it could be valid as a measure of some other variable.

While measurements can be reliable without being valid for a stated purpose, it is *impossible* for a measurement system to be *valid* without being *reliable*. Recall that an unreliable set of scores, a set of scores that do not correlate with themselves upon remeasurement, is essentially equivalent to assigning the scores to the objects in a purely random manner. How could a randomly assigned set of scores possibly correlate with another set of randomly assigned scores? It would be like trying to correlate two columns of random numbers.

Attempting to correlate an unreliable variable with a reliable one, will produce exactly the same result, a zero correlation, for a chain is no stronger than its weakest link. So just as a measurement system can have no validity if it has no reliability, the degree of validity is *limited* by the degree of reliability. Even if our measurements are somewhat reliable, this sets an upper limit on the degree of validity they can have. Thus, whenever we attempt to relate

variables, the weakest link in the process will be the reliability of the measurements we make.

Identification of surrogate variables. Another important application of correlation analysis is in the selection of a variable that can serve as a *surrogate*, or substitute, for another variable and result in cost efficiencies.

For example, consider the advertiser that over the years developed a consumer testing system in which TV commercials were graded in terms of sales persuasiveness. The system was found to be valid in the sense that the commercial scores were related to sales in controlled field experiments. The weakness of the system, however, was that the scores could not be obtained until the full cost of commercial production had been incurred. The advertiser then began a testing program in which scores were first obtained on rough inexpensive commercials consisting of still photos. After the commercials were produced in final form they were tested once again and it was found that the scores correlated highly with the scores obtained on the rough inexpensive versions. Consequently, the advertiser was able to effect cost savings in the future by aborting low scoring commercials while still in preliminary form.

This function of correlation analysis finds application in many areas of quality control. Measurements made early in a process (e.g., oil refining, wine making, etc.) and which have been found to correlate with measurements made on the final product, can be used to adjust processes to help ensure a satisfactory end product.

Advanced analyses. The correlation coefficient also serves as the basis for many advanced statistical analyses. These are the multivariate studies in which many variables, rather than just two, are studied simultaneously. In these analyses the correlation coefficient r plays a major role. The far-ranging importance of this summary measure of association will become apparent in the balance of the chapter as well as in succeeding chapters.

9. Multivariate Correlation Analysis

The expression *multivariate correlation analysis* is somewhat redundant in the sense that the term correlation, by its very definition, supposes more than one variable. However, we adopt usage of the description *multivariate* to suggest analyses in which many—i.e., more than two—variables are involved, making it more general than the bivariate situation discussed in the first half of the chapter. Also, multivariate correlation analysis is to be distinguished as a broader concept than that of *multiple correlation*, one of several techniques that we will study that is but a specific instance of the broader analytical concept.

The input data. As the name implies, multivariate correlation analysis is concerned with the correlations that exist among several variables. As we

Table 7 Format of the input data matrix for multivariate correlation analysis.

	Variables				
Objects	x_1	x_2	x_3	$\cdots$	x_k
Object 1				$\cdots$	
Object 2				$\cdots$	
Object 3				$\cdots$	
$\vdots$	Observational data: each object measured on each variable				
Object n					

learned in the beginning of the chapter, the prerequisite for determining the relationship between two variables is to obtain measurements on a set of objects with respect to both variables. Extending this concept to multivariate correlation, Table 7 shows the format for the collection of data necessary for assessing the degree of correlation among several random variables; namely, each of a set of objects is measured on each of a number of variables $x_1, x_2, \ldots, x_k$.

There are numerous situations in which this type of data collection scheme might arise. For example, the objects could represent a set of cities, and the variables could include crime rate, unemployment rate, population, average income, ratio of home owners to renters, rainfall level, etc. Or, the objects could be students, and the variables could represent grade point average, reading speed, math aptitude test scores, verbal aptitude test scores, family size, household income, hours of TV viewing, etc. Or, the objects could be a sample of sales agents for an insurance company, with each agent measured on such variables as number of policies sold in a year, value of policies sold, years of education, years of sales experience, age, family size, scores on a personality test, etc. Or, as another possibility, the objects might be a set of corporations and the variables might be price of its stock, earnings per share, number of employees, gross sales, assets, outstanding loans, etc.

Criterion and predictor variables. In each of the above examples the list of variables could be extended at will. However, when we collect data we usually do it with a purpose. While *all* of the possible relationships among the various variables, as well as their individual characteristics, might be of interest to us, more often we are primarily concerned with one *key* variable, one that has a special degree of importance to us. We refer to such a variable as a *criterion variable*.

We would be interested in the other variables only insofar as they are related to our variable of prime concern. In other words, the criterion variable stands as a standard against which the importance of the other variables are judged. These latter variables are often referred to as *predictor variables* in the sense that they may be found to have variance in common with the criterion variable, and consequently information about them could be used to predict information about the criterion variable. On the other hand, the predictor variables may serve a purely descriptive function with their predictive capacity, *per se*, not being of primary concern.

The terms *dependent variable* and *independent variable* have often been used in place of criterion and predictor variable, respectively. While the term dependent variable, at first glance, might seem a fair substitute for the term criterion variable, it is really more appropriate in the context of controlled experimental studies where it is safe to conclude that the observed variation in the values of the criterion variable is due at least in part to—i.e., *dependent on*—variation in the manipulated experimental variable(s). However, in the correlational analysis of strictly observational data, as discussed at length in the beginning of the chapter, we cannot infer causality. In a correlation analysis the values of the predictor variables can just as well be thought of as being dependent on the criterion variable, as opposed to vice versa, since we have no basis on which to infer the direction of causality between two variables, if indeed causality exists at all.

The use of the expression *independent variable* in place of predictor variable has even less to speak for itself. It even has questionable desirability when used in the context of experimental studies where its use originated, but referring instead to the fact that the multiple experimental variables were *orthogonal* to each other. Consequently, the expression "independent variable" used in either a correlational or experimental context does not have the meaning we have attached to *variables being independent of one another*; i.e., that do not covary. In fact, in correlation studies the predictor variables are most often *not independent* of one another, more likely being correlated with one another to some extent. And they are certainly *not independent* of the criterion variable; for the objective of correlation analysis is to identify variables which covary with our criterion variable.

This imprecision of descriptive terms is unfortunate. To avoid confusion we will adopt usage of the terms criterion variable and predictor variable when dealing with the correlation analysis of observational data. When referring to controlled experimental studies we will use the terms criterion variable and dependent variable interchangeably, and refer to the experimental variables, those under our manipulation, *as* experimental variables. We will reserve the use of the expression independent variables to mean variables that *are independent of one another*; i.e., that do not covary.

As emphasized throughout the text, every field of study has its own key criterion variables. To the extent that we can identify predictor variables related to these criterion variables, we will have come closer to understanding their underlying natures, and hopefully in the long run how to bring them under our control. In the beginning of the chapter we learned how to relate a single predictor variable to a criterion variable, but since the real world is so complex, it is rare that the identification of a single predictor variable will account for a large share of the observed variation in any given criterion variable; and then, even when a single predictor variable is identified which is highly correlated with our criterion variable, we are not sure that the relationship is not in fact due to some other confounding variables. For these reasons it is desirable to have techniques available which can simultaneously consider the relationships among a large set of variables. The following sections will provide an introduction to the most useful of these techniques.

10. The Correlation Matrix

Once we have made observations on each of a set of objects with respect to each of a number of variables, as shown in Table 7, we are in the position to calculate a correlation coefficient r between *each pair* of variables. Any two columns of the multivariate data matrix shown in Table 7 can be treated in exactly the same way as the solitary two columns portrayed in Table 1 at the beginning of the chapter, and which provided the basis for calculating the correlation coefficient between the two variables.

With the multivariate data matrix we single out every possible pair of variables and calculate a correlation coefficient for each pair, after assuring ourselves that the data has met the necessary assumptions as discussed earlier. We can calculate a correlation coefficient between variables x_1 and x_2, x_1 and x_3, x_1 and x_4, etc. Also, between x_2 and x_3, x_2 and x_4, x_2 and x_5, and so on.

Since so many coefficients are possible among a large set of variables, it is convenient to arrange the coefficients in a systematic and orderly fashion. This is done in the form of a *correlation matrix*, an example of which is shown in Table 8 for a hypothetical set of five variables. It is nothing more than the tabular arrangement of all the correlation coefficients that can be calculated from all the possible pairings of the five variables x_1, x_2, x_3, x_4, and x_5.

Notice that the correlation matrix is square, with as many rows as columns. Each of the row headings stands for a different variable, as do each of the column headings. The first row represents variable x_1, the second row represents variable x_2, and so on. Similarly, the first column stands for variable x_1, the second column stands for variable x_2, and so on until the final column stands for the final variable.

Each cell of the matrix is occupied by a correlation coefficient between the variables represented by the particular row and column that the cell occupies.

Table 8 A sample correlation matrix based on five variables.

Variables	Variables				
	x_1	x_2	x_3	x_4	x_5
x_1	1.00	.36	.49	−.83	.41
x_2	.36	1.00	.57	.16	.37
x_3	.49	.57	1.00	−.51	−.19
x_4	−.83	.16	−.51	1.00	−.09
x_5	.41	.37	−.19	−.09	1.00

For example the cell in Table 8 that is two rows down and four columns over is occupied with a correlation coefficient of $r = .16$, representing the correlation between variables x_2 and x_4. The cell that is located by going three rows down and five columns over is occupied by the coefficient of $r = -.19$, which is the correlation between the variables x_3 and x_5. Similarly, the correlation between any other pair of variables can be found by locating the cell which is the intersection of the variables' respective row and column.

Diagonal coefficients. Inspection of Table 8 will reveal two important features of the correlation matrix. The first characteristic to be noticed is that the diagonal coefficients, those going from the upper left hand corner to the lower right hand corner of the matrix, and outlined in bold borders, are each equal to 1.00 in value; i.e., perfect correlations. But this should be no surprise since they represent the correlations of each of the variables with themselves: x_1 with x_1, x_2 with x_2, x_3 with x_3, etc. Recall that a perfect correlation will occur *only* when the paired values on the respective variables have exactly the same z score value, and this is exactly how the data would look were we to set the values for a variable beside itself—exactly the same values side by side.

Symmetry. The second thing to be noticed about the correlation matrix is that it is symmetrical about the diagonal; that is to say, the portion above the diagonal is a mirror-image of the portion below the diagonal. A bit of reflection will show why this must be the case. Consider the coefficients at the extreme corners of the matrix—at the upper right and lower left: They both equal .41. And they should, for the first represents the correlation between variables x_1 and x_5, while the other represents the correlation between x_5 and x_1, exactly the same pair of variables. Similarly, it will be found that the correlation between variables x_2 and x_4 is identical to the correlation between x_4 and x_2, .16 in both instances. Studying each of the other corresponding pairs of variables above and below the diagonal will show that they too are equal.

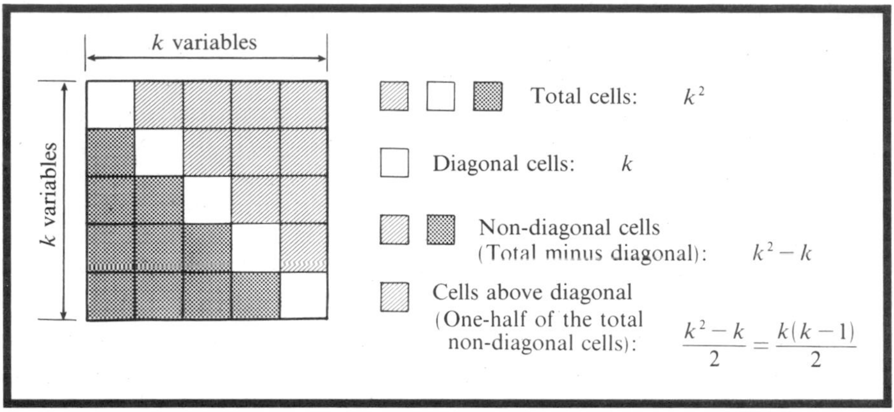

Figure 10 The number of potentially unique correlation coefficients in a correlation matrix based on k variables.

The number of correlations. The symmetry of the correlation matrix results in a substantial degree of redundancy. We could discard either the portion of the matrix above the diagonal or below the diagonal and not lose a bit of information. For that matter, we could also discard the diagonal, since we know on logical grounds that the correlation of a variable with itself must equal 1.0. With how many correlation coefficients are we then left? Figure 10 presents an easy procedure for determining how many potentially unique coefficients are possible among a given set of variables. *We simply discard the diagonal cells and divide the remaining cells in half*. For example, in a 5×5 correlation matrix, representing the correlations among five variables, there are a total of 25 cell entries (5×5). If we throw out the five diagonal cell entries which we know must each be equal to 1.0 in value, we are left with 20 cell entries. Half of these, or 10, will be above the diagonal, and the other half will be below the diagonal. Tossing out either set leaves us with 10 potentially different correlation coefficients.

Despite the fact that the set of coefficients above and below the unit-valued diagonal elements are duplicates of one another, it is of practical value to retain them in the presentation of the matrix, since it makes the reading of the matrix much more convenient than if only the upper or lower triangular sections were presented. For example, if we are interested in knowing the extent to which each of the variables in Table 8 correlates with variable x_3, we merely have to read down the third column (or across the third row). If only half the matrix were presented we would have to do a bit more searching to locate the same coefficients.

Analysis of the matrix. Visual inspection of the matrix can tell us quite a bit about the relationships that exist among our many variables. We can

quickly identify which two variables are most highly correlated with each other, we can identify which variables correlate most highly with each of the individual variables, and we can perhaps even identify clusters of variables that are highly correlated with each other, or those which are relatively independent of one another. What we cannot do by a mere inspection of the matrix is assess the *joint effects* of two or more variables on another variable; nor can we know to what extent the correlation between two variables is due to the effects of a third, fourth, etc., confounding variable. For this task we need the analytical techniques described in the following sections which are designed to cope with these multivariate data analysis problems.

11. Multiple Correlation

In the world at large there are so many possible predictor variables that might be correlated with our criterion variables of interest, and while it is desirable to know the degree to which each of the individual predictor variables correlates with the criterion variable, it would be even more desirable to know the *total* explanatory power of a set of predictor variables *combined*.

The analytical procedure that allows us to determine how much of the variation in a criterion variable is associated with variation in a *set* of predictor variables is known as *multiple correlation*. In the multiple correlation procedure our aim is to construct that *weighted combination* of the values on the respective predictor variables that will correlate more highly with the criterion variable than any other weighted combination. As a result, the *derived variable*, which represents the weighted combination of the values on the various predictor variables, will correlate more highly with the criterion variable than any single predictor variable alone.

Every student is aware of the nature of the weighting procedure whenever final grades are administered. A final grade is composed of a weighted combination of scores on individual variables such as performance on quizzes, midterms, final exams, theme papers, etc. In the same spirit, the predictor variables in a multiple correlation analysis are weighted; not on personal judgment, but on the purely statistical standard that the resulting *composite* scores have the maximum correlation with the values on the criterion variable.

At this point we might object that there are an infinite number of possible weights that could be applied to the values of a set of predictor variables. Even if there were only two predictor variables, the choice of weights that could be associated with the variables is very large. Consider the number of possibilities when the number of predictor variables is five, ten, or more.

How, then, can we possibly identify that particular set of weights that when applied to the values of the predictor variables will result in that set of composite scores, a derived variable, which yields the highest correlation with the criterion variable? Even a purely trial and error procedure with the aid of a

high-speed computer would not be expected to achieve the desired result very easily. The solution to the problem requires advanced mathematical techniques of calculus and matrix algebra, but we need not understand the underlying solution in order to understand the meaning of the resulting weights.

Beta weights. The weights that are applied to the *z scores* on the various predictor variables, those which result in composite scores that correlate most highly with the scores on the criterion variable, are referred to as *beta weights* or *beta coefficients*. They should not be confused with the so-called "*b* coefficients" which are their counterparts when dealing with the raw scores as opposed to the standardized *z* scores.

Each predictor variable is associated with its own beta weight. While we cannot delve into the actual mathematics of how the weights are derived, it is sufficient for our purposes to know that their values are a function of two things: (1) the correlations of the individual predictor variables with the criterion variable, and (2) the correlations that exist among the predictor variables themselves.

It is reasonable that these two factors should determine the value of the beta weights. First of all, we would expect that a predictor variable that correlated highly with the criterion variable should have a high weight associated with it, and predictor variables with low correlations with the criterion variable should have low weights. But this is not enough. We also need to take into account the correlations among the predictor variables themselves. To understand why, consider the situation in which two predictor variables are each highly correlated with the criterion variable, and in addition are highly correlated with *each other*. Contrast this with the situation in which the two predictor variables are again each highly correlated with the criterion variable, but *are not* highly correlated with each other. Which pair of predictor variables would account for more of the variation in the criterion variable? The second pair would. In the first case, the predictor variables, by virtue of their high correlation with one another, would each be accounting for essentially the same variance in the criterion variable. In the second situation, however, with little correlation between the predictor variables, they each account for a unique portion of the variance of the criterion variable, and consequently more of its total variation. Therefore, we would expect a different pair of beta weights in these two situations, despite the fact that the two pairs of predictor variables were individually correlated with the criterion variable to an equal degree. So, in determining the beta weights, not only do we need to take into account the correlation of the predictor variable with the criterion variable, but also the correlations that exist among the predictor variables themselves. That is why a full correlation matrix is needed as input for determining the weights.

Multiple correlation coefficient *R*. The correlation between the set of composite scores, derived from an application of the beta weights, and the scores on the criterion variable is referred to as the *multiple correlation*

coefficient. It is usually designated with the capital letter R to distinguish it from the correlation coefficient r that exists between *two* random variables. Actually, R also measures the correlation between two variables; it is just that one of the variables is *derived* as a weighted combination of several others. The similarities and differences of interpretation of the multiple R and the bivariate r will be discussed after an example is presented which will clarify the overall multiple correlation procedure.

Consider, for instance, the researcher interested in correlates of student grade point average in a given school system. The predictor variables chosen for study include reading speed, verbal aptitude test score, and a musical aptitude score. Based on the correlations that exist among the predictor variables themselves, as well as the correlations between each of the predictor variables and the criterion variable, beta weights are calculated for each of the predictors. Each student then receives a composite score based on the weighted combination of scores on each of the respective predictor variables. These composite scores—comprising a derived variable—are in turn correlated with the grade point averages of the students, yielding a multiple correlation coefficient R. The overall procedure is outlined in Table 9. In short, *a weighted combination of the values on the predictor variables is correlated with the values on the criterion variable*. The weights are those which result in the maximum value of R.

Table 9 An example of the overall multiple correlation procedure.

Objects (Students)	Score on criterion variable (Grade point average z)	Composite score (The derived predictor variable)		z scores on the predictor variables and their respective weights: $\left(\text{beta}_1 \times \text{reading speed}\right)$		$+\left(\text{beta}_2 \times \text{verbal aptitude}\right)$		$+\left(\text{beta}_3 \times \text{musical aptitude}\right)$
1	1.50	1.23	=	(.27)(1.72)	+	(.51)(.98)	+	(−.22)(−1.21)
2	−.78	−.49	=	(.27)(−.97)	+	(.51)(−.15)	+	(−.22)(.68)
3	.05	−.06	=	(.27)(−.26)	+	(.51)(.17)	+	(−.22)(.33)
⋮	⋮	⋮		⋮		⋮		⋮

The multiple correlation coefficient R is the correlation between these two sets of scores.

The beta weights are determined in such a way as to maximize the correlation between the resulting composite scores and scores on the criterion variable.

Interpretation of multiple correlation. The two key sources of interpretation of multiple correlation analysis are the beta weights and the square of the multiple correlation coefficient, R^2. Intuitively, we suspect that the size of the beta weights should somehow reflect the relative importance of the variables to which they are attached. A variable with a high beta coefficient should account for more of the variance in the criterion variable than a predictor variable with a small beta coefficient. And this is so. The squares of the various beta coefficients tell us the *relative* contributions of the respective variables to the variation in the criterion variable. Again, not "contribution" in a causal sense, but simply the extent to which their variation is associated with variance in the criterion variable. The squared beta weights say nothing about the absolute contributions of the separate predictor variables, only their relative importance. For example, in the preceding example, and as shown in Table 9, the beta weight for the reading speed measure was found to be .27, while that for the verbal aptitude measure was found to be .51. Squaring these values yields .073 and .260, respectively. Thus we can conclude that verbal aptitude accounted for somewhat more than three times as much variance in the criterion measure as reading speed, as these variables were measured in the study. Other measures may have produced other results. Notice also in Table 9 the negative beta weight for musical aptitude score, suggesting that it is negatively correlated with grade point average, perhaps reflecting the particular school systems' de-emphasis on right brain hemisphere functions such as music and art in its curriculum.

The second major interpretation of multiple correlation analysis rests on the value of the multiple correlation coefficient R. More precisely, it is upon the value of R^2 that we place our interpretation. Just as the square of the simple correlation coefficient, r^2, signifies the proportion of variance in one variable attributable to, or predictable by a second variable, the value of R^2 signifies the proportion of variance in the criterion variable predictable from variation in the *derived variable* of composite scores. For example, in our preceding example of student grade point average a multiple R of .69 yielding an R^2 of .48 would suggest that about 48% of the variance in grade point average could be predicted from (attributable to, accounted for, explained by) variation in the set of predictor variables taken as a whole. Again, no interpretation of causality is appropriate since we did not systematically manipulate the predictor variables ourselves, but only observed their values "as they fell."

All of the data assumptions and problems of interpretation associated with the bivariate correlation coefficient discussed earlier also accompany the multiple R^2. However, in addition, multiple R^2 has some problems of interpretation that are unique to itself. We will touch upon the issues in the following paragraphs.

Adjusted R^2. If all the predictor variables in a multiple correlation analysis were either all positively or all negatively related to the criterion variable, then the multiple R could turn out to be either positive or negative in value. However, since the set of predictors is likely to be a mixture of variables which are either positively or negatively related to the criterion variable, beta weights must take on positive and negative values to distinguish between them and to properly emphasize their respective values. Consequently, the resulting composite scores will always be positively correlated with the criterion variable.

Because multiple R always takes on a positive value, its initially calculated value must be adjusted downward. To understand why this must be so, imagine that in fact our set of predictor variables are completely unrelated to the criterion variable in the population at large—i.e., each with a $\rho = 0$. Consequently, we should expect a multiple R^2 of zero, since how can the predictor variables *collectively* correlate with the criterion variable if they *individually* do not. But, and this is always our concern, when we sample from a population our observations are subject to sampling error—chance deviations from the population values. Thus, even though our predictor variables in the population are uncorrelated with the criterion variable, our sample will surely yield r's that are chance departures from zero. The multiple correlation technique then blindly takes these chance r's (some $+$, some $-$), each accounting for some variation in the criterion variable, and arrives at beta weights which lead to composite scores which yield a multiple R^2 that is greater than zero, but is really only representing an accumulation of chance relationships. That is, it is biased.

The degree to which the expected value of R^2 will exceed zero, when in fact it is zero in the population, is dependent on two things: The size of our sample of objects n and the number of variables k. Approximately, the *expected* value of R^2 when it is in fact zero in the population will be the ratio of the number of variables to the number of objects on which the variables are measured. More precisely, it is the ratio of the number of variables *less one*, to the number of objects *less one*, $(k-1)/(n-1)$. It follows, then, that when the number of variables is equal to the number of objects we will expect a perfect correlation of $R^2 = 1.0$. This is easiest to see in the case of *two* variables measured on only *two* objects. Imagine the scatter diagram: With only two points they cannot help but fall on a straight line. Analogously for three objects and three variables, four objects and four variables, etc. In short, if the number of objects we study is not large relative to the number of variables, we are bound to get an artificially inflated value of R^2.

The important thing to remember from this discussion is that a multiple correlation analysis should be based on a large sample of objects relative to the

number of predictor variables, and that whenever a multiple R^2 is presented to us that we be assured that it has been adjusted based on the number of variables and objects studied. In any case, we should be very dubious of an R^2 that is reported based on a set of variables that is not much larger than the set of objects upon which they were measured. For example, an R^2 of .9 which would normally delight any investigator, would mean nothing if it were obtained from a set of *thirty* predictor variables measured on a sample of *forty* objects. For such a sample size we would expect a value of $R^2 = .74$ on the basis of chance alone ($29 \div 39 = .74$).

Validation of R^2. In addition to employing a large sample of objects relative to the number of variables in studies of multiple correlation, there are other steps we can take to assure that our results have the meaning we intend. Once we have determined the beta weights from a sufficiently large sample, we can apply the weights to objects in a completely separate sample, obtain a set of composite scores and correlate them with the values of the criterion variable for that sample. To the extent that the resulting R^2 is the same as the R^2 calculated on the initial sample we can be confident that we are indeed measuring a real relationship between our predictor variables and the criterion variable.

Stepwise procedures. Given a set of predictor variables it is not necessary to utilize every single one in the determination of a multiple R^2. Rather, we can begin by selecting the one predictor variable that correlates most highly with our criterion variable, and then introduce a second predictor variable, the one that accounts for the most of the remaining or residual variance in the criterion variable. We can continue this *stepwise* procedure, each time adding that variable that accounts for the most variance in the criterion variable not already explained by the earlier variables, continuing until the inclusion of another variable would account for only an insignificant amount of variance in the criterion variable.

In contrast to the step-up procedure outlined above, the step-down procedure works in the opposite direction. Beginning with the total set of predictor variables, we begin eliminating variables one by one, beginning with the one accounting for the least amount of variation in the criterion variable. We eliminate variables until the elimination of another would sacrifice too much in the way of explained variance. In both the step-down and step-up procedures, the option exists to drop or add predictor variables which had been added or dropped, respectively, in earlier steps.

The stepwise procedures described above provide us with a more parsimonious account of the variance of the criterion variable, and can result in cost savings inasmuch as it is impractical to collect data on a large set of variables when a smaller set contains virtually the same amount of information.

12. Partial Correlation

In the case of multiple correlation we are concerned with combining the values of individual predictor variables to determine their joint contribution to the variation of a criterion variable. We can also consider the reverse possibility—measuring the correlation between two variables when the common variance of other variables is extracted. This type of statistical control of variance can be accomplished by a technique known as *partial correlation*. In the simplest application of partial correlation we can extract the common effects of one variable from the relationship between two other variables. The *partial correlation coefficient* is defined

$$r_{12\cdot 3} = \frac{r_{12} - r_{13}r_{23}}{\sqrt{(1-r_{13}^2)(1-r_{23}^2)}} \tag{5}$$

where

$r_{12\cdot 3}$ is the correlation between variables 1 and 2, with variable 3 held constant,
r_{12} is the correlation between variables 1 and 2,
r_{13} is the correlation between variables 1 and 3, and
r_{23} is the correlation between variables 2 and 3.

Consider the example at the beginning of the chapter in which a correlation was observed between spice sales and shelf space. The correlation between these two variables, measured over a set of ten stores, was $r = .833$. It was conjectured, though, that this relationship might be spuriously high due to the presence of a confounding variable, namely, the size of the stores. If larger stores—defined either in terms of total floor space or total sales—allocated more space to the spice line, much of the relationship between shelf space and spice sales could be attributed to the size of the store rather than to the amount of shelf space, *per se*.

We can ask, then, what would be the correlation between shelf space and spice sales if all the stores were exactly of the same size. If we had a large enough sample of stores we could presumably measure that correlation by selecting a sample of stores that were identical in size. This is not necessary, though, since the problem has a purely statistical solution. All we need to know is the correlations that exist between each pair of variables as shown below:

- Correlation between spice sales and shelf space: $r_{12} = .833$
- Correlation between spice sales and store size: $r_{13} = .859$
- Correlation between shelf space and store size: $r_{23} = .792$

Substituting into the preceding formula we have

$$r_{12\cdot 3} = \frac{.833 - (.859)(.792)}{\sqrt{(1 - .859^2)(1 - .792^2)}} = .489$$

This partial correlation coefficient of .489 between spice sales and shelf space, when store size is held constant, is considerably lower than the correlation of .833 between those variables when store size is overlooked. Entering Appendix Table VII with $df = n - 3$, one less than we would for a simple bivariate r, we find that a coefficient of .582 is needed to reject the null hypothesis at the $\alpha = .05$ significance level for a one-tailed test. Consequently, we cannot reject the null hypothesis that spice sales are uncorrelated with shelf space, when store size is held constant.

Even after the effect of a confounding variable such as store size has been extracted from the relationship between our two key variables of concern, we still cannot conclude that any residual significant relationship is causal in nature. While we may have statistically controlled for store size, what about the possibility of a third or fourth confounding variable, such as age of store or average neighborhood income. And even if we apply the partial correlation technique to extract their effects, we are still not aware of a multitude of other confounding variables. Again, the only way we will be able to determine if shelf space, in and of itself, influences spice sales is to conduct a controlled experiment in which we randomly assign stores to alternative shelf space values and thereby eliminate the presence of confounding variables.

13. Serial Correlation

If we take a column of numbers and set it beside itself, the correlation between the two sets of numbers would naturally be $r = 1$, since each number is paired with itself. However, we could shift one of the columns upward or downward by one value, and then each value would be paired with its preceding value. We could then correlate these *lagged* values to determine if dependencies exist among the successive values. Such a correlation coefficient is referred to as a *serial correlation*. It is also referred to as an *autocorrelation*, in the sense that it involves the correlation of a variable with itself. It can be calculated for lags of any size, but it must be realized that the effective sample size of paired values will decrease by one for each lag, since when the columns of numbers are shifted the values at the ends of the columns have no paired values. The technique is applicable for any variable with values observed sequentially—e.g., crime rates, unemployment levels, sales, stock prices, etc.

The results of a serial correlation analysis are often portrayed graphically in the form of a correlogram as shown in Figure 11. The value of the

correlation coefficient is shown for lags of varying size. In Figure 11*a* sizeable correlations exist up to a lag of four, which tells us that the values of the variable are dependent on preceding values as far as four values back. Figure 11*b* shows a situation in which negative correlations are uncovered for lags of 3 and 4 values, suggesting the variable might follow a cyclical pattern—i.e., current high values are associated with low values 3 to 4 observations back, and vice versa.

If no serial correlation is found between the lagged values of a variable, then we can be fairly confident that the series of observations are independent of one another, a basic assumption of many time series analyses, as well as in all cases of random sampling. In other instances, the existence of correlations between successive values of a variable help us to predict future values.

14. Canonical Correlation

In the section on multiple correlation, we learned how a correlation coefficient could be obtained between a criterion variable and a weighted combination of predictor variables. There are instances, however, when we might not have just one criterion variable in mind, but several. Would it be possible to correlate *two sets* of variables—a weighted combination of predictor variables with a weighted combination of criterion variables? Such a technique does exist; what the mind can conceive, the mind can accomplish.

Canonical correlation is the name given to the procedure for correlating *two derived* variables, each representing a weighted combination of other variables. If the number of possible weights for the predictor variables in the multiple correlation procedure seemed large, imagine the possibilities when two sets of variables are involved. Again, we are dependent upon the procedures of matrix algebra and the high-speed computer for the solution to this problem.

Table 10 provides an overall outline of the canonical correlation procedure. Scores on a set of criterion variables are weighted, then added together to obtain composite scores comprising a *derived criterion variable*. Similarly, the

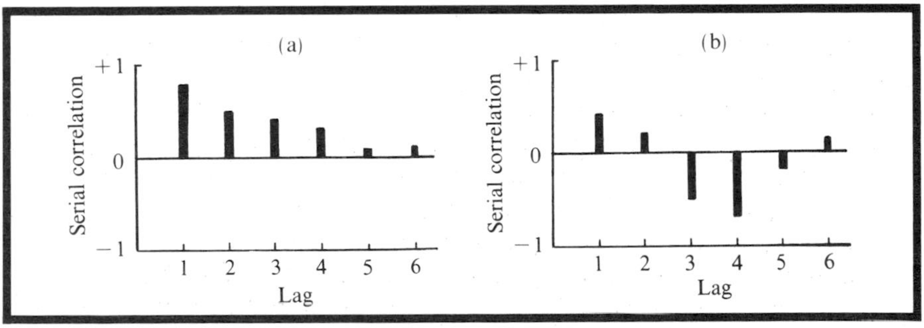

Figure 11 Sample correlograms of the serial correlation coefficient.

scores on a set of predictor variables are weighted, then added together to form the composite scores of a *derived predictor variable*. These two sets of composite scores, the two derived variables, are then correlated with each other just as in the bivariate case. The *canonical* weights, similar to the beta weights of multiple correlation, are derived in such a way that the correlation between the two derived variables is maximized. The square of the resulting canonical correlation coefficient, R_c^2, indicates the proportion of variance of one of the *derived* variables that is associated with the variance of the other *derived* variable. The squares of the canonical weights, as the squares of the beta weights in multiple correlation, tell us the *relative* contributions of the individual variables to the respective *derived* variables. Indeed, we can view multiple correlation as a special case of canonical correlation, a situation in which there is only one variable in one set and a number of variables in the other.

Applications of canonical correlation are most likely to occur in instances in which we are not absolutely sure that a single observed variable can serve as a measure of our criterion dimension of interest. For example, in the measurement of the abstract criterion variable of "job success" there is often no one empirical variable that could adequately measure it. Rather, we might consider

Table 10 An outline of the canonical correlation procedure.

	Variable set I		Composite scores on:		Variable set II	
	Variable R	Variable S	Set I derived variable	Set II derived variable	Variable x	Variable y
Objects	⋯ + Weight × Score	+ Weight × Score =			= Weight × Score	+ Weight × Score + ⋯
1	⋯ + (.19)(−.21)	+ (.46)(1.27) =	.86	.71	= (.39)(.98)	+ (.67)(1.07) + ⋯
2	⋯ + (.19)(.49)	+ (.46)(1.51) =	1.27	1.09	= (.39)(1.21)	+ (.67)(.99) + ⋯
3	⋯ + (.19)(1.12)	+ (.46)(−.53) =	−.11	−.26	= (.39)(−.42)	+ (.67)(−.17) + ⋯
⋮	⋮	⋮	⋮	⋮	⋮	⋮

The canonical correlation coefficient is the correlation between these two sets of composite scores.

The canonical weights are determined in such a way as to maximize the correlation between the resulting sets of composite scores. Once this canonical function is determined, a second function —orthogonal to the first—can be derived from whatever residual variance. Similarly for higher order functions, but the first usually accounts for the bulk of the data's variance.

Note: Values on the respective variables are shown in standardized z score form.

as separate criterion variables, mean supervisor ratings on a set of evaluative variables such as "self-starter," "problem solver," "supervisory skill," etc. A weighted combination of scores on these variables could then serve as the derived criterion variable which could then be correlated with a weighted combination of predictor variables that had been measured earlier; e.g., school grade point average, verbal aptitude test score, mathematical aptitude test score, etc.

As we might suspect, all of the problems of interpretation that multiple correlation analysis is heir to, canonical correlation is doubly so. We need not reiterate the many issues concerning data assumptions, the effects of sample size and the number of variables, interpretations of causality and predictability, etc. It will be worthwhile, though, to consider a unique problem associated with canonical correlation. While the technique helps us avoid depending solely on a single criterion variable, there is the problem that the weights are derived strictly on a mathematical basis, despite the fact that our judgment might tell us that criterion variable A is really more important than criterion variable B or C, which quite possibly might not agree with the results of the canonical analysis. You may object that the same criticism could be leveled at the weights obtained in the multiple correlation procedure, but there is a fundamental difference between the multiple correlation and canonical correlation techniques. In the multiple correlation procedure we have selected a single criterion variable, and for better or for worse it is in our judgment the most valid measure of whatever more elusive dimension we are trying to capture. That being the case we will accept whatever predictor variable weights result from the analysis. In canonical correlation, on the other hand, we have chosen more than one criterion variable, but that is not to say that we feel they are equally important as measures of our "real" criterion dimension. Consequently, we need not accept the resulting canonical weights for the criterion variables, nor the weights for the predictor variables. Instead, we might apply *judgmental* weights to the scores on the respective criterion variables, add them together and create our own derived criterion variable, and then enter it into a multiple correlation analysis.

The extent to which this approach resulted in a multiple R^2 different from the canonical R_c^2, and any accompanying differences in the weights of the various predictor variables obtained under the two approaches, would at the very least shed light on the nature of the real underlying problem; i.e., how do we define a valid measure of our *theoretical* criterion variable.

The above problem is especially manifested in attempts at defining the multivariate trait of "intelligence," which in western society might more appropriately be called "academic aptitude," for the trait is measured on a set of subtests—e.g., short-term memory span, reasoning ability, vocabulary, etc. —for which the composite scores are then correlated with various measures of academic performance, including grade point average and scores on various

achievement tests. Under such a validation system, it is clear that as our school curriculum changes, so must our definition of intelligence, and vice versa.

15. Concluding Comments

We have now been introduced to the third major function of statistical analysis—the identification of associations among the values of two or more variables. At the same time we have seen how the other functions—data reduction and inference—are also relevant to the concept of association.

Correlation analysis, we discovered, is concerned with the associations among random variables, and therefore does not allow causal interpretations for such relationships, playing rather a mainly descriptive and hypothesis generating role. We have also seen how the simple correlation coefficient r between two variables finds application in every facet of higher order multivariate correlation analyses such as multiple correlation, partial correlation, serial correlation, and canonical correlation. The interpretations of the coefficients resulting from these techniques were found to be essentially the same as for the simple correlation between two empirical variables—the main difference being that the interpretation of the coefficients resulting from the multivariate analyses involved a *derived* variable, rather than strictly observational variables. Nonetheless, the more complicated the makeup of the derived variables, the more difficult it becomes to relate the correlation coefficients back to the world which gave rise to the data. In the process, we should have learned that the simplest analyses are often the best analyses; for the further removed we get from our primary data, the greater the chances for statistical and analytical distortions to occur.

At this point it is also worth noting the evolutionary development of our basic statistical concepts. Beginning with a distribution of raw scores, operations were performed to obtain the *mean*. The *mean* was then used to perform further operations on the raw data resulting in the *standard deviation*. The *mean* and *standard deviation* were then used in combination to perform further operations on the data to yield standardized *z scores*. These *z scores* in turn were used to define the *correlation coefficient r*. All simply defined measures but with far-reaching usefulness.

In the following chapters we will study additional techniques for assessing the relationships which occur among variables, and we should not be too surprised if we encounter concepts very similar to those introduced in this chapter.

Chapter 11

Regression Analysis

1. Introduction

Whereas correlation analysis provides us with a summary coefficient of the *extent* of relationship between two variables, *regression analysis* provides us with an equation describing *the nature of the relationship* between two variables. In addition, regression analysis supplies variance measures which allow us to access the accuracy with which the regression equation can predict values on the criterion variable, making it more than just a curve-fitting technique.

While the basic model underlying regression analysis is designed for experimental data in which the levels of the predictor variable are selected or fixed by the investigator, with objects then assigned at random to these levels, the technique can be, and usually is, used to describe the relationship between correlated random variables, where the investigator has no control over the values assumed by the objects on the predictor variable; e.g., the relationship between student grade averages and intelligence test scores, or between the crime rates and unemployment rates of cities, or between crop yields and rainfall levels. There are no severe consequences of this type of application of the basic regression technique, provided the predictor variables are measured with high accuracy.

Examples of the types of *experimental* relationships that can be studied with regression analysis are many. For example, we might want to determine the nature of the relationship between crop yield and levels of fertilizer application, between student test performance and hours of instruction, between disease duration and drug dosage, between blood pressure and controlled dietary salt levels, between product sales and systematically varied advertising levels, between maladaptive behavior frequencies and hours of therapy, between bacteria growth and culture medium concentrations, etc.

While regression analysis can be used with both correlational and experimental data, we will concentrate in this chapter on its application to the former

type, in order to capitalize on the discussions of the preceding chapter. The treatment of experimental data will be addressed in the following chapter on Analysis of Variance, an analytical approach related to regression analysis in its purpose.

2. Overview of Regression Analysis

The recurrent theme of *prediction* appeared throughout our discussion of correlation analysis. While it seemed intuitively clear that the greater the degree of correlation between two variables, the more likely we would be to accurately predict values on one from a knowledge of values on the other, we never learned a specific procedure for accomplishing the prediction. In regression analysis we have such a technique.

The concept of regression analysis—which could well be called prediction analysis—will be easy to understand since much of the spade work has already been done in our study of correlation analysis. Not only will correlation analysis help us in our understanding of regression analysis, but regression analysis will deepen our understanding of correlation analysis.

Just as there is the simple correlation coefficient to measure the degree of relationship between two variables, and the multiple correlation coefficient to measure the degree of relationship between a set of predictor variables and a criterion variable, there is both *simple* and *multiple* regression analysis. In simple regression we are interested in predicting an object's value on a criterion variable, given its value on *one* predictor variable. In the case of multiple regression we are interested in predicting an object's value on a criterion variable when given its value on each of *several* predictor variables. We will begin our study with simple regression, and then discover how we can generalize the concept to the multiple regression situation.

Objectives. The overall objectives of regression analysis can be summarized as follows: (1) to determine whether or not a relationship exists between two variables, (2) to describe the nature of the relationship, should one exist, in the form of a mathematical equation, (3) to assess the degree of accuracy of description or prediction achieved by the regression equation, and (4) in the case of multiple regression, to assess the relative importance of the various predictor variables in their contribution to variation in the criterion variable.

We will touch upon each of these issues, though not in the stated order, since for expositional reasons it is more logical to start with the second named objective—identifying the mathematical equation which describes the relationship between the variables in question.

What might a regression equation look like? Let us begin by looking at the values of several objects on two variables to see if we can gain some insight

into the nature of the desired equation:

	Value on variable x	*Value on variable y*
Object 1	8	31
Object 2	5	22
Object 3	11	40
Object 4	4	19
Object 5	14	49

The objects and variables x and y in this example are left unidentified but could well correspond to any of those named in the introductory section. Before reading on, it will be worthwhile to study the above pairs of values to see if you can identify a systematic relationship between an object's value on variable y and its value on variable x.

It could be noticed, for instance, that the value on variable y *is greater than* the corresponding value on variable x. But we can be more precise. We can say that the value on variable y is *more than triple* the value on variable x. A closer examination will show that the value on variable y is exactly *three times* the value on variable x *plus seven*. In short-hand notation we can specify the relationship as $y = 3x + 7$, which is nothing more than to say that an object's value on variable y is equal to three times its value on variable x plus seven. By convention, though, we often write the equation with the constant term first, $y = 7 + 3x$.

Linear equations. For those who have not already recognized it, based on a study of elementary algebra, $y = 7 + 3x$ is the equation of a specific straight line. Figure 1 presents a refresher course on the characteristics of a straight line, or *linear function*. The equation $y = 7 + 3x$ is but a specific instance of the general equation of a line

$$y = a + bx$$

The value of b is referred to as the *slope* of the line; i.e., its inclination, or the rate of change in the value of the variable y for a unit change in value of the variable x. The higher the value of b, the steeper the slope.

The value of a, or the constant term, represents the value of y when $x = 0$, or in other words the value of y where the line intercepts the y axis. It is, in fact, referred to as the *y-intercept*. In the equation $y = 7 + 3x$, the slope is 3 and the y-intercept is 7. Notice, however, in Figure 1b that the physical appearance of the slope of the line is very arbitrary, depending to a great extent on how stretched out or compressed we make the scale on the y or x axes.

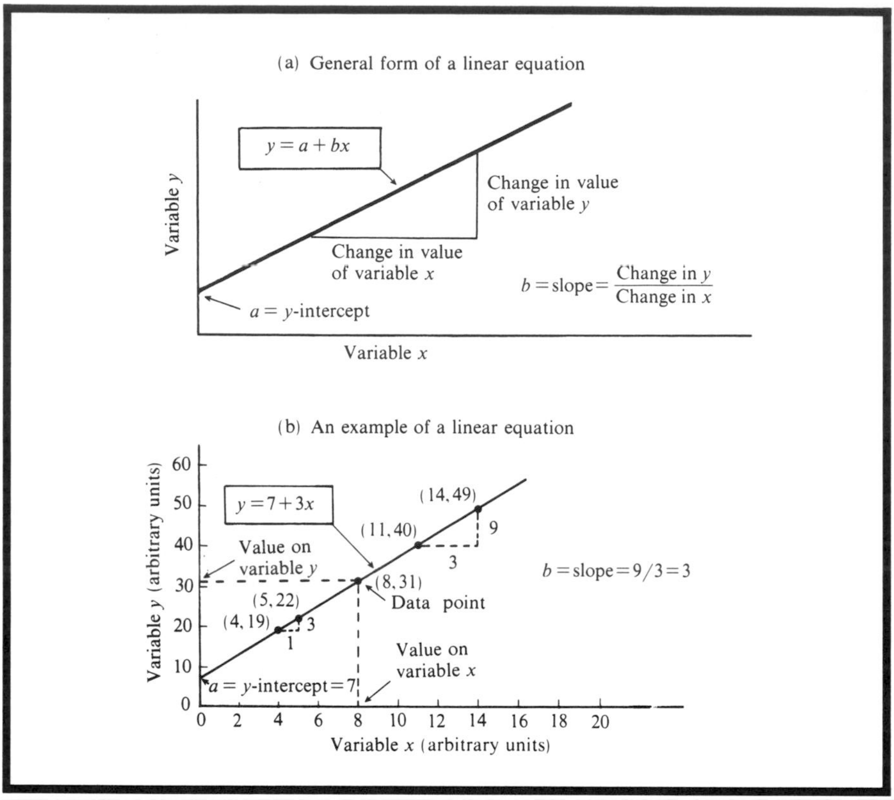

Figure 1 Characteristics of a linear equation.

Returning now to the five pairs of values that were found to be related exactly by the equation $y = 7 + 3x$, note in Figure 1*b* how they lie exactly on the line representing the equation. In real life, however, rarely do we find data that are so perfectly related. More often, we will only be able to say that the value on variable y is *approximately equal* to $7 + 3x$, as illustrated by the pairs of values shown below:

	Value on variable x	*Value on variable y*
Object 6	7	26
Object 7	3	18
Object 8	6	30
Object 9	8	28
Object 10	7	30

As an exercise, plot these five data points onto Figure 1*b* to see how they deviate from the given line; and yet see how the line does represent a fair description of the relationship between the pairs of values.

The importance of the linear equation is that we will be limiting our discussion to data that are linearly related, excluding such non-linear relationships as shown earlier in Figure 5*b* of Chapter 10. But this is not too severe a limitation since many relationships that we encounter are linear in form, and of those that are not, many can be made linear with appropriate data transformations. For example, the values of a variable y may not be linearly related to the values of variable x, but they may turn out to be linearly related to, say, the logarithm of x, or maybe the square root of x, or perhaps the reciprocal of x, or any of a number of other possible transformations.

3. The Regression Line

Given a scatter diagram as shown in Figure 2, how do we go about choosing the "best" of all the possible lines that could pass through the cluster of data points? In other words, compared to the line that is shown, why could we not choose one that was a little steeper or perhaps one not so steep. Or what about one that was positioned a little higher in the cluster of points, or a little lower? There are an infinite number of possible lines, $y = a + bx$, differing in slope b and/or y-intercept a, that could be drawn through the points in Figure 2.

We could say that the line shown passes right through the middle of the cluster of points. But what is the middle? Do we have an objective criterion for determining the middle? Perhaps we should choose the line that passes through the point $(\bar{x}, \bar{y})$ that represents the means of the two variables. This is a good

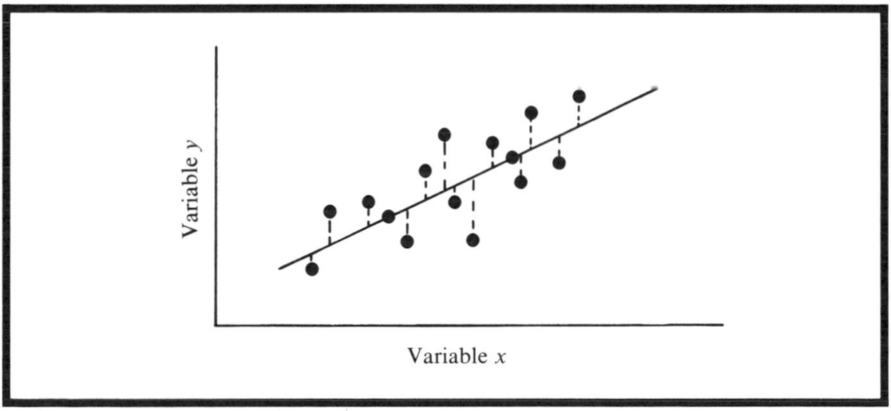

Figure 2 Deviations of data points from a linear function.

start, but there are still an infinite number of lines that can pass through that point, each differing in slope. What would be the best inclination or slope of the line? The fact is, that on the basis of inspection alone, no two individuals are likely to agree on exactly the same line. For this reason we need a well-defined procedure for choosing a "best fitting" regression line

$$\boxed{y' = a + bx} \qquad (1)$$

where y' (read "y prime") represents the predicted value of the criterion variable for a given value of the predictor variable x, and a and b are the y-intercept and slope, respectively, of the line and must be determined from the data. The prime sign ($'$) is used distinguish a predicted value y' from an observed value y.

Least squares criterion. While there are a number of plausible criteria for choosing a best-fitting line, one of the most useful is the *least squares criterion.* In Figure 2 the data points deviate from the given line by varying amounts, the extent of deviation indicated by the dashed lines. With the least squares criterion we choose that particular line, among all possible, that results in the *smallest sum of squared deviations* of the data points from the line. Since the general form of the regression line is $y' = a + bx$, our task is to identify the values of a and b which will minimize $\Sigma(y_i - y_i')^2$, where y_i is an observed value and y_i' is the predicted value.

Slope and y-intercept. It would seem to be a very laborious procedure if we were to determine such a best-fitting line by trial and error alone—positioning and repositioning the line, each time tabulating the squared deviations of the sample data points from the line, until we discovered that line which resulted in the smallest sum of the squared deviations. But as we have seen so often before, the problem of identifying this best-fitting line has a purely mathematical solution. The slope b of the best-fitting line, based on the least squares criterion, can be shown to be

$$\boxed{b = \frac{\Sigma(x_i - \bar{x})(y_i - \bar{y})}{\Sigma(x_i - \bar{x})^2}} \qquad (2)$$

where the summation is over all n pairs of (x_i, y_i) values.

The value of a, the y-intercept, can in turn be shown to be a function of b, $\bar{x}$, and $\bar{y}$; i.e.,

$$\boxed{a = \bar{y} - b\bar{x}} \qquad (3)$$

The derivations of (2) and (3) require calculus techniques and can be found in any advanced mathematical statistics text.

To get a better idea of the numerical calculations involved in determining the values of b and a, we will reintroduce the spice sales vs. shelf space example of the preceding chapter. Table 1 shows the paired values of spice sales and shelf space occupied by the spice line in ten randomly selected stores, as well as the calculations for determining a and b. First we obtain the deviation of each shelf space measure from the mean of that variable; $x_i - \bar{x}$. Then we obtain the deviation of each spice sales measure from the mean of that variable; $y_i - \bar{y}$. Next we take the product of these pairs of deviations; $(x_i - \bar{x})(y_i - \bar{y})$. Finally, we square the deviations of the x_i measures from their mean; $(x_i - \bar{x})^2$. Now, summing the latter two sets of calculations, found in the last two columns of Table 1, across all ten stores, we have

$$b = \frac{\Sigma(x_i - \bar{x})(y_i - \bar{y})}{\Sigma(x_i - \bar{x})^2} = \frac{4{,}490}{33{,}400} = .1344$$

Using this value of the slope b, along with $\bar{x}$ and $\bar{y}$, we can easily determine a, the y-intercept, as

$$a = \bar{y} - b\bar{x} = 70.0 - (.1344)(300.0) = 29.68$$

These calculations, then, demonstrate the applications of the definitional formulas (2) and (3) for the slope and y-intercept, respectively.

Substituting these values of a and b into the general form of the regression line, $y' = a + bx$, we have

$$y' = 29.68 + .1344x$$

as the best-fitting line through the set of ten data points according to the least squares criterion.

The line is shown graphically in Figure 3a, and it does appear to fit the data nicely. The reason the line does not intercept the y-axis at $a = 29.68$, as expected, is due only to the fact that we have curtailed the x and y axes.

If we now wanted to predict the sales of the spice line in a store in which it occupied, say, $x = 250$ in^2 of shelf space, we would simply plug that value into the regression equation $y' = 29.68 + .1344x$ and get

$$y' = 29.68 + .1344(250) = 63.28 \text{ dollars}$$

as our prediction of that store's spice sales. We must be cautioned, though, against applying the equation for values of x which are beyond those used to

Table 1 Sample calculations for obtaining the slope b and intercept a of the best-fitting regression line using the definitional formulas.

x_i Shelf space	y_i Spice sales	$x_i - \bar{x}$	$y_i - \bar{y}$	$(x_i - \bar{x})(y_i - \bar{y})$	$(x_i - \bar{x})^2$
340	71	40	1	40	1600
230	65	−70	−5	350	4900
405	83	105	13	1365	11025
325	74	25	4	100	625
280	67	−20	−3	60	400
195	56	−105	−14	1470	11025
265	57	−35	−13	455	1225
300	78	0	8	0	0
350	84	50	−14	700	2500
310	65	10	−5	−50	100
Sums: 3000	700	0	0	4490	33400
$\bar{x} = 300.0$	$\bar{y} = 70.0$				

$$b = \frac{\Sigma(x_i - \bar{x})(y_i - \bar{y})}{\Sigma(x_i - \bar{x})^2} = \frac{4490}{33400} = .1344$$

$$a = \bar{y} - b\bar{x} = 70.0 - (.1344)(300) = 29.68$$

$$y' = a + bx$$

$$y' = 29.68 + .1344x$$

develop the equation, for the relationship may not be linear for those values of x.

Alternative formulas for b. It is interesting to note that the slope b can also be expressed

$$\boxed{b = r\left(\frac{s_y}{s_x}\right)} \tag{4}$$

where r is the correlation coefficient between the x and y variables, while s_x and s_y are their respective standard deviations.

Yet another equivalent formula for the slope b is given by

$$b = \frac{n\Sigma x_i y_i - \Sigma x_i \Sigma y_i}{n\Sigma x_i^2 - (\Sigma x_i)^2} \tag{5}$$

where the summation is across the $i = 1, 2, \ldots, n$ pairs of observations. The application of this formula to the spice sales example is shown in Table 2, and again we find that $b = .1344$, although by a different route. Computationally, this is sometimes an easier method for calculating b.

Standardized regression equation. It is also interesting to note that the raw score regression equation $y' = a + bx$ simplifies to the standardized form

$$\boxed{z'_y = r z_x} \tag{6}$$

when the x and y variables are expressed as standardized z scores; i.e., with means of zero and standard deviations of one. The value of r is again the correlation coefficient between x and y, and corresponds to the slope of the

Table 2 Sample calculations for obtaining the slope b of the best-fitting regression line using an alternative computational formula.

	x_i Shelf space	y_i Spice sales	x_i^2	$x_i y_i$
	340	71	115,600	24,140
	230	65	52,900	14,950
	405	83	164,025	33,615
	325	74	105,625	24,050
	280	67	78,400	18,760
	195	56	38,025	10,920
	265	57	70,225	15,105
	300	78	90,000	23,400
	350	84	122,500	29,400
	310	65	96,100	20,150
Sums:	3,000	700	933,400	214,490
	$\bar{x} = 300$	$\bar{y} = 70$		

$$b = \frac{n\Sigma x_i y_i - \Sigma x_i \Sigma y_i}{n\Sigma x_i^2 - (\Sigma x_i)^2} = \frac{10(214{,}490) - (3{,}000)(700)}{10(933{,}400) - (3{,}000)^2} = .1344$$

$$a = \bar{y} - b\bar{x} = 70.0 - (.1344)(300) = 29.68$$

$$y' = a + bx$$

$$y' = 29.68 + .1344x$$

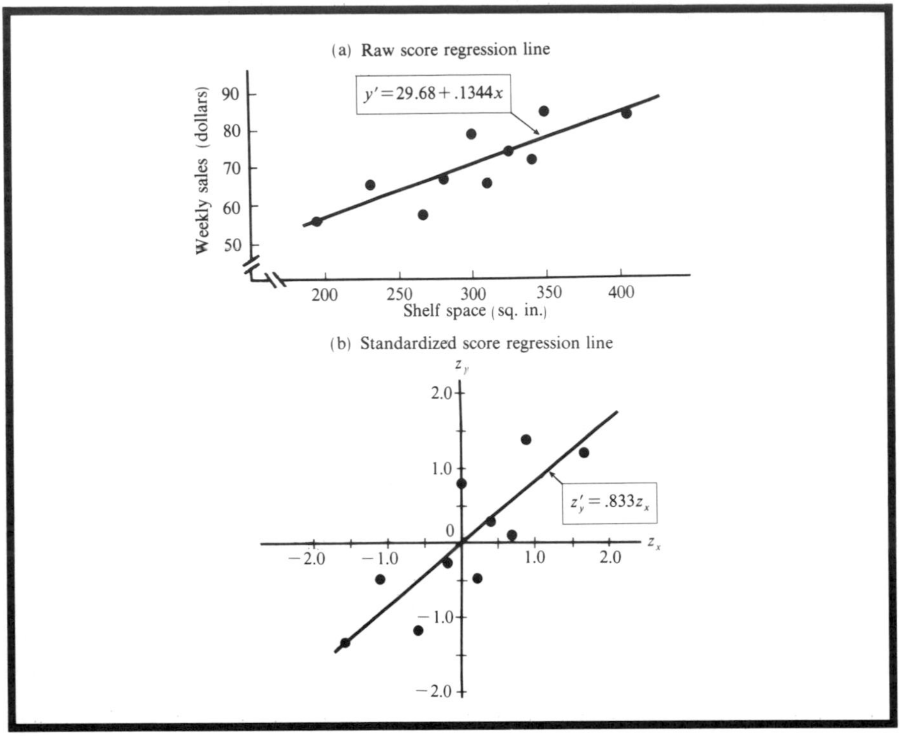

Figure 3 The best-fitting regression lines for the spice sales example.

line relating their standardized scores. There is no intercept term since the equation passes through the origin (0,0) corresponding to the means of the respective z variables.

The standard score form of the regression equation for the spice sales example is portrayed in Figure 3*b*. The data points and the slope $r = .833$ were taken from Table 5 of Chapter 10. The difference in appearance of the slope and dispersion of data points in parts *a* and *b* of Figure 3 is due to the difference in scales for the two graphs; i.e., raw vs. standardized scores.

4. The Regression Model

In the preceding example we mechanically applied the formulas for the slope b and y-intercept a to obtain an equation for the best-fitting line through the given set of data points. However, for the resulting regression equation to be properly interpreted, a number of assumptions must be met concerning the populations of data we are studying.

The essence of the linear regression model is shown graphically in Figure 4. Specifically, the assumptions of the model are as follows:

1. For each value of the predictor variable x, there is a probability distribution of independent values of the criterion variable y. From each of these y distributions, one or more values is sampled at random.

2. The variances of the y distributions are all equal to one another, a condition referred to as *homoscedasticity*.

3. The means of the y distributions fall on the regression line $\mu_y = \alpha + \beta x$; where μ_y is the mean of a y distribution for a given value of the predictor variable x, β (beta) is the slope of the line, and α (alpha) is the y-axis intercept of the line.

What we see, then, is that for any given value of the predictor variable x, the values of the criterion variable y vary randomly about the regression line. Consequently, any individual observation of the criterion variable, y_i, will deviate from the population regression line by a certain amount, call it e_i, where the value of e_i can be either positive or negative, depending upon whether the observation falls above or below the true regression line. Since these e_i's represent deviations from the mean of a y distribution, their average value will be zero.

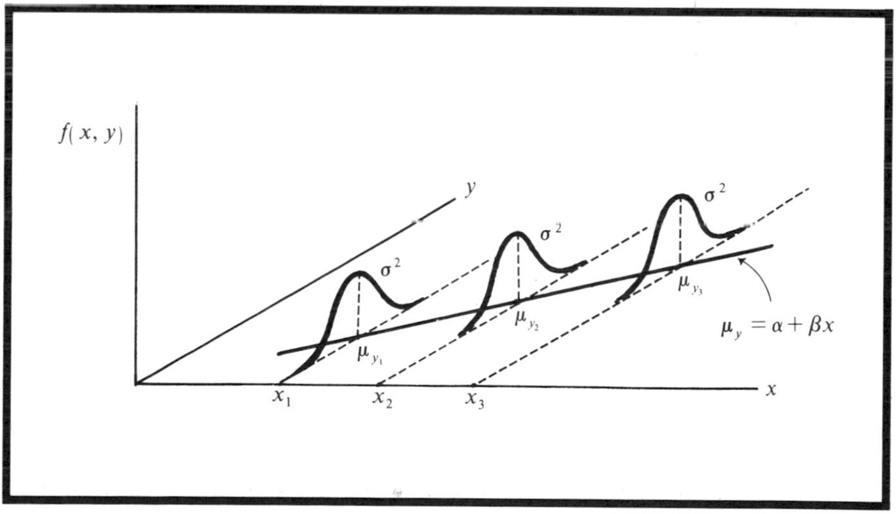

Figure 4 The regression model of independent y populations with equal variances, and with means falling on a straight line.

Based on the above assumptions, we can characterize an individual observation of the criterion variable, y_i, as being equal to

$$y_i = \alpha + \beta x + e_i$$

That is, the observed value, y_i, is the sum of a fixed part dictated by the true regression line, $\alpha + \beta x$, plus a random part, e_i, due to the natural variation of the y values about the regression line. It follows, then, that for any given value of the predictor variable x, the variation of the y_i values is identical to the variation of the e_i's, and it is assumed that this variation is the same regardless of the value of x.

The importance of the e_i's lies in the fact that they represent the primary source of error in trying to predict values of the criterion variable y. In the following section we will learn how to estimate the variance of these deviations from the true regression line.

5. Accuracy of Prediction

If the assumptions of the above regression model are met, we can be assured that the least squares method will yield a sample regression line, $y' = a + bx$, which is an unbiased estimate of the true, but unknown, population regression line, $\mu_y = \alpha + \beta x$. However, the a and b estimates of α and β are subject to sampling error just like any other sample statistics, so they will be sources of error in trying to predict the criterion variable value from a given value of the predictor variable. But, by far the one greatest source of error in attempting to predict individual values of the criterion variable y does not lie in the errors of estimating the slope and y-axis intercept of the regression line, but in the random variation of the y_i's about the regression line—the e_i's of the preceding section.

Standard error of estimate. The variation of the y_i values about the population regression line can be estimated by assessing their variation about the sample regression line. The standard deviation of the observed y_i values about the predicted values y_i' is referred to as the *standard error of estimate*, designated $s_{y \cdot x}$, and is given by the formula

$$s_{y \cdot x} = \sqrt{\frac{\Sigma(y_i - y_i')^2}{n-2}} \qquad (7)$$

where the summation is across the $i = 1, 2, \ldots, n$ sample observations. The reason we divide by $n-2$, instead of $n-1$ as was customary with other sample

Table 3 Calculations for obtaining the standard error of estimate.

x_i Shelf space	y_i Spice sales	$y_i' = 29.68 + .1344x_i$	$y_i - y_i'$	$(y_i - y_i')^2$
340	71	75.38	−4.38	19.184
230	65	60.59	4.41	19.448
405	83	84.11	−1.11	1.232
325	74	73.36	.64	.410
280	67	67.31	−.31	.096
195	56	55.89	.11	.012
265	57	65.30	−8.30	68.890
300	78	70.00	8.00	64.000
350	84	76.72	7.28	52.998
310	65	71.34	−6.34	40.196
$\bar{x} = 300$	$\bar{y} = 70$	$\bar{y}' = 70$	0.00	266.466
$s_x = 60.92$	$s_y = 9.83$	$s_{y'} = 8.19$		

$$s_{y \cdot x} = \sqrt{\frac{\Sigma(y_i - y')^2}{n-2}} = \sqrt{\frac{266.466}{8}} = 5.77$$

standard deviations, is due to the fact that there are two constraints on the data—the slope and y-axis intercept which were used to obtain the predicted values y_i'.

Although $s_{y \cdot x}$ is referred to as the standard error of estimate, it is not a standard error in our conventional use of the term as a measure of the standard deviation of the sampling distribution of a statistic. Rather, it is an estimate (when squared) of the variance of the y populations about the true regression line, as shown in Figure 4. It might, more appropriately, be called the "standard deviation about regression." In this terminology the subscript notation of $s_{y \cdot x}$ is also more meaningful, signifying the standard deviation of y for a given x.

Table 3 shows the calculations involved in determining the value of $s_{y \cdot x}$ for our spice sales vs. shelf space example. First we substitute each value of x into the sample regression equation $y' = 29.68 + .1344x$ to arrive at the y' estimates. The differences between the observed and predicted values, $y_i - y_i'$, are then squared and summed across the $n = 10$ stores, and finally divided by $n - 2$ before taking the square root; specifically,

$$s_{y \cdot x} = \sqrt{\frac{266.466}{8}} = 5.77$$

This, then, is our *estimate* of the variation of the y populations about the true regression line.

Confidence bands. At first glance we might think that $s_{y \cdot x}$ could be used to create a confidence band about the sample regression line, reflecting the maximum expected error in predicting y from x, with a given probability, say .95 or .99, similar to other confidence intervals based on sample statistics. But this is not so, since the errors in predicting y are not only due to $s_{y \cdot x}$, which estimates the random variation of y about the true regression line, but there are also two other sources of error: (1) the error in estimating the overall elevation or y-axis intercept of the true regression line, α, and (2) the error in estimating the slope β of the true regression line.

Furthermore, the error due the second of the above two factors—the error in estimating the slope—becomes more pronounced the more the predictor value x_i deviates from the average x value under study. The consequence of this ever-increasing error the further we move from the average value of the predictor value, is a *bowed* confidence band about the sample regression line. This can be seen most easily from a study of Figure 5. In part *a* of the figure, a sample regression line is superimposed upon the true but unknown regression line. Imagine, instead, that an infinite number of such sample regression lines

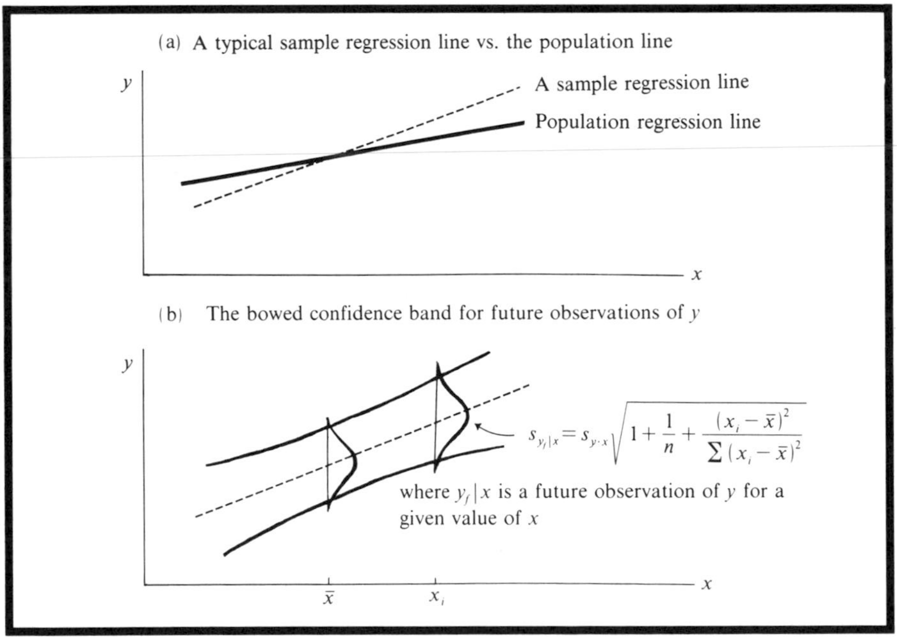

Figure 5 Errors of prediction resulting from a sample regression equation (See text).

were in the figure. Each would vary in slope due to sampling error, but the net effect would be the same: The further we move from the mean of the x variable, the larger the discrepancy between the sample and true regression line.

Figure 5*b* shows the resulting confidence band with its bowed feature. Its width, measured vertically at any given value of the predictor variable x, is a function of the three sources of error outlined above: the natural variation of y about the true regression line, $s_{y \cdot x}$; the error in estimating the y-axis intercept, α; and the error in estimating the slope β of the line. We have studied the first component, $s_{y \cdot x}$, in some detail, but to adequately probe the formulas and interpretations of the other two error sources would require extensive discussion, and is best left for advanced study. However, for reference purposes, the relevant expression appears in Figure 5*b*.

Although $s_{y \cdot x}$ alone is not sufficient to precisely estimate the expected magnitudes of our prediction errors, it is fortunate that as n becomes large, say greater than 100, it can provide approximate confidence intervals, since the errors in the estimation of α and β become small relative to $s_{y \cdot x}$. This will be apparent from a study of the formula in Figure 5*b*: As n becomes large, the latter two terms become negligible. Under these large sample conditions, we can then expect that approximately 95% of our prediction errors are within $\pm 1.96 s_{y \cdot x}$ of the sample regression line, and that approximately 99% are within $\pm 2.58 s_{y \cdot x}$ of it, provided, of course, we make the further assumption that the y populations for each predictor value x are normally distributed in addition to having equal variance.

Reduction of prediction errors. An understanding of how well the regression equation predicts the criterion variable y—compared to simply predicting its overall mean value $\bar{y}$, regardless of the value of x—can be had by studying the graph of the equation among the data points of a correlational scatter diagram.

Parts *a* through *d* of Figure 6 show progressively higher degrees of correlation between two hypothetical variables. Beginning with a zero correlation, we see that the variation in prediction errors (indicated by the bold arrows) is exactly equal to the variation of the criterion variable itself (indicated by the double arrow).

This state of affairs can be contrasted with the situation portrayed in parts *b*, *c*, and *d* of Figure 6, where we see that with increasing degrees of correlation the deviations of the observed scores from the regression line get smaller and smaller. That is, the errors of prediction are reduced as the degree of correlation between the variables increases. The limiting situation, of course, would be the case of a perfect correlation in which all the observed points would lie right on the regression line and consequently there would be no errors of prediction.

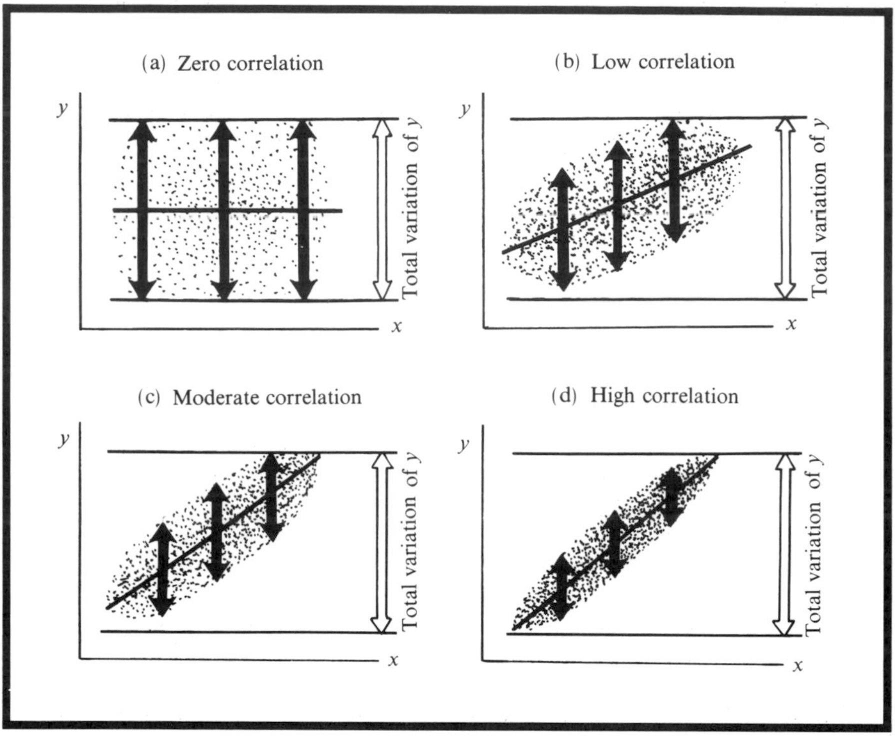

Figure 6 The reduction of prediction errors with increasing correlation.

If it is recognized that the bold arrows in Figure 6 reflect the variance of the y values about the regression line, $s^2_{y \cdot x}$, and that the double arrows reflect the overall variance of the y values, s^2_y, then the preceding relationship can be summarized concisely as

$$\frac{s^2_{y \cdot x}}{s^2_y} \doteq 1 - r^2$$

where $\doteq$ is the symbol for "is approximately equal to" and r^2 is the square of the correlation coefficient between variables x and y. The relationship would be exact were it not for the fact that $n-1$ is used in the definition of s_y, whereas $n-2$ is used in the definition of $s_{y \cdot x}$.

The spice sales vs. shelf space example will illustrate the above relationship. Recalling that $r = .833$, and obtaining the needed variance estimates from

Table 3, we have

$$\frac{(5.77)^2}{(9.83)^2} \doteq 1-(.833)^2$$

or

$$.345 \doteq .306$$

In situations where n is large the approximation will be much closer. In this example the various statistics were based on a sample of only $n=10$. Notice that if we multiply the left side of the equation by $(n-2)/(n-1)$—i.e., 8/9—we have an exact relationship $(8/9)(.345)=.306$. Thus, we conclude that the knowledge of the regression equation between x and y has reduced the variance of our errors of prediction to just over 30% of what it would be if we simply predicted the average value of y all the time, regardless of the value of x, or if we did not know the value of x. This is also one context in which meaning is given to the interpretation of r^2 as a measure of the proportion of variance in one variable accounted for by variation in the other.

Proportion of variance explained. Another interpretation of the efficacy of a regression equation, and its relation to r^2 as a measure of the amount of variance in one variable accounted for by the variance in another variable, can be seen graphically in Figure 7. For clarity of illustration only a few data points are shown, rather than the swarm of data used in Figure 6.

For each observed data point in the figure there is a corresponding predicted point. The observed data points are shown as open circles, while the predicted points as solid circles. Notice that if we project the observed data points and the corresponding predicted points against the y axis, we can compare their respective variations. In part *a* of Figure 7, portraying a high degree of correlation, the variation of the predicted y values is almost the same as the variation of the observed y values. In other words, nearly all the variation in the y variable is accounted for, or predictable by, the variation in the x variable which gave rise to the predicted scores through the regression equation. We see further in parts *b*, *c*, and *d* of the figure that the variance of the predicted scores compared to the variance of the observed scores gets smaller and smaller as the degree of correlation decreases, until we reach the limiting case of zero correlation, in which case none of the variance in the y variable is predictable from the variance in the x variable, because there is absolutely no variation in the predicted y values; for when there is a zero correlation between two variables, the best we can do in terms of prediction is to predict the mean value of the criterion variable regardless of the value of the

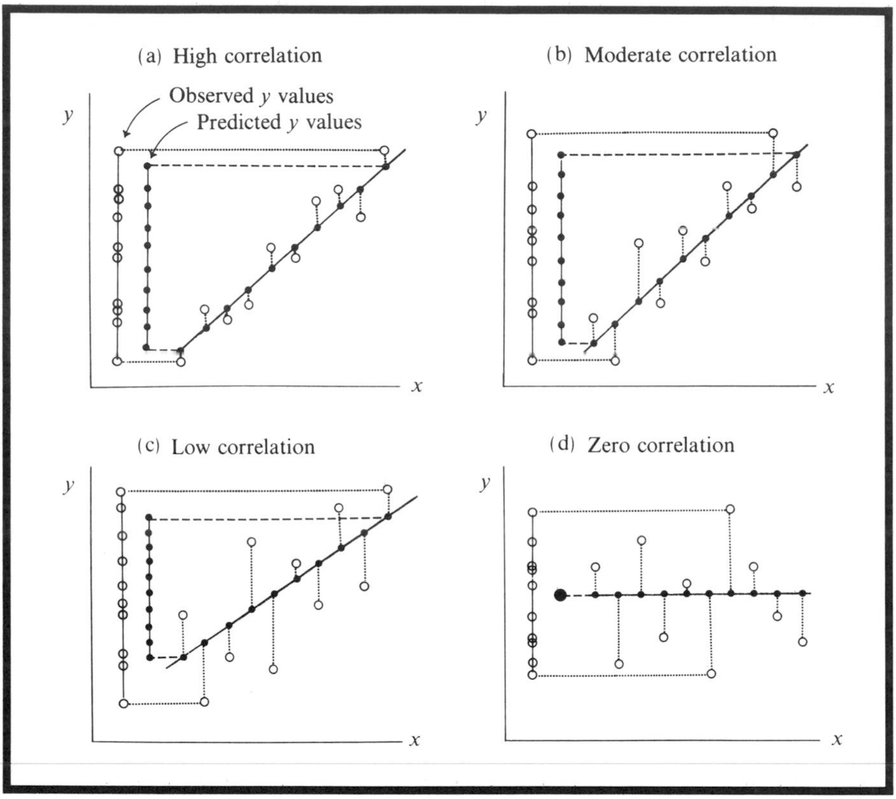

Figure 7 Variation in the criterion variable accounted for by variation in the predictor variable.

predictor variable. This is the situation of a horizontal regression line, one with a slope of zero.

The relationship described above can be described in mathematical terms. If we calculated the variance of the observed and predicted y values, the ratio of the latter to the former would be nothing other than the value of r^2. That is,

$$\frac{s_{y'}^2}{s_y^2} = r^2$$

where the numerator of the ratio is the variance of the *predicted* y values, while the denominator is the variance of *observed* y values.

Again, we can illustrate the above relationship using the spice sales vs. shelf space data from Table 3. Substituting the appropriate values, we can

confirm that

$$\frac{(8.19)^2}{(9.83)^2} = (.833)^2$$

does in fact satisfy the equality.

We see, then, that r^2 represents the proportion of variance in the criterion variable accounted for by variance in the predictor variable which gives rise to the predicted y' values via the regression equation. In the above example $r^2 = .69$, signifying that 69% of the variation in spice sales is accounted for by variation in shelf space occupied by the product.

It should be noted that if the data were experimental rather than correlational in nature, then the value of r^2 is no longer the square of the "correlation coefficient"—i.e., an estimate of the population ρ^2—but only resembles it in its mathematical definition. In such instances, r^2 is often referred to as the *coefficient of determination*, and its value will be influenced by the particular values of the predictor variable chosen by us for study. Remember that in the case of correlational data, we have no influence over the values of the predictor variable, while in the case of experimental data, we choose the values of x. An example of the latter situation would be if we purposefully and randomly varied the amount of shelf space occupied by the spices in a random sample of ten stores, rather than simply observing how the two variables covaried in a natural setting.

Figures 6 and 7 should be studied to see the complementary relationship between these two interpretations of r^2 vis-a-vis the regression line and the predicted scores. In one case, as the correlation increases, the errors of prediction decrease. Alternatively, as the correlation increases, we are able to account for more of the variation in the criterion variable with values predicted from the regression equation. It should be clear, then, that the task of regression analysis does not end with the development of the regression equation, but further involves an assessment of the accuracy with which the relationship is described.

6. Significance Test of the Slope

Although its discussion has been reserved until now, one of the first things we want to do upon obtaining the sample regression equation is to test its slope b.

If there is no relationship between the variables x and y, then the slope of the regression equation would be expected to be zero. To test the hypothesis H_0: $\beta = 0$, we need to know the standard error of the sample slope b, which is

the estimate of β, and it is given by

$$s_b = \frac{s_{y \cdot x}}{\sqrt{\Sigma(x_i - \bar{x})^2}} \tag{8}$$

where $s_{y \cdot x}$, the standard error of estimate, is given by formula (7).

Choosing a significance level α beforehand, we can then test the null hypothesis with the t variable

$$t = \frac{b - \beta}{s_b}$$

which is distributed with degrees of freedom $df = n - 2$.

For the spice sales vs. shelf space example the standard error of the slope b is given by

$$s_b = \frac{s_{y \cdot x}}{\sqrt{\Sigma(x_i - \bar{x})^2}} = \frac{5.77}{\sqrt{33{,}400}} = .032$$

where the denominator and numerator values were obtained from Tables 1 and 3, respectively.

The significance test of the sample slope $b = .1344$ is then

$$t = \frac{b - \beta}{s_b} = \frac{.1344 - 0}{.032} = 4.20$$

which with degrees of freedom $df = 8$ is well beyond the critical value of $t = 1.86$ needed to reject the null hypothesis at the $\alpha = .05$ significance level using a one-tailed test. Consequently, we can be confident that the observed linear equation was not simply a chance departure from a horizontal line, the situation when there is no relationship between the two variables. To use the t test, however, we must make the more stringent assumption that the y populations not only have equal variances, but are also normal in form.

7. Analysis of Residual Errors

If violations of the above assumptions of the regression model are not evident from a knowledge of the data source or from an inspection of the plot of the y values against the x values, then a graph of the prediction or *residual errors*, $y - y'$, will help to point out possible deviations from the assumptions.

Figure 8 presents four examples of such residual plots. In part *a* of the figure the residual errors are evenly distributed and not related to the value of

x, suggesting that the relationship between y and x is indeed linear, as required, and that the variance of y for each value of x is the same, as required by the homoscedasticity assumption.

In part b of Figure 8 the residual errors increase in variance as x increases, suggesting that the homoscedasticity assumption has been violated by the data.

Figure 8c shows a curvilinear pattern for the residual errors, reflecting a curvilinear relationship between the x and y variables themselves, invalidating the linear regression model.

Part d of Figure 8 shows the residual prediction errors increasing as x increases and also becoming more dispersed. Such a pattern indicates a violation of either the linearity or homoscedasticity assumptions, or quite possibly both.

This type of residual analysis, along with an inspection of the graph of the original data, will prevent to a large extent the misapplication of the linear regression model and help us to avoid incorrect conclusions based on a purely mechanical application of the technique to a body of data.

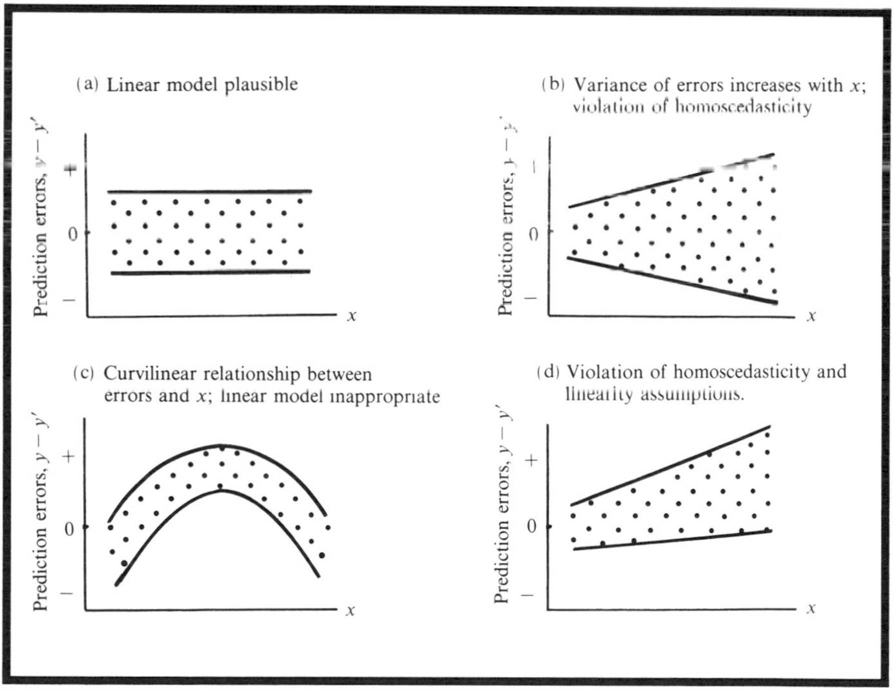

Figure 8 Examples of residual prediction error plots and likely interpretations.

As for the assumption of independent y distributions for the various values of x, it is best verified from a logical analysis of the data source. If the same or related objects contribute to more than one data point—either within or between the y distributions—then the observations, and consequently the prediction errors, are not likely to be statistically independent.

8. Multiple Regression

Multiple regression is an extension of the concept of simple regression. Rather than using values on one predictor variable to estimate values on a criterion variable, we use values on *several* predictor variables. In using many predictor variables instead of just one, our aim is to reduce even further our errors of prediction; or, equivalently, to account for more of the variance of the criterion variable.

The input data for a multiple regression analysis is similar to that for a multiple correlation analysis; namely, a random sample of objects measured on some criterion variable of interest, as well as on k predictor variables. While the multiple correlation analysis requires that the predictor variables are random variables—as opposed to being determined by the researcher—there are multiple regression models to cover both types of situations.

An example of the type of problem to which multiple regression analysis lends itself would be the prediction of college grade point average based on predictor variables such as high school grade point average, aptitude test scores, household income, scores on various entrance exams, etc. A number of other examples of the application of multiple regression analysis will be introduced at the end of the chapter, after the basics of the technique have been studied.

Multiple regression equation. The multiple regression equation will be recognized as similar to the simple regression equation, but instead of a single predictor variable x we have several predictor variables $x_1, x_2, \ldots, x_k$. The general form of the equation is

$$\boxed{y' = a + b_1x_1 + b_2x_2 + \cdots + b_kx_k} \tag{9}$$

where y' is the predicted value of the criterion variable and the values of a and the b coefficients must be determined from the sample data. Since it is based on sample observations, equation (9) must be thought of as an estimate of the true but unknown population equation

$$\mu_y = \alpha + \beta_1x_1 + \beta_2x_2 + \cdots + \beta_kx_k \tag{10}$$

Equations (9) and (10) do not represent straight lines as in the case of simple regression where we have only one predictor variable, but rather represent *planes* in multi-dimensional space, a concept admittedly difficult to conceive and virtually impossible to portray graphically. However, its application is easy enough.

As in simple regression, the least squares solution is used to determine the best multiple regression equation; i.e., the values of $a, b_1, b_2, \ldots, b_k$ that will yield values of y' such that the sum of the squared deviations of the predicted y' values from the actual observed y values—$\Sigma(y - y')^2$—is at a minimum. Alternatively, we can think of the least squares solution as that *weighted sum of values on the various predictor variables* that correlates most highly with the values on the criterion variable. For example, if the least squares criterion yielded the following equation for a three-predictor variable problem

$$y' = 24.3 + 7.1x_1 + 6.2x_2 + 91.5x_3$$

we would know that no other equation would yield predictions y' which would correlate more highly with the observed values of y; or, equivalently, no other equation would result in a smaller value of $\Sigma(y - y')^2$, the sum of the squared differences between the actual and predicted values of y.

Regression coefficients. The values of $b_1, b_2, \ldots, b_k$ in the regression equation $y' = a + b_1x_1 + b_2x_2 + \cdots + b_kx_k$ are alternatively referred to as *b coefficients* or as *regression coefficients*. In the following section, we will get a better idea of the interpretation, and limits on the interpretation, of the *b* coefficients; where they will be contrasted with *beta* coefficients based on the regression equation in standardized *z* score form.

9. Importance of the Predictor Variables

The multiple correlation coefficient R tells us the correlation between the weighted sum of the predictor variables and the criterion variable. Consequently, the squared multiple correlation coefficient R^2 tells us what proportion of the variance of the criterion variable is accounted for by *all the predictor variables combined*.

Still, it would also be worthwhile to know how much each of the *individual* predictor variables contributes to the total explained variance, or, alternatively, to the total reduction in prediction errors. For example, a multiple R^2 of .70 signifies that 70% of the variance of the criterion variable is accounted for, or predictable by, a given set of predictor variables. If, say, there were five predictor variables in this particular situation, would we be able to determine

how much of that 70% could be attributed to each of the five predictor variables?

The fact is, there is no satisfactory method for determining the *absolute* contributions of individual predictor variables to their combined effect in accounting for the variance of a criterion variable, when we are dealing with correlational rather than experimental data. The problem lies in the fact that the predictor variables are usually correlated among themselves. We could sooner unscramble an omelette.

If the predictor variables were *uncorrelated* with each other, the problem would be simple. We would merely take the square of the correlation coefficient of a predictor variable with the criterion variable, r^2, as the measure of that predictor variable's contribution to the multiple R^2. In this situation of *independent* predictor variables, the r^2 of the individual predictor variables with the criterion variable will sum to R^2, and then it would be simple arithmetic to determine their percentage contribution to the sum.

When the predictor variables are correlated among themselves, however, the sum of the individual r^2 will be greater than R^2, since most of the predictor variables are *duplicating* the predictive power contained in another predictor variable. In other words, much of the explained variance of the criterion variable would be counted more than once.

What about the possibility of *eliminating* a particular predictor variable from our regression analysis and observe the extent to which R^2 drops in value. While this procedure seems appealing on the surface, it will not accomplish our purpose. If we added the decrements in the value of R^2 resulting from the withdrawal of each predictor variable, the sum would again exceed R^2. And for the same reason as before. Each time we remove a predictor variable we are removing some predictive ability that is in common with other predictor variables, and consequently we end up tabulating it more than once. This procedure, furthermore, can result in the gross misinterpretation of the predictive capacity of a variable. Imagine, for example, that the removal of a particular predictor variable results in a negligible decrease in the value of R^2. Are we to conclude that this variable is unrelated to the criterion variable? Not necessarily. It may be highly correlated with another predictor variable, and in that sense was superfluous for the analysis, but in the absence of that other variable may well have resulted in a substantial drop in the value of R^2.

Beta coefficients. About the best we can do in assessing the relative importance of the various predictor variables is to look at their coefficients in the multiple regression equation when all variables are in their *standardized z score* form; i.e., each with a mean of zero and a standard deviation of one. The coefficients of the standardized predictor variables are referred to as *beta*

coefficients or *beta weights*, and the general form of such a prediction equation can be written

$$\boxed{z'_y = beta_1 z_1 + beta_2 z_2 + \cdots + beta_k z_k} \tag{11}$$

where z'_y is the predicted standardized score on the criterion variable. The beta's are spelled out in (11) so as not to confuse them with the β's in equation (10) where there they refer to the theoretical parameters of the population equation in raw score form. The beta's in equation (11) on the other hand, are actually empirical "beta estimates" of the corresponding coefficients of the population equation in standardized z score form. Nonetheless, through common usage they have come to be called simply beta coefficients or beta weights, and are the same as those discussed in conjunction with multiple correlation in the preceding chapter.

The beta coefficients are also sometimes referred to as *partial regression coefficients*. The term "partial" derives from the fact that these regression coefficients are related to the partial correlation coefficients (see Chapter 10) between the respective predictor variables and the criterion variable. That is, the value of the coefficient of each predictor variable x is a function of the correlation between that predictor variable and the criterion variable *as well as the correlations that exist among the predictor variables themselves*. As we have seen in our study of the partial correlation coefficient, it expresses the correlation between two variables under the condition that all other concommitantly measured variables are held constant; that is, it statistically extracts the effects of other variables which correlate with the two variables with which we are concerned—the criterion variable and a given predictor variable.

Since each variable in the standardized form of the multiple regression equation (11) has exactly the same standard deviation and mean, the absolute values of the beta coefficients will tell us the *rank order* of importance of the predictor variables. For example, in the equation

$$z'_y = .44z_1 + .09z_2 + .27z_3$$

predictor variable z_1, with a beta coefficient of .44, has a more important contribution to the criterion variable than the other two predictor variables. The *relative* importance of any two predictor variables can be obtained by taking the ratio of the squares of their respective beta's. For example, $.44^2 = .194$ for predictor z_1, versus $.27^2 = .073$ for predictor z_3, tells us that the first predictor variable accounts for over two and a half times as much of the variance in the criterion variable as does the third predictor variable—i.e.,

$.194 \div .073 = 2.66$. Had the prediction equation been formed in the raw score form—i.e., with the actual observed values of y and the predictor variables x_1, x_2, and x_3—we might have found it to be

$$y' = 15.2x_1 + 17.7x_2 + 46.9x_3$$

a form in which the b coefficients 15.2, 17.7, and 46.9 tell us *nothing* about the relative contributions of the three predictor variables to the criterion variable. And that is because the means and standard deviations of the predictor variables have not been taken into account. Hence, the raw score form of the regression equation is fine for predicting actual values of the criterion variable y, but the beta coefficients from the standardized score form of the equation are needed to interpret the relative importance of the various predictor variables.

It must be stressed that the beta regression coefficients from the standardized form of the prediction equation can inform us only of the *relative* importance of the various predictor variables, not the *absolute* contributions, since there is still the joint contributions of two or more variables taken together that cannot be disentangled. Also, we must be cautioned that the relative importance of any two predictor variables is dependent upon *which other predictor variables have been included in the analysis.* If we had included additional predictor variables, or fewer, the relative importance of two variables vis-a-vis one another could well be quite different.

Despite these difficulties of interpreting the contributions of individual variables, we can often be content with the knowledge that the variables taken as a group account for such and such of the total variance in the criterion variable, as revealed by R^2. Also, it is often enough to know the *rank-ordering* of the predictor variables in terms of their efficacy in accounting for the variance of the criterion variable. The beta regression coefficients can provide us this information at a glance.

10. Selection of Predictor Variables

For reasons of simplicity of explanation, or cost efficiency, we may want to construct a regression equation with as few predictor variables as possible. While we would begin our analysis with a large set of variables we would like to eliminate those which accounted for only a trivial amount of the variation in the criterion variable.

Stepwise procedures. One method of choosing a smaller set of predictor variables from among a larger set is the stepwise procedure described in our discussion of multiple correlation in Chapter 10. In the step-up procedure we begin with the predictor variable accounting for the most variance in the

criterion variable, and then, one at a time, add the variables which account for the most of the remaining or *residual* unexplained variance. We continue introducing predictor variables until the resulting increase in R^2 becomes insignificant.

A general idea of how this stepwise procedure is accomplished can be found in Figure 9. We begin with the predictor variable most highly correlated with the criterion variable. The errors of prediction resulting from that regression equation are then correlated with the values of each of the remaining predictor variables to identify the one which accounts for the most of this unexplained *residual* variance. Then, the errors of prediction resulting from the regression equation incorporating both predictor variables are correlated with the values of the remaining predictor variables to identify the one that can account for most of this residual variance. This stepwise procedure is stopped at the point where the introduction of another variable would account for only a trivial or statistically insignificant portion of the unexplained variance.

This step-up or *forward addition* procedure can be contrasted with the step-down or *backward elimination* procedure in which we begin with *all* of the predictor variables and eliminate one by one, always the least predictive, until we reach the point where the elimination of another would sacrifice a significant amount of explained variance in the criterion variable. It should be noted that the step-up and step-down procedures will not necessarily result in the same regression equation. It is possible for each to result in the same R^2 but with completely different sets of variables eliminated from the analysis.

All regressions. An alternative method of identifying a concise regression equation is to consider *every possible* regression equation that could be constructed from our set of predictor variables. For this *all-regressions* approach it can be shown that if we have k predictor variables there are 2^k possible regression equations. For example, assume we had five predictor

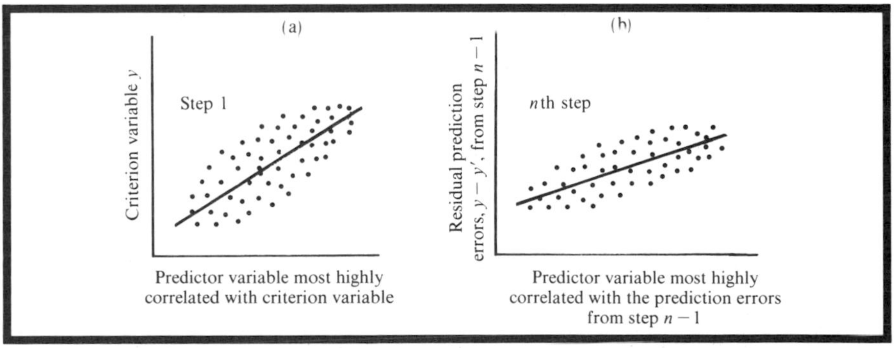

Figure 9 The selection of predictor variables in the stepwise regression procedure.

variables: The first one *could* or *could not* appear in the equation, the second one *could* or *could not* appear in the equation, etc. The two possibilities for each of the five variables results in $2\times2\times2\times2\times2$ or $2^5 = 32$ possibilities.

Modern computers are so fast that every one of the possible regression equations can be computed within seconds, provided the number of variables is not too large. Rather than looking at every single equation though, which would number over 1,000 with as few as ten variables, we could ask the computer center statistician to provide us with the "best" equation when using *one* variable, when using *two* variables, when using *three* variables, and so on. By "best" we mean the ones associated with the largest R^2's.

The stepwise and all-regressions procedures allow us to identify a regression equation based on relatively few predictor variables, yet which accounts for virtually all the variance that could be explained if we used the entire set of predictor variables. This is desirable from the standpoint of parsimony of explanation and economy of data collection.

On the other hand, there is the danger that we might select variables for inclusion in the regression equation based purely on chance relationships. Therefore, as stressed in our discussion of multiple correlation, we should apply our chosen regression equation to a fresh sample of objects to see how well it does in fact predict values on the criterion variable. This validation procedure is absolutely essential if we are to have any faith at all in future applications of the regression equation.

11. Applications of Regression Analysis

As stressed throughout the chapter, two key benefits to be derived from the application of regression analysis include (1) the prediction of values on a criterion variable based on a knowledge of values on predictor variables, and (2) the assessment of the relative degree to which each predictor variable accounts for variance in the criterion variable.

In terms of specific applications of the technique, the possibilities are near limitless. In business and economics there is interest in identifying predictors of sales, productivity, unemployment rates, inflation rates, strike activity, etc. Researchers in education and psychology are interested in predictors of academic achievement, career success, aptitudes, personality traits, mental health, etc. Sociologists, psychologists, and anthropologists are interested in predictors of crime, marriage, divorce, and birth rates. Predictors of crop yield, animal behaviors, disease durations, and bodily functions such as blood pressure or skin temperature are of interest to researchers in biology and medicine. These are just a few of the criterion variables which might be studied in various fields, and many more could be identified, each with a long list of potential predictor variables.

What is important to realize is that alternative regression analyses can be applied to the same basic analytical problem, depending upon the objects we choose to study. Consider, for example, in the world of business, which perhaps has the widest range of objects for study, an automobile manufacturer interested in the criterion variable "sales of Model *A*." Now this is a very broadly stated problem and needs sharper definition. To be more specific, the manufacturer is interested in knowing which variables are related to the sales of Model *A*, information which can then be used to possibly increase its sales. While we recognize that on the surface this problem lends itself to a regression analysis, we must formulate the problem more specifically in order to apply the analytical technique. It is at this point that the ingenuity of the investigators comes into play.

A number of options are available to the planners of the study. The criterion variable has been identified as the sales of Model *A*, so the next task is to identify a set of *objects* on which to measure this variable. We might consider the individual *dealerships*, since they surely vary with respect to the criterion variable. Next, we need to identify a set of potential predictor variables, characteristics of the dealerships that might be related to our criterion variable of Model *A* sales.

A management, sales, and research team will brainstorm to come up with a comprehensive set of potential predictor variables: e.g., population density in a fifteen-mile radius, distance to nearest dealer of competitor *B*, local advertising expenditure, number of feet of street frontage, average gasoline price in a fifty-mile radius, number of sales personnel, number of service personnel, number of years at existing site, average trade-in allowance, number of autos on hand, etc.

As can be easily seen, the list could go on and on. Notice, also, that the nature of the objects chosen for the analysis, dealerships, more or less dictates the nature of the predictor variables. Note further that some of the predictor variables are under the manufacturer's control (e.g., number of sales personnel, trade-in allowance, number of autos on hand, etc.) while others are not (e.g., average gasoline price, years at existing site, population density, etc.). However, even in those instances in which the manufacturer has no direct control over the predictor variable, the regression analysis could still be beneficial in selecting future dealership sites. In the case of the predictor variables which are under the manufacturer's control, they can later be manipulated on a test basis to determine if they do indeed *cause* changes in sales.

The auto marketer can attack the very same problem from a completely different angle. With sales of Model *A* still the criterion variable, we may choose as objects not individual dealerships, but individual *sales persons*. The predictor variables could then include dimensions such as years of experience, age, scores on a personnel test, grade in a training course, height to weight

ratio, etc. The regression analysis will then tell us how these characteristics are related to Model *A* sales for individual sales persons.

Or, alternatively, the manufacturer could use marketing *territories* as objects, which would dictate predictor variables such as population, number of Model *A* dealers, number of dealers of competitor Model *B*, advertising expenditure, and many of the same variables that were applicable in the dealership analysis.

Yet another possibility is to use *time periods* as the objects of our analysis. Sales during the various time periods would be related to such variables as number of used cars sold, advertising expenditure, rainfall, weeks since new model introduction, unemployment rate, sales of Model *B*, size of payroll, total hours open to public, etc.

We can even use *prospective car buyers* as objects. The criterion variable could be a numerical rating of interest in purchasing Model *A*, and the predictor variables could be ratings on a series of image characteristics such as "attractive styling," "roomy interior," comfortable ride," "good resale value," "economical to drive," etc. The extent to which the ratings on the various image dimensions predict ratings of purchase interest would shed further light on the factors that influence sales of Model *A*.

What we have seen in the above examples is that a single criterion variable can be studied with a number of alternative regression analyses. By understanding the extent to which characteristics of dealerships, sales personnel, sales territories, time periods, and consumers are related to Model *A* sales, the marketing efforts of the automobile manufacturer can be adapted to improve the sales of Model *A*, the criterion variable.

This type of analysis—identifying a criterion variable of interest, selecting an appropriate set of objects on which to measure it, and identifying a set of potential predictor variables—is applicable to the widest possible range of analytical problems, whether the criterion variable is sales, crop yields, attitudes, academic achievement, job success, strike activity, disease levels, crime rates, or life span. It should also be clear from the preceding examples that regression analysis is more than the mechanical application of a statistical technique to a matrix of data. The formulation of the problem, the identification of criterion and predictor variables and the objects upon which they are measured, and the interpretations of the resulting regression equation and the accompanying R^2 and beta weights, will determine how useful the analysis will be. And, of course, we must be satisfied that the raw data conforms to the statistical assumptions of the given regression model.

Also, we have seen that there is no one regression analysis that is most appropriate for understanding a criterion variable, but rather the greatest understanding is most likely to result from a number of alternative analyses, each viewing the problem from a different angle.

To round out our discussion of regression analysis the balance of the chapter will touch briefly on several special topics related to the application and interpretation of the technique.

12. Collinearity Problem

A particularly vexing problem in the application of multiple regression analysis arises from the situation in which two or more predictor variables are very highly correlated with each other. This is referred to as the *multicollinearity* problem, or simply as *collinearity.* Under such conditions the computer attempting to analyze the data according to its stated instructions is likely to go awry. Exactly when this will happen is not always identifiable, otherwise it could be prevented. We should be forewarned, though, to use some common sense in our selection of predictor variables so as not to include groups of variables that we know on logical grounds must be highly correlated with each other. For example, we would not include the variables of sales, costs, and profits into a regression analysis since any two will automatically determine the third.

Related to the collinearity problem is the situation in which we include a predictor variable that is really not a predictor variable as such but rather a slight variation of the criterion variable. For example, if our criterion variable was defined as the sales performance of a set of sales reps, and among our predictor variables we included the commissions earned by the reps, it is unlikely that we would gain any information on the other predictor variables we studied, since commissions would account for virtually all the variance in the criterion measure. If there is no variance left to account for, how can we assess the importance of the remaining predictor variables? Either we should have left sales commissions out of the problem or let it stand as a criterion variable. In situations such as this, we cannot expect the computer to think for us.

13. Dummy Variables

In order to use qualitative predictor variables (e.g., gender) in a regression analysis, we can transform the variables into quantitative *dummy variables.* Essentially what we do is convert each level of a qualitative variable into a binary variable. For example, the qualitative variable *gender* (male vs. female) could be made into a dummy variable representing *maleness*—i.e., male vs. not male—with the respective numerical values 1 and 0. Or we could construct the dummy variable representing *femaleness*—i.e., female vs. not female—again with the respective numerical values 1 and 0. The three levels of the qualitative political affiliation variable—Democrat, Republican, and Independent—could be made into three dummy variables: Democrat vs. not

Democrat, Republican vs. not Republican, and Independent vs. not Independent. In each case, one level of the dummy variable could take on the value of 1 and the other level a value of 0.

The benefit of such a transformation is that the quantitative dummy variables can now be introduced as predictors into a regression analysis. For example we could determine if student gender is related to their test performance. Or we could determine if the dominant political affiliation of a voter district is related to the district's crime rate.

However, when we use dummy variables we must keep the collinearity problem in mind. We cannot introduce dummy variables for every level of the qualitative variable, since they are not independent of one another. If we know that an individual has a value of 0 on the dummy variable of "maleness" we can predict perfectly the individual's value on the dummy variable "femaleness"—namely, it must be 1. Therefore we need to include only one of these two dummy variables. In the case of dominant political affiliation of a voter district, if we know that a district scores 0 on "Democratness", and 1 on "Republicanness" we know for sure that it must score 0 on "Independentness." So, we need to include only two of these three dummy variables. Knowing a district's value on any two of the three dummy variables will automatically inform us of its value on the remaining dummy.

In general, when we construct dummy variables from a qualitative variable, we will always want to use *one less* than the number we can create. For example, if we have classified voters into ten occupation categories, we can create nine dummy variables for use in a regression analysis for predicting frequency of voting based on occupation.

14. Autoregression

An interesting application of regression analysis is to predict values on the criterion variable based on values of the same criterion variable obtained earlier in time. Imagine the price of Stock A on each of 100 trading days. Now let us pair each of these prices with the price on the immediately preceding day, as explained in our discussion of serial correlation. We can now try to predict the price of Stock A on a given day based on its price the previous day. In fact, we could turn it into a multiple regression equation by introducing as additional predictor variables the stock's price two days back, three days back, etc.

While we should be so lucky as to be able to predict the future in the stock market, the autoregression technique is useful in identifying dependencies among data collected sequentially which we may wish to extract before submitting the data to further analysis. The technique is also useful for projecting time series data such as crime rates, fertility rates, strike activity, etc.

15. Regression to the Mean

The expression "regression" originated from the observation that exceptionally tall fathers tended to sire sons who, when matured, tended to be shorter than their father's height. Similarly, exceptionally short fathers had sons who tended to be taller than their fathers. While a full interpretation of such a finding would require theories of genetics and the dynamics of mate selection, we can attribute it partly to the phenomenon of *regression to the mean*.

To understand the concept consider that any empirical measurement of a characteristic is composed of two parts—the *true value* of the characteristic plus or minus some *error*. On repeated measurements the true value remains the same but the error component fluctuates. We know that when we measure a large number of objects with respect to a characteristic, some of the objects will score high, some low, and some in between. Now wherever the object scores, part of the score is due to an error component. Thinking in terms of conditional probabilities, we can ask ourselves whether those objects that scored exceptionally high were not benefiting from a large error component; and, similarly, those that scored exceptionally low, were in the receipt of a large negative error component. Cast in a different light, suppose we knew only the size of an object's error component: What would we predict as the object's total score if we knew it had an exceptionally large positive error component—would it tend to be above or below the mean. The dynamics of this phenomenon become apparent when we *remeasure* our set of objects on the same characteristic and compare their respective values on the two measurements. What we find is that those that scored exceptionally high (or low) on the first measurement score closer to the mean on the second measure; that is, there is a *regression to the mean*. The greater the error component, the greater will be the regression or "turning back" to the mean.

This phenomenon is worth bearing in mind whenever exceptional scores on a single measurement are singled out for attention; especially when the objects possessing these scores are to receive special due, as in academic, medical, or business settings. For example, a year after introduction of a new product, two cities are singled out as having exceptionally high sales. During the next year these cities receive all manner of special attention and marketing expenditure. After the second year it is found that their performance has dropped compared to their first year performance. The cities which had the best second-year performance gain were those with lackluster first-year performance. In other words, much of the initial variation in sales from city to city was due purely to chance variation, and the cities that performed best during the first year just happened to be recipients of a larger positive error than the other cities. On the other hand, this need not be so: It could well be that the

variation was not due to error at all, but to fundamental causal factors operative in each city. This is the variance that regression analysis tries to tap.

Consider as another example a mutual money market fund that boasts having the best performance of all the leading investment funds during the most recent year. We should not be too impressed with this performance. After all, of the many funds, and of the many starting each year, one of them *had* to do better than all the others. This is a truism. Again, we want to know the reliability of this performance. Will it duplicate its performance next year, or will another fund claim the leadership role, while the other regresses to the mean.

16. Self-Fulfilling Prophesy

The true validity of the predictions arising from a regression analysis cannot always be ascertained. Since the analysis is not purely an intellectual exercise but the basis for action, the outcome of the analysis may often provoke activities that make the predictions come true—the phenomenon of *self-fulfilling prophesy*. If sales for certain stores in a chain are predicted to be above average, these stores may enjoy special promotions and other attention they might not normally receive, and consequently live up to expectations but for the wrong reasons. The same is likely to happen when students are placed into special classes based on achievement or aptitude test scores.

On the other hand, we might experience a *self-negating prophesy* in which dire predictions are forestalled through corrective actions. For example, predictions of falling sales, poor academic achievement, or disease onset may result in special remedial efforts to avoid such possibilities. In instances such as these, it becomes a very philosophical question as to whether our regression analysis has validity, for while our dire predictions did not come true, we surely benefited from the analysis.

17. Concluding Comments

In this chapter we have touched upon the basic concepts of regression analysis, a technique for describing the mathematical relationship between a criterion variable and one or more predictor variables. We also discovered how r^2, beta coefficients, and the measures of prediction error help us to interpret the practical value of a regression equation. While regression analysis can be applied to problems in which the predictor variables are either random variables or fixed experimental variables, we concentrated our examples on the former type since they are so commonplace. Experimental variables, as they are related to a criterion variable, will be discussed at greater length in the following chapter on Analysis of Variance, although it should be recognized that they can also be described by the regression approach.

Chapter 12

Analysis of Variance

1. Introduction

Although regression analysis is useful for identifying and describing a linear or other systematic relationship between quantitative variables, there are analytical situations in which it is not easily applied. For instance, we might confront two variables that are related to each other, but only over part of the range of studied values, and not necessarily in a simple linear or polynomial fashion. Nonetheless, we are interested in uncovering the fact that the two variables *are* related, albeit in a complex and perhaps undescribable manner.

A second class of situations in which regression analysis is cumbersome, involves experimental or predictor variables which are *qualitative* in nature. For example, rather than being interested in the effects of the *level* of fertilizer application, we might be interested in the effects of alternative *kinds* of fertilizer; or instead of drug dosage, *type* of drug; or instead of hours of instruction, *type* of instruction; or instead of advertising levels, *kind* of advertising creative execution; etc. Similarly, we might be interested in whether different cities differ in academic achievement of their students, or whether different political candidates differ in public acceptability, or whether men and women differ in performance on a particular task. In all of these situations the predictor variable is composed of values which differ in *kind*, rather than in *quantity*.

To handle the above types of analytical problems we have available a versatile approach known as *analysis of variance*. It can be used for identifying relationships between predictor and criterion variables, whether the predictor variables are quantitative or qualitative in nature. Although the above types of relationships can also be handled by a multiple regression analysis using dummy variables for the various levels of the qualitative predictor variable, as discussed in the preceding chapter, the *identification* of the relationships is typically done by the analysis of variance techniques outlined in this chapter.

In fact, the regression analysis and analysis of variance approaches are often used hand in hand, dealing as they do with the identical data. We can think of the analysis of variance technique as testing hypotheses about the *presence* of relationships between predictor and criterion variables, regression analysis as describing the *nature* of those relationships, and r^2 as measuring the *strength* of the relationships.

Analysis of variance techniques will also be found to be useful in guiding the efficient *design* of our data collection schemes, especially complex experiments, since the design and analysis of experiments go hand in hand.

2. Overview of Analysis of Variance

Analysis of variance—often abbreviated with the acronym *anova*—is a broad class of techniques for identifying and measuring the various sources of variation within a collection of data. Taken literally, "analysis of variance" could be synonymous with the field of statistical analysis itself, for we have found that all statistical analyses are essentially concerned with analyzing the variation inherent in data collections. But in addition to this more general meaning, analysis of variance refers to a set of well-defined procedures for partitioning the total variation of a data collection into its component parts.

To even touch upon the variety of analysis of variance techniques would require a book again this size, so we will only deal with the fundamental concepts in this introductory chapter, as well as some primary applications. Since in most analysis of variance problems we are interested in the differences in the mean values of a criterion variable which are associated with different values of the experimental or predictor variable, we should not be surprised that we must consider an appropriate sampling distribution to determine whether those sample means differ more than would be expected by pure chance. An understanding of the relevant sampling distribution, to which we turn next, is fundamental to the understanding of all analysis of variance models.

3. The F Distribution

Imagine a normally distributed population with mean μ and variance σ^2. Now suppose we draw two successive samples of size n_1 and n_2, and calculate the respective sample variances s_1^2 and s_2^2. Further, let us define a new statistic F which is the ratio of the first sample variance to the second sample variance:

$$F = \frac{s_1^2}{s_2^2}$$

What would we expect the value of F to be? More specifically, what would we expect the *sampling distribution* of F to look like? Since s_1^2 and s_2^2 are each unbiased estimates of the population variance σ^2, we would expect that in the long run the mean value of F would be 1; i.e.,

$$E(F)=E\left(\frac{s_1^2}{s_2^2}\right)=\frac{E(s_1^2)}{E(s_2^2)}=\frac{\sigma^2}{\sigma^2}=1$$

That is, if we drew an infinite number of sample pairs from a normally distributed population, and each time formed the ratio of the two sample variances, $F=s_1^2/s_2^2$, the average of these F ratios would equal 1.

However, since the two sample variances, s_1^2 and s_2^2, are themselves subject to sampling variation—i.e., each has its own sampling distribution—the ratio $F=s_1^2/s_2^2$ must also exhibit sampling variation. Although the expected value of F is 1, sometimes it will be smaller, and sometimes is will be larger.

Since the value of F is a function of the two sample variances, s_1^2 and s_2^2, it follows that the sampling distribution of F should be dependent upon the nature of the sampling distributions of s_1^2 and s_2^2. Although we have not studied the characteristics of the sampling distribution of a sample variance s^2, as we have that of the sample mean $\bar{x}$, it should be intuitively clear that the variation of the sampling distribution of s^2 should depend to an extent upon the sample size n. The larger the sample size, the smaller will be the variation of the sampling distribution of the sample variance s^2.

Returning then to the nature of the F distribution, it should be apparent that its shape and variation should be dependent upon the sample sizes n_1 and n_2 upon which the sample variances s_1^2 and s_2^2 are based. For as the sampling distributions of s_1^2 and s_2^2 depend upon n_1 and n_2, respectively, so too must the distribution of $F=s_1^2/s_2^2$ depend on those sample sizes. As a result, we have a different F distribution for each possible pair of sample sizes upon which it could be based. More specifically, we have a different F distribution for each pair of *degrees of freedom* associated with the respective sample variances. In each case, the degrees of freedom equal one less than the sample size; i.e., n_1-1 for the numerator of the F ratio, and n_2-1 for the denominator. For example, the F distribution based on samples of $n_1=10$ and $n_2=15$, would have the pair of degrees of freedom $df=9,14$. Figure 1 shows F distributions for various degrees of freedom. Notice that as the degrees of freedom get large, the F distribution approaches the shape of a normal distribution.

Since the F distribution is a well-defined probability distribution, we can determine the probability that a given F will be above a certain value. Or,

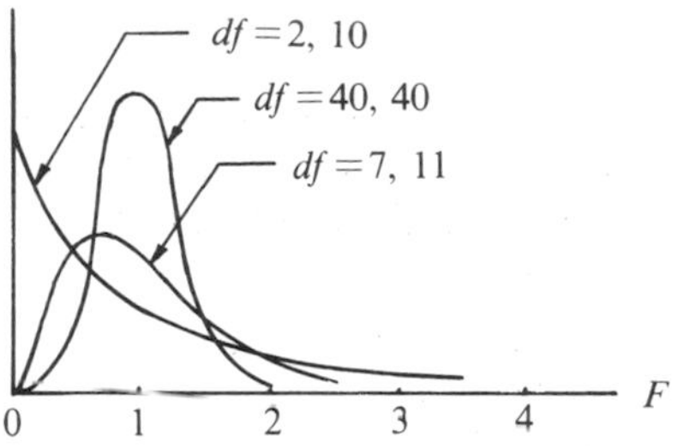

Figure 1 Some *F* distributions for different pairs of degrees of freedom.

conversely, and more useful for our purposes, we can determine the value of *F* which will be exceeded, say, 5% or 1% of the time by chance alone. These critical *F* values, for various pairs of degrees of freedom, are presented in Appendix Table V. In that table we see, for example, that for degrees of freedom of 5 and 17 ($df = 5, 17$) an *F* of 2.81 will be exceeded with a probability of .05, and an *F* of 4.34 will be exceeded with the probability of .01, due to chance alone. This is shown graphically in Figure 2.

While at the moment the relevance of the *F* distribution—defined as the ratio of two *variances*—to the testing of the difference among two or more sample *means* may seem obscure, its application to such situations will become evident in the following sections. The important things to remember from this discussion are that the *F* ratio is a statistic defined as the ratio of two independent estimates of a normally distributed population's variance σ^2—i.e., $F = s_1^2/s_2^2$—and that the sampling distribution of *F* is a function of the sample sizes upon which s_1^2 and s_2^2 are based.

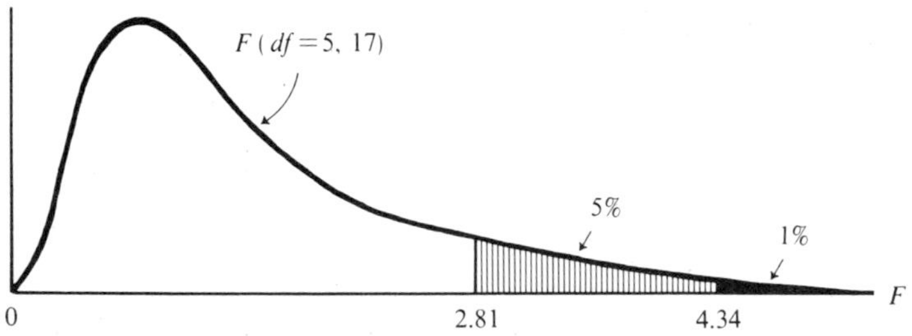

Figure 2 Critical *F* values for the .05 and .01 significance levels for $df = 5, 17$.

4. One-Way Analysis of Variance

As stated in the beginning of the chapter, there are countless situations in which we would be interested in evaluating whether two or more sample means differ more than would be expected by chance. If we let k represent the number of sample means with which we are concerned, we can state the null hypothesis

$$H_0: \mu_1 = \mu_2 = \mu_3 = \cdots = \mu_k = \mu$$

which supposes that the sample means are all estimates of the same population mean μ. Through an appropriate statistical model we must determine the tenability of this hypothesis. The alternative hypothesis, H_1, is that at least one of the means differs from the others.

To understand the application of the analysis of variance procedure to the above type of hypothesis testing problem, let us consider a concrete example. A marketer of grocery products is test marketing a new product in a random sample of 12 supermarkets. Included in this market test is a test of three alternative packages—identical in graphic design but differing in the predominant color. The three colors are Blue, Red, and Green. Of the total of 12 stores chosen for the sales test, four are *randomly chosen* to carry the Blue package, four the Red package, and four the Green package. We could just as well be testing three alternative drugs, teaching methods, fertilizers, etc.

An audit of first-week unit sales for the product is presented in Table 1. We see that the mean unit sales for the stores carrying the Blue, Red, and Green packages are $\bar{x}_B = 14.0$, $\bar{x}_R = 18.0$, and $\bar{x}_G = 11.5$, respectively; and that the corresponding population variance estimates are $s_B^2 = 32.7$, $s_R^2 = 26.0$, and $s_G^2 = 21.7$, respectively.

Table 1 Unit sales of a new product in 12 stores as a function of package color

	Unit sales of stores carrying · · ·			
	Blue package	Red package	Green package	
	6	18	7	
	14	11	11	
	19	20	18	
	17	23	10	
Sample means:	14.0	18.0	11.5	(overall mean = 14.5)
Sample variances:	32.7	26.0	21.7	

We want to test the null hypothesis

$$H_0: \mu_B = \mu_R = \mu_G$$

which assumes—rightly or wrongly—that there is no difference in the sales engendered by the Blue, Red, and Green packages, and that our sample means $\bar{x}_B = 14.0$, $\bar{x}_R = 18.0$ and $\bar{x}_G = 11.5$ are merely chance fluctuations from a common population mean μ. This is essentially the same as hypothesizing that there is no relationship between package color and the product's sales. To test the credibility of the hypothesis, we will take advantage of the F distribution.

Within groups variance. To obtain a value of F we need *two* estimates of the population variance. Notice in Table 1 that the set of Blue, Red, and Green package stores each supply an estimate of the population variance; namely, $s_B^2 = 32.7$, $s_R^2 = 26.0$, and $s_G^2 = 21.7$, respectively. Since no one of these estimates is any better than any of the others, our best bet is to *pool* them and create a single estimate based on this "within" groups variation. Since the sample size is the same for each group, we can merely take the mean of the three independent variance estimates to get our *pooled within groups variance estimate*, s_w^2, which turns out to be

$$s_w^2 = \frac{32.7 + 26.0 + 21.7}{3} = 26.8$$

If the various within groups variance estimates had been based on different sample sizes, then we would have had to take their weighted average to arrive at the pooled value.

The estimated population variance based on the pooled within groups variance should not be confused as being the same as the variance of the total set of 12 sales figures. In the latter situation, the variation of the 12 scores would be taken about the overall mean, whereas in the case of the within groups variance, the variation is measured about the respective group means. The pooled within groups variance s_w^2 is really a measure of random variability or random "error" among the measured objects, stores in this instance. It will serve as one of the two variance estimates needed to form an F ratio.

Between groups variance. A second source for estimating the population variance is the variation between the group means $\bar{x}_B = 14.0$, $\bar{x}_R = 18.0$, and $\bar{x}_G = 11.5$, measured about the total mean of all 12 measures, $\bar{x}_T = 14.5$. True, we are obtaining a variance estimate based on only a sample of three numbers (14.0, 18.0, 11.5), but nothing in our study of the definition of variance estimates said we had to have a large set of numbers to calculate a variance; in fact, we can calculate a variance with as few as two numbers.

The variance of the three sample means about the total mean, which will later yield the *between groups* or "between treatments" variance estimate, turns

out in this example to be

$$\frac{(14.0-14.5)^2+(18.0-14.5)^2+(11.5-14.5)^2}{2}=10.75$$

But this variance estimate is not really an estimate of the population variance σ^2, but an estimate of the variance of the mean, $\sigma_{\bar{x}}^2$—the square of the standard error of the mean. But we know that

$$\sigma_{\bar{x}}^2=\frac{\sigma^2}{n}$$

and

$$s_{\bar{x}}^2=\frac{s^2}{n}$$

so it follows that

$$s^2=ns_{\bar{x}}^2$$

where s^2, in our example, is the estimate of the population variance based on the between groups variation. To distinguish it from the within groups variance estimate we will subscript it with *bg* for "between groups," and we have

$$s_{bg}^2=ns_{\bar{x}}^2=4(10.75)=43.0$$

as our second estimate of the population variance. But keep in mind that this between groups estimate of the population variance, unlike the within groups variance, reflects not only random variation between stores but also any variation in the data due to the possible treatment effects—namely, the package colors.

***F* ratio.** Now that we have two independent estimates of the population variance, we can form an *F* ratio

$$F=\frac{s_{bg}^2}{s_w^2}$$

which is simply the ratio of the between groups variance estimate to the within groups variance estimate. In the present example we have

$$F=\frac{43.0}{26.8}=1.60$$

which is associated with a pair of degrees of freedom, $df=2,9$.

Degrees of freedom. Each of the two variances forming the F ratio is associated with its own degrees of freedom. For s_{bg}^2, the between groups variance estimate, the degrees of freedom equal $k-1$, the number of groups less one. In this example, the degrees of freedom for s_{bg}^2 equals 2. That is, once the overall mean is calculated—which depends on the three group means—only two of the three means are free to vary about the overall mean; i.e., knowing two of the deviations automatically dictates the third deviation, since, by definition, the deviations must sum to zero.

The degrees of freedom associated with the within groups variance estimate are equal to $\Sigma n_j - k$, where n_j equals the sample size for the jth group; in other words, the sum of the individual group sample sizes—which comprises the total sample size—less the number of groups. We lose a degree of freedom for each group mean that is calculated. In the present example, the within groups variance estimate s_w^2 is associated with nine degrees of freedom—which can be thought of as $12-3$ (total sample size less the number of groups) or as 3×3 (3 degrees of freedom from each of the three groups). The F ratio in our example, then, has degrees of freedom $df = 2, 9$.

Significance test of *F*. Since the between groups variance $s_{bg}^2 = 43.0$ reflects variation due to random differences between stores *plus* variation due to the treatments—i.e., package colors—while the within groups variance estimate $s_w^2 = 26.8$ reflects only random store differences, our F ratio looks like this:

$$F = \frac{\text{(Variance due to store differences) } \textit{plus} \text{ (package color effects)}}{\text{(Variance due to store differences)}}$$

As a result, the extent to which F exceeds a value of 1, is indicative of a possible real package color or treatment effect. But remember that an F ratio can exceed a value of 1 by chance alone, even if there were *no* treatment effects. Is it likely, we must ask, that we would get an $F=1.60$ as above, if in fact package color had absolutely no effect. In other words, is it likely that we would have gotten the same sales pattern among the stores if *all* packages were, say, Blue. Notice that this is essentially the same hypothesis testing question we are faced with when evaluating a z or t ratio. Consequently, we can select a significance level α which corresponds to an "unlikely" outcome given that the expected value of F equals 1—corresponding to the hypothesis of no treatment effect differences.

To determine if an $F=1.60$ is likely to have occurred by chance alone, we consult Appendix Table V. The degrees of freedom associated with our F are 2 and 9 for the numerator and denominator, respectively. For $df = 2, 9$ and a significance level of $\alpha = .05$ we find a critical value of $F_c = 4.26$. Consequently, our $F=1.60$ must be considered simply a chance deviation from its expected value of 1.0, given that the null hypothesis were true.

So we see, that while a casual inspection of the store sales in Table 1 suggested that the package colors did affect sales, our statistical analysis of the data tells us that we could well have expected those same sales figures if the stores all carried the same color package.

Our failure to reject the null hypothesis is no guarantee that we have not reached an erroneous conclusion. Just as when we test hypotheses involving one or two means, we can still make type I and type II errors. As before, the probability of a type I error—rejecting the null hypothesis when it is in fact true—is equal to α, the significance level used to assess the credibility of the hypothesis, often .05 or .01. The probability of committing a type II error—failing to reject a false null hypothesis—is a more complicated probability to determine, for it depends among other things on the true population values for all k means, which could assume an infinite variety of values.

Finally, it should be noted that a rejection of the null hypothesis could be due not to differences between the population means but to differences between the population *variances*. It is for this reason that we must assume *homoscedasticity*—or homogeneity of population variances—when evaluating an F ratio, for then if the F ratio is statistically significant, it must be attributed to unequal population means, rather than to unequal population variances—assuming, of course, that the populations are normally distributed.

Formulas. In the preceding examples we did not show all of the intermediate calculations to arrive at our final F ratio value. Details of the calculations can be found in Table 2 which traces the analysis of variance from raw data through to a summary analysis of variance table. Also, mathematical notation is introduced for the sake of generality. In the notational scheme, x_{ij} refers to the ith observation in the jth group, where j varies from 1 to k, the number of groups in the analysis, and i varies from 1 to n_j, the number of observations in the jth group. Also, $\bar{x}_{.j}$ is the notation for the mean of the jth group, and $\bar{x}_{..}$ is the mean of the total number of observations.

Note that the two variance estimates—also referred to as the *mean sum of squares*—are arrived at in two steps. First the sum of the squared deviations around the appropriate means are obtained, and then the sum is divided by the relevant degrees of freedom. Also note that in the calculation of the between groups sum of squared deviations, each deviation of a group mean from the total mean is weighted by the number of observations in that group. Also worthy of note is that the within groups and between groups sum of squared deviations should add to the "total" sum of squares. This relationship can serve as a good check on the arithmetic calculations. It also points out how the total variation of the data is partitioned into two components—within groups and between groups. And it is this type of examination of the sources of variation in a body of data that earns "analysis of variance" its descriptive name.

Table 2 Outline of steps in one-way analysis of variance.

I. The raw data and group means.

Individual store sales for alternative package colors:

	Blue	Red	Green	
	6	18	7	
	14	11	11	
	19	20	18	
	17	23	10	
$\bar{x}_{.j}$	14.0	18.0	11.5	$\bar{x}_{..} = 14.5$

II. Calculation of "within groups" sum of squared deviations SS_w.

Sum of squared deviations · · ·

within Blue group $= (6-14.0)^2 + (14-14.0)^2 + (19-14.0)^2 + (17-14.0)^2 = 98.0$
within Red group $= (18-18.0)^2 + (11-18.0)^2 + (20-18.0)^2 + (23-18.0)^2 = 78.0$
within Green group $= (7-11.5)^2 + (11-11.5)^2 + (18-11.5)^2 + (10-11.5)^2 = 65.0$

Pooled sum of squared deviations within groups $= \Sigma_{j=1}^{k} \Sigma_{i=1}^{n_j} (x_{ij} - \bar{x}_{.j})^2 = 241.0$

(*Note*: The sum of squared deviations within each group, when divided by their degrees of freedom, will yield the Table 1 variances.)

III. Calculation of "between groups" sum of squared deviations (weighted by group sample size).

$$SS_{bg} = \sum_{j=1}^{k} n_j(\bar{x}_{.j} - \bar{x}_{..})^2 = 4(14.0-14.5)^2 + 4(18.0-14.5)^2 + 4(11.5-14.5)^2 = 86.0$$

IV. Calculation of "total" sum of squared deviations.

$$SS_T = \sum_{j=1}^{k} \sum_{i=1}^{n_j} (x_{ij} - \bar{x}_{..})^2 = (6-14.5)^2 + (14-14.5)^2 + \cdots + (10-14.5)^2 = 327.0$$

V. Summary analysis of variance table.

Source of variation	Sum of squared deviations	Degrees of freedom	Variance estimate	F ratio
Between groups (Colors)	86.0	2	43.0	1.60
Within groups (Error)	241.0	9	26.8	—
Totals	327.0	11		

Table 3 Short-cut computational formulas for the sums of squared deviations and variance estimates in a one-way analysis of variance, derivable from the definitional formulas in Table 2.

Source of variation	Sum of squared deviations (SS)	*df*	Variance estimates
Total	$SS_T = \sum_{j=1}^{k}\sum_{i=1}^{n_j} x_{ij}^2 - \dfrac{\left(\sum_{j=1}^{k}\sum_{i=1}^{n_j} x_{ij}\right)^2}{\sum_{j=1}^{k} n_j}$	$\sum_{j=1}^{k} n_j - 1$	—
Between groups	$SS_{bg} = \sum_{j=1}^{k} \dfrac{\left(\sum_{i=1}^{n_j} x_{ij}\right)^2}{n_j} - \dfrac{\left(\sum_{j=1}^{k}\sum_{i=1}^{n_j} x_{ij}\right)^2}{\sum_{j=1}^{k} n_j}$	$k-1$	$s_{bg}^2 = \dfrac{SS_{bg}}{k-1}$
Within groups	$SS_w = SS_T - SS_{bg}$	$\sum_{j=1}^{k} n_j - k$	$s_w^2 = \dfrac{SS_w}{\sum_{j=1}^{k} n_j - k}$

Table 3 presents alternative "short-cut" computational formulas for the determination of the two key variance estimates in the simple analysis of variance.

5. Factorial Designs

In the preceding section we studied the application of the analysis of variance technique to an experiment involving the effects of a single experimental variable, package color, on a criterion variable of unit sales. That type of analysis is often referred to as a "one-way" or "one-factor" analysis of variance, because only a single experimental variable or "factor" is being assessed. Usually, however, we are interested in the effects of several factors on our criterion variable of interest. For example, in addition to package color, we might be interested in the effects of package *shape*, or *price*, or product *aroma* on product sales. Rather than repeat our experiment for each of the other experimental variables with other sets of stores, we can design a more efficient multi-factor experiment in which the effects of two or more experimental variables are assessed simultaneously with the same set of objects.

There are many forms that multi-factor data collections can assume, and each has its appropriate analysis of variance model. We will look at the most common of these experimental designs, begining with the *factorial* design, in

which each level of each factor studied is combined with each level of every other factor under consideration, to form *composite* treatment conditions. To understand the logic of the approach we will begin with a general discussion of a two-factor factorial design, followed by a numerical example, and then see how the approach can be extended to a three-factor or higher level design.

Two-factor factorial design. As our introduction to factorial experimental designs, let us reconsider the package color experiment of the preceding section. Instead of testing only the effects of package color on sales, let us introduce a second experimental variable—the *shape* of the package, with the levels of this factor being square or rectangular. In total, then, we have six different treatment combinations based on the respective levels of the two experimental factors—three levels of package color times two levels of package shape.

Now let us take our 12 randomly selected stores and assign them at random to the six experimental conditions so that we have two stores measured under each of the six treatment conditions. Ideally, we would want to have more than two independent observations per treatment condition group, or *cell*, to improve the power of our experiment, but two is the minimum we need to obtain a "within treatment groups" error variance estimate, as we did in the one-way analysis of variance. The design of this experiment is portrayed in Figure 3. This type of design can be described as a two-factor 2×3 factorial design with two replications (observations) per cell, in that there are two experimental factors—shape and color of package—with two levels of the first and three levels of the other factor, resulting in 2×3 treatment combinations, each with two independent observations. It is further described as a *completely randomized design*, as is the one-way analysis of variance of the preceding section, in that the objects comprising the total sample are randomly assigned to the treatment groups.

Notice in Figure 3 that if we compare the means of the four stores under each package color condition, we have exactly the same comparison as in the earlier example of a one-way analysis of variance design. But in the present factorial design, our objects—the stores—are doing double duty for us, for we can now compare the means of the six stores under each package *shape* condition to see if that variable has an effect on the product's sales. But again, we need an analytical procedure to statistically evaluate whatever sample differences we find.

The analysis of variance procedure for the above type of factorial design—in which there is more than one observation per cell or group—is not that different from the type we used for the single-factor experiment. As before, we can calculate a "between groups" variance estimate, based on the variation of the *six* group means about the overall mean; and we can also

		Package color			
		Blue	Red	Green	Totals
Package shape	Square	2 stores	2 stores	2 stores	6 stores
	Rectangular	2 stores	2 stores	2 stores	6 stores
	Totals	4 stores	4 stores	4 stores	12 stores

Figure 3 Schematic diagram of a 2×3 factorial experimental design with two independent randomly assigned objects per cell.

calculate a pooled "within groups" variance, based on the variation of the two observations in each group about their respective group means (remember that all we need is a minimum of two observations for a variance estimate). The key difference in this design is that the between groups variance can be further subdivided into component sources of variation. This should be clear from an examination of the six groups in Figure 3. Some of the variation between the six groups is due to package color, some is due to package shape, and then there is a left-over or *residual* amount of variation which is attributable to the "interaction" between these two factors, a concept discussed later in this section.

The various sources of variation and their associated degrees of freedom are shown in Table 4. For package color there are 2 degrees of freedom (3 minus 1), for package shape there is 1 degree of freedom (2 minus 1), for the color × shape interaction there are 2 degrees of freedom (2×1—the product of degrees of freedom for color and shape, respectively), and there are 6 degrees of freedom for the pooled within groups source of variation, since each of the six groups has $2-1$ degrees of freedom. Finally, the total degrees of freedom is equal to 11, the total number of observations less 1.

After the sum of squared deviations has been calculated for each source of variation and then divided by their respective degrees of freedom, F ratios can be formed to test the effects of package color, package shape, and the interaction between the two. In each case, the denominator of the ratio will be the variance estimate based on the pooled within groups sum of squared deviations, for as in the one-way analysis of variance, the within groups variance is a measure of random "error" or chance differences among the

Table 4 Summary analysis of variance table for the 2×3 factorial experiment portrayed in Figure 3.

Source of variation	Degrees of freedom	Variance estimate	F ratio
Package color (C)	2	s_C^2	$\frac{s_C^2}{s_w^2}$
Package shape (S)	1	s_S^2	$\frac{s_S^2}{s_w^2}$
$C \times S$ interaction	2	s_{CS}^2	$\frac{s_{CS}^2}{s_w^2}$
Within groups (w)	6	s_w^2	—
Total	11		

stores. The variance estimate based on differences among the *means* for the various package colors, on the other hand, includes not only this random error but any systematic effects of the package color; similarly for the package shape source of variation. Consequently, the extent to which the F ratio exceeds a value of 1 will reflect the existence of treatment effects—e.g., color, shape, or the color$\times$shape interaction. Again, each F ratio must be evaluated by consulting Appendix Table V for the appropriate pair of degrees of freedom associated with the numerator and denominator of the F ratio, and for the chosen significance level.

Numerical example. We can now look at an actual data analysis of a two-factor factorial experiment to see the ways in which it is similar to and different from the one-way analysis of variance design.

A school system is interested in evaluating three alternative textbooks, A, B, and C. They are also interested in knowing the effect of alternative amounts of classroom instruction for the semester course in question; 60 hours vs. 100 hours are chosen for the test. So, there are two experimental variables, *textbook used* and *hours of classroom instruction*, with three and two levels respectively, resulting in $3 \times 2 = 6$ treatment combinations.

From across the school system, a sample of 18 classes are selected at random. These 18 classes are assigned completely at random to the six treatment combinations, so that there will be three classes taught under each of the six conditions. Now it is true that the sample of 18 classes differ on a variety of variables including instructor ability, average student ability, class size, etc.—but since the classes are assigned at random to the six alternative

treatment conditions, every treatment has received a "fair shake" and no systematic bias favors one over another. Our statistical test will account for these random differences.

At the end of the semester course a standardized achievement test is administered to the students, and each class average is the data of interest. The average scores for the 18 classes are presented in Table 5, arranged according to the six treatment groups. For example, the three classes assigned to the *textbook A/60 hour* treatment condition had average scores on the achievement test of 80, 86, and 80—the averages reported to the nearest whole number for the sake of simplicity of exposition. Also presented in Table 5 are the means for the six treatment conditions, the means for the three textbook conditions, and the means for the two hours of instruction conditions, as well as the overall mean of all 18 classes.

We know from our previous discussion that the total variation of the 18 data points can initially be broken down into two parts—variation *between the six treatment groups* and variation *within the six treatment groups*. But the variation between the six treatment groups can subsequently be partitioned into separate components: (a) variation between textbook conditions, (b) variation between hours of instruction conditions, and (c) a left-over or *residual* variation which is a measure of the textbook by hours of instruction *interaction*, which is a reflection of whether the textbook effect varies depending upon hours of instruction, or vice versa.

The within groups variation, on the other hand, pooled from all six treatment combination groups, reflects the random variability of the classes (the entire *milieu* of teacher, students, physical environment, etc.) in the school system— that is, the classes within any one treatment group would be expected to exhibit the same variability as the classes in the school system at large, since applying a treatment condition to them is only assumed to add some constant effect to them, and we know that adding a constant value to a group of scores will affect the mean of the group but not the variation.

As can be seen in Table 5, these various sources of variation are first computed as the sum of squared deviations about the appropriate means, then divided by their respective degrees of freedom to yield variance estimates. The within groups error variance estimate is then used as the denominator of the F ratios which have as their numerators the variance estimates based on the various sources of variation. The resulting F ratios can then be tested for statistical significance against a prespecified α level by consulting Appendix Table V, with degrees of freedom conforming to those of the variance estimates in the F's numerator and denominator. The probabilities of the various F's, the p values, assuming the null hypothesis, can also be reported.

With the above overview of the analysis we can now consider the computations in Table 5 in more detail. In part I of the table, the raw data is organized

Table 5 Outline of steps in the analysis of variance of a 2-factor 3×2 factorial experimental design.

I. The raw data, x_{ijk}, along with cell, column, and row means. ($i = 1, 2, 3$ observations per cell, $j = 1, 2, 3$ columns, and $k = 1, 2$ rows of the data matrix.)

		Factor I: Textbook used			
		Text *A* ($j = 1$)	Text *B* ($j = 2$)	Text *C* ($j = 3$)	Row factor means, $\bar{x}_{..k}$
Factor II: Hours of Instruction	60 hours ($k = 1$)	80 86 80 $\bar{x}_{.11} = 82.$	75 74 79 $\bar{x}_{.21} = 76.$	79 80 78 $\bar{x}_{.31} = 79.$	$\bar{x}_{..1} = 79.$
	100 hours ($k = 2$)	90 86 88 $\bar{x}_{.12} = 88.$	76 80 78 $\bar{x}_{.22} = 78.$	86 81 82 $\bar{x}_{.23} = 83.$	$\bar{x}_{..2} = 83.$
Column factor means, $\bar{x}_{.j.}$		$\bar{x}_{.1.} = 85.$	$\bar{x}_{.2.} = 77.$	$\bar{x}_{.3.} = 81.$	$\bar{x}_{...} = 81.$ (Total mean)

II. Total sum of squared deviations.

$$SS_T = \sum\sum\sum (x_{ijk} - \bar{x}_{...})^2 = (80-81)^2 + (86-81)^2 + \cdots + (82-81)^2 = 346$$

III. Between groups (cells) sum of squares.

$$SS_{bg} = \sum\sum n_{.jk}(\bar{x}_{.jk} - \bar{x}_{...})^2 = 3(82-81)^2 + 3(76-81)^2 + \cdots + (83-81)^2 = 276$$

Partitioning of the between groups (cells) sum of squares:

A. Between columns sum of squared deviations (between textbook conditions).

$$SS_C = \sum n_{.j.}(\bar{x}_{.j.} - \bar{x}_{...})^2 = 6(85-81)^2 + 6(77-81)^2 + 6(81-81)^2 = 192$$

B. Between rows sum of squared deviations (between hours of instruction conditions).

$$SS_R = \sum n_{..k}(\bar{x}_{..k} - \bar{x}_{...})^2 = 9(79-81)^2 + 9(83-81)^2 = 72$$

C. Residual columns $\times$ rows interaction (textbook $\times$ hours of instruction interaction).

$$SS_{CR} = \sum\sum n_{.jk}(\bar{x}_{.jk} - \bar{x}_{.j.} - \bar{x}_{..k} + \bar{x}_{...})^2$$
$$= 3(82-85-79+81)^2 + 3(76-77-79+81)^2 + \cdots + 3(83-81-83+81)^2 = 12$$

Or, more easily:

$$SS_{CR} = \textit{Between groups} - \textit{Columns} - \textit{Rows} = SS_{bg} - SS_C - SS_R = 276 - 192 - 72 = 12$$

IV. Within groups (cells) sum of squares (error variation).

$$SS_w = \textit{Total} - \textit{Between groups} = SS_T - SS_{bg} = 346 - 276 = 70$$

V. Summary analysis of variance table.

Source of variation	Sum of squared deviations	Degrees of freedom	Variance estimate	*F* ratio
Between treatment groups	276	5	55.20	9.47**
Textbooks	192	2	96.00	16.47**
Hours of instruction	72	1	72.00	12.35*
Text × Hours interaction	12	2	6.00	1.03
Within groups (error)	70	12	5.83	—
Totals	346	17		

$^*p < .01$ $\quad$ $^{**}p < .001$

by treatment group, and means are calculated for all six treatment conditions, as well as for the three textbook and two hours of instruction conditions. In addition, the total mean of all observations is calculated.

In part II of Table 5 the total sum of squared deviations is obtained by summing the squared deviations of the 18 raw data observations from the overall mean of 81.

In part III the *between* treatment groups sum of squared deviations is obtained by subtracting the overall mean from each treatment group mean, squaring the results, weighting by the sample size in each treatment group ($n = 3$), and summing the results. But, as mentioned above, the resulting between treatment groups sum of squared deviations, 276, can be partitioned into sub-components; namely, that due to textbook conditions, that due to hours of instruction conditions, and the left-over residual textbook by hours of instruction interaction. The textbook sum of squared deviations is obtained by squaring the deviations of the three textbook condition means from the overall mean, weighting by the sample size(6) upon which each of the textbook means is based, and summing, to get 192. The hours of instruction sum of squared deviations is obtained similarly, except the squared deviations of the two hours of instruction conditions' means from the overall mean are now weighted by a sample size of 9, the number of observations upon which each is based, and the resulting sum is 72. The residual sum of squared deviations, the textbook by hours of instruction interaction, is then most easily obtained by subtracting the results of the *textbook* and *hours of instruction* computations from the between treatment group sum of squares; namely, $276 - 192 - 72 = 12$. It can also be obtained independently by the calculations shown in part III*C* of the table, which is highly recommended as a check on the accuracy of the other computations.

The *within* treatment groups sum of squared deviations, reflecting random error, is obtained in part IV of Table 5 by subtracting the between groups sum of squared deviations from the total sum of squared deviations. As a check, it can be computed independently in the same manner as in Table 2, part II, in which the squared deviations of each cell observation about the cell mean are summed, then pooled from all the treatment cells.

The sum of squared deviations for the various sources of variation are summarized in part V of Table 5. The degrees of freedom for the between treatment groups' source of variation is one less than the number of groups, or $6 - 1 = 5$; for textbooks it is $3 - 1 = 2$; for hours of instruction it is $2 - 1 = 1$; and for the textbook by hours of instruction interaction, $df = (2)(1) = 2$, the product of the df for textbooks and hours of instruction. Notice that the df for the latter three sources sum to the df for the between treatment groups, just as do the sum of squared deviations. The degrees of freedom for the within groups error variation is $df = 6(3 - 1) = 12$; which corresponds to 2 for each of the 6 cells, or alternatively, $df = 18 - 6 = 12$, which corresponds to the total

sample of observations less 1 for each of the 6 treatment groups. As a check, the degrees of freedom should total $df = 18 - 1 = 17$, the total number of observations less 1, which they do.

The various variance estimates are then obtained by dividing the various sum of squared deviations by the respective degrees of freedom. The within groups error variance estimate is then used as the denominator of each F ratio, the numerator in each case corresponding to the variance estimate based on the experimental variable of interest. Although we have shown a line for the between treatment groups' source of variation in Table 5, it is optional, since in any event we will proceed to test the component effects of textbook, hours of instruction, and their interaction. The degrees of freedom associated with each F correspond to those possessed by the numerator and denominator of the F ratio. For example, for the textbook F test, we have $df = 2, 12$, and the critical F value at the $\alpha = .05$ significance level can be found in Appendix Table V to be 3.29. Since the observed F ratio is 16.47 we can reject the null hypothesis that there is no difference between the tested textbooks. For hours of instruction, we have $df = 1, 12$ and the critical F for $\alpha = .05$ can be found in Appendix Table V to be 4.75. Since it is surpassed by the obtained value of 12.35 we can also reject the hypothesis that 60 vs. 100 hours of instruction have no differential effect on achievement test scores. The textbook by hours of instruction interaction with $df = 2, 12$ proved nonsignificant, so we cannot reject the hypothesis that the textbook effect is the same for the alternative hours of instruction. As is customary, we have added "p values" to the summary table to indicate just how improbable each F ratio was, if the null hypothesis were true.

In summary, we conclude that there is a difference among the textbooks, and between the two levels of hours of instruction, and that these two variables do not interact with one another. The natural question of which textbooks differed from which will be postponed to a later section on *multiple comparisons*. Table 6 presents the general form of a summary analysis of variance table for the type of two-factor factorial design that we have just considered.

The individual student data. The question should arise as to why, in the above textbook by hours of instruction experiment, we were only interested in the overall mean performance of each of the 18 classes and not in the individual performance of the students within each class. It would seem at first glance that we are sacrificing important information by not taking into account that each class is based on a large number of students, and nowhere do we take these unspecified "sample sizes" into account. Later we will consider the usefulness of the individual student data, but first the reasons why it was not considered as a matter of course in the present experimental design.

The experimental objects (the basic analytical units) in this study consisted of the entire class *milieu* (the full complex of teacher, set of students, physical environment, etc.) and *not the individual student*. The six experimental

Table 6 General summary table for the analysis of variance of a 2-factor factorial experimental design with n independent observations per cell (treatment group).

Source of variation	Sum of squared deviations	Degrees of freedom	Variance estimate	F ratio
Between groups	SS_{bg}	$rc-1$	$s_{bg}^2 = SS_{bg}/(rc-1)$	s_{bg}^2/s_w^2
c Columns (C)	SS_C	$c-1$	$s_C^2 = SS_C\ /(c-1)$	s_C^2/s_w^2
r Rows (R)	SS_R	$r-1$	$s_R^2 = SS_R/(r-1)$	s_R^2/s_w^2
$C \times R$ interaction	SS_{CR}	$(c-1)(r-1)$	$s_{CR}^2 = SS_{CR}/(c-1)(r-1)$	s_{CR}^2/s_w^2
Within groups (error)	SS_w	$cr(n-1)$	$s_w^2 = SS_w/cr(n-1)$	—
Totals	SS_T	$ncr-1$		

treatment combinations were *assigned at random to the 18 classes*, and not to the individual students. The difference between studying the class *milieu* and studying individual students, is the difference between studying cornstalks and ears of corn, or between ears of corn and kernals of corn. Since our experimental treatments were assigned at random and applied to the 18 classes, our concern is with the variation of the performance of those 18 classes, and we do not need the detailed information for each student except to produce a total score for each class

In fact, the class sizes *have* contributed to the reliability of the inter-class data, in that the given classes resulted in less inter-class variability than if the classes were smaller in size. And this *empirical* effect is of benefit to us, since the variation between classes (in the absence of treatment-produced differences) is the error variance in this study, as measured by the variation of the three classes in each treatment cell about the cell mean, pooled across the six treatment cells, with degrees of freedom $df = (3-1)(6) = 12$.

Still, we might persist in wondering whether we could somehow benefit from the much larger degrees of freedom associated with the individual student scores, reasoning as follows: If we *assume* that the various classes represent random samples of students from a larger population, then the variance s^2 of each class when divided by its sample size, s^2/n, would be nothing other than $s_{\bar{x}}^2$, an estimate of the variance of the sampling distribution of the class means; which, in turn, would correspond to the error variance estimate obtained in the analysis of variance of the experimental data—the pooled within cells variation of the class means about their respective cell mean. Now *if* the student-based and class milieu-based variance estimates are one and the same, why not take advantage of the much larger degrees of freedom associated with the student-based estimates, taking a weighted average of the 18 separate

estimates, which, if the average class size equalled 30 students, would have $df = (18)(30) - 18 = 522$, which is essentially ∞, infinity, in terms of the critical F required to declare statistical significance. To see the effect of this *assumption*, turn to the F table in Appendix Table V, for $\alpha = .05$, and compare the critical value of F for an error variance denominator with $df = 12$ vs. $df = \infty$: The respective F values are 4.75 and 3.84 (for a numerator $df = 1$), 3.89 and 3.00 (for a numerator $df = 2$), 3.49 and 2.60 (for a numerator $df = 3$), etc. Taking the ratios of these critical F's, for $df = 12$ vs. $df = \infty$, we get values of $3.84/4.75 = .81$, $3.00/3.89 = .77$, $2.60/3.49 = .74$, etc. So, if the *assumption* that the student-based and class-based estimates of the error variance were equivalent, then the larger degrees of freedom for the student-based estimate would require critical F's about 20% to 30% lower than if the experimental within-cells class-based estimate (*of the same value*) were used with its $df = 12$, resulting in increased power for identifying treatment effects. Unfortunately, this approach is not valid, since the fundamental assumption is incorrect.

The problem with the above approach is the premise that the class compositions represent random samples of a larger population, each with comparable expected mean and variance estimates. It implies that *student differences* fully account for *class differences*, which is not the case; there are classes of the gifted and classes of the retarded, classes in the ghetto and classes in Beverly Hills, classes of males and classes of females, etc., each occurring more often than would be expected by random formation of the classes, and the within-class variances are likely to be systematically different (usually smaller) than if completely random classes were formed. Correlated to varying degrees with these student differences are differences in teacher characteristics, supervisory personnel characteristics, budgets, instructional materials, physical plant differences, etc., all of which contribute to between-class differences. It should be clear, then, that variation among students within a class cannot fully account for variation *among classes* of students.

Using the class milieu as the experimental object, their random assignment to the alternative treatments effectively randomizes their many differences among the various treatment groups so that each experimental group of classes is expected to be like every other group, and the variation among the classes in each treatment group provides us with the needed error variance estimate of how *classes* vary in the population at large.

In the final section of this chapter, Analysis of Covariance, we will consider using the individual student information for estimating regression equations relating student predictor characteristics with criterion variables. Such relationships are sometimes useful for increasing the power of our analyses, by decreasing the size of the error variance estimate.

In the following discussion we will take a closer look at the meaning of an *interaction* effect among variables.

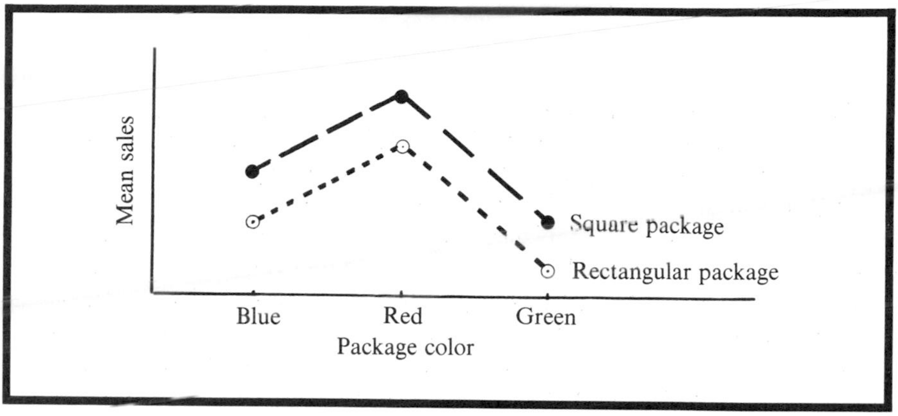

Figure 4 Hypothetical results of a two-factor experiment with no interaction effect present. (Lines connect the data points to emphasize differences.)

Interaction. Whenever we measure the effect of two or more experimental or predictor variables on a criterion variable, we must be on the alert for possible *interaction effects*—i.e., whether or not the effect of one variable on the criterion variable is the same regardless of the existing levels of a second variable. This is more easily seen graphically. Figure 4 shows some hypothetical results from the *package color* × *shape* experiment. We see that both package color and package shape have an effect on sales. The red package outsells the other colors, and the square package outsells the rectangular. Also, we see that the relative effect of package color on sales is the same *regardless of whether the package is square or rectangular*; or, equivalently, the square vs.

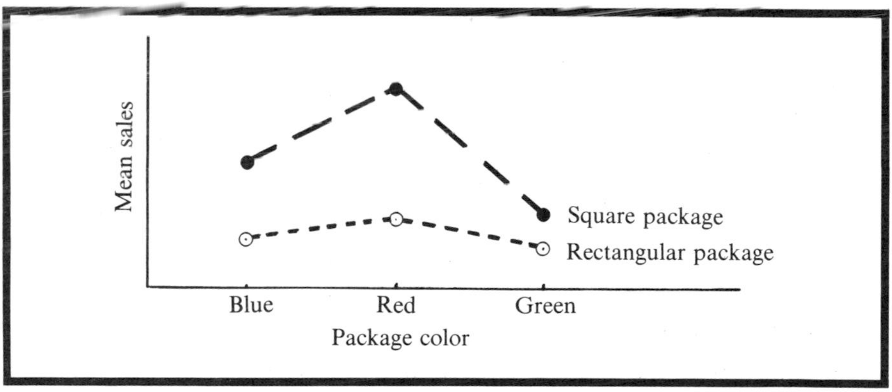

Figure 5 Hypothetical results of a two-factor experiment with an interaction effect present.

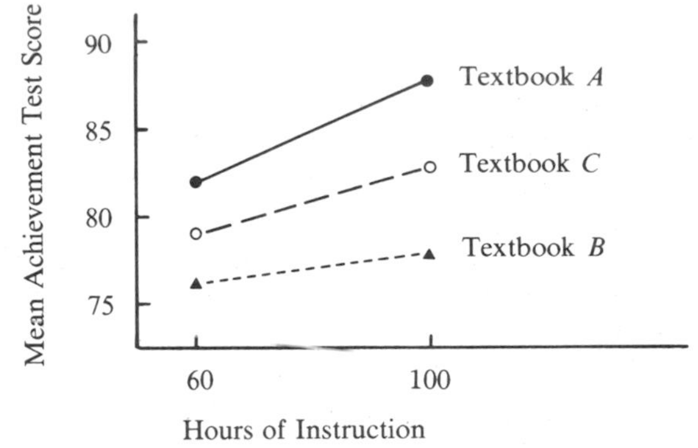

Figure 6 Mean achievement test scores, by hours of instruction and type of textbook.

rectangular difference is the same regardless of color—i.e., there is no interaction between package color and package shape.

Figure 5, on the other hand, shows a hypothetical example in which there *is* an interaction between the effects of the two experimental variables, package color and shape. In this instance, the color of the package (the first experimental variable) has an effect on sales (the criterion variable) *depending on the shape of the package* (the second experimental variable). Specifically, the different color packages lead to different sales only when the package is square in shape. When it is rectangular, color has no effect. Alternatively, we could say that package shape has a different effect on sales depending on the color of the package. Notice in Figure 5 that the difference in sales engendered by the square and rectangular package is much more pronounced when the package is red in color. It is the departure from parallel effects that signals an interaction.

Figure 6 presents the results of the *textbook* × *hours of instruction* experiment. Although the effects are not perfectly parallel—since it seems that the 60 vs. 100 hours of instruction effect differs from textbook to textbook—the departure is not more than would be expected by chance, as evidenced by the insignificant F ratio in the analysis of variance of Table 5.

There are two key reasons for identifying the existence of interaction effects among experimental or predictor variables. The first is for strictly statistical purposes. The validity of most multi-factor analytical models rests on an assumption of no interaction effects among the experimental or predictor variables. For example, if the data in Figure 7 were subjected to an analysis of variance, it is likely that the F test for the difference in effectiveness of drugs A and B would prove nonsignificant, since that source of variation would be

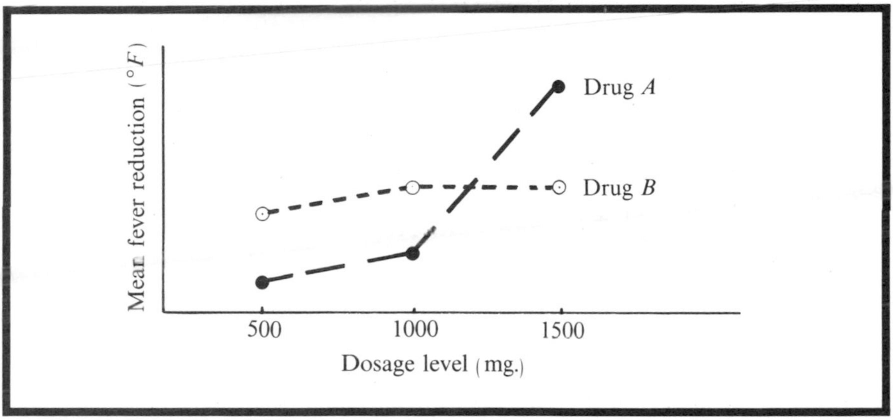

Figure 7 An interaction between the effects of drug type and dosage.

based on their overall performance, *averaged over dosage level.* But we see that there is a difference between drugs A and B *depending upon dosage level.* That is, there is an interaction between *type of drug* and *dosage level.* And that exemplifies the second key reason why we want to detect interaction effects. Simply looking at overall effects, without taking into account the levels of other variables, may lead us to make generalizations from our data that are misleading or drastically incorrect.

In many types of experimental designs, such as the one testing the effects of package color and shape, or textbook type and hours of instruction, it is possible to test for interaction effects. If they prove to be insignificant, we can proceed with confidence to test the "main effects" of package color and package shape, or textbook and hours of instruction. If the interaction does prove to be significant, we must then qualify our statements about the effects of either experimental variable—e.g., "The various package colors have a differential effect on sales only when the package is square in shape," or "drug A is superior to drug B, but only at the high dosage level." Figure 8 shows several experimental results with and without interactions. How we go about identifying which sub-effects are significant and which are not will be the subject of the later section on multiple comparisons.

Three-factor factorial design. Factorial experimental designs are not limited to two experimental variables as in the preceding examples. We could expand the two-factor package test into a three-factor design by introducing a third variable, say price. The expanded design is shown in Figure 9 in a three-dimensional schematic. With price at two levels, we now have a total of $2 \times 3 \times 2 = 12$ experimental groups. Since we need at least two stores per group

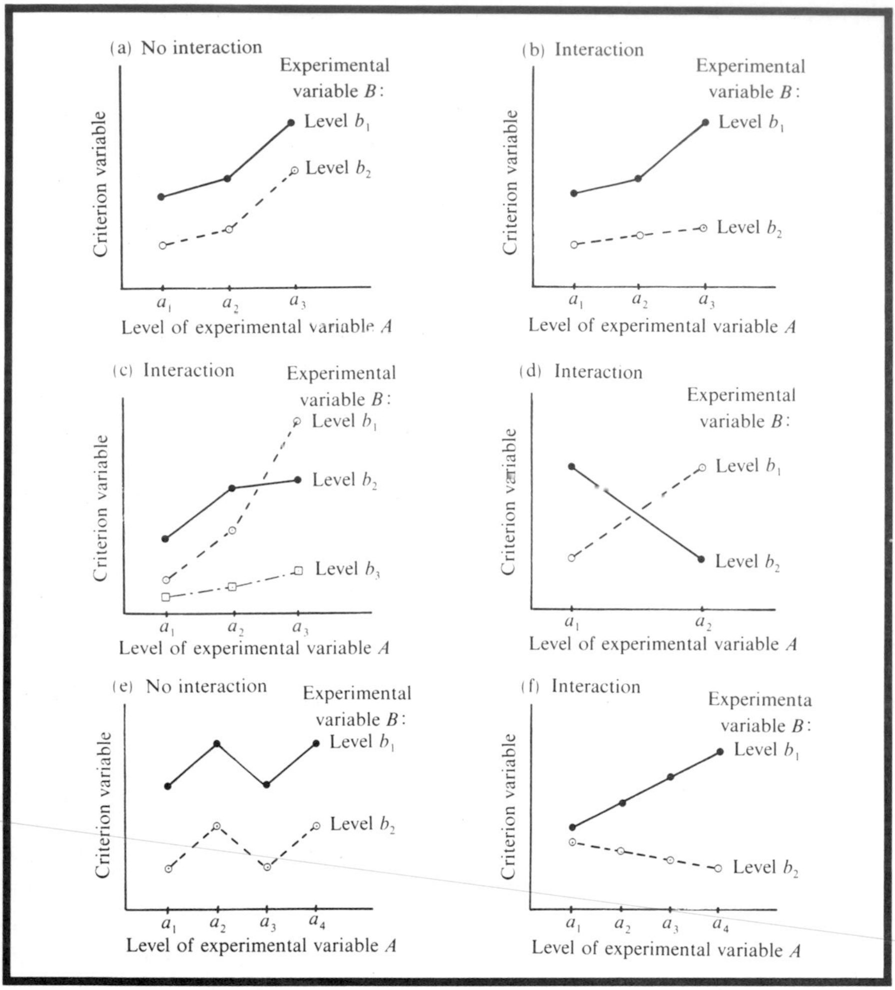

Figure 8 Examples of two-factor experimental results with and without the presence of interaction effects.

in order to obtain a within groups error variance estimate, we would have to increase the number of stores in our test to at least 24. In such a design we must be on the alert for several possible interaction effects, for now there are three experimental variables that can interact with each other—e.g., color with shape, color with price, shape with price, and then there can be a three-way interaction which involves a situation in which a two-way interaction itself differs depending on the level of the third variable.

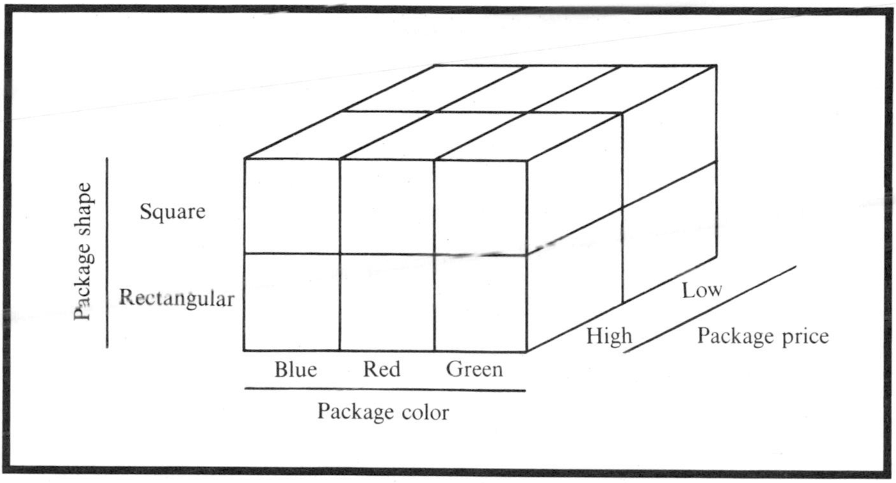

Figure 9 Schematic diagram of a $2 \times 3 \times 2$ three-factor experimental design.

In the study of the alternative textbooks and the different levels of hours of instruction, we could expand it to a three-factor design by introducing a third variable, perhaps class size, wherein we would purposefully manipulate the class enrollment to, say, either 20 or 40 students. As a result we would have a $3 \times 2 \times 2$ factorial design for a total of 12 treatment combinations. To have a minimum of two classes per condition we would need a total sample of 24 classes to assign at random to the 12 treatment groups.

As we might expect, factorial designs can be expanded to as many variables as we like. On the one hand, the advantage of these multi-factor designs is that several experimental variables can be assessed more efficiently than if separate one-factor experiments were conducted, since the same experimental objects do multiple duty, being measured at levels of a number of variables simultaneously; and furthermore, the results of such experiments can be generalized more confidently than if each variable were measured independently of other factors.

On the other hand, if the experimental design gets too complicated, with too many experimental variables, the administration and strict control of the experiment becomes troublesome, and the existence of too many higher-order interactions may obstruct the forest for the trees. Still, in most experimental work, we do not begin with complicated experiments, but with simple one or two variable designs and then expand upon them a little at a time, as we learn more about the fundamental variables in our particular area of study. Or, as in

Table 7 General summary table for the analysis of variance of a 3-factor factorial experimental design with n independent observations per cell (treatment group).

Source of variation	Degrees of freedom	Variance estimate	F ratio
Factor A (a levels)	$a-1$	s_A^2	s_A^2/s_w^2
Factor B (b levels)	$b-1$	s_B^2	s_B^2/s_w^2
Factor C (c levels)	$c-1$	s_C^2	s_C^2/s_w^2
AB interaction	$(a-1)(b-1)$	s_{AB}^2	s_{AB}^2/s_w^2
AC interaction	$(a-1)(c-1)$	s_{AC}^2	s_{AC}^2/s_w^2
BC interaction	$(b-1)(c-1)$	s_{BC}^2	s_{BC}^2/s_w^2
ABC interaction	$(a-1)(b-1)(c-1)$	s_{ABC}^2	s_{ABC}^2/s_w^2
Within groups (error)	$abc(n-1)$	s_w^2	—
Total	$abcn-1$		

the case of many correlation and regression analyses, we may study many variables at once strictly as a "fishing" or hypothesis generation expedition.

Table 7 presents the general format for the summary analysis of variance table for a 3-factor factorial design.

6. Randomized Block Design

In all of the examples discussed so far, independent groups of objects were measured at each treatment level or combination; i.e., unrelated sets of stores were used for the various package color and shape conditions, unrelated sets of classrooms were used for the various textbook and hours of instruction treatment conditions, and unrelated sets of patients were used for the various drug type and dosage conditions. In other words, our *total* sample of objects was randomly assigned to the various conditions without regard to individual characteristics of those objects, knowing that our randomization procedure would assure that all groups would be the same in their *expected value* for these extraneous characteristics. Nonetheless, if they are related to our criterion variable, the *variation* in these uncontrolled characteristics will appear in the within treatment groups variation and thus contribute to the size of the error variance, which we want to keep as small as possible.

This unwanted error variance can be reduced experimentally if we form several sets or *blocks* of homogeneous objects, with as many objects per block

as there are treatments. If we have m treatments, then we would form several blocks each containing m homogeneous objects. *The m objects in each block are then assigned at random to the m treatment conditions.* We have already seen a version of this design in Chapters 8 and 9, when we considered the case of $m = 2$ treatments, and the blocks conformed to twins, room-mates, spouses, neighbors, or other naturally occurring homogeneous pairs. Now we want to expand that concept to the situation in which we have more than two treatments under study. Consequently, we must either look for, or create, homogeneous blocks with as many members as we have treatment conditions, so that the members within each block can be assigned at random to the treatments. The result, then, is a rectangular data matrix in which the columns represent the various treatments, and the rows represent the various blocks (or vice versa). This type of experimental design is known as a *randomized block design*, or more descriptively as a *treatments × blocks design*. *The members within each block are randomly assigned to the treatment conditions*, thus the *randomized block* name.

As mentioned above, we need to locate naturally occurring blocks of homogeneous objects, or create them. Naturally occurring blocks would include animal litters, siblings, partitioned land plots, cliques, neighbors, peas in a pod, etc. If we cannot identify naturally occurring blocks of homogeneous objects, we can always create them. The important thing is that we form them on the basis of a variable that is logically related to the criterion variable of interest, otherwise the formation of the blocks will serve no useful purpose and can even reduce the power of our experiment instead of increasing it. For example, if we wanted to test the effects of four drugs on learning ability, we would form blocks of subjects based on their aptitude scores. Suppose 28 subjects were sampled for the study. We could rank them in aptitude score, and then take each four in turn as comprising a block, resulting in seven blocks of subjects. Within each block, the subjects are assigned at random to the four treatment conditions. Thinking of this procedure for a minute should make it clear how it differs from the *completely randomized* one-way design in which the 28 subjects would have been assigned at random to the four treatment conditions without regard to their aptitude scores, resulting in greater random error variation. The important thing is that the formation of the blocks is based on some variable which is logically related to the criterion variable under study, otherwise why control for it, for its variation would not affect variation in the criterion variable. In a learning study, for example, it would do no good to form blocks based on homogeneity of some unrelated variable such as blood cholesterol level, and in fact such irrelevant blocking can decrease the power of our experiment to below what it would be with a regular randomized one-way design, since some degrees of freedom for the error variance are lost by virtue of the blocking feature of the design.

Whenever possible, the randomized block (treatments × blocks) design is to be preferred over a completely randomized design such as those discussed earlier in the chapter. The increased power of the treatments × blocks design is due to the fact that effective block formation will reduce our error variance term—the denominator of the F ratio—and consequently we are more likely to get a significant F ratio for a given difference between treatment populations. To see this in a different light, recall that in the one-way analysis of variance the total variation is partitioned into *between treatments* and *residual error*, while in the treatments × blocks design it is split into *between treatments, between blocks*, and *residual error*. It follows that the extraction of the blocks variation from the data will leave a smaller residual error variation. This will become clearer in the following numerical example, where we will see not only how the error variance is decreased, but how we simultaneously lose degrees of freedom for the error variance by virtue of the blocking. Consequently, a larger F is required for significance and that effect must be offset by the smaller denominator of F, in order for the blocking operation to be of benefit to us.

Numerical example. Table 8 presents a detailed outline of the analysis of variance for a treatments × blocks experiment. Four treatment conditions and five blocks are used. The four treatments consist of the injection of drug A, drug B, a placebo injection, or a "handling-only" control condition. The five blocks consist of 4-member litters of a laboratory animal species. Within each litter the members are assigned at random to the four treatment conditions. The criterion measure is the number of trials required to learn a maze to a specified criterion.

Part I of Table 8 presents the raw data along with the *treatment* means, *block* means, and *overall* mean. In part II we obtain the total sum of squared deviations by the usual method—subtracting the overall mean from each observation, squaring the deviation, and summing all the results. In part III the variation of the treatment means about the overall mean is obtained, and each squared deviation is weighted by the sample size upon which each treatment mean is based. Similarly, in part IV, the variation of the block means about the total mean is obtained, again weighted by the sample size upon which each block mean is based. Finally, in part V, the residual error is obtained, either by the indicated calculations or by subtracting out the treatment and block sum of squared deviations from the total sum of squared deviations.

The various sums of squared deviations are presented in the summary analysis of variance table in part VI of Table 8. Beside them are their respective degrees of freedom. For treatments we have $df = 4-1 = 3$, for blocks we have $df = 5-1 = 4$, and for residual error we have $df = (4-1)(5-1) = 12$, the product of the degrees of freedom for the treatments and blocks. Again, each variance estimate is obtained by dividing each sum of squared deviations by their respective degrees of freedom. The residual error variance

Table 8 Outline of steps in the analysis of variance of a randomized block (treatments × blocks) experimental design.

I. The raw data, x_{ij}, along with treatment and block means.

Blocks ($i = 1,2,3,4,5$)	Treatment conditions ($j = 1,2,3,4$) Control	Placebo	Drug A	Drug B	Block means $\bar{x}_{i.}$
Litter 1	22	25	39	34	$\bar{x}_{1.} = 30.0$
Litter 2	26	35	42	37	$\bar{x}_{2.} = 35.0$
Litter 3	14	15	28	19	$\bar{x}_{3.} = 19.0$
Litter 4	19	24	33	24	$\bar{x}_{4.} = 25.0$
Litter 5	14	16	23	11	$\bar{x}_{5.} = 16.0$
Treatment means, $\bar{x}_{.j}$	$\bar{x}_{.1} = 19.0$	$\bar{x}_{.2} = 23.0$	$\bar{x}_{.3} = 33.0$	$\bar{x}_{.4} = 25.0$	$\bar{x}_{..} = 25.0$ (Total mean)

II. Total sum of squared deviations.

$$SS_T = \sum\sum(x_{ij} - \bar{x}_{..})^2 = (22-25)^2 + (25-25)^2 + \cdots + (11-25)^2 = 1590$$

III. Column (treatments) sum of squared deviations.

$$SS_C = \sum n_{.j}(\bar{x}_{.j} - \bar{x}_{..})^2 = 5(19-25)^2 + 5(23-25)^2 + \cdots + 5(25-25)^2 = 520$$

IV. Row (blocks) sum of squared deviations.

$$SS_R = \sum n_{i.}(\bar{x}_{i.} - \bar{x}_{..})^2 = 4(30-25)^2 + 4(35-25)^2 + \cdots + 4(16-25)^2 = 968$$

V. Columns × rows (treatments × blocks) residual error sum of squared deviations.

$$SS_{CR} = \sum\sum(x_{ij} - \bar{x}_{.j} - \bar{x}_{i.} + \bar{x}_{..})^2 = (22-19-30+25)^2 + (25-23-30+25)^2 + \cdots$$
$$+ (11-25-16+25)^2 = 102$$

Or, more easily:

$$SS_{CR} = \textit{Total} - \textit{Columns} - \textit{Rows} = SS_T - SS_C - SS_R = 1590 - 520 - 968 = 102$$

VI. Summary analysis of variance table.

Source of variation	Sum of squared deviations	Degrees of freedom	Variance estimate	F ratio
Treatments	520	3	173.3	20.39*
Litters	968	4	242.0	28.47*
Residual error	102	12	8.5	—
Totals	1590	19		

$^*p < .001$

VII. Summary analysis of variance table (in general).

Source of variation	Sum of squared deviations	Degrees of freedom	Variance estimate	F ratio
c Column treatments (C)	SS_C	$c-1$	$s_C^2 = SS_C/(c-1)$	s_C^2/s_{CR}^2
r Row blocks (R)	SS_R	$r-1$	$s_R^2 = SS_R/(r-1)$	s_R^2/s_{CR}^2
$C \times R$ residual error	SS_{CR}	$(c-1)(r-1)$	$s_{CR}^2 = SS_{CR}/(c-1)(r-1)$	—
Totals	SS_T	$cr-1$		

estimate is then used as the denominator of the F ratio for testing the treatment variance. With $df = 3,12$ the obtained $F = 20.39$ is well beyond the critical F needed at the $\alpha = .05$ significance level. Indeed, its p value is less than 1 in a thousand. For the litter blocks we have $F = 28.47$ with $df = 4,12$ which is also highly significant, as we would expect it to be.

Since we have rejected the null hypothesis that the four treatments conditions are the same, we naturally want to know which means differ from which. This is a question of *multiple comparisons*, and since it is a tricky subject it deserves discussion in a separate section of its own, later in the chapter.

The general form of the summary analysis of variance table for the treatments × blocks (randomized block) experimental design is presented in part VII of Table 8.

7. Other Designs

It is a rare research problem that cannot be successfully attacked with one or another of the designs already introduced, whether a one-way or factorial completely randomized design, or a randomized block design. Nonetheless, there are a multitude of other designs—usually special cases or combinations of the above designs—which strive to improve the power and efficiency of an experiment for a given cost of data collection. The full discussion of the many variations would fill a book again this size, and therefore is best left for a separate course of study. We will touch upon the more common of these alternative experimental designs.

Perhaps the best of all possible designs from the standpoint of power and efficiency is to use individual objects as blocks, such that *each object* receives all k treatments, as shown in Table 9. This is not always feasible, for we cannot always test a patient under k different drug conditions, nor could we test a student under k different instructional treatment conditions. But there are many situations where the alternative experimental treatments can be administered to each individual. For example, if we wanted to test the persuasiveness of three alternative anti-smoking TV commercials, we could draw a random sample of smokers and show each smoker all three messages, each time obtaining a numerical rating of the commercial's persuasiveness. We would be sure to randomize the order in which the three commercials were viewed so as to *counterbalance* or "wash out" any order-of-presentation effects. This type of design is often known as a *repeated measures design* or *subjects-as-their-own-control*, but it will be recognized as a special case of the randomized block, or treatments × blocks design as we prefer to call it, in which an individual serves as a block. In order for this special case to be strictly valid we need to assume that the inter-correlations between the columns of treat-

Table 9 An experimental design in which each object is measured under all levels of the experimental variable.

Objects	Experimental variable A			
	Level a_1	Level a_2	...	Level a_k
Object 1	———	———	———	——→
			...	
Object 2	———	———	———	——→
			...	
⋮	⋮	⋮		⋮
			...	
Object m	———	———	———	——→

ment observations are the same, a safe assumption if the order of presentation of the treatments is randomized from individual to individual. Otherwise the correlation between the first and second treatment, say, is not likely to be the same as between the first and final treatment, if the treatments are always presented in the same order. To appreciate the power and efficiency of this type of design, it needs to be recognized that measurements on the same individual are much less variable than measurements on twins, which are less variable than measurements on siblings, which are less variable than measurements on "matched" individuals, which are less variable than measurements on random individuals; the latter condition representing the completely randomized experimental design.

An interesting application of the repeated measures and factorial design involves the technique of *conjoint measurement*, or *trade-off analysis*, in which ratings or rankings are obtained for several alternative concepts, or stimuli, which are generated as factorial combinations of the levels of the variables under study. For example, to help in the design of an investment vehicle we could consider *three* levels of return on investment, *four* levels of liquidity, and *two* levels of downside risk, which would generate $3 \times 4 \times 2 = 24$ different investment descriptions. Presented in scrambled order, potential investors could then rate or rank the 24 concepts in terms of their appeal. The analysis of the mean ratings or rankings for each level of each variable would then provide an idea of the relative importance of, or trade-offs between, the various input variables, which would then guide product development. Or, as another example, a local government could formulate developmental plans by having a sample of residents rate a large number of proposals consisting of the factorial combination of alternative property tax rates, income tax rates, bond offerings, services to be offered, major new construction, etc. The analysis or "unfolding" of the data would then provide the importance or elasticities of the individual variables, providing input for the design of the optimum plan to serve the

Table 10 A two-factor experimental design in which different groups of objects are measured under the levels of one experimental variable (B), but all objects are measured under all levels of the other experimental variable (A).

Experimental variable B	Objects	Experimental variable A			
		Level a_1	Level a_2	$\cdots$	Level a_k
Level b_1	Object b_{11}	———	———	———	——→
				$\cdots$	
	Object b_{12}	———	———	———	——→
				$\cdots$	
	$\vdots$	$\vdots$	$\vdots$		$\vdots$
				$\cdots$	
	Object b_{1m}	———	———	———	——→
Level b_2	Object b_{21}	———	———	———	——→
				$\cdots$	
	Object b_{22}	———	———	———	——→
				$\cdots$	
	$\vdots$	$\vdots$	$\vdots$		$\vdots$
				$\cdots$	
	Object b_{2n}	———	———	———	——→

community, recognizing that the most appealing concept is not always the most feasible.

Another type of useful design is a combination of a randomized design and the repeated measures design, and is shown in Table 10. The experimental objects are randomly assigned to the alternative levels of one experimental variable (variable B), but each object is measured under *all* levels of the other experimental variable (variable A). This type of design could be used for testing consumer interest in alternatively worded new product descriptions (variable A) at alternative prices (variable B). Different groups of consumers would be exposed to the alternative prices, but at a given price level each consumer would be exposed to all of the alternative product descriptions or positionings.

Or, if the objects could not logically receive all levels of the variable A treatment, then the objects in Table 10 could represent *blocks* of homogeneous objects. For example, the blocks could consist of tumored rat litters, the B variable could be radiation vs. chemical therapy, and the A variable could represent alternative daily diets. The available litters would be randomly assigned to the radiation or chemical therapy condition, and within each litter the members would be randomly assigned to the alternative daily diets. In such designs, though, it must be recognized that two different error variance terms

must be called into play, since the B variable effect is one *between independent groups*—just as in the one-way randomized design—while the A variable effect is akin to the randomized block treatment effect.

These are just a sampling of the various types of experiments that can be conducted. Each has its own analysis of variance model, with an accompanying set of statistical assumptions. In each case, however, the basic principle is the same. Data is collected in such a way that the total variation can be broken down into its component sources, including the treatment effects in which we are interested and an estimate of the relevant error variance so that we can form the appropriate F ratio with which to test those treatment effects.

Also, it should be recognized that the interpretation of the results of an experiment depends upon the manner in which the levels of the experimental variables were selected. If the levels under investigation were chosen at random from a population, then what is called a *random effects* model is appropriate and we may generalize the results to the population at large. On the other hand, if the levels of the experimental variable were purposefully selected because they were of special interest, then a *fixed effects* model is appropriate and the results of the experiment may not be generalized beyond those particular experimental values. Finally, there are *mixed effects* models in which both random effects and fixed effects are present. Within each of these three classes of analytical models, there are numerous types of specific designs. Their details can be found in advanced texts on analysis of variance and experimental design.

Other subjects for advanced study include designs which allow us to study many variables at once—e.g., drug type, drug dosage, intake regimen, stage of symptoms, etc.—without having to study every possible combination of the levels of the various variables, which is often impractical or uneconomical. These types of designs, though, require stringent assumptions about the absence of interaction effects. Details of these kinds of designs can be found in advanced texts under the headings of *fractional factorial designs*, *nested designs*, *latin square designs*, *graeco-latin square designs*, and *orthogonal array designs*, to name the most common ones.

8. Multiple Comparisons

In both Chapters 8 and 9 we learned a straightforward method for making inferences about the difference between *two* population means, either using the interval estimation or the equivalent hypothesis testing approach. However, now we are confronted with the problem of testing the differences among *many* means. We establish the null hypothesis of equal population means

$$H_0\colon \mu_1 = \mu_2 = \cdots = \mu_m$$

which we test with the F ratio. Either we reject the hypothesis or we fail to reject it. But then, in either case, what do we do next? Do we go ahead and test individual pairs, or combinations, of means? For the possibility that the null hypothesis H_0 is *not rejected*, namely that the hypothesis of equal population means is tenable, there are basically three schools of thought. One states that no further tests are warranted—we failed to reject the null hypothesis and so it would be illogical to continue pursuing the issue. Another says that tests of pairs of means or combinations of means are okay provided that they had been *planned* prior to the collection of the data. Yet another approach endorses making any and all comparisons of means that look interesting.

Now even if the above issue could be solved satisfactorily, those who opt to make the comparisons are faced with still another problem; namely, what type of significance test and probability level is appropriate for declaring a significant difference between means, especially when we make a large number of such comparisons, many of which may not be independent of (orthogonal to) one another. This is a problem since we know that if we make a large number of tests of the null hypothesis at, say, the $\alpha = .05$ significance level, we will declare significant differences 5% of the occasions by chance alone when the null hypothesis is indeed true. We could make these type I errors even more frequently if we make tests that are not orthogonal to one another. For example, if we compare $\bar{x}_1$ with $\bar{x}_2$, it will not be orthogonal to the comparison of $\bar{x}_2$ with $\bar{x}_3$, since if $\bar{x}_2$ was exceptionally high, by chance alone, both significance tests would reflect this same extraordinary sampling error.

Now let us backtrack and consider the problem of what to do if our F test *has rejected* a null hypothesis H_0: $\mu_1 = \mu_2 = \cdots = \mu_m$. We can conclude that at least one of the population means differs from the others, but there are an infinite number of other possible deviations from the $\mu_1 = \mu_2 = \cdots = \mu_m$ situation. Should we go ahead and test every possible pair or combination of means? And how should the significance level be determined for so many comparisons?

That there is no easy solution to the problems outlined above is evidenced by the wide variety of techniques that have been proposed to attack the issues. The various approaches to the problem are named after pioneer workers in the field, including *Fisher*, *Scheffé*, *Tukey*, *Duncan*, *Newman*, *Keuls*, and *Dunnett*, to name the most common approaches. It would be impossible to review these many approaches in a few pages. Suffice it to say that each approach is better than the others for a given departure from the hypothesis of equal means, $\mu_1 = \mu_2 = \cdots = \mu_m$. But since there are an *infinite* number of departures from the null hypothesis, the exact nature of which we are never aware, it is difficult to choose one approach over the other. However, an emphasis on the nature of the problem, rather than on the variety of proposed solutions, should clarify the overall issue and provide a general strategy for addressing the problem. It is one of the most difficult subjects of all.

Planned vs. unplanned comparisons. As pointed out above, some researchers feel that pairs of means from among m means in an analysis of variance can be tested for significance by the traditional t test regardless of whether the overall F was significant, provided the comparison had been *planned* before the data was collected. In practice, the difficulty is in verifying which had been planned and which had not. Researchers rarely write them down and have them notarized.

Furthermore, researchers are likely to differ in the logic of their preplanned comparisons. At the extreme, a researcher could render the concept meaningless, by "preplanning" every conceivable comparison among the m means in the data collection, falsely believing that that is the most efficient way to get the maximum information from the experimental data.

The distinction, then, between planned vs. unplanned comparisons is not as clearcut or meaningful a criterion as it may seem as a basis for performing individual t tests among means when the F test is not significant; even if such tests could be justified in light of an insignificant overall F test of the entire set of means.

When tests of means are made contingent on a *significant* overall F test, the significance levels used in those subsequent comparisons are termed "protected," since the significant F acts as a gate-keeper preventing unwarranted comparisons. The most accepted approach for going directly to comparisons of interest, regardless of the overall F test, is if we are testing *a set of orthogonal comparisons* among the means; for that, as we will see, is planning in a truer sense.

Orthogonal vs. non-orthogonal comparisons. We have repeatedly referred to comparisons among means, and whether they are orthogonal or non-orthogonal comparisons. To arrive at a more general definition of these concepts let us consider some examples.

If we wanted to compare $\bar{x}_1$ with $\bar{x}_2$, we would write the difference between the means as $\bar{x}_1 - \bar{x}_2$. This is equivalent to the sum, or linear combination, $(+1)\bar{x}_1 + (-1)\bar{x}_2$. Notice that the respective coefficients of the means are

$$+1 \quad -1$$

and further notice that these coefficients sum to 0.

Or, if we wanted to compare $\bar{x}_1$ with the average of $\bar{x}_2$ and $\bar{x}_3$, we would have

$$\bar{x}_1 - \frac{\bar{x}_2 + \bar{x}_3}{2} = \bar{x}_1 - \frac{\bar{x}_2}{2} - \frac{\bar{x}_3}{2}$$

with the respective coefficients of the three means equal to

$$+1 \quad -\frac{1}{2} \quad -\frac{1}{2}$$

and notice again that the coefficients sum to 0.

Now let us say we want to compare the average of $\bar{x}_1$ and $\bar{x}_2$ with the average of $\bar{x}_3$ and $\bar{x}_4$, in which case we have

$$\frac{\bar{x}_1 + \bar{x}_2}{2} - \frac{\bar{x}_3 + \bar{x}_4}{2} = \frac{\bar{x}_1}{2} + \frac{\bar{x}_2}{2} - \frac{\bar{x}_3}{2} - \frac{\bar{x}_4}{2}$$

with respective coefficients of the four means being

$$+\frac{1}{2} \quad +\frac{1}{2} \quad -\frac{1}{2} \quad \frac{1}{2}$$

and again they sum to 0.

Each of the above comparisons of means is also referred to as a *contrast of means*. The essential characteristic is that the sum of the *coefficients*, or *weights*, equals 0. If we symbolize each weight with the letter w_i, then we can specify the sum of the weights as

$$\Sigma w_i = 0 \tag{1}$$

as the condition that must hold for a contrast of means. The reason for this should be apparent. If we wanted to test whether the difference between two means $\bar{x}_1$ and $\bar{x}_2$ is 0, it would not be logical to use coefficients of, say, $+3$ and -2.

Sometimes it is more convenient to multiply the coefficients by a constant so that we do not have to deal later with fractions. For instance, multiplying by 2, the coefficients $+\frac{1}{2}$ $+\frac{1}{2}$ $-\frac{1}{2}$ $-\frac{1}{2}$ can be written alternatively as

$$+1 \quad +1 \quad -1 \quad -1.$$

Similarly, multiplying by 3, the coefficients $+1$ $-\frac{1}{3}$ $-\frac{1}{3}$ $-\frac{1}{3}$ can be written as

$$+3 \quad -1 \quad -1 \quad -1$$

and the $\Sigma w_i = 0$ condition still holds. However, when using these *convenient* coefficients we should keep in mind the original defining coefficients for the contrast in question, and the multiplicative factor used to transform the original set to the convenient set. This is because the transformed coefficients will yield a contrast value that is a multiple of the original. We will see the

implications of this later in the section when we consider interval estimates for the contrasts.

With the above definition of a contrast of means, we can now define *orthogonal contrasts of means*: *Two contrasts of means are orthogonal to one another if the sum of the products of their respective coefficient weights equals zero*. Symbolically, if we let w_{1i} represent the coefficient weights for the first contrast, and w_{2i} represent the coefficient weights for the second contrast, then the two contrasts are orthogonal to each other only if:

$$\boxed{\Sigma w_{1i} w_{2i} = 0} \tag{2}$$

As an example, consider the following set of four means and three selected contrasts:

	$\bar{x}_1$	$\bar{x}_2$	$\bar{x}_3$	$\bar{x}_4$
Contrast 1:	+1	−1	0	0
Contrast 2:	+1	0	−1	0
Contrast 3:	+1	+1	−1	−1

Each of the three comparisons above is a legitimate contrast of means in that $\Sigma w_i = 0$ holds in each case, but are they all orthogonal to one another? Testing the first two contrasts against the orthogonality criterion $\Sigma w_{1i} w_{2i} = 0$, we have

$$(+1)(+1) + (-1)(0) + (0)(-1) + (0)(0) = 1$$

so the first two contrasts are not orthogonal, or independent of one another. We had an explanation of this earlier, in that any error reflected in $\bar{x}_1$ will be reflected in both contrasts.

To test whether the first and third contrasts are orthogonal, we have

$$(+1)(+1) + (-1)(+1) + (0)(-1) + (0)(-1) = 0$$

so we are assured that the first and third contrasts are orthogonal to one another. Verify that the second and third contrasts are not orthogonal, and describe each contrast in words.

In general, if we have a set of m means, we can create a set of $m-1$ orthogonal contrasts; however, different sets of $m-1$ contrasts are possible. Consider a set of four means for which we can create $4-1=3$ orthogonal

contrasts, such as:

	$\bar{x}_1$	$\bar{x}_2$	$\bar{x}_3$	$\bar{x}_4$
Contrast 1:	+3	−1	−1	−1
Contrast 2:	0	+2	−1	−1
Contrast 3:	0	0	+1	−1

An alternative set of orthogonal contrasts for the four means would be:

Contrast 1:	+1	+1	−1	−1
Contrast 2:	+1	−1	0	0
Contrast 3:	0	0	+1	−1

As an exercise, verify that the contrasts within each of the above sets are in fact orthogonal, and convince yourself that you cannot create a fourth orthogonal contrast within each set. Also recognize that there is an infinite number of *non-orthogonal* contrasts that you can create.

The distinction between sets of orthogonal and non-orthogonal contrasts also makes the planned vs. unplanned distinction more transparent, when we consider the following four possibilities:

	Orthogonal	*Non-orthogonal*
Planned	✓	?
Unplanned	?	✓

"Planned non-orthogonal" contrasts and "unplanned orthogonal" contrasts each seem to be a contradiction in terms. For if our contrasts are planned, we would logically choose orthogonal rather than non-orthogonal contrasts, for reasons which will become apparent; and if the contrasts are unplanned, it is unlikely that we would stumble upon a set of orthogonal contrasts. The two genuine categories, then, would seem to be *planned orthogonal contrasts* and *unplanned non-orthogonal contrasts*, or simply orthogonal vs. non-orthogonal contrasts, since these latter terms tend to imply whether they were planned or not.

Part of the importance that we place on a set of *orthogonal* contrasts rests on the fact that all the research information contained in the set of means is captured by a set of orthogonal contrasts. While this may not seem obvious at present, recall that we could not think of any additional contrasts in the above examples that were not dependent on one or another of the given ones. It is true, however, that we can look at the data with *alternative* sets of orthogonal

contrasts, but this is akin to looking at the data from different angles, it does not increase the information available to us. It is analogous to the "rotation of reference axes" that is discussed later in the Factor Analysis chapter. Another key advantage of orthogonal contrasts is presented in the following discussion.

Comparison-wise vs. set-wise significance levels. Since we are considering both *individual* contrasts of means and *sets* of such contrasts, two different significance levels are at issue. When considering individual comparisons, the significance level of *each* is of interest, and these are referred to as *comparison-wise* significance levels. When considering an entire set of comparisons, we are also interested in the significance level for the *full set* of comparisons being correct (i.e., not a single type I error in the set), and that is referred to as the *experiment-wise* or *set-wise* significance level.

The importance that we place on a set of orthogonal comparisons is that both of these significance levels are known to us. Since the nature of the comparisons within an orthogonal set are planned, they will be associated with whatever significance level α we choose. And since the comparisons are orthogonal to one another, we know that the *set-wise significance level* will be the *product of the respective comparison-wise significance* levels, by virtue of the multiplication rule of independent probabilities. For example, for a set of three orthogonal comparisons, each with a significance level $\alpha = .05$, the probability that the entire set of three comparisons is without a type I error will be $.95 \times .95 \times .95 = .86$, since .95 is the probability of correctly failing to reject the null hypothesis for each of the three independent comparisons. The probability of at least one type I error in the set will then be $1 - .86 = .14$.

On the other hand, when we deal with sets of unplanned non-orthogonal comparisons, these probabilities are not generally available to us, because of the unplanned nature of the comparisons, and because of the non-independence among them.

If we want to establish a *set-wise* rather than comparison-wise significance level at, say, $\alpha = .05$, this can be accomplished approximately by dividing the .05 value among the $m - 1$ orthogonal comparisons, to obtain the necessary comparison-wise significance levels. For example, for a set of 4 orthogonal comparisons we have $.05/4 = .0125$ as the required significance level for each, in order that the *set-wise* level is .05. As a check, we see that $(.9875)^4$ is very nearly the .95 value desired for the set. In general, the approximate relationship is

$$(1 - \alpha/n)^n \doteq 1 - \alpha,$$

where α is now the *set-wise* significance level and n is the number of orthogonal comparisons, and α/n is the comparison-wise significance level. More often than not, though, researchers typically give priority to the compari-

son-wise significance level at the usual .05, .01, or .001 values, letting the set-wise level fall where it may.

Interval estimates of contrasts. We will now consider popular approaches to creating interval estimates for preplanned orthogonal contrasts and unplanned non-orthogonal contrasts.

For the discussions that follow it will be useful if we identify any given contrast with the letter l. This is simply the value that the weighted combination of means assumes. In general, for a set of m means, any given *contrast of means* can be expressed

$$l = w_1\bar{x}_1 + w_2\bar{x}_2 + \cdots + w_m\bar{x}_m, \qquad \text{where } \Sigma w_i = 0$$

which is nothing other than we have been doing in the examples above; namely, summing the products of the various means with their respective weights. This can be shortened to the form

$$\boxed{l = \Sigma w_i\bar{x}_i, \qquad \text{where } \Sigma w_i = 0} \tag{3}$$

as the general definition of a linear contrast of means.

When evaluating a contrast of means we usually establish the null hypothesis

$$H_0\colon E(l) = 0$$

where $E(l)$, the *expected value* of the contrast l, is hypothesized to be 0. Since we are usually working with small samples, we can test the observed value of l against the hypothesized value with

$$t = \frac{l - E(l)}{s_l}$$

or, since $E(l)$ is assumed to be 0, we can write more simply

$$\boxed{t = \frac{l}{s_l}} \tag{4}$$

This will be recognized as another instance of the general t ratio

$$t = \frac{\hat{\theta} - \theta}{s_{\hat{\theta}}}$$

where θ is any given parameter, $\hat{\theta}$ is a sample estimate of that parameter, and $s_{\hat{\theta}}$ is the standard error of the sampling distribution of the estimate. To this point we have considered a number of identities for $\hat{\theta}$, including a single mean, $\bar{x}$; a mean difference between paired measures, $\bar{D}$; the difference between two independent sample means, $\bar{x}_1 - \bar{x}_2$; a proportion, $\hat{p}$; the difference between two independent proportions, $\hat{p}_1 - \hat{p}_2$; a regression coefficient, b; and now we consider yet another statistic, l, a contrast of means.

Instead of using

$$t = \frac{l}{s_l}$$

to test a preplanned hypothesis about l, we can equivalently create a *confidence interval* for $E(l)$, the expected value of l:

$$\boxed{E(l) = l \pm t_c s_l} \tag{5}$$

where t_c is the critical value of t for a given confidence level. Since we already know how to calculate l from (3), as a weighted sum of the means in question, all we need to know is the value of s_l, which is the standard error of the sampling distribution of l, and it is given by

$$s_l = \left(\frac{s}{\sqrt{n}}\right)\sqrt{\Sigma w_i^2}$$

and we have as our confidence interval:

$$\boxed{E(l) = l \pm t_c(s/\sqrt{n})\sqrt{\Sigma w_i^2}} \tag{6}$$

where

$E(l)$ = the expected value of contrast l.
l = the observed value of contrast l.
n = sample size upon which each mean is based (if equal sample sizes, otherwise see modification below).
Σw_i^2 = sum of the squared contrast coefficient weights.
s = the square root of the error variance estimate from the analysis of variance.
t_c = the critical value of t for a given confidence level, with degrees of freedom corresponding to those of the error variance estimate from the analysis of variance.

Notice especially that the standard error of l, s_l, is very similar to the standard error of an individual mean, except that it is increased by a factor of $\sqrt{\Sigma w^2}$. Also be aware that if the various means under study were based on samples of different size, then we would calculate the standard error as

$$s_l = s\sqrt{\frac{\Sigma w_i^2}{n_i}}$$

This may occur in a completely randomized design, but in a randomized block design the sample sizes are necessarily equal for the various means.

To see how the above confidence interval is applied, we will use the data from the earlier *control/placebo/drug A/drug B* experiment which was in the form of a randomized block (treatments × blocks) design. The four group means are reproduced in Table 11 and two alternative sets of orthogonal contrasts are considered. The procedure is outlined in detail, showing all the necessary calculations, so they need not be recreated here. Notice that once the confidence intervals are created, we inspect them to see whether the null hypothesis value $E(l) = 0$ is *included in the interval.* If not, we reject the null hypothesis. The only contrast for which the null hypothesis was not rejected was the handling-only control vs. placebo injection condition, and that missed by the narrowest of hairs, the interval being $-8.02 < E(l) < .02$, which just does include the 0 value. It is at this point that statistical theory breaks down and human nature takes over, for the investigator will surely want to think about possible real differences between the control and placebo conditions, which narrowly missed statistical significance.

In the above example the researcher was making preplanned orthogonal contrasts, each with a significance level $\alpha = .05$, so it is possible to determine the *set-wise* significance level for *no* type I errors in the entire set of three contrasts. The probability of correctly failing to reject the null hypothesis is .95 for *each* of the three contrasts, so $.95 \times .95 \times .95 = .86$ is the joint probability of correctly failing to reject the null hypothesis in all three instances, again by the multiplication rule for independent probabilities. As a result, the probability of at least one type I error in the set is $1 - .86 = .14$. In a strict sense, these are approximate probabilities, since the contrasts are not fully independent, because we used an error variance estimate s^2, instead of its unknown population value σ^2, in each comparison.

In the above example the researcher was making preplanned orthogonal contrasts, permitting an estimate of the set-wise significance level. But now let us consider an opposite extreme in which we have a large collection of unplanned non-orthogonal contrasts; or mixture of planned and unplanned, since who is to say what was planned. For this type of situation, we need to

Table 11 Interval estimates for two alternative sets of orthogonal contrasts of means (data from Table 8).

A set of orthogonal contrasts	Factor†	Treatment Means (n = 5 each): Control $\bar{x}_C = 19$	Placebo $\bar{x}_P = 23$	Drug A $\bar{x}_A = 33$	Drug B $\bar{x}_B = 25$	Σw	Σw^2	Value of contrast $l = \Sigma w\bar{x}$	Interval estimates for testing H_0: $E(l) = 0$, with a $\alpha = .05$ *comparison-wise* significance level. $l \pm t_c s_l$; $l \pm t_c(s/\sqrt{n})(\sqrt{\Sigma w^2})$ (see note); $l \pm (2.18)(\sqrt{8.5}/\sqrt{5})(\sqrt{\Sigma w^2})$; $l \pm 2.84\sqrt{\Sigma w^2}$
$\overline{CP}$ vs. $\overline{AB}$	(×2)	+1	+1	−1	−1	0	4	−16	$-16 \pm 2.84\sqrt{4}$, $-21.68 < E(l) < -10.32$*
C vs. P	(×1)	+1	−1	0	0	0	2	−4	$-4 \pm 2.84\sqrt{2}$, $-8.02 < E(l) < .02$
A vs. B	(×1)	0	0	+1	−1	0	2	8	$8 \pm 2.84\sqrt{2}$, $3.98 < E(l) < 12.02$*
An alternative set of orthogonal contrasts									*Note*: s is the square root of the error variance estimate from Table 8; t_c is based on df for the error variance.
C vs. $\overline{PAB}$	(×3)	+3	−1	−1	−1	0	12	−24	$-24 \pm 2.84\sqrt{12}$, $-33.84 < E(l) < -14.16$*
P vs. $\overline{AB}$	(×2)	0	+2	−1	−1	0	6	−12	$-12 \pm 2.84\sqrt{6}$, $-18.96 < E(l) < -5.04$*
A vs. B	(×1)	0	0	+1	−1	0	2	8	$8 \pm 2.84\sqrt{2}$, $3.98 < E(l) < 12.02$*

*Reject null hypothesis, H_0: $E(l) = 0$, with comparison-wise $\alpha = .05$

† The convenient coefficients used will produce intervals for hypothesis testing purposes. For intervals centered on the indicated mean differences, divide the resulting value of l and the upper and lower confidence limits by the (×) factor.

Sample computations for the C vs $\overline{PAB}$ contrast:

- First verify that $\Sigma w = 0$: $\Sigma w = +3 + (-1) + (-1) + (-1) = 0$
- Calculate Σw^2: $\Sigma w^2 = (+3)^2 + (-1)^2 + (-1)^2 + (-1)^2 = 12$
- Calculate contrast l: $l = \Sigma w\bar{x} = (+3)(19) + (-1)(23) + (-1)(33) + (-1)(25) = -24$
- Calculate confidence interval for $E(l)$ and determine if the null hypothesis value, H_0: $E(l) = 0$ falls within the interval (see computations in final columns above). If not, reject H_0.
- To confirm that two contrasts within the set are orthogonal to one another, multiply their respective weights and the sum of the products should equal 0. For example, for the C vs $\overline{PAB}$ and the P vs $\overline{AB}$ contrasts we have:

$(+3)(0) + (-1)(+2) + (-1)(-1) + (-1)(-1) = 0 - 2 + 1 + 1 = 0$ (Confirm for every other pair of contrasts in the set.)

have a *wider* confidence interval for each comparison, making it more difficult to reject the null hypothesis, since we are making comparisons willy-nilly, yet we want to control the size of the set-wise significance level.

For a set of such unplanned non-orthogonal contrasts, interval estimates can be created which provide us with a conservative level of confidence that the entire set of intervals is correct—that is, assurance that we will not commit *any* type I error with a probability greater than α. The interval estimate—often referred to as a *Scheffé-type* interval—is given by

$$\boxed{E(l) = l \pm \sqrt{(m-1)F_c}\,(s/\sqrt{n})\sqrt{\Sigma w_i^2}} \qquad (7)$$

where $E(l)$, l, s, n, and Σw_i^2 are the same as in (6). But now, instead of t_c we have the larger value $\sqrt{(m-1)F_c}$, where m is the total number of means under consideration, and F_c is the critical F value associated with degrees of freedom $m-1$, v, where v represents the degrees of freedom associated with s^2, the error variance estimate from the analysis of variance. Again, if the means are based on variable sample sizes, we would use

$$E(l) = l \pm \sqrt{(m-1)F_c}\left(s\sqrt{\Sigma w_i^2/n_i}\right)$$

instead of (7).

The application of this broader interval estimate to a large set of unplanned non-orthogonal contrasts is shown in Table 12. It is the same control/placebo/drug A/drug B data, but now instead of just three orthogonal contrasts we are looking at a full dozen contrasts. Even with the wider confidence interval, the null hypothesis is rejected for 9 of the 12 contrasts studied, which should not be surprising since they are *not* all independent of one another. Compare this *haphazard* approach with that in Table 11.

A word of caution about the choice of coefficient weights, w_i: If the primary purpose of creating the interval estimates is to test whether the null hypothesis value H_0: $E(l) = 0$ falls within the interval, then it makes no difference what multiple of values we choose for the coefficients, so long as $\Sigma w_i = 0$. But if we are also interested in the interval estimate as it relates to the actual means (and not a multiple of them), then we need to select the coefficients more systematically. For example, in Tables 11 and 12, the coefficients for the $\overline{CP}$ vs. $\overline{AB}$ contrast should be $+\frac{1}{2}\ +\frac{1}{2}\ -\frac{1}{2}\ -\frac{1}{2}$, instead of the computationally more convenient $+1\ +1\ -1\ -1$. Similarly, the coefficients for the C vs. $\overline{PAB}$ contrast should be $+1\ -\frac{1}{3}\ -\frac{1}{3}\ -\frac{1}{3}$. However, this distinction in choice of weights only affects the obtained value of l and the limits of the confidence intervals, and *not* the significance conclusions. As an

Table 12 Interval estimates for a set of non-orthogonal contrasts of means (data from Table 8).

Contrasts	Factor[†]	Control $\bar{x}_C = 19$	Placebo $\bar{x}_P = 23$	Drug A $\bar{x}_A = 33$	Drug B $\bar{x}_B = 25$	Σw	Σw^2	Value of contrast $l = \Sigma w\bar{x}$	Interval estimates for testing H_0: $E(l) = 0$, with an $\alpha = .05$ *set-wise* significance level. $l \pm \sqrt{(m-1)F_c}\, s_l$; $l \pm \sqrt{(m-1)F_c}\,(s/\sqrt{n})\sqrt{\Sigma w^2}$ (see note); $l \pm \sqrt{(4-1)3.49}\,(\sqrt{8.5}/\sqrt{5})\sqrt{\Sigma w^2}$; $l \pm 4.22\sqrt{\Sigma w^2}$	
C vs. P	(×1)	+1	−1	0	0	0	2	−4	$-4 \pm 4.22\sqrt{2}$,	$-9.97 < E(l) < 1.97$
C vs. A	(×1)	+1	0	−1	0	0	2	−14	$-14 \pm 4.22\sqrt{2}$,	$-19.97 < E(l) < -8.03$*
C vs. B	(×1)	+1	0	0	−1	0	2	−6	$-6 \pm 4.22\sqrt{2}$,	$-11.97 < E(l) < -.03$*
P vs. A	(×1)	0	+1	−1	0	0	2	−10	$-10 \pm 4.22\sqrt{2}$,	$-15.97 < E(l) < -4.03$*
P vs. B	(×1)	0	+1	0	−1	0	2	−2	$-2 \pm 4.22\sqrt{2}$,	$-7.97 < E(l) < 3.97$
A vs. B	(×1)	0	0	+1	−1	0	2	8	$8 \pm 4.22\sqrt{2}$,	$2.03 < E(l) < 13.97$*
$\overline{CP}$ vs. A	(×2)	+1	+1	−2	0	0	6	−24	$-24 \pm 4.22\sqrt{6}$,	$-34.34 < E(l) < -13.66$*
$\overline{CP}$ vs. B	(×2)	+1	+1	0	−2	0	6	−8	$-8 \pm 4.22\sqrt{6}$,	$-18.34 < E(l) < -2.34$*
$\overline{CP}$ vs. $\overline{AB}$	(×2)	+1	+1	−1	−1	0	4	−16	$-16 \pm 4.22\sqrt{4}$,	$-24.44 < E(l) < -7.56$*
A vs. $\overline{CPB}$	(×3)	−1	−1	+3	−1	0	12	−32	$-32 \pm 4.22\sqrt{12}$,	$-46.62 < E(l) < -17.38$*
B vs. $\overline{CPA}$	(×3)	−1	−1	−1	+3	0	12	0	$0 \pm 4.22\sqrt{12}$,	$-14.62 < E(l) < 14.62$
C vs. $\overline{PAB}$	(×3)	+3	−1	−1	−1	0	12	−24	$-24 \pm 4.22\sqrt{12}$,	$-38.62 < E(l) < -9.38$*

[†] The convenient coefficients used will produce intervals for hypothesis testing purposes. For intervals centered on the indicated mean differences, divide the resulting value of l and the upper and lower confidence limits by the (×) factor.

* Reject null hypothesis, H_0: $E(l) = 0$, with *set-wise* $\alpha = .05$.

Note: m = number of means in the set.
s = square root of the error variance estimate from the *anova* (Table 8).
F_c = the critical F for $\alpha = .05$, with $df = m - 1 = 3$ for the numerator, and df for the denominator equal to those for the error variance estimate—in this example $(m-1)(n-1) = 12$.

exercise, recompute the above intervals. How are the original confidence interval limits related to the new ones?

Finally, it is worth comparing the results from the alternative interval estimation approaches that we have considered. Notice that the interval estimates in Tables 11 and 12 are not directly comparable, since the set-wise confidence levels differ—86% in Table 11 and 95%, at least, in Table 12. To make a more direct comparison of the interval estimates under the alternative approaches, we can recompute the interval estimates for the orthogonal contrasts in Table 11 so that its set-wise confidence level is 95%, as for the non-orthogonal set. Dividing $\alpha = .05$ by 3, we have .0167 as the required significance level for each contrast test. From Appendix Table IV we can obtain the critical value of t (with $df = 12$, still, corresponding to the error variance estimate) through interpolation to be 2.81. Choosing the drug A vs. B contrast for illustration, we replace $t_c = 2.18$ with $t_c = 2.81$ in Table 11, and the interval estimate becomes

$$2.82 < E(l) < 13.18$$

for a set-wise confidence level of 95%, and a comparison-wise level of 98.33%. This compares with the narrower interval from Table 11

$$3.98 < E(l) < 12.02$$

for the set-wise confidence level of 86%, and a comparison-wise level of 95%. This, in turn, compares with the broader interval from Table 12

$$2.03 < E(l) < 13.97$$

when the comparison was among many unplanned non-orthogonal contrasts, with a set-wise confidence level of 95%. In this example, the drug A vs. B effect was so large, that the $E(l) = 0$ hypothesis is rejected no matter which of the alternatives was pursued. In borderline situations, conclusions could differ.

In summary, we have presented the two extreme approaches for creating interval estimates for contrasts among many means. At one end we considered preplanned orthogonal contrasts, the much preferred approach, and at the other end unplanned non-orthogonal contrasts, with wider intervals required for the latter in order to achieve a comparable set-wise confidence level. Between these two extremes are several alternative approaches mentioned at the start of the section, which formulate interval widths somewhere between the two extremes that we have studied; their full validity, though, depending upon the accuracy of their assumptions about the true but unknown popula-

tion means. As an exercise, these alternative methods can be researched and applied to the data of Tables 11 and 12 to determine the extent of difference among the various approaches.

Our general conclusion, then, is that we should try to avoid sets of non-orthogonal comparisons with our *experimental data.* Aside from being able to infer causality, the great advantage of experimentation over a *correlational* regression analysis is that we can deal with *orthogonal predictor variables*, allowing us to assess their individual contributions to the criterion variable; and so it is counterproductive, a step backward, to *introduce* non-orthogonality into our analysis. It is a rare research question that cannot be answered with one or two well chosen sets of orthogonal comparisons.

Significance tests of contrasts. Instead of the interval estimation approach above, we could perform equivalent significance tests. For the data in Table 11 we could use

$$t = \frac{l}{s_l} = \frac{l}{(s/\sqrt{n})\sqrt{\Sigma w_i^2}}$$

and arrive at exactly the same conclusions as with the interval estimate

$$l \pm t_c(s/\sqrt{n})\sqrt{\Sigma w_i^2}.$$

However, instead of t it is advantageous to use t^2, for reasons which will become clear. It so happens that an F with $df = 1, v$ (i.e., the two-group situation) is equal to t^2 with $df = v$. Verify this by consulting Appendix Tables IV and V, and comparing values of t^2 and F for various df. Consequently, we can write

$$F = t^2 = \frac{l^2}{s_l^2} = \frac{l^2}{(s^2/n)(\Sigma w_i^2)}$$

or

$$\boxed{F = \frac{nl^2}{s^2 \Sigma w_i^2}} \qquad (8)$$

where the values of n, l, s, and Σw_i^2 are identical to those in (6) and (7). Again, the above is for the situation with a constant sample size n for the various means. For variable sample sizes from mean to mean, n_i, we would

replace (8) with

$$F = l^2/s^2\Sigma(w_i^2/n_i).$$

Not only can the F in (8) be checked for significance for a given contrast, but the relationship points a way for us to obtain the contribution of the contrast to the total or between treatments sum of squared deviations. We can rewrite (8) as

$$\boxed{F = \frac{nl^2/\Sigma w_i^2}{s^2}} \tag{8a}$$

and with the error variance estimate s^2 in the denominator, we recognize that since the numerator has only one degree of freedom, it corresponds not only to the *mean sum* of squared deviations for l in the F ratio, but also the *sum* of squared deviations. Thus, the values of $nl^2/\Sigma w_i^2$ corresponding to the various orthogonal contrasts will add to the sum of squared deviations between treatments (or $l^2/\Sigma(w_i^2/n_i)$ for variable n_i).

The above approach to testing a set of planned orthogonal contrasts is presented in Table 13. It is the data from the *textbook* × *hours of instruction* factorial experiment considered earlier in the chapter. Notice how the sum of squared deviations in the final column of the table sum to the total sum of squared deviations between treatment groups. Also notice that the first two contrasts have sum of squared deviations which add to that for the *textbook* sum of squared deviations—the latter having been broken down into two orthogonal contrasts, each with one degree of freedom. Similarly, the textbook ×hours of instruction interaction sum of squared deviations is also broken down into two orthogonal components.

Whereas the initial analysis of variance in Table 5 informed us that the alternative textbooks were a significant source of variation, the orthogonal contrasts in Table 13 tell us further that textbook B performed more poorly than the average of textbooks A and C, and that textbook A performed better than textbook C. No new information is added about the hours of instruction, since we knew from the initial analysis that they differed, and with one degree of freedom cannot be broken down any further. The absence of detailed interaction effects is also confirmed. For example, the final contrast in Table 13 compares the difference between textbook A and C at the two alternative hours of instruction conditions; i.e.,

$$(T_AH_{60} - T_CH_{60}) - (T_AH_{100} - T_CH_{100}),$$

yielding the coefficients shown in the bottom row of the table.

Table 13 Significance tests and r^2 for a set of orthogonal contrasts of means (data from Table 5).

	The six treatment means ($n = 3$ each)									Significance tests, where each F has $df = 1,12$; and s^2 is the error variance estimate from Table 5.		
Textbook (T):	Textbook A		Textbook B		Textbook C							
Hours instruction (H):	60 hrs	100 hrs	60 hrs	100 hrs	60 hrs	100 hrs				$F = t^2 = \frac{l^2}{s_l^2} = \frac{l^2}{(s^2/n)\Sigma w^2} = \frac{nl^2}{s^2\Sigma w^2}$	Sum of squares	$r^2 = \frac{SS_l}{SS_{\text{Total}}}$
Group code:	T_AH_{60}	T_AH_{100}	T_BH_{60}	T_BH_{100}	T_CH_{60}	T_CH_{100}		Value of contrast				
Group means:	82.	88.	76.	78.	79.	83.	Σw^2	$l = \Sigma w\bar{x}$	l^2	$= \frac{3l^2}{5.83\Sigma w^2} = \frac{.515l^2}{\Sigma w^2}$	$SS_l = \frac{nl^2}{\Sigma w^2}$	$= \frac{SS_l}{346}$
Contrasts												
$\overline{T_AT_C}$ vs. T_B	+1	+1	−2	−2	+1	+1	12	+24	576	.515(576)/12 = 24.72***	144	.416
T_A vs. T_C	+1	+1	0	0	−1	−1	4	+8	64	.515(64)/4 = 8.24*	48	.139
H_{60} vs. H_{100}	+1	−1	+1	−1	+1	−1	6	−12	144	.515(144)/6 = 12.36**	72	.208
$(\overline{T_AT_C}$ vs. $T_B)\times H$ *interaction*	+1	−1	−2	+2	+1	−1	12	−6	36	.515(36)/12 = 1.54	9	.026
$(T_A$ vs. $T_C)\times H$ *interaction*	+1	−1	0	0	−1	+1	4	−2	4	.515(4)/4 = .52	3	.009
											$SS_{bg} = \overline{276}$	$R^2 = \overline{.798}$

$^*p < .05$
$^{**}p < .01$
$^{***}p < .001$

Geometric view of the contrasts:

Group means

	T_A	T_B	T_C
H_{60}	82.	76.	79.
H_{100}	88.	78.	83.

Contrasts

$\overline{T_AT_C}$ vs. T_B

	T_A	T_B	T_C
H_{60}	−1	−2	+1
H_{100}	+1	−2	+1

T_A vs. T_C

	T_A	T_B	T_C
H_{60}	+1	0	−1
H_{100}	+1	0	−1

H_{60} vs. H_{100}

	T_A	T_B	T_C
H_{60}	−1	+1	+1
H_{100}	−1	−1	−1

$(\overline{T_AT_C}$ vs. $T_B)\times H$

	T_A	T_B	T_C
H_{60}	+1	−2	+1
H_{100}	−1	+2	−1

$(T_A$ vs. $T_C)\times H$

	T_A	T_B	T_C
H_{60}	+1	0	−1
H_{100}	−1	0	+1

When making significance tests, multiplying the contrast coefficients by a constant factor, to obtain convenient whole numbers, does not affect the resulting F ratio since both the numerator and denominator are increased by the same factor and they cancel each other.

Coefficient of determination, r^2. The above approach to obtaining the sum of squared deviations for each contrast, can be further used to compute the value of r^2, the square of the correlation coefficient, but in this experimental data context referred to as the *coefficient of determination*. In essence, it is the squared correlation between the criterion scores and the contrast coefficients, and can be shown to equal

$$r^2 = \frac{nl^2/\Sigma w_i^2}{SS_{Total}} = \frac{SS_l}{SS_{Total}} \tag{9}$$

where n, l, and w_i are as in (6), (7), and (8), and SS_{Total} is the total sum of squared deviations from the Table 5 analysis of variance.

The values of r^2 are shown in the final column of Table 13, for the data of the textbook × hours of instruction experiment. Since the numerator of (9) is the sum of squared deviations for an individual contrast, we see again how we can interpret r^2 as the proportion of total variation accounted for by a predictor variable. Indeed, in the example of Table 13 we can think of each of the five contrasts as a *dummy variable* in a multiple regression analysis. Since the contrasts are independent of one another, their respective r^2 can be summed to provide the multiple R^2 which provides the proportion of the total variation accounted for by all six treatment combinations, which can be thought of as a qualitative variable. Recall that when we have a qualitative variable with m levels, we use only $m-1$ dummy variables in a multiple regression equation, which happens to correspond to the $m-1$ orthogonal contrasts from among a set of m means. As an exercise, perform the Table 13 analysis for the data in Tables 11 and 12.

We can now see how *regression analysis*, *analysis of variance*, and *correlation analysis* are all related to one another, each serving a different function. Whereas analysis of variance tests for the existence of relationships between predictor and criterion variables, regression analysis further provides an equation describing that relationship, and correlation analysis, using r^2, measures the extent of these relationships.

Trend analysis. When our predictor variable is quantitative rather than qualitative in nature, we are not only interested in whether there is a significant F ratio with regard to the differences among the respective means on the

criterion variable, but we are more interested in whether those means follow a systematic pattern or trend; say, linear, quadratic, cubic, etc.

To test hypotheses about the various forms that the relationship between the predictor and criterion variables assumes, we need a special set of orthogonal contrasts to assess the trends. To understand why, suppose that we have three means, A, B, and C. Now consider the variety of ways in which those three means might be arranged: ABC, ACB, BAC, BCA, CAB, and CBA, a total of six different orderings. But the F ratio for testing the hypothesis that the three means are identical does not take into account the *order* of the three means, only the differences among them. However, we can use a set of orthogonal contrasts that will take into account the actual order of the means that is observed, and consequently will distinguish between, say, an ABC vs. ACB ordering of the means.

As always, with m means, we can have a set of $m-1$ orthogonal contrasts, each with $df=1$. The particular contrasts in which we are usually interested involve those measuring linear, quadratic, cubic, and sometimes higher-order trends in the series of means. If we have three means, then we have only two orthogonal contrasts to consider, the linear and quadratic. The coefficients for the various types of trends are shown in Table 14, for the situations involving three, four, or five means. In most instances we hypothesize a *linear* relationship, *with the remaining* components considered a part of the error variance.

To see how the trend contrast coefficients take the ordering of the sample means into account, suppose we obtained means of 7.1, 9.6, and 12.9 on our criterion variable, corresponding to *equally spaced* values on a quantitative

Table 14 Sets of coefficients for orthogonal trend contrasts of means.

(a) Trend coefficients for three means.

Linear:	−1	0	+1
Quadratic:	−1	+2	−1

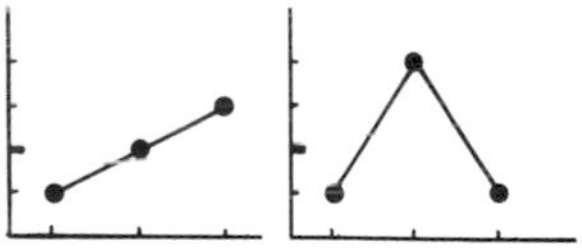

(b) Trend coefficients for four means.

Linear:	−3	−1	+1	+3
Quadratic:	−1	+1	+1	−1
Cubic:	−1	+3	−3	+1

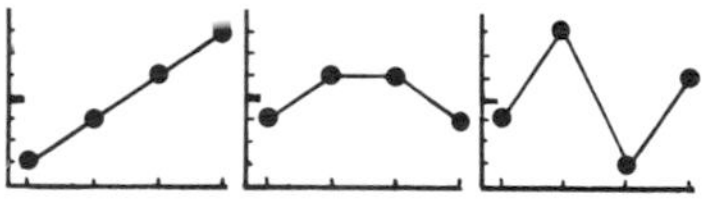

(c) Trend coefficients for five means.

Linear:	−2	−1	0	+1	+2
Quadratic:	−2	+1	+2	+1	−2
Cubic:	−1	+2	0	−2	+1
Quartic:	+1	−4	+6	−4	+1

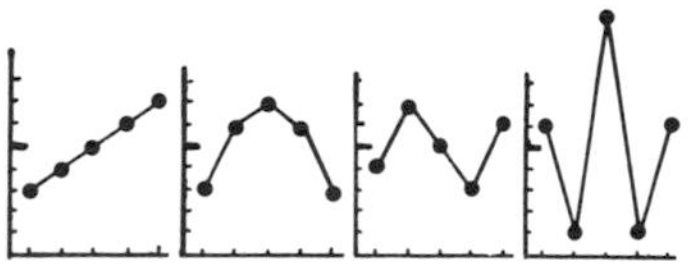

predictor variable (e.g., three levels of drug dosage, or hours of instruction, or number of learning trials, etc.). Using the linear trend coefficients from Table 14, namely

$$-1 \quad 0 \quad +1$$

and applying them to our sample means, we have

$$\textit{Linear trend contrast} = (-1)(7.1) + (0)(9.6) + (+1)(12.9) = 5.8$$

which like any other contrast of means will have $df = 1$, and can be tested for statistical significance by the same procedures as in the preceding sections, utilizing the error variance estimate from the analysis of variance of the original data, with its associated df.

The coefficients for the second contrast, the quadratic component, are shown in Table 14 to be

$$-1 \quad +2 \quad -1$$

which will test a parabolic up-then-down pattern of the three means. The value of the contrast is then

$$\textit{Quadratic trend contrast} = (-1)(7.1) + (+2)(9.6) + (-1)(12.9) = -.8$$

which also has $df = 1$. The virtually zero quadratic trend reflects the fact that most of the variation in means is due to the linear component.

Table 15 presents a summary of the tests of the *linear* and *quadratic* trends for the three means described above, based on the additional information about the sample sizes upon which each mean is based (6, 8, and 5, respectively) and the value of the error variance estimate, in this instance the pooled within-groups variance, $s^2 = 11.2$, which would be available from the analysis of variance of the primary data, and has $df = 16$ (representing $5 + 7 + 4$).

Notice that one additional wrinkle is introduced in this worked example, due to the different sample sizes for the various treatment groups. Whereas when our sample size was a constant n from group to group we could use

$$F = \frac{l^2}{s_l^2} = \frac{l^2}{(s^2/n)\Sigma w_i^2} = \frac{nl^2/\Sigma w_i^2}{s^2}$$

Table 15 Test of the linear and quadratic trend for $m = 3$ means.

	Treatment groups					F test using pooled within-groups error variance estimate, $s^2 = 11.2$ with $df = 16$.
	I	II	III			
Sample sizes (n_i):	6	8	5			
Group means $(\bar{x}_i)$:	7.1	9.6	12.9	$\Sigma(w_i^2/n_i)$	$l = \Sigma w_i \bar{x}_i$	$F = \frac{l^2}{s_l^2} = \frac{l^2}{s^2\Sigma(w_i^2/n_i)} = \frac{l^2/\Sigma(w_i^2/n_i)}{s^2}$
Contrasts						
Linear trend	−1	0	+1	.367	5.8	8.18*
Quadratic trend	−1	+2	−1	.867	−.8	.07
						*$p < .05$

to test each contrast, the presence of variable sample sizes requires that we use the slightly modified

$$F = \frac{l^2}{s_l^2} = \frac{l^2}{s^2\Sigma(w_i^2/n_i)} = \frac{l^2/\Sigma(w_i^2/n_i)}{s^2}$$

to test the contrasts.

The F tests are shown in the final column of Table 15, where we see that the linear trend is statistically significant, while the quadratic component is virtually zero. To fully understand how the trend coefficients take into account the actual order of the means, recompute the linear and quadratic contrasts for the identical means, but now in the order 7.1, 12.9, and 9.6.

Not only can the above type of analysis be used to measure the trend of criterion means associated with equally spaced values of a quantitative variable, but it can also be used to assess the trend components of means obtained at *fixed intervals of time*, as in ongoing surveys of population characteristics. Further analyses of such *time series* are discussed in Chapter 18.

9. Experimental vs. In-Tact Groups

In the preceding discussions we looked at the analysis of variance technique as it applies to experimental data; i.e., data collected from objects that are *randomly assigned to the alternative experimental treatments*. This random assignment of objects to experimental treatments is necessary for the validity of our analysis and any inferences of *causality* between the experimental variable and the criterion variable.

There are many instances, though, when analysis of variance is applied to groups of data that are not generated by objects randomly assigned to experimental treatments, but by *in-tact* groups of objects which vary on some dimension of interest. For example, with regard to the variable of gender, males and females are in-tact groups; with regard to political affiliation, Democrats, Republicans, and Independents are in-tact groups; with regard to educational achievement, high school and college graduates are in-tact groups; and the various levels of the variables of race, religion, and nationality, to name just a few, are represented by in-tact groups. Consequently, whenever we perform an analysis of in-tact groups with respect to some criterion variable of interest, we cannot infer any causal link between the variable defining the groups and the criterion variable. For example, if we discover a difference between smokers and non-smokers in their performance on a standardized test, we cannot attribute that difference to the smoking behavior *per se*. For example, there could be hormonal, genetic, personality, socio-cultural, or attitude variables which were at the root of the test performance differences and which at the same time predisposed individuals to the smoking behavior. Only if the individuals had been assigned to the smoker and non-smoker groups at random by the researchers, admittedly an impractical procedure, could differences between the groups be interpreted as being caused by the smoking behavior. Fortunately, such experiments can be performed with various animal species.

Such in-tact groups as discussed above will be recognized as analogous to the quantitative random variables studied in correlation analysis. While the distinction in the interpretation of experimental and correlational data has been treated at length in earlier chapters, it is re-introduced here to emphasize its importance. Research results appearing in the popular press, as well as many analyses performed in business and government, and even many academic research studies, emphasize relationships between variables which are based on strictly correlational or in-tact group data, often with the implication that the variables are causally related.

Whenever we come across statements relating two variables, the first thing we should ask is whether the supporting data is *experimental* or *correlational*. For only if the data came from a properly conducted and analyzed experiment in which objects were *randomly assigned to the treatment conditions* can we be confident that a causal link exists between the variables. This is not to say that correlational data cannot uncover variables that are very probably causally related—e.g., the incidence of tobacco smoking and the incidence of cancer—but we must realize that the relationship may also be due, either in whole or in part, to other unidentified confounding variables, as discussed above. The same holds true for differences found between such in-tact groups as males and females; Protestants, Catholics, Jews, Moslems, Buddhists, and

Hindus; Whites and Blacks; liberals and conservatives; circumsized and intact males; bottle-fed and breast-fed individuals; heterosexuals and homosexuals; murderers and non-murderers; rich and poor; etc.

The importance of this discussion is not that in-tact groups are without real differences, but rather that the causes of these differences are not always what we might think at first glance. Instead, they are due to a complex of factors including both *environmental* and *biological* sources, to varying degrees; and these sources, in turn, can be either *systematic* (i.e., controllable) or primarily *random* in nature. While differences between relatively transient groups such as users of Brand *A* vs. Brand *B* are likely to be mostly environmentally based, differences between more intransigent populations such as nationality, racial, religious, aptitude, sexual preference, and criminal groups are more likely—bottomline—to be predominantly biological than socio-cultural in nature, often due indirectly to sanctions against inter-group mating which tends to preserve unique genetic pools, originating perhaps from random gene sampling patterns thousands upon thousands of years ago. This, in turn, implies that efforts to erase (or address) differences between groups by naïve social means are destined to failure. A further consideration of this very important planetary issue is beyond the domain of statistical analysis *per se*.

10. Multivariate Analysis of Variance

Until now we have been concerned with analyses that have had only one criterion variable. It is often the case, however, that we are interested in not just one criterion variable, but perhaps two, three, four or more such characteristics. These situations might arise because we are unsure which treatment effect should be of prime concern to us, or we are legitimately concerned with a number of alternative effects, guided by some theory or by pure speculation.

For example, in a laboratory study of alternative drugs we might be interested in their respective effects on various behavioral and biological measures; say, trials to learn a maze, performance on an activity wheel, various blood chemistry measures, brain tissue characteristics, longevity, etc.

Another instance in which we might be interested in the multiple effects of our experimental variables would involve a communications study in which the stimulus material (perhaps alternative learning material or advertisements) differ in content and exposure rate, and we want to study their effects on the various criterion variables of comprehension, recall, persuasiveness, as well as overall and specific attitudes toward the subject matter.

Yet another situation would involve an experiment on the effects of alternative fertilizer and irrigation treatments. We would not only be interested in the key criterion measure of crop yield, but also in the size of the resulting produce, their chemical composition, and their taste, among other things.

In a study of educational treatments, our interest may not be restricted to scores on a single achievement test but on several, plus scores on attitude scales, delayed recall measures, absenteeism, etc.

The analysis of these types of studies in which we measure a number of alternative criterion variables, as opposed to a single key one, is referred to as *multivariate analysis of variance*, known by the abbreviation *manova*. It would seem that the analyses discussed previously in this chapter should also be called multivariate analyses of variance inasmuch as they involve more than one variable—at least one experimental predictor variable plus one criterion variable—but since they involve only *one criterion variable* they are usually labelled *univariate* analysis of variance, no matter how many experimental treatment variables are employed. Every discipline has its semantic conventions. Table 16 summarizes the four basic classes of analysis that we have considered. In addition to this nomenclature we will continue to use the term "multivariate" in its more general sense, as involving the simultaneous study of two or more variables.

Analysis of multiple criterion variables. From the drug, communication, agriculture, and education examples given above, it should be obvious that the analysis and interpretation of studies with multiple criterion variables will be somewhat more complicated than the situation in which we have only one criterion measure. Especially difficult is when we have both a large number of

Table 16 Characteristics of major classes of analysis of variance (*anova*) and multivariate analysis of variance (*manova*).

Name of Analysis	*Number of Predictor Variables*	*Number of Criterion Variables*
One-factor *anova*	**One** (e.g., drug dosage)	**One** (e.g., white blood cell count)
Multi-factor *anova*	**Many** (e.g., advertising expenditure, creative execution, allocation among media, allocation over time)	**One** (e.g., awareness of product)
One-factor *manova*	**One** (e.g., diet type)	**Many** (e.g., blood pressure, weight, cholesterol measure, mood)
Multi-factor *manova*	**Many** (e.g., fertilizer type, fertilizer amount, seed pretreatment, time of sowing)	**Many** (e.g., crop yield, produce size, protein content)

predictor variables and a large number of criterion variables, for not only does the possibility arise of uncovering higher order interactions among the treatment variables but these interactions could well differ depending upon which criterion variable we are looking at.

Indeed, a major problem of multivariate analysis of variance is how to analyze the many criterion variables: Should we look at each one separately, performing a univariate analysis of variance for each criterion variable? Should we consider all the criterion variables taken as a whole? Should we consider one or another linear combination of the criterion variables, and if so, should we add together their raw score or standardized z score form? If we combine criterion variables, should we use one set of weights or another? Should we look at selected subsets of the criterion variables, or what? The answer is, *all of the above* are done, plus more, depending upon the investigators' motivations.

This situation in which we have a plethora of criterion variables, while efficient from a data-gathering point of view, should impress upon us the importance of approaching an experiment with some kind of guiding theory as to the expected effects of our experimental variables, rather than treating the research as a shotgun with the hope that some significant effects will be identified in some variable or another. Furthermore, these criterion variables ideally should be identified before the experiment is conducted, so that we do not start making additional unplanned measurements *ad infinitum*, once the data begins to be analyzed—although in some instances initial data analyses might provide insights that can be checked out with the gathering of further information from our experimental objects, provided those treatment effects have not dissipated over time.

We must realize, too, that the sheer number of criterion variables that we measure will lead to finding statistically significant effects by chance alone—the type I error. To avoid this spurious identification of significant effects, an analytical technique is necessary which takes into account the number of comparisons we are making, just as in the multiple comparisons problem associated with a single criterion variable. How to address this problem of assessing multiple criterion variables is a difficult issue and the impetus for much theoretical work.

Significance tests. A number of test statistics have been formulated for determining whether treatment effects are present when considering *all* criterion variables as a whole.

Recall that in the univariate case, when we have but a single criterion variable, we are able to test (1) whether a mean obtained from a single sample differs significantly from some hypothesized value, or (2) whether the difference between the means from two independent samples differs significantly from a hypothesized zero difference; in each instance using Student's t statistic, or the large sample z ratio. When we deal with more than two sample groups, as

earlier in this chapter, we can use the univariate F statistic to test the hypothesis that the population means are all equal to one another. Corresponding tests exist for the multivariate case where we have several criterion variables, and consequently several means and variances for each group, as well as a *covariance* and *correlation matrix* for each group, reflecting the extent of relationship *among the many criterion variables within each group*.

When we have a single sample measured on several criterion variables we can test whether the *set* of means on those variables differs from a hypothesized set, using *Hotelling's T^2 statistic*, which is a generalization of Student's t statistic to more than one criterion variable. For example, we could test whether the set of means on various blood and urine measures for a sample of patients with a given disease differs from a *set* of normative population values. Or, if we took measurements on a number of variables before and after some experimental treatment administered to a single sample, we could test whether the *set* of mean *before-after differences* differed significantly from the hypothesized *set* of zero differences.

When we measure two samples on a number of variables, we can again use *Hotelling's T^2 statistic* or *Mahalanobis' D^2 statistic* to determine whether the *set* of means for one group differs significantly from the *set* of means for the other. If they differed significantly, it would reflect the differential effects of the alternative treatments that the two independent samples received.

In situations where we measure more than two independent groups, each on a number of criterion variables, then a variety of test statistics are in use—most commonly, *Wilks' Lambda*, *Hotelling-Lawley Trace*, and *Roy's Largest Root Statistic.*

All of the above test statistics are based on matrix algebra techniques and their mathematical definitions are subjects for advanced study. However, most multivariate computer statistical software packages report them, as well as alternative statistics, plus their associated probabilities, allowing users to identify significant effects.

Multivariate analysis of in-tact groups. Just as a univariate analysis of variance can be performed on data originating from in-tact groups, so too can a multivariate analysis of variance be performed on such non-experimental groups—but again, with no conclusions of causality. For example, we might compare voters for proposition A with those for proposition B (or males and females, or classroom A and B, etc.) on a number of variables in an attempt to ascertain if and how they differ. This type of problem of discriminating between in-tact groups can also be addressed by a related technique known as *discriminant analysis*, the subject of Chapter 14.

Multivariate normal distribution. In the Correlation Analysis chapter, while studying the linear relationship between *two* continuous random variables, we were introduced to the *bivariate normal distribution*. It appeared as a

bell-shaped mound, such that for each value of each of the two variables, the other variable was normally distributed. Furthermore, the variables were linearly related; i.e., a linear regression equation could be fitted to the data, the deviations from the regression line also being normally distributed. The extent of relationship between the two variables could be seen by taking a horizontal cross-section of the mound, which could range from a circle (reflecting zero correlation between the two variables) to an ellipse of varying narrowness (depending upon the tightness of the relationship).

When we study objects measured on many variables, not just two, the concept of a *multivariate normal distribution* is introduced. Precise mathematical definitions of the properties of the distribution are complicated, and like so many multivariate topics, depend upon higher mathematics and would seem a symbol salad to the majority of individuals who will eventually use multivariate analytical techniques. It is worthwhile, though, to try to visualize the properties of the multivariate normal distribution, even if it is only an intuitive understanding.

A set of variables is said to have a *multivariate normal distribution* if each of the variables has a linear regression on each of the others (simple regressions) *and* on every possible subset of the other variables (multiple regressions), and where the deviations from the regression equations are normally distributed. Is this of much help? We learned earlier that a multiple regression equation could be viewed as a plane in multi-dimensional space, a *hyperplane*, which itself is difficult enough to visualize, so we may be begging the issue to expect the above description to help us visualize the multivariate normal distribution.

Perhaps better would be to imagine each variable as having a univariate normal distribution for each and every possible combination of values of the remaining variables; and further imagine a bivariate normal distribution existing for each value of a third variable; and this latter situation in turn existing for each value of a fourth variable; and so on and so forth. Or simply extrapolate the bivariate normal distribution to multi-dimensional space.

The importance of the multivariate normal distribution is primarily theoretical in nature, since it is the underlying data assumption of virtually all multivariate analytical techniques. Still, the violation of the assumption has never stopped anyone from gathering and analyzing data in which they were interested. Techniques for identifying departures from the multivariate normal distribution are still a subject of early developmental work. However, just as the Central Limit Theorem lessened our concerns when univariate samples departed from normality—provided our sample sizes were large—it correspondingly adds confidence to our sometimes cavalier use of multivariate techniques.

11. Analysis of Covariance

The data analyses that we have considered so far involved one or more *predictor* variables and one or more *criterion* variables. But there are many more variable characteristics that our analytical objects possess—whether measured or not—and they are variously referred to as *uncontrolled* variables, *extraneous* variables, *covariables*, or more commonly as *concomitant* variables.

The existence of these extra variables presents both *problems* and *opportunities* in the analysis of our data—to the extent that they might be correlated with the predictor and criterion variables of prime concern. A methodology that takes into account the relationships among the concomitant, predictor, and criterion variables is referred to as *analysis of covariance*, often abbreviated *ancova*.

There are several practical applications of the approach, some more controversial than others. We will begin by considering the questionable use of the technique with *experimental* data, followed by applications which are more universally accepted.

Adjustment of experimental results. When conducting an experiment, the reason we randomly assign the alternative treatment conditions to our objects of study is to ensure that there are no biasing influences on the subsequently measured criterion variable(s). That is, going into the experiment we want our experimental groups to be essentially alike, so that any observed group differences in the criterion variable can be attributed to the *alternative treatments* received by the experimental groups, as opposed to differences which existed between the groups even before the treatments were administered.

In some unfortunate situations, however, substantial differences might be present among the experimental groups in some concomitant variable which we know or suspect is related to our criterion variable of interest. For example, in a learning experiment, two randomly formed groups of individuals could be found to differ in mean academic aptitude, despite the fact that they were used in an experiment in which they were assigned alternative self-instructional educational software and then had their performance measured on an achievement test. In this example, the alternative software represent the levels of the experimental *predictor* variable, the achievement test scores represent the *criterion* variable, and the aptitude scores represent the *concomitant* variable.

When such a concomitant variable is discovered to have a different mean value for the various levels of the predictor variable, it is further described as a *confounding variable*. To the extent that such a confounding variable is also related to the criterion variable (e.g., as aptitude is related to achievement), we want to eliminate its influence on the criterion variable, so that any observed differences in the means on the criterion variable can be attributed to the

predictor variable itself, rather than to the concomitant variable which is confounded with it.

The above scenario is portrayed in Figure 10, in its general form. If it looks similar to the *spice sales/shelf space/store size* diagram in the Correlation Analysis chapter (Figure 8 of Chapter 10), it is because they are conceptually the same; the difference being in the nature of the predictor variable. In the correlational study, the predictor variable of shelf space is a continuous random variable (not under our control), whereas in the type of experimental study we are currently considering the predictor variable of software type represents an experimentally manipulated variable, which also happens to be discrete rather than continuous in nature. In other respects, the confounding problem is the same.

In terms of our learning experiment example, then, the problem is this: How can we be sure that the observed difference between the treatment groups in achievement test scores (the criterion variable) is due to the alternative educational software (the experimental variable), as opposed to the ingoing difference in aptitude scores (the confounding variable)?

The analysis of covariance technique is often used to cope with the above type of situation, in which ingoing differences are evident among the experimental treatment groups. The procedure is also sometimes known loosely as "weighting the data." By using *correlational* regression techniques—which identify the relationship between the confounded concomitant variable and the criterion variable—the imbalances between the treatment groups with respect to the confounded variable are taken into account and an *adjustment* is made in the group means on the criterion variable. The mean criterion score of the group going in with a "disadvantage" is increased, while that of the group entering the experiment with an "advantage" is decreased.

Suppose, for example, that in our learning experiment Groups *A* and *B* had mean achievement scores of 75 and 86, respectively, after exposure to the experimental treatment of alternative self-instructional software. But further suppose, that going into the experiment, the mean aptitude scores of Groups *A* and *B* on a standardized aptitude test were 105 and 112, respectively. In summary:

	Group A	*Group B*
Experimental variable:	Software *A*	Software *B*
Mean on criterion variable (achievement score):	75	86
Mean on confounded variable (aptitude score):	105	110

Did Group *B* outperform Group *A* on the achievement test, in whole or in part, because of the experimental treatment (the software used) or because of their superior aptitude?

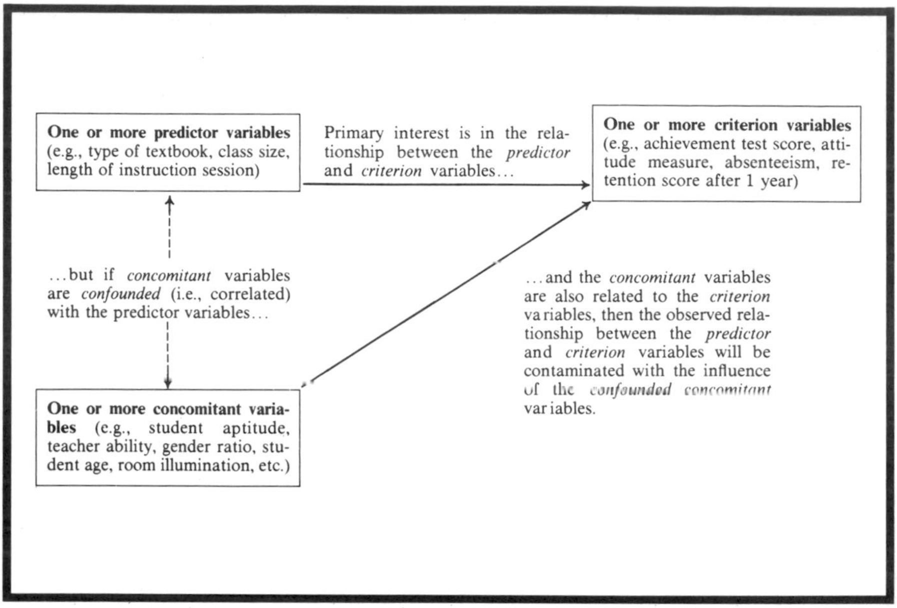

Figure 10 General paradigm of a data problem addressed by analysis of covariance.

The analysis of covariance technique makes adjustments in the mean criterion scores for the alternative treatment groups by taking into account the extent of correlation between the *confounded* variable and the *criterion* variable, measured separately *within each group* (where inter-object variation due to the treatment variable is not present). If the correlation between the concomitant variable and the criterion variable is essentially the same within each treatment group, as required by the technique, then the separate estimates can be pooled to form a single regression equation describing the relationship between the concomitant and criterion variables—aptitude and achievement, in this instance. This regression equation can then be used to adjust the group means on the criterion variable, recognizing that the groups would have differed in their predicted criterion scores by virtue of their differences in the concomitant variable alone.

For instance, in the present example where Group *B* had superior aptitude, we would expect a downward adjustment in their mean achievement score, while that of Group *A* would be adjusted upward, with any remaining difference between the two groups then tested for statistical significance, and attributed to the alternative experimental treatments; i.e., software *A* vs. *B*. We might find, for example, that the adjusted mean achievement scores of

Groups *A* and *B* are 77 and 84, respectively, and the significance test of this result could well be different than if the means had not been adjusted by an analysis of covariance. Note that the difference between the adjusted means on the criterion variable would have been *increased* rather than decreased if Group *B* had had the lower mean aptitude score, since Group *B* would then have been at a disadvantage, yet outperforming Group *A* despite the handicap, consequently deserving an even larger adjusted mean score.

The actual magnitude and direction of the adjustment in the treatment group means will depend upon the means and variances of the variables, as well as their inter-correlations. Depending upon the strength of the relationship between the confounded and criterion variables, and how far out of balance the treatment groups are with respect to the confounded variable, the observed difference in group means could either be increased, diminished, wiped out, or reversed, by the adjustment procedure.

Controversy over adjusting experimental data. We see in the above application of the analysis of covariance technique the use of *correlational* regression techniques with *experimental* data. As such, this hybrid approach is subject to controversy. The main problem lies in the interpretation that the ingoing bias which was present between the treatment groups with respect to the confounded variable has been removed, and that the experimental nature of the data remains intact. However, this interpretation does not address the fact, or reason why, the bias was present in the first place. Keep in mind that we do not expect (statistically significant) differences between our experimental treatment groups with respect to *any* extraneous concomitant variable, if we have randomly assigned our objects to the alternative treatment groups. If between-group differences in one or more concomitant variables are present, it is due either to a rare instance of chance sampling or to some systematic bias introduced in the experiment's execution, or by a faulty non-random object assignment procedure.

If the between-groups difference in a concomitant variable is due to normal sampling variation, we do not need an analysis of covariance adjustment since our traditional statistical test takes this chance variation into account. Natural variation in data is why we perform statistical tests, rather than accept observed group differences at face value. Furthermore, any chance differences are likely to be small and efforts to correct for them by a covariance analysis will probably be ineffective, since the resulting adjustments in the criterion means will be slight, and though the technique simultaneously reduces the error variance estimate (as explained later in the section), the approach also sacrifices degrees of freedom, requiring a larger critical F value to claim significance; so the trade-off between reduced error variance and lost degrees of freedom may not pay off in greater power as intended, especially when the adjustment in treatment means is small.

If small imbalances between groups do not warrant a covariance analysis adjustment, what about the situation where a *large* difference is observed between experimental groups with respect to a concomitant variable. By definition, such a difference, if due to chance alone, will be rare. A statistically significant difference between groups for any given concomitant variable, under the null hypothesis, will have a probability equal to α, which we usually set at .05 or .01, an unlikely event. If it does happen, and if it is uncovered prior to the administration of the experimental treatments, the most expedient solution would be to scrap the group assignments and redo it, although this tactic is also open to question. Still, it seems a more prudent approach than proceeding with a costly experiment that will be fraught with interpretation problems due to the ingoing bias. But what if the imbalance in the concomitant variable is first noticed *after* the experiment has been conducted. In such an instance we may have no recourse but to use an analysis of covariance adjustment, but we can no longer consider it a purely experimental study, for once we acknowledge a bias and employ correlational techniques, we can no longer be fully confident of causal interpretations vis-a-vis our experimental predictor variable. And critics may view the results with scorn.

Furthermore, if the concomitant variable is first measured after the experimental treatments have been administered, how does it differ formally from a criterion variable, and how do we know that the "imbalance" was not an *effect* of the treatments? True, in some instances we can logically rule out the possibility that the treatments caused the group differences in a concomitant variable which is measured after the treatments have been administered (e.g., group differences in the ages or educational backgrounds of the subjects), but in other situations we cannot be so sure (e.g., "aptitude" test scores could well be influenced by the treatments). In the latter situation, making an adjustment in a criterion variable based on a concomitant variable which itself reflects treatments effects could result in highly misleading results. For example, if treatment conditions affected both "aptitude" and achievement test scores, then treating the former as a concomitant variable in an analysis of covariance would tend to eliminate the observed group differences in achievement. Would we then dare conclude that the treatment variable had no effect?

Perhaps the most important argument against the application of the analysis of covariance technique to *experimental data* is that by adjusting for an out of balance concomitant variable, we are lulled into believing that we have removed all bias from the treatment means. This is unlikely, for even though we adjusted for the one concomitant variable that we happened to measure, there are still a *million* other concomitant variables that we did not measure and which could just as well be out of balance, or which may have been *in balance* going into the experiment, but which were then thrown out of whack once the adjustment is made on our measured concomitant variable

—this because of the various types of relationships that exist among all possible concomitant variables. The very fact that the groups exhibit a difference on one concomitant variable is reason enough to suspect that some systematic bias was present in the assignment of treatments to objects, and consequently the groups would be expected to differ in any number of other unmeasured concomitant variables. In such a situation, trying to balance the groups with respect to all variables would be like trying to solve a Rubik's Cube in which the moveable squares of each color were unequal in number.

In the learning experiment example considered above we had but *one* experimental predictor variable, *one* criterion variable, and *one* confounded concomitant variable. Imagine instead that we had performed a multi-factor experiment with *many experimental predictor variables* and *many criterion variables*—difficult enough to analyze—and then we discovered imbalances in *several concomitant variables*, some out of balance for some of the treatment groups, and others out of balance for other groups, some in all groups, and all in some groups, with interactions present. At this point we would be in serious trouble, and even the most dedicated researcher might start looking toward a career in the *arts* after the frustration of trying to reconcile the tangle of inter-related variables, for with such a morass of data we could "massage" it with covariance techniques for half a lifetime and still not be confident in our conclusions, nor could we expect others to be. We must remember that the goal of our research is to simplify, not to complicate our understanding of life.

The lesson to be learned, then, is that we should take special care in the initial design of an experiment, especially in the sample selection and the *random assignment of the treatments to the experimental objects*. If after the assignment of objects has been made, small differences are observed between the groups in some concomitant variable(s), we need not automatically resort to an analysis of covariance. Nor should we start tampering with the random assignment by juggling objects back and forth between the groups, subjectively "balancing" them, a practice that has been observed all too often. Small differences are to be expected between groups, and the theory underlying our statistical tests of significance take this random variation into account. In most cases, it is unlikely that a concomitant variable will be so highly related to the criterion variable, and be so out of balance from one treatment group to the other, that it will have any appreciable effect on the final analysis. If the difference is huge, something is wrong and little of the experimental nature of the study can be salvaged. What is done, cannot be undone.

Reduction of error variance. Irrespective of whether the experimental treatment groups are out of balance with respect to a concomitant variable, the analysis of covariance technique can still be used to reduce the size of the *error variance estimate* used as the denominator of the F ratio.

To the extent that a concomitant variable is correlated with the criterion variable *within treatment groups*, the contribution of the concomitant variable to the variation of the criterion variable can be extracted, resulting in a smaller residual error variance estimate. Essentially, it leaves the variation in the criterion variable that would be expected if the level of the concomitant variable were held fixed. It is a statistical method of accomplishing what is done experimentally in a *randomized block design*, in which the blocks of homogeneous objects are formed on the basis of some concomitant variable known to be related to the criterion variable. As we have seen earlier, by extracting the variation between the blocks, the residual error variance is reduced below the level it would be if a purely randomized one-way experiment had been performed. The analysis of covariance procedure uses correlational regression techniques instead of experimental blocking to accomplish the error variance reduction.

However, as in the randomized block design, the control of a concomitant variable by the analysis of covariance methodology will not routinely be of benefit to us. For although the technique reduces the error variance estimate, it simultaneously sacrifices *degrees of freedom*, requiring a larger critical value of the F ratio in order to declare statistical significance. Thus, there is a *trade-off* between reduced error variance (the denominator of the F ratio) and the increased value of the critical F required for significance. Whether the technique pays off will depend on the strength of the relationship between the concomitant and criterion variables.

This hit-or-miss nature of the technique's effectiveness in increasing the power of our analysis—our ability to detect real treatment effects—raises a number of questions. If the analysis of covariance is tried and then discarded because it does not result in a good trade-off between error reduction and the needed size of the critical F ratio, then we might be presented with analyses that have capitalized on chance relationships. Also, by statistically removing all the variation due to the concomitant variable, we will be capitalizing on chance empirical relationships. Just as there is regression to the mean on the remeasurement of single characteristics, there would be a regression to the mean with the remeasurement of the relationship between the concomitant and criterion variables. Consequently, we will often reduce the error variance much more than is warranted by the *true* relationship between the concomitant and criterion variables.

The question is also raised as to why, if data on the concomitant variables was available, a randomized block experiment was not conducted, using it as a blocking variable. If the concomitant variable was first considered or measured *after* the experiment was conducted, then another opportunity to capitalize on chance has been introduced. And further, if the covariance analysis is decided

upon only after the data has been examined and is called into use only when there are *borderline* significance situations, we will again be capitalizing on chance conditions.

In summary, the *reported* use of the analysis of covariance technique to reduce error variance in an experimental study may be so selective that it overcompensates for a concomitant variable, increasing the chance of making a type I error, *incorrectly rejecting* the null hypothesis. This type of selective use and reporting is unlikely in a randomized block experiment where we have committed ourselves to a concomitant blocking variable at the *design stage* of the experiment, rather than holding several options on it.

Analysis of in-tact groups. Despite its controversial use with experimental data, the analysis of covariance technique is ideally suited for analyzing differences between in-tact groups. In many disciplines it is not always possible to conduct experiments, and research must be confined to data that is correlational in nature.

We have already studied correlational relationships between *continuous* random variables in the Correlation Analysis chapter. We can now consider the relationship between the levels of *discrete* in-tact groups and various continuous variables of interest. For example, for any given criterion variable, we may be interested in understanding the differences among various racial, cultural, community, political, religious, social, economic, or medical diagnostic groups. Or, differences between alternative aptitude, attitude, and achievement groups. Or between rock, soil, fossil, water, or air samples drawn from different geographic or time periods. Behavioral groups are always an interesting topic of a study—vegetarians vs. non-vegetarians, smokers vs. non-smokers, voters for alternative A vs. B vs. C, users vs. non-users of a given product, criminals vs. non-criminals, etc.

When studying such in-tact groups we usually measure each member of the sample on a large number of variables. We may find that the groups differ on some variables and not on others, and we can try to deepen our understanding of the nature of the groups by performing an analysis of covariance, selecting any particular variable or combination of variables as a "confounding" variable, and extracting its contribution to the variation of the remaining variables. We might find, for example, that a difference between males and females in average income for a given profession disappears when we control for years of experience, or the difference might become more pronounced, or the direction of the difference might change. Or, we might study various causes of death, comparing diet, behavioral, and physiological characteristics of individuals who have died from various causes; sometimes treating these characteristics as explanatory variables and other times as confounding variables, in an attempt to isolate which variables actually contributed to the deaths. For instance, those who have died from heart disease could be

discovered to have eaten less fish than other individuals, but was it the fish diet *per se* or a concomitant decrease in the consumption or red meat that was related to the cause of death, or some entirely different confounded variable?

This covariance approach to analyzing in-tact groups is analogous to the partial correlation procedure for analyzing the correlational relationships among continuous random variables presented in Chapter 10. It is the challenge to the researcher to identify which variables to study from the multitude available, to formulate some kind of theory which will provide a parsimonious explanation of the vast amount of data, and to consider the implications of the findings. We will return to the subject of identifying differences among in-tact groups in the *Discriminant Analysis* chapter.

Analysis of quasi-experiments. *Psuedo-* or *quasi-experiments* are often conducted in which alternative experimental treatments are administered to in-tact rather than randomly formed groups, and then analysis of covariance is used to statistically control for the concomitant variables upon which we know the groups to differ.

For example, practical considerations or budgetary contraints might require that we test two alternative teaching methods with only two existing classrooms of students. Since the classes differ on a number of concomitant variables, including student aptitude, attitude, absenteeism, etc., we could resort to an analysis of covariance in an attempt to remove their effects from observed differences in the criterion measure of achievement score, *provided* we know the correlation of the confounding variables with the criterion variable. But again, since the subjects were not assigned to the treatment groups at random, observed differences in even the adjusted group means on the criterion variable can only be suggestive of the efficacy of the treatment variable. While within-group estimates of the correlation between some concomitant variables (e.g., aptitude) and the criterion variable can be obtained, the effects of other confounded variables such as teacher experience, classroom size, gender ratios, time-of-day, and physical environment cannot be assessed or controlled with the available data; requiring, instead, experimentation in which there is both *random assignment* and *replication* of the classroom *milieu* (the complex of student, teacher, and environment variables, as in the Table 5 study).

Another example of the above type of application of covariance analysis to a quasi-experiment would involve testing two alternative employee incentive plans at two different factories. Differences in productivity, the criterion variable, could be adjusted for factory differences in prior productivity, as well as other employee characteristics known to differ from one plant to the other, and which are related to productivity. Similar quasi-experiments are also performed among in-tact hospital, prison, community, and military populations.

However, in all cases, the existence of a multitude of unassessed confounding variables makes the results of such studies suggestive at best. When such investigations produce interesting results, funds could become available for conducting experiments, in which there is both replication and random assignment of the experimental objects (whether classrooms, hospital wards, prison cell blocks, or corporate offices) to the experimental treatments. Nothing is more dangerous than trying to make an inference from a sample of one.

Interpretation of experimental effects. The analysis of covariance technique can be used in experimental studies involving *many* criterion variables in an attempt to assess the treatment effects on one (or more) of the criterion variables when the effects on other criterion variables are statistically held constant. Such an analysis can often provide insights into the manner in which the treatments have their effect. We might discover, for example, that the differential effects of two fertilizers on crop yield is due largely to their parallel effect on germination rates.

Or, the different sales effectiveness of two alternative advertising campaigns conducted in randomly selected sets of markets could be attributed to differences in the awareness (or intrusion) level of the two campaigns, rather than to differences in their persuasiveness.

Or, in an experimental study of the effects of class size, resulting differences in achievement might be found to be correlated with differences in absenteeism, student participation, and classroom disruptions, *resulting* from the different levels of the experimental variable.

This type of application of the covariance technique should be recognized as different in theory and purpose from the situation where the effects of an ingoing confounding variable are extracted from the variation in the criterion variable. In this present application, we are examining between-group differences in the mean values on the *criterion* variables which have been *caused by the experimental treatments*. Our goal is to understand the nature of the treatment effects, not to remove spurious effects as in the confounding variable paradigm. It is the difference between absenteeism as an experimental effect, and absenteeism as an ingoing bias between treatment groups.

Estimation of missing data. Yet another use of the analysis of covariance technique is in supplying substitute data for an *occasional missing observation* in a research study—data that has been lost, was not gathered, or is unusable for one good reason or another. Information on a concomitant variable can be used to predict the value of the missing observation.

In summary, there is a wide variety of problems to which analysis of covariance can be applied, some more controversial than others. It is a natural type of analysis for correlational data, where confounding relationships are the rule, but it is of more questionable benefit when used to adjust experimental data, where its very use might raise more questions than it solves. Because of

the many potential pitpalls in its interpretation, whether with correlational or experimental data, the analysis of covariance technique should always be approached with caution.

12. Concluding Comments

The analysis of variance methodology is an alternative to the regression analysis approach to identifying relationships among predictor and criterion variables. Although the techniques differ in actual execution, approaching the data from different angles, they are equivalent in terms of the assumed underlying model of the data and arrive at the same conclusions. We have found how the analysis of variance F ratio tests for the *existence* of a relationship, the regression coefficient b describes the *nature* of the relationship, and r^2 measures the *magnitude* of the relationship.

We progressed from the situation of a *single predictor* variable and a *single criterion* variable, to that where there were *many* of each, and then further to the introduction of a third set of variables—*concomitant variables*—potentially related to the criterion and predictor variables, generating the analysis of covariance technique which attempts to utilize the information inherent in these extra variables.

In the following chapter we will continue to pursue our primary theme of the identification of relationships among variables, but for the special situation in which all the variables are *categorical* in nature; e.g., qualitative variables such as gender, race, religious affiliation, and diagnostic category.

Chapter 13

Analysis of Category Data

1. Introduction

Very often our data collection consists exclusively of *category data*; i.e., frequency counts made on nominally scaled qualitative variables such as religion, political affiliation, diagnostic category, race or species membership, cause of death, blood type, and the like.

When we study only one such variable we are interested in how frequently each categorical value of the variable occurs among our set of objects. We might want that information for purely descriptive purposes, or we might want to test a *hypothesis* about the *expected frequency* of each category value.

In the multivariate situation—when each object is measured on two or more such variables—we also want to determine whether the variables are *related to one another* or are independent. This is exactly the same problem we have been pursuing all along, whether in our study of correlation analysis, regression analysis, or the analysis of variance techniques of the preceding chapters. The key difference is in the nature of the variables that we now want to study.

In fact, the issue of qualitative variables was partially addressed in the preceding chapters, where we saw that the alternative values of such variables could be used as regression predictor variables when converted into individual binary dummy variables, or used as levels of a predictor variable in an analysis of variance, as in the package color experiment. However, in those instances, the criterion variable was always quantitative in nature. Now we want to consider the situation where we have *only* category data.

2. The χ^2 Statistic and Goodness-of-Fit Test

Suppose we present a random sample of 150 voters with three alternative tax proposals—call them A, B, and C—and ask each voter to express a preference. Now if the three proposals were equally appealing, we would

expect one-third of the sample of 150 voters to choose each of the alternatives, or 50 apiece. Instead, we observe that proposals A, B, and C are selected by 43, 56, and 51 voters, respectively. Keeping in mind that we are observing only a sample from a larger population, we want to know whether we can reject the hypothesis that the *expected frequencies* for the three alternatives are the same; namely, 50, 50, and 50.

More generally, we can state the above problem as follows: How can we determine if the *observed frequencies* of occurrence of the categorical values of a qualitative variable allow us to reject a hypothesis about the *expected* or theoretical frequencies of occurrence?

To determine whether the observed frequencies differ significantly from the expected frequencies of occurrence, we can use a very useful statistic known as the *chi-squared statistic*, which is denoted χ^2 and is pronounced *ki squared*. It is defined as follows:

$$\chi^2 \text{ statistic} = \sum_{i=1}^{k} \frac{(O_i - E_i)^2}{E_i}$$

where

O_i is the *observed* frequency in category i
E_i is the *expected* or hypothesized frequency in category i
k is the total number of categories

and the summation is across the $i = 1, 2, \ldots, k$ categories, or "cells" as they are often called.

In words, the formula states that for each cell we should subtract the *expected* frequency from the *observed* frequency, square the result, divide by the *expected* frequency, and then sum these results from each cell. For the present example, the observed and expected frequencies for the $k = 3$ cells are as follows:

Favored tax proposal:	A	B	C
Expected frequencies (E_i):	50	50	50
Observed frequencies (O_i):	43	56	51

Substituting into the formula for the χ^2 statistic and performing the

computations, we have:

$$\chi^2 = \sum_{i=1}^{k} \frac{(O_i - E_i)^2}{E_i} = \frac{(43-50)^2}{50} + \frac{(56-50)^2}{50} + \frac{(51-50)^2}{50} = 1.72$$

Now just as the t and F test statistics are associated with *degrees of freedom* and have their accompanying distributions, so too does the χ^2 statistic. In this instance of k values of a qualitative variable, the degree of freedom (df) for the χ^2 statistic equals $k-1$, or one less than the number of values of the categorical variable. For our tax proposal example we have $df = 3 - 1 = 2$.

Now although the χ^2 statistic is discrete in the values that it can assume, it happens to follow very closely a continuous distribution also known as the chi-squared distribution, and also denoted χ^2, so we must study the context of usage to determine whether χ^2 stands for the test statistic defined above, or the continuous distribution which it approximates for given degrees of freedom. The approximation is very good provided all expected cell frequencies are 5 or greater. This is a conservative rule, and even smaller expected frequencies have resulted in good approximations.

The χ^2 distribution differs in shape depending upon its degrees of freedom. Figure 1 shows some examples of the distribution. The tabled values of χ^2 for various degrees of freedom and for various significance levels can be found in Appendix Table VI. For the present example, for $df = 2$ and a significance level $\alpha = .05$, we find in Appendix Table VI that a χ^2 of 5.99 must be exceeded in order to reject the hypothesis about the expected frequencies. We obtained a value of $\chi^2 = 1.72$ from our data above, so we cannot reject the hypothesis of equal preference among the tax alternatives. The departures of the observed from the expected frequencies were well within the bounds of

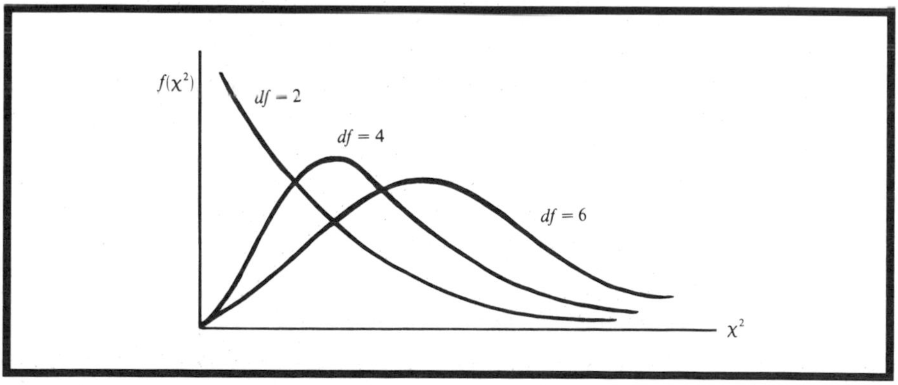

Figure 1 Some χ^2 distributions with their associated degrees of freedom, df.

what would be expected by chance alone. This statistical comparison of the observed and expected frequencies of the values of a variable is referred to as a *goodness-of-fit* test.

In the preceding example the expected cell frequencies were all equal to one another, but it should be clear that this need not be the case. Suppose, for instance, that a botanist theorizes that a given population of seeds should yield pink, red, and white flowered plants in a ratio of 3:2:1. That is, in the long run, out of every 6 flowering plants (3+2+1), 3 would be expected to be pink, 2 red, and 1 white, or respective percentages of 50.0%, 33.3%, and 16.7%. So if a total of 250 seedlings were studied, the expected frequencies for pink, red, and white flowered plants would be 125.00, 83.25, and 41.75, respectively (obtained by applying the percentages against the total of 250 seedlings). Not only can the expected cell frequencies differ, being dictated by our particular hypothesis, but the expected frequencies, unlike the observed frequencies, need not be in the form of *whole* numbers. But what *is* essential in the use of the χ^2 statistic is that we are dealing with *frequencies* and not percentages. Also, the frequencies in the various cells must be independent of one another—an observation cannot appear in more than one cell. For example, we could not set up pink, red, and white flower cells, if some plants had, say, red and pink flowers. The cells must be *mutually exclusive* and *exhaustive* with respect to the possible values of the variable we are studying.

In the following sections we will look at the use of the χ^2 statistic in determining whether two qualitative variables are *related* to one another, and we will find that we must again look at observed vs. expected frequencies.

3. Test of k Independent Proportions

Very often we want to know whether alternative qualitative treatments are differentially related to "success" vs. "failure" on another variable. For example, are four different teaching methods related to the number of students passing vs. failing a standardized test? Are three different insecticide treatments related to the number of pests that survive vs. die? Are five different nationalities related to whether or not death is from heart disease? Or, in general, are the k values of a qualitative variable related to success vs. failure on a criterion variable?

To consider a concrete example, suppose a direct-mail marketer wishes to determine whether four alternative mailing pieces differ in their ability to pull responses. Four random samples of 125 prospects each are drawn from a large list of potential customers. Each sample is then mailed one of the four alternative mailing pieces, and after a predesignated cutoff date a tabulation is made of the number of prospects responding to the solicitations. It was found that of those receiving mailing pieces A, B, C, or D, the numbers responding

were 34, 19, 27, and 20, respectively, or proportions .272, .152, .216, and .160, since each sample contained 125 prospects. The results of the experiment are summarized in Table 1.

Since χ^2 is defined in terms of frequencies, rather than proportions, it is necessary that we evaluate the deviations of the observed frequencies from the expected frequencies. For that reason *results in the form of percentages or proportions must first be transformed back into frequencies* as in Table 1.

Since we have no *a priori* theory as to what the expected cell frequencies should be, we have to base them on the data itself. We note in the far right marginal totals of Table 1 that the total number responding to the mailing piece is 100 out of 500, or a proportion of .20. Since our working hypothesis is that the four mailing pieces do not differ in their effectiveness, we would expect that each should result in the same proportion of response. Since 125 prospects received each mailing piece, we would expect $.20 \times 125$ or 25 prospects in each sample to respond to the solicitation they received. Consequently, 100 of the 125 would be expected to not respond. These expected frequencies are presented in the upper right-hand corners of the outcome cells in Table 1. The values of χ^2 for the data is found to be 7.10.

To determine whether $\chi^2 = 7.10$ is greater than would be expected by chance, we need to refer to Appendix Table VI for the critical value of χ^2 at, say, the $\alpha = .05$ significance level, for the appropriate degrees of freedom for the χ^2 at issue. But while we have 8 cells in our example, the cell frequencies are not all free to vary, given the marginal frequency constraints. In fact, only

Table 1 Frequency of response to four mail-order solicitations.

Response	Mailing Piece *A*	Mailing Piece *B*	Mailing Piece *C*	Mailing Piece *D*	Totals
Ordered	25 34	25 19	25 27	25 20	100
Did not order	100 93	100 106	100 96	100 105	400
Sample sizes:	125	125	125	125	500

$$\chi^2 = \frac{(34-25)^2}{25} + \frac{(19-25)^2}{25} + \frac{(27-25)^2}{25} + \frac{(20-25)^2}{25}$$

$$+ \frac{(93-100)^2}{100} + \frac{(106-100)^2}{100} + \frac{(96-100)^2}{100} + \frac{(105-100)^2}{100} = 7.10, \ df = 3$$

3 of the 8 cell frequencies are free to vary, with the marginal totals fixed as they are. Imagine, for example, that the frequencies in the first three cells of the top row of Table 1 were provided. Then, by subtraction from the marginal totals, the remaining cell frequencies are automatically determined. Consequently, we have $df = 3$ for the present example. In general, for such a two-way frequency classification, the number of degrees of freedom will equal

$$(r-1)(c-1)$$

where r is the number of *rows*, and c is the number of *columns* in the table. In the example of Table 1 we have $(2-1)(4-1)$ or 3 degrees of freedom.

At the $\alpha = .05$ significance level, the critical value for χ^2 with $df = 3$ is $\chi^2 = 7.82$. Since the value of χ^2 for the present example is 7.10, we barely fail to reject the null hypothesis that the four mailing pieces are equivalent in their effectiveness. We may be committing a type II error, failing to detect a real difference, but unless we were willing to increase the probability of a type I error by changing our significance level to $\alpha = .10$, then the only way we can increase the power of our test is to increase the sample sizes. But these are considerations that should have been taken into account during the design of the experiment, and not afterwards, as is too often the case.

In summary, the above approach can be used to test the difference between k *independent* proportions—independent in the sense that objects contributing to one proportion cannot in any way, directly or indirectly, contribute to the other proportions. Although the test is concerned with proportions, the actual mechanics of the test must deal with the *frequencies* upon which the proportions are based.

4. The $r \times c$ Contingency Table

The data matrix for testing the difference between k independent proportions, as discussed in the preceding section, will be recognized as a $2 \times k$ *contingency table*, where the 2 rows correspond to the particular dichotomous variable of interest and k refers to the various levels of the experimental or predictor variable. The task was to determine if the two variables were contingent upon each other, which was the alternative to the null hypothesis. The $2 \times k$ contingency table will also be recognized as a special case of the more general $r \times c$ *contingency table* in which there is no restriction on the number of rows r or the number of columns c. Again, following the same logic as presented in the preceding section, the degrees of freedom for such a contingency table will be $(r-1)(c-1)$. For example, a 5×3 contingency table will have $df = (5-1)(3-1) = 8$.

Table 2 The joint frequency distribution of geographic residence and political affiliation among a sample of 500 voters.

Political affiliation	Geographic residence				Totals		
	East	*Midwest*	*South*	*West*	f	p	
Democrat	92 (73.2)	39 (47.1)	49 (43.3)	53 (69.4)	233	.456	These proportions are applied to the column totals to yield the expected cell frequencies in each column.
Republican	41 (60.3)	58 (38.8)	25 (35.7)	68 (57.2)	192	.384	
Independent	24 (23.5)	4 (15.1)	19 (14.0)	28 (22.4)	75	.150	
Totals	157	101	93	149	500	1.00	

$$\chi^2 = \frac{(92-73.2)^2}{73.2} + \frac{(39-47.1)^2}{47.1} + \cdots + \frac{(28-22.4)^2}{22.4} = 43.14, \; df = 6$$

The χ^2 evaluation of $r \times c$ contingency tables is useful for identifying an association between two *qualitative* variables, examples of which were provided in the introduction. Table 2, for example, presents the results of a survey to determine the existence of an association between geographic region of residence and political affiliation, both qualitative variables. The procedure for calculating the expected frequencies is essentially the same as that outlined for the $2 \times k$ contingency table. As shown in Table 2 the row totals at the extreme right of the table are converted into proportions based on the total sample, in this case 500. As a null hypothesis it is then assumed that the expected frequencies in each column of the table should follow *the same pattern as the marginal totals*. That is, since the overall proportions of Democrats, Republicans, and Independents are .466, .384, and .150, respectively, we assume that these same proportions should hold in the East, Midwest, South, and West regions of the country. Therefore, we apply these proportions to the total samples in each of these regions to obtain the individual *expected cell frequencies*. For example, of the 157 persons living in the East region, $.466 \times 157 = 73.16$ would be expected to be Democrats under the null hypothesis, $.384 \times 157 = 60.29$ would be expected to be Republicans, and $.150 \times 157 = 23.55$ would be expected to be Independents. Similarly, the overall proportions would be multiplied times the sample sizes in each of the other regions, in order to determine the remaining expected frequencies.

The analysis of the observed and expected frequencies in Table 2 results in $\chi^2 = 43.14$, which for 6 degrees of freedom—$(4-1)(3-1)$—is well above the critical value of $\chi^2 = 12.59$ needed at the $\alpha = .05$ significance level, as found in Appendix Table VI. Consequently, we conclude that political affiliation *is* contingent or dependent upon geographic region of residence. It is in this context that χ^2 is considered a *test of independence* between two variables. Had the data resulted in an insignificant value of χ^2 we would have concluded that the two variables were independent or "uncorrelated". Again, the fact that we did discover dependence between the two variables does not imply a causal relationship that someone will somehow change their political affiliation merely by changing geographic location. The interpretation is the same as that for the observed correlation between two random quantitative variables as discussed in Chapter 10. On the other hand, if the individuals had been *assigned at random* to the alternative levels of one of the variables, as in the mailing-piece example of the preceding section, then any dependency discovered between the two variables could be interpreted in a causal way.

The applications of χ^2 in the preceding example, in which political affiliation can be viewed as a qualitative criterion variable, will be recognized as analogous to the one-way analysis of variance of in-tact groups in which the criterion variable is quantitative in nature.

5. The 2×2 Contingency Table

The χ^2 evaluation of 2×2 contingency tables is an especially useful analytical device. In fact, if we had but *one* statistical tool, this would be it. Nothing could be more basic than the 2×2 contingency table when it comes to testing for the relationship between variables. This is because *two values* is the minimum number of levels that a variable can have, and *two variables* is the minimum number that we can relate. So in the 2×2 contingency table, the two columns represent the values of one variable, and the two rows represent the values of the second variable. Nothing could be more fundamental.

Aside from the many variables which have only two values—e.g., male vs. female, success vs. failure, organic vs. inorganic, vote for vs. against, live vs. dead, etc.—*any* variable can be made into a dichotomous variable, either by partitioning its values around some convenient value such as the median, in the case of quantitative variables, or by creating dummy variables in the case of multi-valued qualitative variables. The following sections will show the appeal of the 2×2 contingency table.

Testing two independent proportions. The χ^2 distribution can be used to test the difference between *two* independent proportions. Such an application is, indeed, the limiting case of the 2× k contingency table analysis for assessing the difference between k independent proportions, as discussed earlier.

Suppose, for example, that a pollster is interested in knowing whether males and females differ in their endorsement of a given attitudinal statement. Of a sample of 100 males, 33% agree with the statement, while only 17% of a sample of 100 females agree with it. Is this difference greater than would be expected by chance alone, if in fact males and females in the sampled populations did not differ in their response to the attitude? Table 3 presents the relevant 2×2 contingency table for the problem, in terms of frequencies as required, and it yields $\chi^2 = 6.83$, with $df = 1$—i.e., $(2-1)(2-1) = 1$. Since the

Table 3 Agreement with attitude *A* by gender.

Response to Attitude *A*	Gender: *Male*	Gender: *Female*	Totals
	25	25	
Agree	33	17	50
	75	75	
Disagree	67	83	150
Totals:	100	100	200
	$\chi^2 = 6.83,\ df = 1$		

critical value of χ^2 at the $\alpha = .05$ significance level for $df = 1$ is equal to 3.84, we can reject the null hypothesis that males and females are alike with respect to the given attitude; or, equivalently, we can alternatively say that the attitudinal variable and gender are related, dependent, or correlated.

Again, the above χ^2 test is only approximate in nature and requires that the expected cell frequencies are at least equal to five for the approximation to the real probabilities to be satisfactory. Also, the test can be shown to be equivalent to the approximate normal distribution z test studied in Chapter 9 for exactly the same type of problem. Recall that in that approach we required that np and nq be 5 or greater for both samples. This corresponds to the requirement that the expected frequencies in the χ^2 contingency table be at least 5 in value. To further appreciate the equivalence of the z and χ^2 tests when $df = 1$, note that the critical value of $\chi^2 = 3.84$ is equal to the square of the critical z value—i.e., $(1.96)^2 = 3.84$. It is left as an exercise to show that the value of χ^2 obtained for this example, is equal to the square of the value of z obtained by the method of Chapter 9, using the pooled value of $\hat{p}_1$ and $\hat{p}_2$ to define the standard error of the difference between the two proportions.

Assessing validity. Suppose a medical researcher has come upon a blood test purported to diagnose the presence of cancer. To assess its validity the test is applied blind to blood specimens from a sample of 750 hospital patients—120 with cancer, and 630 without cancer. The results of the experiment are presented in Table 4. An analysis of the observed and expected numbers of "hits" and "misses"—correct and incorrect diagnoses, respectively—results in $\chi^2 = 1.55$, which with $df = 1$ does not approach the critical value of $\chi^2 = 3.84$ needed at the $\alpha = .05$ significance level. The researcher must conclude, therefore, that the test is without merit. There is no strong evidence that the test did any better in identifying cancer patients than could have been done by a completely random selection procedure.

Table 4 Accuracy of a diagnostic test.

	Actuality		
Diagnosis	*Cancer*	*No Cancer*	Totals
Cancer	24 29	126 121	150
No Cancer	96 91	504 509	600
Totals:	120	630	750
		$\chi^2 = 1.55$, $df = 1$	

Notice that the blood test resulted in 121 "false positives"— predicting a cancerous condition when it was not present—and 91 "false negatives"—predicting the absence of cancer when in fact it was present. On the other hand, 29 correct diagnoses of cancer were made, and 509 correct diagnoses of a cancer-free condition were made. But this level of performance is no better than chance, given the overall incidence of the condition in the sample.

Assessing reliability. A sample of 100 consumers are asked to express their preference between two cola formulations, *A* and *B*: 50 choose *A*, and 50 choose *B*. The marketing manager wonders how to interpret these results—does it mean that consumers cannot tell the two formulations apart, or does it mean that two consumer segments exist in the population, one favoring formulation *A* and the other formulation *B*?

To answer the question the same 100 consumers are presented with the same choice a day later, and 52 choose *A* while 48 choose *B*, hardly any difference from the original test. But this data alone does not determine whether the choices between *A* and *B* were random or reliable. Rather, we must know the extent to which the 50 consumers who preferred formulation *A* in the first test are represented among the 52 who favored it in the second test.

The association between the *original* and *retest* preferences is presented in the 2×2 contingency matrix of Table 5. The χ^2 evaluation of the data results in $\chi^2 = 23.08$, which is well above the critical value for $df = 1$ at the $\alpha = .05$ significance level. We can be confident, then, that two taste preferences exist in the sampled population. Had the association not been found between the test and retest preferences, either formulation could have been marketed with equal success. The actual results, however, prompted the marketing manager to recommend the marketing of both formulations, under different brand names, in order to appeal to the broadest base of consumers.

Another advantage of the 2×2 contingency table is that we can often take the position that if *a relationship does not show up* in such an analysis, it is not strong enough to be of *practical significance* to us.

Table 5 Test-retest reliability for a cola preference experiment.

	Retest		
First Test	*Prefer A*	*Prefer B*	Totals
Prefer A	38 (26)	12 (24)	50
Prefer B	14 (26)	36 (24)	50
Totals:	52	48	100

$\chi^2 = 23.08$, $df = 1$

6. Test of Two Non-Independent Proportions

Many occasions arise where we want to test the difference between two proportions *based on the same sample*, or related samples—for example, the proportion of a sample agreeing with attitude *A* versus the proportion agreeing with attitude *B*. Since the two proportions are based on the same set of individuals, they are not independent of one another, and consequently we cannot use the techniques outlined earlier in the chapter and in Chapter 9, designed for independent proportions. However, there is a relatively simple application of the 2×2 contingency table that can be used to attack the problem.

Suppose, for example, that we wish to determine whether there is a difference in the level of endorsement for two attitudinal statements, *A* and *B*, among a sample of 50 individuals, where it is found that *A* is endorsed by 25 individuals (50%), and *B* is endorsed by 17 individuals (34%). The first step is to determine the amount of overlap or association between the responses to the two statements. This is presented in the 2×2 contingency matrix of Table 6a, again in frequency form. The value of $\chi^2 = 10.78$ for the data, significant at the $\alpha = .05$ level, confirms that there is an association between the responses to the two attitudinal statement; i.e., those who agree with statement *A* are more likely to agree with *B* than those who disagree with *A*.

Table 6 Analysis of two non-independent proportions.

(a) Test of the association between responses to items *A* and *B*.

Response to item *A*	Response to item *B*: *Agree*	Response to item *B*: *Disagree*	Totals
Agree	14 (8.5)	11 (16.5)	25
Disagree	3 (8.5)	22 (16.5)	25
Totals	17	33	50

$\chi^2 = 10.78$, $df = 1$

(b) Test of the levels of response to items *A* and *B*.

Agree with B, but not A	*Agree with A, but not B*	Total
3 (7)	11 (7)	14

$\chi^2 = 4.57$, $df = 1$

While the above analysis tells us that there is an association between the responses to the two items, it does not tell us whether the *level* of response to statement A is different from the level of response to statement B. To answer this question we must first eliminate from our analysis the 14 individuals who *agreed with both A and B*, and the 22 who *disagreed with both A and B*—since these 36 individuals tell us nothing about whether the levels of endorsement for A and B differ.

Of the 14 remaining individuals, 11 agreed with statement A, but disagreed with B; while 3 agreed with B, but disagreed with A. If, in fact, A and B were equal in their levels of endorsement in the sampled population, then it would be expected that the 14 (3+11) individuals who differed in their response to the items would split 7 in favor of A and 7 in favor of B. The question is, does the 3:11 split differ more than chance from the hypothesized 7:7 split. Part *b* of Table 6 presents the evaluation of the observed and expected frequencies, yielding $\chi^2 = 4.57$, which for $df = 1$ is significant at the $\alpha = .05$ level*. We can conclude, then, that the level of agreement for items A and B do differ. Had we inappropriately treated the two proportions in question as though they were independent, we would not have detected this difference; for, as a rule, the stronger the positive correlation between two variables, the smaller the difference required for statistical significance. It is for this reason that we always strive to measure two or more variables among the same or related samples of objects, rather than among independent samples.

7. Multivariate Contingency Tables

Moving now from the elegant simplicity of the 2×2 contingency table, we can wonder whether we can identify associations among *many* qualitative variables, when there are no restrictions on the number of variables or the number of values each possesses. We are speaking now about contingency tables in *three* dimensions, when dealing with *three* variables, and in k dimensions when studying k variables. For example, we might want to identify the inter-relationships among the variables of gender, political affiliation, religion, and referendum vote.

This would be analogous to the multivariate analyses previously studied, involving many continuous quantitative variables. In the case of such quantitative variables, however, each could be expressed as an additive combination of the other variables. With categorical variables, however, we do not have numerical values available. Instead, we can think in terms of contributions to the *expected frequencies* within each cell of the multivariate contingency table.

Recall that in the two dimensional tables that we have studied, the expected cell frequencies were obtained using the marginal proportions. In the

*The *exact binomial* probability is .0574.

multivariate contingency table, any given cell will represent the intersection of many marginal proportions. Just as the probability of the joint occurrence of independent events is equal to the product of their respective probabilities, we can think of the expected cell frequencies as being a joint function of the various marginal proportions. If it is further recalled from elementary algebra that the logarithm of a product of terms is equal to the *sum of the logarithms* of the terms being multiplied, then at least a vague notion will be had of the *loglinear models* which are used to analyze multivariate contingency tables.

Whereas the general loglinear models are used to model the cells of entire multivariate contingency tables, a special case occurs when we are primarily interested in the probability of membership in *one* of the many categories as a function of the others. For example, we might want to predict membership in the "yes" referendum vote category, as a function of membership in the remaining categories of gender, political affiliation, and religion. The equation which produces the probability p of an individual belonging to the category of interest is referred to as a *logistic regression* equation. It is based on the *log odds* of membership in the category of interest—$\log(p/q)$—which we learned in Chapter 6 is called the *logit of* p. If we let

$$\log\left(\frac{p}{q}\right) = y$$

and solve for p based on what we learned in high school algebra, we have

$$p = \frac{1}{1 + e^{-y}}$$

where e is the base of the natural logarithm system. The value of the exponent y is of the form

$$y = \alpha + \beta_1 x_1 + \beta_2 x_2 + \cdots + \beta_k x_k$$

which looks like an ordinary regression equation, except now the x's represent the 0 or 1 binary dummy values for the category predictor variables, and the α and β's (estimated from the data) are a function of the logarithms of the odds of belonging to the various predictor categories.

This type of analysis is one of the most recently developed areas of study in statistics, and is attracting much attention from researchers and theoreticians in the field, and this activity should promote the study of categorical variables which are so common in our lives and need to be understood through the identification of their many complicated inter-relationships.

8. Concluding Comments

In the preceding chapters on correlation analysis, regression analysis, and analysis of variance, our emphasis was on the identification of relationships among quantitative variables, or between categorical predictor variables and quantitative criterion variables. In this chapter we studied relationships that were strictly between categorical variables. This was approached through the analysis of contingency tables, using the χ^2 statistic to determine whether observed cell frequencies differed significantly from hypothesized or expected cell frequencies. We also considered the possibility of analyzing multivariate contingency tables in which relationships could be drawn between many categorical variables.

In the following chapter we will look at another approach which strives to relate predictor variables to categorical criterion variables, just as logistic regression, but which is a more effective approach if the data meets certain assumptions. And this is to be expected, for the more orderly the data, the more precise we can be in our inferences.

Chapter 14

Discriminant Analysis

1. Introduction

Very often our criterion variable of interest is dichotomous in nature, and we are interested in predictor variables which are related to the two criterion values. For example, we might be interested in predictor variables which discriminate between voters for and against an issue, smokers and non-smokers, breast-fed and bottle-fed individuals, schizophrenics and non-schizophrenics, diseased and healthy cells, users and non-users of a given product, credit risks and non-risks, dry and productive well drillings, murderers and non-murderers, school dropouts and non-dropouts, etc.

There are still other situations in which our criterion variable may have more than two values but be purely qualitative in nature. For example, we might be interested in differences between Protestants, Catholics, and Jews; or between Democrats, Republicans, and Independents; or between members of four different nationalities; or between users of five different brands of a given product; or between several plant species; or between various illnesses; etc. In each case we would like to know whether values on various predictor variables are related to the alternative values on the qualitative criterion variable.

Discriminant analysis is a procedure for identifying such relationships between qualitative criterion variables and quantitative predictor variables. Alternatively, we can think of discriminant analysis—or discrimination analysis as it is also known—as a procedure for identifying boundaries between groups of objects, the boundaries being defined in terms of those variable characteristics which distinguish or discriminate the objects in the respective criterion groups.

The benefits of such a technique are the same as for regression analysis. First of all, we can learn which variables are related to the criterion variable, and secondly we will be able to predict values on the criterion variable when given values on the predictor variables. A lending institution would be able to

distinguish credit risks from non-risks; medical specialists would be able to differentiate one disease from another; geologists would be able to discriminate good and poor well-drilling sites; personnel departments would be able to discriminate between successful and unsuccessful job trainees; etc.—just a few of the specific benefits that might be derived from the application of discriminant analysis. This chapter will serve as a non-mathematical introduction to the technique, with emphasis on its rationale and interpretation.

2. Overview of Discriminant Analysis

In this and the following sections we will learn how discriminant analysis is essentially an adaptation of the regression analysis technique, designed specifically for situations in which the criterion variable is qualitative rather than quantitative in nature. We will first study the procedure as it applies to dichotomous criterion variables, and then as it applies to the situation in which the criterion variable is multi-valued but qualitative in nature.

The input data. Table 1 presents the most general form of the input data matrix used in discriminant analysis. A number of objects classified into two or more criterion groups are measured on each of a number of predictor variables.

Table 1 Input data matrix for a discriminant analysis.

Groups to be discriminated	Objects within the groups	Scores on the predictor variables			
		x_1	x_2	...	x_k
Group A	Object A_1			...	
	Object A_2			...	
	⋮				
	Object A_m			...	
Group B	Object B_1			...	
	Object B_2			...	
	⋮				
	Object B_n			...	
Group C	Object C_1			...	
	Object C_2			...	
	⋮				
	Object C_l			...	

To say that the objects are *classified* into two or more groups is equivalent to saying that each object possesses one of the *values* on the associated qualitative variable. For example, credit applicants can belong to the credit *risk group* or the *non-risk group*; or they can be thought of as having the label, or value, of *risk* or *non-risk* on the "credit worthiness" variable. Throughout our discussion of discriminant analysis it will be useful to keep these two alternative interpretations in mind—on the one hand, *objects belonging to groups*; and on the other hand, *objects having values on a qualitative variable*.

A number of aspects of the data matrix in Table 1 should be noted. Firstly, the groups of objects are *mutually exclusive*; i.e., an object belonging to one group cannot belong to another group. Secondly, every object, regardless of group membership, is measured on the same set of predictor variables. Thirdly, the number of objects in each group need not be the same.

It will be noticed that the input data matrix for discriminant analysis is not that different from that used in multiple correlation and regression analysis. Each object is measured on several quantitative variables; the main difference being that the objects in discriminant analysis are further grouped according to some meaningful criterion. This categorization of the objects is analogous to the quantitative criterion variable found in the input data matrix for multiple correlation and regression analysis.

The criterion variable. As noted above, the classification labels attached to the objects constitute the criterion variable in discriminant analysis. As indicated in the introduction, the criterion variable can have a minimum of two values—e.g., voter vs. non-voter, smoker vs. non-smoker, productive vs. dry oil wells, etc.—or it may have several values—e.g., Protestant, Catholic, or Jew; Democrat, Republican, or Independent; users of brand *A*, *B*, *C*, or *D*; etc.

In either case, whether the criterion variable is dichotomous in nature or a multi-valued qualitative variable, the task of discriminant analysis is to classify the given objects into groups—or, equivalently, to assign them a qualitative label—based on information on various predictor or classification variables.

Predictor variables. As with regression analysis, the particular objects chosen for our analysis, as well as the particular criterion variable, will dictate the nature of the predictor variables. If we are interested in buyers of different models of automobiles, we will measure the buyers on such characteristics as age, income, family size, annual miles driven, and we might also consider their image ratings of the various models on dimensions such as safe to drive, comfortable, good trade-in value, etc. If we are interested in discriminating diseases we will want to measure patients in terms of body temperature, blood pressure, pulse rate, weight, diet, age, as well as in terms of various blood and urine variables. If we are interested in discriminating between voters and non-voters for a given election, we might want to measure the individuals in

terms of demographic characteristics such as age, income, length of residence in the community, etc. as well as their responses to various attitudinal and belief statements.

In short, we will want to measure our objects on those variables which we believe to be *related* to the objects' membership in one or another of the criterion groups. In this respect, our choice of predictor variables is exactly the same as in regression analysis, which should not surprise us since the task of discriminant analysis is essentially the same as that of regression analysis, and only the specifics of the techniques differ.

Key assumptions. While the effectiveness of discriminant analysis will be found to rest on the existence of predictor variables which differ in *mean value* from one criterion group to another, there are two key assumptions with regard to the *variances* of the predictor variables and their *inter-correlations*. Firstly, it is assumed that the variance of a given predictor variable is the same in the respective populations from which our groups of objects have been drawn. Different predictor variables can have different variances, as they most often do, but the variance of a given variable must be the same in each criterion group population.

Secondly, it is necessary that the *correlation* between any two predictor variables is the same in the respective populations from which our alternative criterion groups have been sampled. In other words, the correlation *matrix* of predictor variables must be the same in each group.

The reasons for these assumptions will become more evident in the following discussion of how discriminant analysis goes about classifying objects into groups.

3. The Discriminant Function

We have seen in regression analysis how a regression equation is developed which involves a weighted combination of values on various predictor variables. This equation allows us to predict or estimate an object's value on a quantitative criterion variable when given its values on each of the predictor variables.

In discriminant analysis we employ a concept very similar to the regression equation, and it is called the *discriminant function*. Whereas the regression equation uses a weighted combination of values on various predictor variables to predict an object's value on a continuously scaled criterion variable, the discriminant function uses a weighted combination of those predictor variable values to *classify* an object into one of the criterion variable *groups*—or, alternatively, to assign it a value on the qualitative criterion variable.

The discriminant function, then, which for convenience we will designate with the letter L, is nothing more than a *derived variable* defined as a weighted

sum of values on individual predictor variables. Each object's score on the discriminant function—its *discriminant score*—will depend upon its values on the various predictor variables.

In symbolic form, the discriminant function can be expressed as follows:

$$\boxed{L = b_1x_1 + b_2x_2 + \cdots + b_kx_k}$$

where $x_1, x_2, \ldots, x_k$ represent values on the various predictor variables and $b_1, b_2, \ldots, b_k$ are the weights associated with each of the respective predictor variables, and L is an object's resultant discriminant score.

An example of a specific discriminant function would be

$$L = .7x_1 + .2x_2 - .4x_3$$

Each object in our analysis would have a value on this discriminant function depending upon its values on the predictor variables x_1, x_2, and x_3. The above equation will be recognized as essentially the same as a multiple regression equation. In both cases we have a weighted sum of predictor variable values creating a composite score—L in the case of discriminant analysis, y' in the case of regression analysis. When the criterion variable is quantitative in nature, as in regression analysis, it is easy to understand how the composite y' scores can be estimates of the actual y values, but when the criterion variable is qualitative in nature how can the numerical composite L scores be used to predict an object's group membership?

The cutoff score. The answer to the preceding question lies in the fact that associated with the discriminant function is a *cutoff score*. Objects with discriminant scores greater than the cutoff score are assigned to one of the criterion groups, and objects with discriminant scores less than the cutoff score are assigned to the other criterion group.

Expanding on our earlier example we might have as our discriminant function

$$L = .7x_1 + .2x_2 - .4x_3$$

with the stipulation:

If L is greater than 16, assign object to Group A
If L is less than or equal to 16, assign object to Group B

Each object's discriminant score would be judged against the cutoff score of 16 to determine its group membership. Consider an object with scores of 15, 40,

and 11 on variables x_1, x_2, and x_3, respectively. Its value on the discriminant function would be calculated as

$$L = .7(15) + .2(40) - .4(11) = 14.1$$

Since a discriminant score of 14.1 is less than the cutoff score of 16, this particular object would be assigned to Group *B*.

On the other hand, consider an object with values of 25, 30, and 7 on the variables x_1, x_2, and x_3, respectively. Its score on the discriminant function would be calculated as

$$L = .7(25) + .2(30) - .4(7) = 20.7$$

This object, having a discriminant score of 20.7, exceeding the cutoff score of 16, would be assigned to Group *A*.

Following the same procedure, each object in our analysis would be classified into one criterion group or the other, depending upon its values on the individual predictor variables. As for the *accuracy* of our classifications, that will be the subject of a later section.

4. Understanding the Discriminant Function

It is apparent from the above examples that the defining characteristics or *parameters* of the discriminant function are (1) the *weights* associated with each predictor variable, and (2) the critical *cutoff score* for assigning objects into the alternative criterion groups. These parameters of the discriminant function are determined in such a way as to *minimize the number of classification errors*. Note the analogy with regression analysis. While we cannot go into the mathematics of the derivations of these defining characteristics, we can gain an insight into their definitions graphically.

One predictor variable. Figure 1 shows the simplest type of discrimination problem: Two groups of objects, *A* and *B*, are measured on a *single* predictor variable *x*. Since in this instance we do not have to worry about alternative weights for alternative predictor variables, having only one, we need only determine a cutoff score on that one predictor variable. This could be accomplished by considering alternative cutoff points along the predictor variable continuum and assessing the number of classification errors resulting at each point. However, if our assumption of equal variance of the predictor variable in each group holds, the problem will have a more direct and less tedious mathematical solution, a fact that will be appreciated when the number of predictor variables grows large.

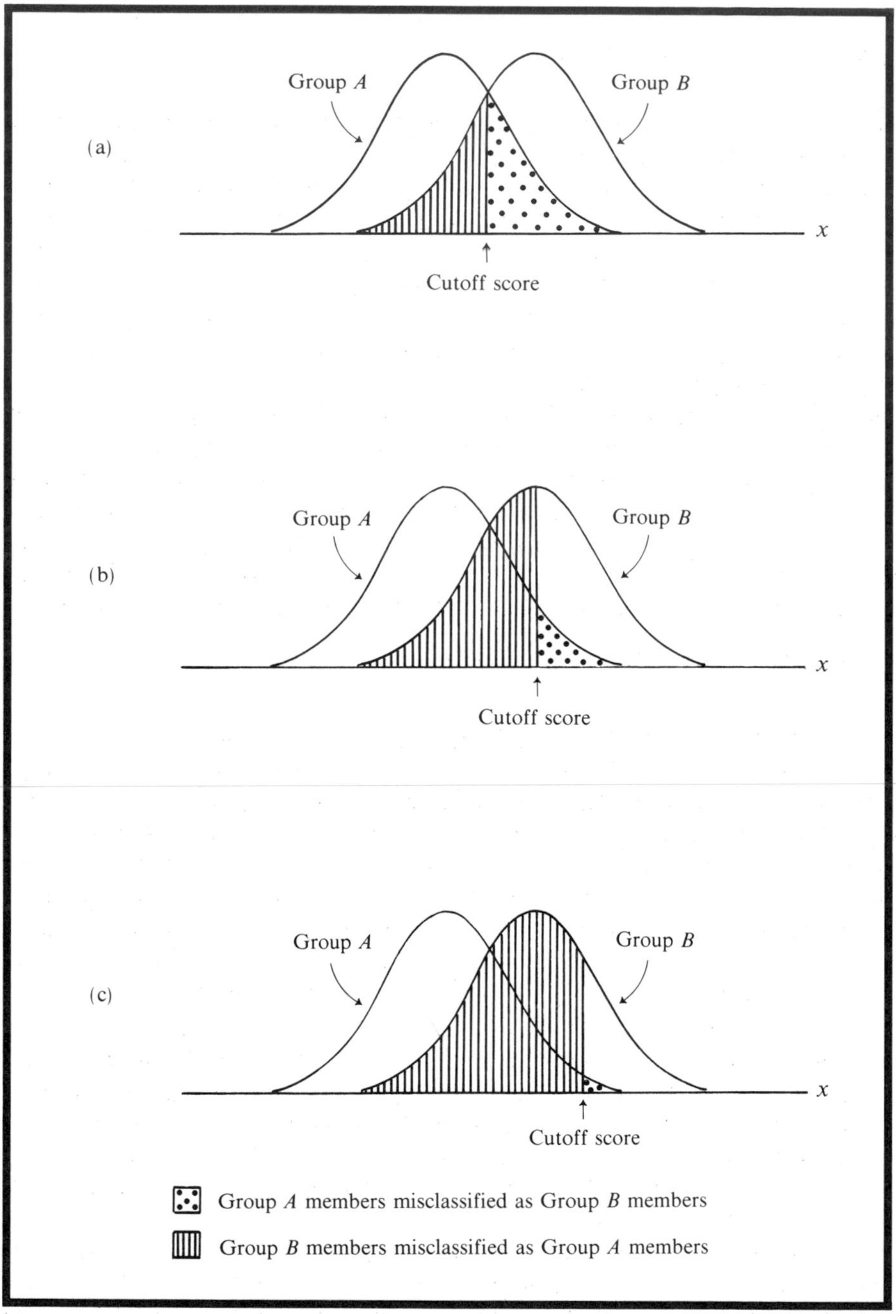

Figure 1 Errors of classification using alternative discriminant cutoff scores for the one-predictor-variable situation.

Alternative cutoff scores for the two-group, one-predictor variable situation can be found in parts *a*, *b*, and *c* of Figure 1. In each case, objects scoring *higher* than the cutoff score are presumed to belong to Group *B*, and objects scoring *below* the cutoff score will be predicted to belong to Group *A*. The resulting errors of classification are indicated by the striped area for misclassifications of Group *B* members, and the dotted area for misclassifications of Group A members.

Of the infinite number of possible cutoff scores along the predictor variable, we choose the one that results in the *fewest errors of classification*. In the example of Figure 1, with groups of equal size and equal variance, the best cutoff score is located midway between the means of the two groups. Notice how the total number of errors increase as the cutoff score is moved further from the ideal point.

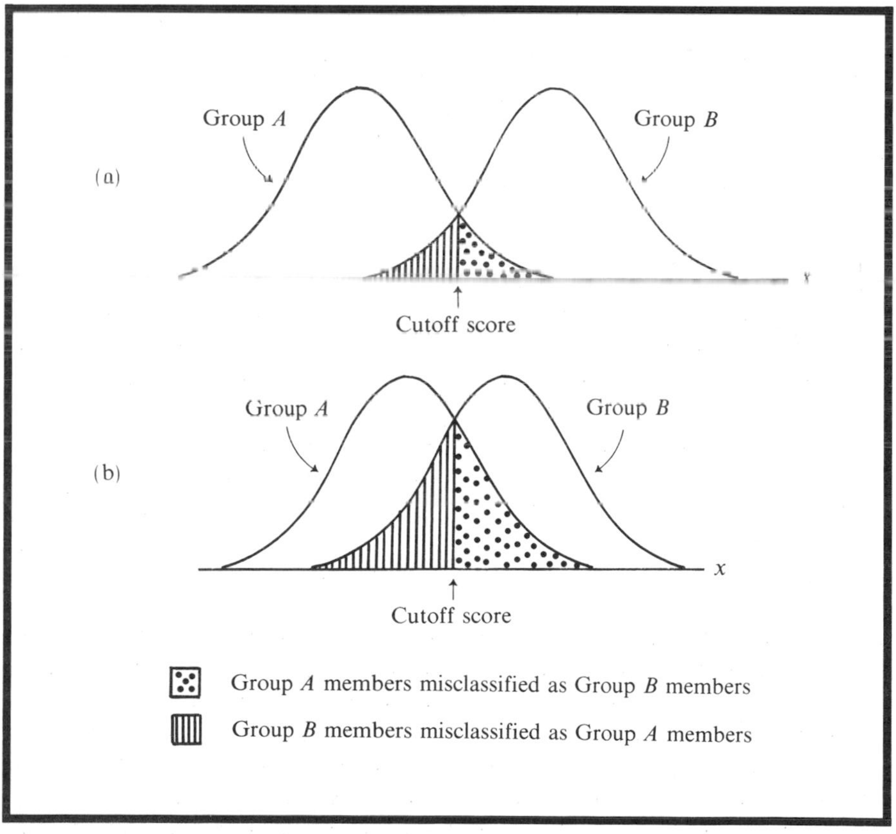

Figure 2 Errors of classification when there is (a) a large vs. (b) a small group difference, for the one-predictor-variable situation.

Unless there is absolutely *no* overlap between the criterion groups with respect to the predictor variable we are bound to make errors of classification. It will be apparent from a comparison of parts *a* and *b* of Figure 2 that the smaller the difference between two groups on the predictor variable—i.e., the greater the overlap—the more errors of classification we are destined to make. We can draw another analogy, then, with regression analysis: Just as a high correlation between predictor variable and criterion variable results in reduced errors of prediction in regression analysis, *a large difference between criterion groups* with respect to the predictor variable results in *fewer errors of classification* in discriminant analysis.

Multiple predictor variables. Instead of using just *one* predictor variable to discriminate between two criterion groups, we can extend the discrimination concept to *several* predictor variables—more precisely, to a weighted combination of several predictor variables.

By weighting the values of various predictor variables we can *derive* a single predictor variable—the *discriminant function*— which can then be treated just as in the one-predictor situation discussed above. Each object will have a single score on the discriminant function in place of its scores on the various predictor variables. At the same time a cutoff score will be determined such that when the criterion groups are compared with respect to the discriminant scores the errors of classification are minimized.

This general procedure of assigning weights to the various predictor variables and selecting a cutoff score so as to minimize the errors of classification can be seen intuitively and graphically in Figure 3, where two groups of objects, *A* and *B*, are measured on *two* predictor variables, x_1 and x_2. The objects in the two groups are plotted in a scatter diagram in part *a* of the figure, revealing a number of characteristics of the data. First we notice that the groups overlap *both* on variables x_1 and x_2. Secondly, we notice that there is a moderate correlation between the two predictor variables, and it appears to be the same in each group as required by the discriminant analysis model. Thirdly, we see that the variation of scores on each variable is about the same in each group, another of our data assumptions. Finally we notice that the groups differ more on variable x_1 than on x_2, so we might expect that x_1 will have the greater weight in the derived discriminant function, assuming their variances are the same.

Figure 3b shows the distribution of *discriminant scores* for the two groups. The particular discriminant function, $L = .9x_1 + .5x_2$, was derived such as to *maximize the difference* between the two criterion groups with respect to their resulting discriminant scores, and consequently *minimize their overlap*. As can be seen by the location of the cutoff score there are many fewer errors of classification than if either x_1 or x_2 alone had been used to discriminate the groups. The weights determined for the two predictor variables result in composite discriminant scores that lead to less overlap between the groups

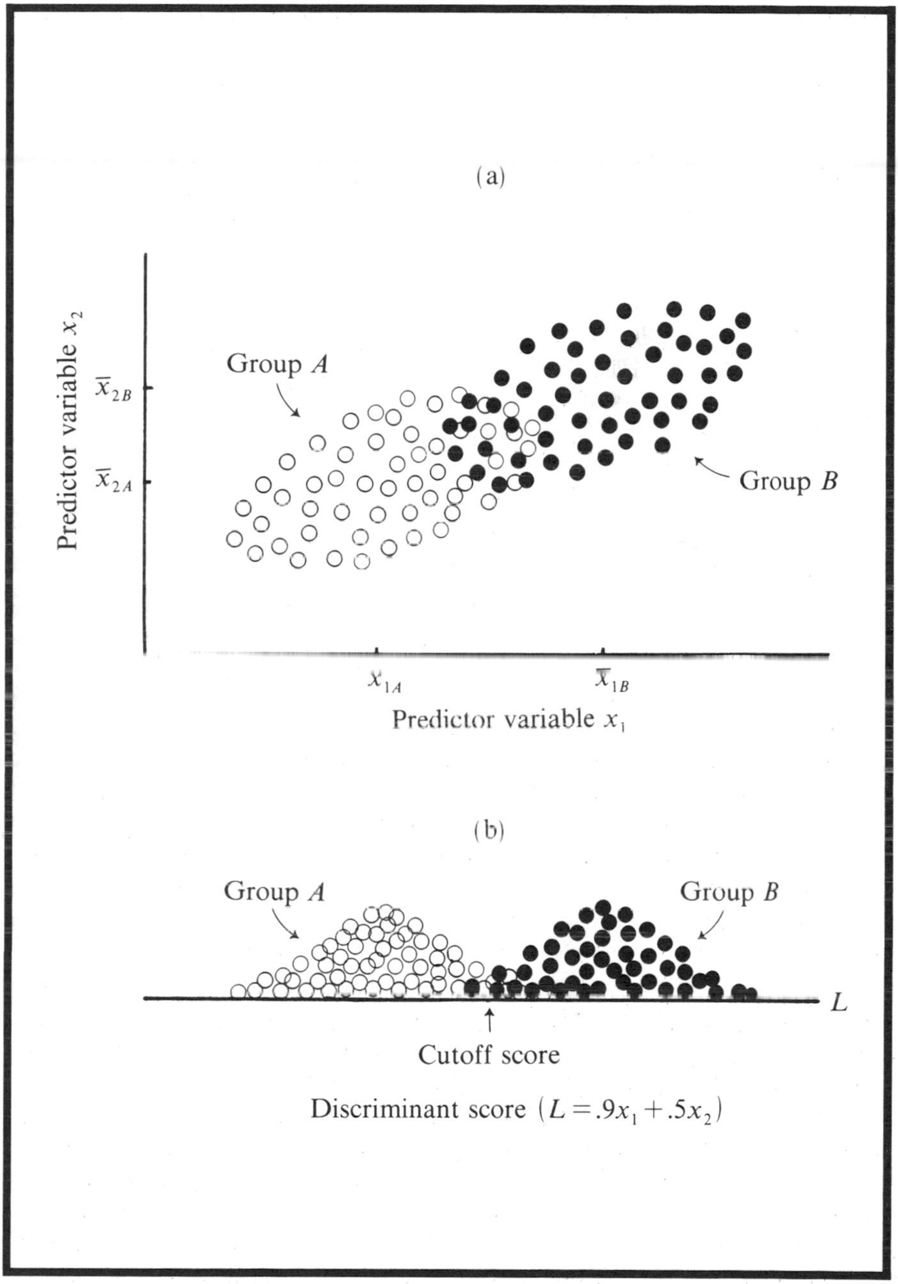

Figure 3 Two groups compared on (a) two predictor variables, and (b) the derived discriminant function.

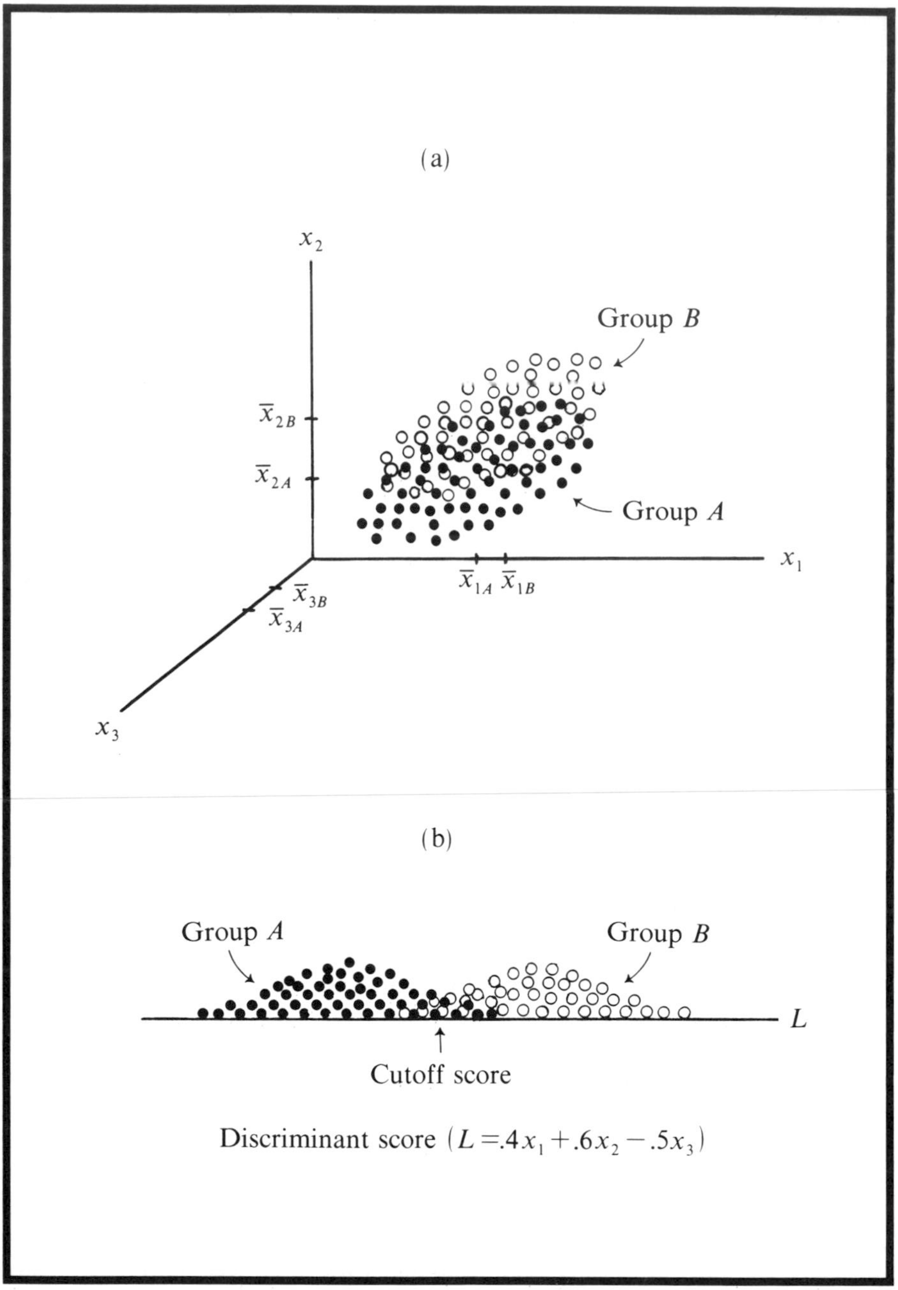

Figure 4 Two groups compared on (a) three predictor variables, and (b) the derived discriminant function.

than any other choice of weights: This is, in fact, the criterion for choosing the weights.

We can now see how the same procedure can be applied to the situation in which the two criterion groups are measured on *three* predictor variables. Figure 4a shows a *three-dimensional* scatter plot of the scores of Group *A* and Group *B* members on predictor variables x_1, x_2, and x_3. Part *b* of the figure shows how the discriminant scores—the weighted sums of the predictor variable scores—are distributed. Again, the weights associated with each predictor variable are chosen such that there is a maximum difference between the criterion groups in their discriminant scores, for the larger the difference between the groups, the smaller the overlap in scores, and consequently the fewer errors of classification when the optimum cutoff score is employed.

This procedure can be generalized to any number of predictor variables. In each case we simply reduce the situation to a one-predictor problem by creating a derived variable—the discriminant function—which is a weighted sum of the values on the individual predictor variables. In determining these weights, as in multiple regression analysis, the *correlations* that exist among the predictor variables are taken into account. Also, in assessing the *difference* between two groups on a particular predictor variable—which will influence the size of that variable's weight—we must evaluate that difference with respect to the *variation* of the variable. A mean difference between two groups of, say, *five* units, on one variable could be substantially greater than a difference of *twelve* units on another variable, depending on the variances of the respective variables.

More than two criterion groups. When we want to discriminate between more than two groups we need more than one discriminant function. Generally, we will need *one fewer* discriminant functions than the number of criterion groups. If we want to discriminate between buyers of model *A*, *B*, and *C* automobile , we will need *two* functions. If we want to discriminate between patients with disease *A*, *B*, *C*, and *D* we will need three functions. The exception to this rule is when we have fewer predictor variables than we have criterion groups, in which case we will have as many discriminant functions as we have predictor variables.

The use of more than one discriminant function in classifying objects into criterion groups can best be conceptualized with an example. Suppose we want to discriminate between buyers of Model *A*, *B*, and *C* automobile, based on measurements of these buyers on four predictor variables, x_1, x_2, x_3, and x_4.

Since we have three criterion groups, we will need two discriminant functions. The first of these functions will discriminate the members of one of the groups from the members of the other two groups. The second discriminant function will then discriminate between the remaining two groups.

For example, our first function might be as follows:

$$L = 3.1x_1 - 4.2x_2 + 16.1x_3 + .7x_4$$

with the stipulation:

If L is greater than 120.1, assign buyer to group B

The values of each buyer on variables x_1, x_2, x_3, and x_4 will be plugged into the function and those with discriminant scores exceeding the cutoff score of 120.1 will be classified as a buyer of Model B. Why were members of Buyer Group B singled out for the first discriminant function? Because they were the easiest to discriminate from members of the other groups.

We now need a second discriminant function to discriminate between Groups A and C. It might be as follows:

$$L = 1.1x_1 + 2.7x_2 - 25.3x_3 - 1.1x_4$$

with the stipulation:

If L is greater than 87.5, assign buyer to Group A

Those buyers not already classified into Group B based on the first discriminant function will have their values on the four predictor variables substituted into the above function. If their resulting discriminant score exceeds the cutoff value of 87.5 they will be classified as a member of Group A. The remaining buyers, those with scores under 87.5 will be categorized as members of Group C.

To gain a more intuitive grasp of how separate discriminant functions are determined when there are three criterion groups, imagine a sequential procedure in which members of Group A are compared against all others; then Group B against all others; then Group C against all others. Whichever of these comparisons results in the best discrimination—i.e., the fewest classification errors—will form the basis of the first discriminant function. The second discriminant function is then based on the remaining two groups. The concept can easily be generalized to four, five, or more criterion groups. In each case, of course, the evaluation of the differences between the groups takes into account the variances of the variables as well as their inter-correlations.

Stepwise procedures. As with regression analysis, discriminant analysis lends itself to the selection of predictor variables through stepwise procedures. Rather than using an entire set of predictor variables in our discriminant

function, we can use a smaller set that discriminates between the criterion groups virtually as well as the entire set itself.

This reduction in the number of predictor variables can be accomplished due to the intercorrelations or redundancies among the predictor variables. However, as with regression analysis, we must be concerned about the collinearity problem, instances in which two or more predictor variables are very highly correlated. (See Chapters 10 and 11 for a fuller discussion of the stepwise techniques.)

5. Evaluation of the Discriminant Function

One of the main things we would like to know about our discriminant function is whether it, in fact, succeeds in discriminating between members of our criterion groups. If the groups do not differ on the individual predictor variables, it is unlikely that they will differ on the derived discriminant function. On the other hand, chance differences between groups on the individual predictor variables could accumulate to produce apparent discrimination between the groups.

Indices of discrimination. There are a number of summary indices of the amount of discrimination achieved in a discriminant function analysis. This is not the place to delve into their mathematical definitions and statistical properties, rather it is enough for our purposes to know that these summary measures exist, and that they provide a statistical basis for determining if the degree of observed differentiation between groups is beyond what would be expected by chance alone.

Among the available indices is R^2, the square of the multiple correlation coefficient, previously encountered in our evaluation of the regression equation. However, we might wonder, can such a correlation be computed when we have a dichotomous criterion variable? It will be recalled from our introductory discussion of correlation in Chapter 10 that reference was made to a variety of correlation coefficients that could be used when the assumptions of the product moment correlation coefficient were not met. One of these, the *point biserial correlation*, is used for correlating values on a continuous variable with the two values of a dichotomous variable. All we need to do is make the dichotomous variable into a dummy variable with arbitrary values of 1 and 0 for the two levels in question. We can then proceed with the same calculations involved in determining the product moment correlation. The value of R^2 will then reflect the correlation between the 0 and 1 values on the dichotomous criterion variable and the scores on the continuous discriminant function. Figure 5 should remove any doubt that such a correlation can be calculated.

There are discrimination indices other than R^2, which are based on the extent of difference between the criterion groups with respect to the intercorre-

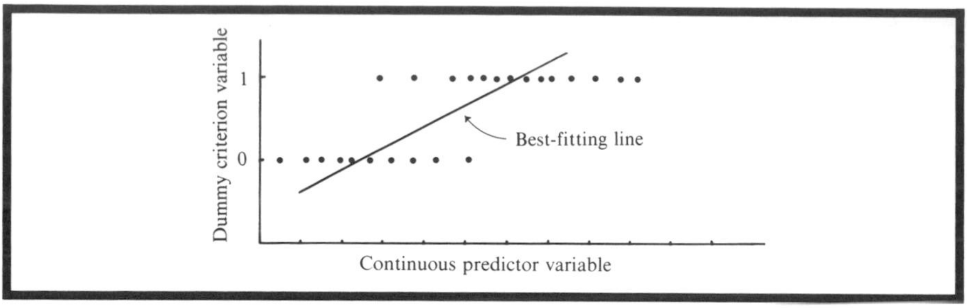

Figure 5 The scatter diagram for a dichotomous criterion variable and a continuous predictor variable.

lated predictor variables and the derived discriminant function based upon them. They include *Mahalanobis'* D^2, *Wilks' Lambda*, and *Rao's V*. The definitions of these indices are topics for advanced study and are mentioned only in passing to promote familiarity with their names allowing those interested to pursue the subject further. For our present purposes, it is important only to know that such indices exist which will help to evaluate a particular discriminant function, just as methods are available for evaluating a regression equation.

6. Accuracy of Classification

The indices named in the preceding section provide summary measures of the discrimination achieved in a discriminant function analysis. Still, perhaps the most meaningful evaluation of the discriminant function will be in terms of the *actual errors of classification*, both in *number* and in *type*.

The accuracy of classification can readily be seen in a matrix such as that shown in Table 2. Often referred to as a *confusion matrix*, it presents a tabulation of the objects' *actual* group membership versus their *predicted* group membership. The table presents the hypothetical results of a discriminant analysis in which 300 insurance applicants were classified as *risks* or *non-risks*—i.e., likely or not likely to make a claim—based on their scores on a number of predictor variables.

Inspecting the marginal totals of Table 2 we see that 100 of the 300 individuals were predicted to be risks, while the remaining 200 were predicted to be non-risks. These figures happen to correspond to the *actual* number of risks and non-risks, but it should be absolutely clear that this should not be taken as a mark of success in our prediction procedure, for we can stipulate beforehand how many of our total number of predictions we want to fall into each group. More important are the frequencies *within* the body of the table,

Table 2 Predicted vs. actual group membership for 300 insurance applicants (developmental sample).

Predicted group	Actual group: Non-risk	Actual group: Risk	Totals
Risk	30 (errors)	70	100
Non-risk	170	30 (errors)	200
Totals	200	100	300 applicants

reflecting the *association* between predicted and actual group membership. It will be recalled from our discussion of contingency tables, of which the confusion matrix is one, that the cell frequencies within the body of the table can vary widely for any given set of marginal frequency totals.

We find in Table 2 that our predictions of risks are correct in 70 out of 100 cases (70%); and our predictions of non-risks are correct in 170 out of 200 (85%). Or, in total, we are correct in 240 out of 300 predictions, for an overall percentage of 80% correct classification. It can be verified that with the given marginal totals, we would expect to be correct 167 cases out of the 300 (56%) on the basis of chance alone; i.e., if we assigned the 100 risk and 200 non-risk labels to our 300 individuals completely at random.

As pointed out above we are not constrained to making our predictions for the two criterion groups in the same proportion as they are observed in the population. For example, we could take the very extreme position and predict that *all* 300 of the insurance applications were *non-risks*. Since two out of three cases will be non-risks in any event, we would be correct two-thirds (67%) of the time without even doing a discriminant analysis. This degree of accuracy, though, is of little solace to the insurance company that is interested in singling out risks. The above approach fails to identify a single risk.

So even though our discriminant function analysis resulted in only 80% accuracy of classification overall, it did succeed in identifying 70 of the 100 risks. In the process it erred in classifying 30 applicants as risks who proved not to be; and made an additional 30 errors in classifying applicants as non-risks when in fact they were risks. It is just like disease diagnosis.

Costs of errors. The above observations of the different *types* of prediction errors raises an important issue in discriminant analysis. *Are all types of errors of equal importance*? In our example, what are the alternative consequences or costs of, on the one hand, predicting someone to be a risk when they are not, versus, on the other hand, predicting that someone is not a risk when they are?

In the first instance we will be foregoing some income by refusing a policy to some actual non-risk applicants. In the second instance, we will be granting a policy to some applicants who are in reality risks. Obviously the second type of error, taken on a one for one basis, is more costly than the first type, but the only way to rule out all such errors is to predict that *everyone* is a risk, and sell no policies at all. This, of course, is no more realistic than predicting everyone to be a non-risk.

Fortunately, the discriminant analysis procedure can be adapted to take into account the differential costs of the alternative types of errors. We cannot expect miracles though. If we place too much weight on one type of error, the only plausible solution may be to predict that *everyone* belongs in the group for which we especially do not want to misclassify its members, and thereby eliminate any possibility of misclassifying them. However, with a little trial and error we may decide upon alternative costs that result in satisfactory discrimination, yet tend to minimize the more costly errors at the expense of a greater number of less costly errors.

Despite our ability to weight the importance of alternative types of errors, the harsh reality of discriminant analysis is that it is not very successful in identifying members of a minority group. For example, if 95% of our total sample falls in one criterion group and 5% falls in the other, it is virtually impossible to do better than chance in discriminating between the groups. The task becomes harder as the deviation from a fifty-fifty split increases. This is not to say that it cannot be done, for if we can identify one or more predictor variables that are highly related to the criterion classification we should expect some degree of success. The difficulty is in identifying such predictor variables which can be reliably measured.

Validation. As with the regression equation, the discriminant function should be *validated* by testing its efficacy with a fresh sample of analytical objects. The observed accuracy of prediction on the sample upon which the function was developed will always be spuriously high, because we will have capitalized on chance relationships. The true discriminatory power of the function will be found when we test it with a completely separate sample.

Table 3 shows the classification of insurance risks using the original discriminant function, but now applied to a *new sample* of 300 applicants. Note how the accuracy of prediction has dropped. Of our 100 risk predictions, 50 are correct and 50 are incorrect. On the basis of chance alone we would expect one-third of our predictions to be risks, so we are still performing above chance but not nearly so well as was apparent, and deceptively so, in our *developmental* sample. Our overall level of accuracy in the *validation* sample, taking into account the 150 correctly classified non-risks, is 67%, compared with 80% accuracy with the development sample.

As with regression analysis, the importance of this type of cross-validation of our discriminant function cannot be overstated. The last thing we want to

Table 3 Predicted vs. actual group membership for 300 insurance applicants (validation sample).

Predicted group	Actual group: Non-risk	Actual group: Risk	Totals
Risk	50 (errors)	50	100
Non-risk	150	50 (errors)	200
Totals	200	100	300 applicants

do is act headstrong upon derived relationships that have no demonstrated validity.

7. Importance of the Predictors

In addition to knowing how well our discriminant function differentiates between members of our criterion groups, we would like to know the importance of the individual predictor variables in contributing to that success. Just as with regression analysis, we would expect some of the predictors to be more important than others, and knowledge of these variables would help us understand and perhaps eventually manipulate membership in the criterion groups.

While we cannot influence whether someone is a credit risk or not, there are some applications of discriminant analysis in which we can hope to affect group membership from a knowledge of the important predictor variables. For example, knowing the discriminating characteristics of smokers and non-smokers might help us to develop special educational programs which could affect whether an individual becomes a smoker. Similarly, if we can identify differences between diseased and healthy cells, we can hope to manipulate those characteristics and thereby control the incidence of the diseased cells. At the very worse we will have identified variables which will lead to further hypotheses and testing.

The *relative* importance of the predictor variables can be determined from the squared coefficient weights associated with each variable in the discriminant function. But, as with regression analysis, the discriminant function must be in the standardized z score form

$$\boxed{L_z = beta_1 z_1 + beta_2 z_2 + \cdots + beta_k z_k} \tag{2}$$

where the beta weights are analogous to those in the multiple regression equation.

Again, as with regression analysis, the squared beta weights reflect only the relative importance of the variables and do not reflect their absolute importance. Also, the relative importance of any given variable, vis-a-vis another, will depend upon which other variables are included in the analysis, and which have been omitted. Eliminating or introducing variables into the discriminant function could completely alter the contributions of individual variables. And when we have more than two criterion groups, and consequently more than one discriminant function, the importance of the predictor variables can vary from one function to another.

8. Discriminant vs. Regression Analysis

Discriminant analysis is sometimes incorrectly used in instances where regression analysis is the more appropriate and powerful technique. This misuse of discriminant analysis involves the dichotomization of a continuous criterion variable for no apparent reason other than to lend the data to a discriminant function analysis.

For example, a sample of students might be measured on the continuous criterion variable of number of hours of TV watched per week. Rather than using these measures as the criterion variable in a regression analysis, the students will be divided for no justifiable reason into two groups, heavy and light viewers, and these two criterion groups would form the basis for performing a discriminant analysis. The detail of the observed data has been discarded —it is like tossing out the watermelon to eat the rind.

Perhaps one justifiable basis for dichotomizing a continuous criterion variable would be if the linearity assumption of the regression model were violated. Even then, a data transformation is preferable to dichotomization, for if ever there is a cardinal sin in statistical analysis it is to use a weaker analysis when a more powerful and efficient analysis is readily available. Rather than applying the technique to the data, some would like to apply the data to the technique. One can only guess that this type of behavior is motivated by the fallacious belief that the more esoteric or less well-known a technique, the more magical will be its benefits.

Tandem use. There are legitimate instances in which discriminant and regression analysis can be applied to the same set of data. Thinking back over the many dichotomous criterion variables mentioned in the chapter it will be evident that they are not, in reality, always of an all or none nature. For example, whereas borrowers either do or do not default on their loans, there are *degrees* of default. Also, while individuals either do or do not smoke, those who do, do so in varying amounts. In other words, the criterion variable can be

thought of as being *continuously* scaled for *one* of the two criterion groups, but *all-or-none* for the other group.

In such situations, what we might do is apply a discriminant analysis to the dichotomous groups, and a regression analysis to the members of the group for which the criterion variable is continuously scaled. It would be interesting to learn, for example, if the variables which are predictive of whether or not an individual smokes tobacco are the same as those which predict how much the person smokes. Or whether the variables which discriminate between dry and productive well sites, are able to predict the level of productivity of the producing wells. It is this type of complementary use of statistical techniques that will prove in the end to be the most beneficial, for empirical data is always richer than the means we have to capture it.

9. Concluding Comments

We have seen how discriminant analysis is similar in principle to regression analysis, finding its special application in situations which involve a criterion variable that is qualitative in nature. The technique allows us to classify objects into groups based on their values on various predictor variables. As with regression analysis, alternative methods are available to assess the degree to which the technique has improved upon chance or uninformed prediction. We also find in discriminant analysis a prime example of how the individual statistical building blocks of mean, variance, and correlation combine to create a higher-order analytical technique. Also, we find in discriminant analysis, as with regression analysis and correlation analysis, a fusion of the three key functions of statistical analysis—data reduction, inference, and the identification of associations among variables.

Chapter 15

Factor Analysis

1. Introduction

The very early stages of most statistical investigations can be characterized as fishing expeditions. Often there is little prior quantitative information about the area of study and only the flimsiest of hypotheses exist as to the relevant variables and their inter-relationships. Consequently, the early research efforts are necessarily of a hit and miss nature—many variables are studied with the hope of identifying some kind of order among them which will deepen our understanding of the subject matter.

We can ask ourselves, for example, how the relevant variables were isolated for inclusion in the correlation analyses, regression analyses, experimental studies, and discriminant analyses discussed in the earlier chapters. Unfortunately, these variables did not leap up and stare the investigator in the face. While we might like to think that we as researchers are so insightful as to be able to conjure the variables that are most relevant to our area of interest, it is more often the case that the identification of these variables is the result of some earlier investigations.

When we conduct such preliminary studies we will usually want to obtain information on the widest possible variety of variables, in order to increase our chances of hitting upon some which, individually or jointly, will expand our knowledge about the subject matter. For the analysis of this type of data, it would be highly desirable to have a technique available that could identify and summarize the many inter-relationships that exist among the individual variables.

While we do have correlation analysis available to identify the inter-relationships among variables, practical problems arise when we have a large number of variables under study. For example, when we are studying the relationships among ten variables, 45 different correlation coefficients are possible. When we study fifty variables, a total of 1,225 correlations

are generated. And the number of coefficients keeps accelerating as the number of variables increases, dictating the need for a higher-order data reduction technique, one that can systematically summarize large correlation matrices.

Factor analysis is just such a data reduction technique. It is a family of procedures for removing the redundancy from a set of correlated variables and representing the variables with a smaller set of "derived" variables, or *factors*. Alternatively, the factor analysis procedure can be thought of as removing the duplicated information from among a set of variables, or, we may think of it loosely as the grouping of similar variables.

2. Overview of Factor Analysis

The essence of *factor analysis* is shown schematically in Figure 1, where nine variables $v_1, v_2, \ldots, v_9$ are clustered into three separate groupings. Variables v_1, v_4, v_5, and v_8 are clustered together, meaning that they are highly correlated with one another and represent a common underlying variable, or *factor* as it is called. Similarly, variables v_3 and v_7 define a separate factor. And the variables v_2, v_6, and v_9 contribute to form a third factor. In each case the subset of variables can be thought of as *manifestations of an abstract underlying dimension*—a factor. So, instead of having to understand nine separate variables, we have simplified matters and have to consider only three factors which contain virtually all the information inherent in the original nine variables.

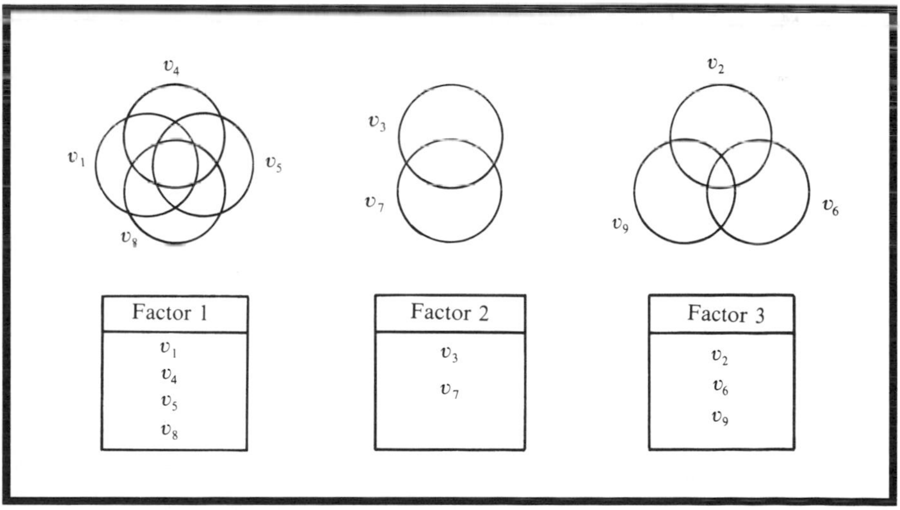

Figure 1 Nine variables reduced to three factors.

In practice, the derived factors are never so clear-cut as the ideal case portrayed in Figure 1. There is usually some amount of overlap between the factors, since each of the original variables defining a factor usually has some degree of correlation with some of the other variables. They are shown as completely separate from one another due to the difficulty of portraying such relationships on two-dimensional surfaces. It could be, for example, that variables v_1, v_4, v_5, and v_8 which define the first factor, could be marginally related to the variables constituting the second and third factors. However, it is the task of factor analysis to form factors that are relatively *independent* of one another. So in the present example, we can be confident that the variables defining Factor 1 are more highly correlated with one another than they are with the variables defining Factors 2 and 3.

Later in the chapter we will learn more about the actual factor analysis procedure and the properties of the derived factors. First, some of the main applications of this powerful technique will be presented.

3. Applications of Factor Analysis

There is a wide range of data analysis problems to which factor analysis can be applied. Several of the key types of applications are described in the following paragraphs.

Identification of underlying factors. One of the most important uses of factor analysis is in the identification of factors underlying a large set of variables. By clustering a large number of variables into a smaller number of homogeneous sets and creating a new variable—a factor—representing each of these sets, we have simplified our data and consequently are more likely to gain insight into our subject matter.

For example, an airline might study fifty different variables that it thinks are important to the flier in choosing an airline—variables such as courtesy of personnel, degree of on-time performance, convenience of schedules, check-in time, etc. Individually, and collectively, the variables would shed little light on the overall consumer dynamics involved in choosing one airline over another. However, if a factor analysis could reduce these fifty variables into a handful of more general and abstract factors, we would be on our way to a better understanding of the fundamental variables underlying a flier's decision process.

Similarly, we could submit social attitudes, mental abilities, personality characteristics, career interests, food preferences, disease histories, or blood characteristics to a factor analysis. In each instance, the large number of possible variables would be reduced to a smaller, more manageable, and interpretable number of *factors*.

Screening of variables. A second important use of factor analysis is in the *screening* of variables for inclusion in subsequent statistical investigations, such as regression analysis or discriminant analysis. Since factor analysis identifies groupings of variables that are highly correlated with one another, we can choose a *single* variable from each factor for inclusion among a set of potential predictor variables, thereby avoiding the collinearity problem.

For example, a personnel department wishing to develop a short screening application for job applicants could perform a factor analysis of information provided by a sample of applicants on a long list of variables ranging from high school grades to interests in various hobbies. Sample variables from each of the resulting factors would then provide the basis on which to develop a shorter screening questionnaire for future use, the scores on which could then be correlated with job performance.

Similarly, a short academic achievement test could be constructed by selecting questions from a large battery of items which were found to be highly intercorrelated and formed underlying factors. The shorter form would then save administration time without sacrificing much of the information inherent in the longer form. Also, the selected items could serve as predictor variables in a follow-up study.

Or, a fast-food restaurant chain wishing to expand its franchise could factor analyze the variables believed to be related to outlet sales—population density, distance to nearest school, average household income in a two-mile radius, etc. The elimination of redundant variables through factor analysis would be a big step toward the development of a regression equation for predicting outlet sales from a handful of neighborhood variables, which in turn would guide the selection of new restaurant sites which would best serve the community.

Summary of data. A third type of application of factor analysis owes to its flexibility in being able to extract as few or as many factors as we desire from a set of variables. Also, in one very common type of factor analysis, the first extracted factor typically accounts for a large portion of the variance of information inherent in the entire set of variables, and each succeeding factor accounts for less and less. This feature allows us to use a few factors, perhaps only one or two, to account for the bulk of the variance contained in the entire set of variables.

For example, a group of voters might be asked to rate some candidates on a large number of image variables—community-minded, good manager, progressive, etc. A factor analysis of this candidate image data could reduce the large number of specific images into a couple of broader, more abstract factors that accounted for most of the data. The candidates could then concentrate their public relations efforts on the key image factors, those accounting for the bulk of the variance in the voters' attitudes.

Or, as another example, a factor analysis could be performed on supervisor ratings of a number of management trainees on a wide variety of variables. Rather than each trainee having a number of different scores on these evaluative variables, factor analysis procedures would enable each trainee to have a *single* score on the largest factor, the one accounting for the bulk of the information inherent in the individual variables.

Sampling of variables. A fourth use of the factor analytic technique is in the selection of a small group of representative, though uncorrelated variables from among a larger set in order to solve a variety of *practical* problems.

For example, an automobile advertiser limited to a thirty-second TV commercial would like to identify perhaps three or four automobile features from among dozens, to stress in its advertising. If these three or four features happened to be communicating the same underlying characteristic—i.e., manifesting the same factor the advertising message would be repetitive and not as broad as if the features had been drawn from separate factors.

Or, to consider another example, an investment manager might want to put together a portfolio of stocks that represents a large group of energy-related companies, but for practical reasons cannot invest in each of the stocks. A factor analysis of the stocks' price movements on each of a large number of trading days could reveal groupings of similar stocks, as far as price movement is concerned. A stock from each of these factors could then form the basis of a portfolio that would be expected to perform very similarly to the larger population of stocks as a whole.

Or, a factor analysis of student food preferences could aid a school cafeteria in designing an efficient menu selection, one that was not encumbered with too many redundant entries, appealing instead to the widest possible range of tastes.

Clustering of objects. In addition to identifying similarities among variables, factor analysis can be used to cluster people (or other objects). In this procedure, often referred to as *inverse* factor analysis, a sample of individuals (in the role of variables) are measured on a number of random variables (in the role of objects), and are clustered into homogeneous groups based on their inter-correlations.

For example, a sample of people could be grouped together based on their similarity of attitudes toward educational needs. After the factor analysis has partitioned the original sample of people into relatively homogeneous subgroups, the subgroups could then be compared in terms of demographics, past education, and on the very attitudes that gave rise to the groups. An educational institution could then use this information to tailor its education programs to fit the needs of each of the consumer groups.

Or, a large sample of individuals could be clustered into homogeneous groups based on the similarities in their incidence of various illnesses. The

resulting groups of individuals might then shed light on the etiology of the diseases.

As can be seen from the preceding list of applications, factor analysis is an extremely powerful and versatile technique. At one and the same time it *summarizes* data and *identifies relationships* among variables, two of the basic functions of statistical analysis. Further, it serves an *inferential* role in those instances in which we generalize the results to a larger population. Also, we see again how the correlation coefficient, one of the fundamental building blocks of statistical analysis, is at the heart of factor analysis. In the following sections we will take a closer look at some of the details of the factor analytic procedure to get a better grasp of the relationship between the input variables and the extracted factors.

4. The Input Data Matrix

Now that we have an idea of the major applications and benefits resulting from factor analysis, we can take a closer look at the overall procedure involved in the technique.

Rather than citing specific variables and objects, which could detract from the understanding of the procedure itself, we will begin by referring to the variables, objects, and factors in the abstract, keeping in mind that the variables could be any of those mentioned in the preceding sections or chapters. After the procedure has been described in general terms, a few specific examples will be presented to add the necessary flavor to the understanding and appreciation of the technique.

As stressed in the preceding sections, factor analysis depends to a great extent on the correlations between variables. It is not too hard to imagine that once we have a correlation matrix, as described in Chapter 10, we might be able to identify "factors" by visual inspection alone. As we scanned the correlation matrix row by row, or column by column, we would find sets of variables that were highly correlated with one another, suggesting that they were all measuring the same fundamental underlying dimension, or factor.

This procedure of visually inspecting the correlation matrix in search of variables that are highly correlated with each other, yet which have low correlations with other variables in the matrix, would be fine if we were dealing with a small matrix of perhaps five or so variables, but once the number of variables starts getting larger, the number of correlation coefficients, as we have seen, begins to accelerate, making it virtually impossible to summarize the matrix by visual inspection alone.

The development of factor analysis helped solve this data reduction need. It is a well-defined procedure for identifying and extracting the redundancy within the correlation matrix, a technique which can avoid the personal

judgments of an investigator attempting to perform this task by inspection alone. Also, the technique, unlike personal judgment, results in factors with quantitative properties that aid in their interpretation and application.

Figure 2 shows the key stages in the factor analysis procedure. The first stage is the original data matrix: A set of objects $O_1, O_2, \ldots, O_n$ are measured with respect to a number of different variables $v_1, v_2, \ldots, v_k$. We have seen this *objects by variables* ($O \times v$) matrix many times before in the earlier chapters, since it is the fundamental input for many multivariate analyses.

The $O \times v$ matrix could represent the ratings of consumers (the objects) for Model *A* automobile on a variety of perceived product benefits (the variables). It could represent scores of job applicants (the objects) on various tests (the variables). It could represent the price movement of various stocks (the variables) on a number of individual trading days (the objects). It could represent the values of cities (the objects) on a number of dimensions such as unemployment rate, cost of living index, crime rate, population, etc. (the variables). Or, it could represent various individuals' (the objects) frequencies of various symptoms (the variables). In each instance, each of the objects has a value or score on each of the variables.

The number of possibilities for the input $O \times v$ data matrix is virtually limitless. The important thing to remember, once again, is that whatever variables we wish to study, we need to identify an appropriate scale and set of objects on which to obtain a variety of values of the variables. And, as we know, the larger the sample of objects we study, the more reliable will be our

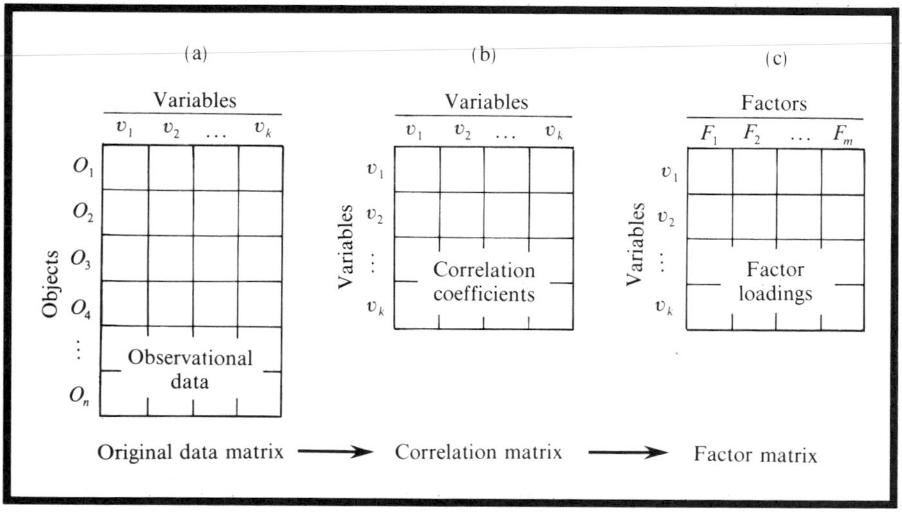

Figure 2 Key stages in the factor analysis procedure.

results. Typically, we would want an input objects by variables matrix in which there are considerably more objects than variables, in order to avoid the identification of relationships between variables which we *believe* to be real, but in actuality are due to chance alone; a circumstance that is more likely to happen as the number of variables grows large.

Various rules of thumb are used in the determination of the size of the sample of objects: e.g., ten times as many objects as variables, or the square of the number of variables, with a minimum of one hundred objects in cases with fewer than ten variables. However, these are only guidelines, cost considerations and the particular purpose of the analysis should be the final determining factors in the decision.

5. The Correlation Matrix

The second key stage in the factor analysis procedure, as shown in Figure 2, is the creation of the *correlation matrix*. We have seen in Chapter 10 how this is accomplished. It is nothing more than the systematic arrangement of the simple correlation coefficients that exist between each pair of variables. A correlation is calculated for each combination of two variables— v_1 with v_2, v_1 with v_3, v_1 with v_4, etc. Also, v_2 with v_3, v_2 with v_4, and so on until a correlation coefficient is obtained for *each possible pair of variables*.

At this point it is almost essential that we enlist the aid of the high-speed computer. For, as we have noted before, as few as twenty variables will generate 190 correlation coefficients. To do these individual calculations with a simple hand calculator and double-checking them for human error would require weeks or months of full-time labor; for a hundred-variable problem, *years*.

6. The Factor Matrix

Once the correlation matrix has been formed, a series of operations are performed upon it employing matrix algebra, operations that are beyond the scope and purpose of our introductory discussion, and are not necessary for understanding the general procedure. Figure 2, however, shows the format of the end product of these operations, the *factor matrix*.

Factor loadings. In a *factor matrix*, the columns represent the derived factors, and the rows represent the original input variables. The cell entries of the factor matrix, which are called *factor loadings*, and which vary in value from −1.00 to +1.00, represent *the degree to which each of the variables correlates with each of the factors*. In fact, these factor loadings are nothing more than the correlation coefficients between the original variables and the newly derived factors, which are themselves variables.

Inspection of the factor loadings will reveal the extent to which each of the variables contributes to the meaning of each of the factors. Within any one column of the factor matrix, some of the loadings will be high and some will be relatively low. Those variables with the high loadings on a factor will be the ones that provide the meaning and interpretation of the factor. Those variables with low or zero loadings on a particular factor will not contribute to the meaning of that factor, but rather will tend to contribute to the meaning of one of the other factors by virtue of their high loadings on those factors. We can liken a factor to a *criterion variable* in a regression analysis and the input variables can be seen as the *predictor variables*, or vice versa.

Factor scores. If a factor is itself a variable, we can ask on what objects is it measured? The answer is that the factors are measured on the same objects upon which the input variables are measured. So, in addition to each of our original objects having a value on each of the input variables, they also have a value on each of the derived factors, and these latter scores are referred to as *factor scores*. An object's score on a factor represents *a weighted combination of its scores on each of the input variables*.

Variables with *high loadings* on a particular factor will be associated with *high weights* in the equation for determining the factor scores. On the other hand, variables with *low loadings* on a particular factor will be associated with *low weights*. It follows logically, then, that an object will tend to score high on a factor only if it scores high on the variables which load most highly on that factor.

It is easier to see now how a variable can be thought of as having a loading on (correlation with) a factor. Since each object has a score on a factor and a score on a given variable, it would be easy enough to determine the correlation between these two sets of scores. This, of course, is a simplified version of how the factor loadings are obtained, since no mention has been made as to how the *weights* are determined for the equations which yield the factor scores in the first place. If the weights depend on the loadings, and the loadings in turn depend on the factor scores resulting from the application of the weights, then what came first, the chicken or the egg!

The *algorithm*, or set of mathematical operations, needed to solve the above problem is again beyond the scope of this presentation. For our purposes it is enough to understand that the factor loadings and the weights for determining the factor scores are determined more or less simultaneously in such a way that there will be little or *no correlation between the resulting factors*; that is, so they will be independent of one another.

For two *factors* to be independent of one another has the same meaning as two *variables* being independent of one another. It is the same to say that no association exists between them: If an object scores above average on one factor, we have no clues as to whether it scores above or below average on the

other factors—that is, the scores on the separate factors do not systematically covary.

It follows, then, that variables which are highly correlated with each other will form a factor, while variables that are not correlated with each other will form separate factors. It is on this basis that the factor analysis procedure is able to remove redundancy from a set of variables.

7. Number of Factors Extracted

In perhaps the most common variation of factor analysis, called *principal components* factor analysis, as many factors are initially extracted as there are variables. If we have nine input variables, we will extract nine factors. If we have twenty input variables, we will extract twenty factors. What, we might ask, is to be gained by extracting as many factors as we have variables if our original aim was data *reduction*.

The answer lies in the fact that the first extracted factor typically accounts for the largest part of the total variance inherent in the data collection. The second factor extracted usually accounts for quite a bit of information also, though less than the first factor. Each succeeding factor accounts for less and less of the total variance, so that by the time we get to the final factors they necessarily account for less variance than an individual variable.

This property of the initial factors accounting for more variance than the final factors is shown in Table 1 for a hypothetical nine-variable problem. Notice that the first factor accounts for a full 41% of the total variance inherent in the data. Extracting the second factor adds another 23%, for a cumulative total of 64%. So these two factors account for nearly *two-thirds* of the data's variation, while two individual variables would only account for about 22%: This is because each variable on average accounts for 11% of the total variation—100% divided by 9 variables equals 11% per variable, on average.

There is, in fact, a basic difference between principal components analysis and the more mainstream factor analysis models. In principal components analysis, each factor or "component" is viewed as *a weighted combination of the input variables*, with as many components derived as there are variables. In the mainstream factor analysis models, on the other hand, each input variable is viewed as *a weighted combination of factors*, with the number of factors being fewer in number than the original set of input variables. Indeed, principal components analysis is often used as a preliminary step to help decide the difficult question of how many factors to solve for in the latter approach. For our purposes, though, it is best to think of the issue more generally, in which factors represent abstractions of the input variables, whether derived by the

Table 1 The percentage of total variance accounted for, and the associated eigenvalues, for the successively extracted factors in a factor analysis of a set of nine variables (principal components solution).

The extracted factors	% of total variance accounted for		Eigenvalues	
	Incremental	Cumulative	Incremental	Cumulative
F_1	41%	41%	3.69	3.69
F_2	23	64	2.07	5.76
F_3	14	78	1.26	7.02
F_4	7	85	.63	7.65
F_5	5	90	.45	8.10
F_6	4	94	.36	8.46
F_7	3	97	.27	8.73
F_8	2	99	.18	8.91
F_9	1	100	.09	9.00

principal components approach or the mainstream factor analytic models. Though mathematically different, they produce highly similar results.

Eigenvalues. Associated with each derived factor is a quantity known as an *eigenvalue*, which corresponds to the equivalent number of variables which the factor represents. For example, a factor associated with an eigenvalue of 3.69 indicates that the factor accounts for as much variance in the data collection as would 3.69 variables, on average. If, for instance, we are dealing with a nine variable problem, as in Table 1, in which each input variable would account on average for $100\% \div 9 = 11\%$ of the total variation, then a factor with an eigenvalue of 3.69 would account for $3.69 \times 11\% = 41\%$ of the total variation in the data collection (e.g., Factor F_1 in Table 1). Compare this with Factor F_9 which has an eigenvalue of only .09 and therefore accounts for only $.09 \times 11\% = 1\%$ of the variance of the data.

The importance of this concept of an eigenvalue also rests on the decision we must make as to *how many* factors we will retain from the analysis. We do not want to retain all nine factors in our example since the latter factors account for only about one or two percent of the total variance. One frequently used rule of thumb is to retain factors to the point where an additional factor would account for less variance than a typical variable; that is, less than one eigenvalue.

In our example, a typical variable would account for 11% of the variance, so based on the figures in Table 1 we would choose to retain only three of the nine factors. To retain a fourth factor would only yield an additional 7%, less

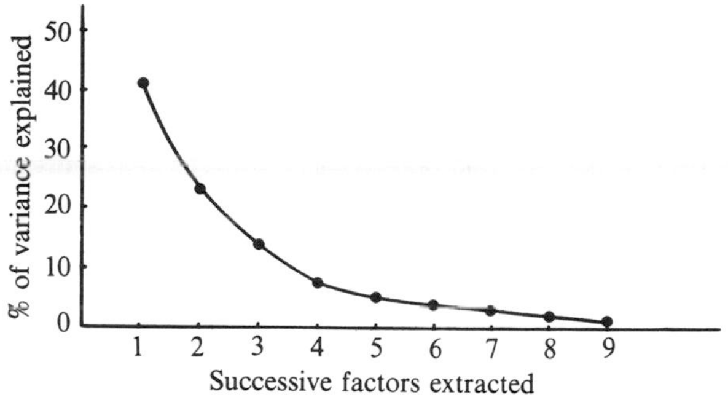

Figure 3 Scree curve showing the percentage of total variance accounted for by each of nine successively extracted factors (Data from Table 1).

than the amount inherent in a single variable. And since our primary concern is data reduction, it would be somewhat illogical to retain a factor that contains even *less* information than an original *input* variable.

Scree test. The eigenvalue criterion for deciding on how many factors to retain is only a guideline, and should not be followed blindly. The plot of the incremental variance accounted for by each successive factor can also help in our decision.

This plot is known as the *scree curve* and is shown in Figure 3. It gets its name due to the similarity of the curve's tail to the stoney rubble accumulated at the base of a slope. The idea of the scree test is that the factors along the tail of the curve represent mostly random error variance—i.e., rubble—and therefore we should select the factor solution just prior to the levelling of the curve. This criterion also suggests the 3-factor solution for the present example, reinforcing the decision based on the eigenvalues.

Variance explained. A third basis on which to choose the number of factors to retain is to consider the total variance accounted for, or "explained", by the factors. While we would want to account for as much of the variance as possible, at the same time we want to do it with as few factors as possible.

The decision then becomes a *trade-off* between the amount of *parsimony* and *comprehensiveness* we can attain. To have more of one is to have less of the other. We must decide on what balance will serve our purpose.

The decision will be largely dependent on the use to which the analysis will be put. In the present example, the 3-factor solution accounts for a respectable 78% of the total variance inherent in the entire set of nine variables. However, depending on the reasons for performing the factor analysis, we may be happy with the 2-factor solution that accounts for only 64% of the total variance; or

we may feel it is necessary to take the 4-factor solution that accounts for 85% of the variance.

Comprehensibility. Another approach to selecting a factor solution is to inspect a number of different solutions with respect to the meanings of the variables loading on the respective factors, and decide which solution makes the most sense in light of what is already known about the subject matter. So both comprehensiveness *and comprehensibility* of the factors can guide our decision. In the present example, we might study the 2-, 3-, and 4-factor solutions before making a final decision on how many factors to retain.

We see, then, that there is a variety of bases on which to select the number of factors that we should solve for. In addition to those named above, there are a number of other purely statistical criteria. In the end, though, the decision will have to rest on the purpose of our analysis. And there is no reason why we cannot retain alternative solutions; one solution for selecting input variables for a regression analysis, for instance; and another solution for understanding the *dominant* factors underlying the set of variables, or for generating hypotheses for further study.

In short, the applications of our analysis, as discussed at the beginning of the chapter, should dictate the criteria for deciding on the number of factors to extract from our original set of variables. And in most cases, our decisions will take into account *all* of the procedures named above.

8. Rotation of Factors

Having decided on how many factors to extract from our original set of variables, we have the option of redefining the factors in order that the explained variance is redistributed among the newly defined factors. Recall that in the principal components form of factor analysis the first extracted factor accounts for the largest portion of the total variance, and each successive factor accounts for less and less. This lopsided distribution of explained variance often poses problems of interpretation. With most of the variables loading high or moderately high on the first factor, and relatively few variables loading high on the other factors, it can be difficult to get an impression of the underlying meaning of the factors.

A redefinition of the factors, such that the loadings on the various factors tend to be either very high (near 1.0 or −1.0) or very low (near zero), eliminating as many medium-sized loadings as possible, would aid in the interpretation of the factors. Such a redefinition of the factors is accomplished by a procedure known as *factor rotation*.

There are many different types of rotation possible, known by the names of *varimax, quartimax, equimax,* and *oblimin*, to mention the most common

procedures. The distinctions between these techniques and their methodologies are not important for our purpose. It is sufficient to know that the rotation techniques *redefine* the factors in order to make sharper distinctions in the meanings of the factors. The techniques named above are merely different criteria for the accomplishment of that task, and their definitions can be pursued in advanced courses of study.

The essence of the factor rotation concept can be seen in Figure 4. It is a graphical portrayal of three of the variables—v_1, v_5, and v_7—from the nine-variable example of the preceding sections. The variables are represented as points on a rectangular coordinate system in which the horizontal and vertical axes represent Factors 1 and 2, respectively. The dashed lines in Figure 4 indicate the projections of the points (the variables) onto the axes (the

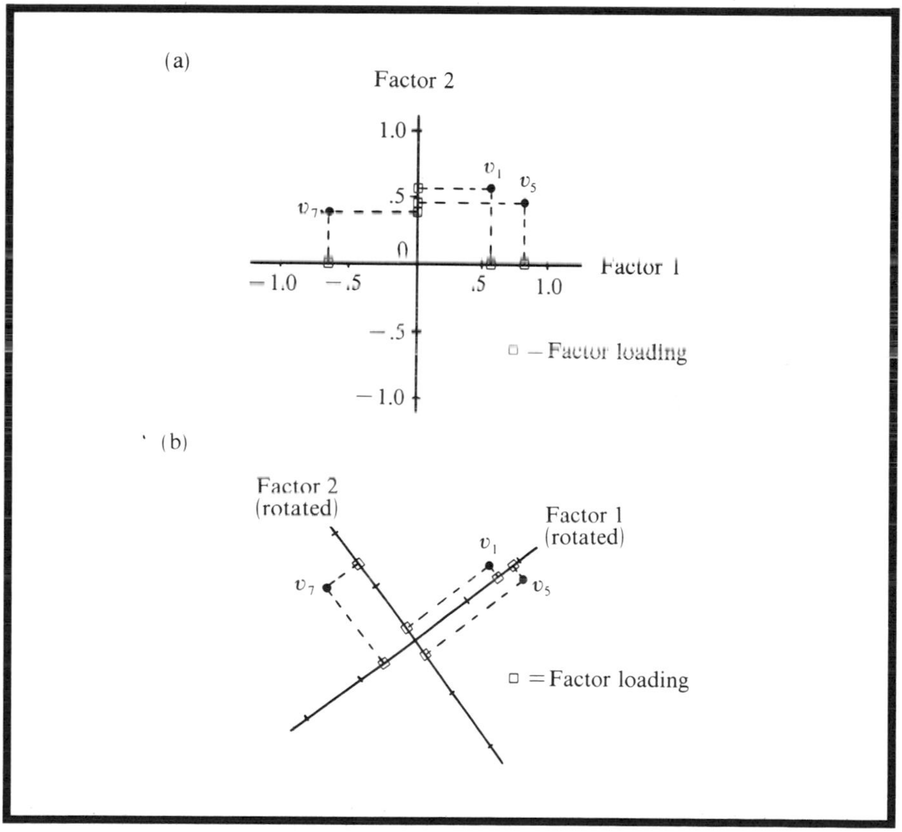

Figure 4 An example of factor rotation and the accompanying changes in factor loadings.

factors), and the point of projection on an axis represents the *loading* of the variable on that factor. Consider variable v_5, for example. It has a very high loading on Factor 1, but only a moderately high loading of Factor 2. Variable v_1 has a moderately high loading on both Factors 1 *and* 2. The same is true with v_7.

In the bottom half of Figure 4 the factor axes have been rotated approximately 45° counterclockwise. The variables, though, have been left to occupy their former positions. However, the projections of the variables on the factor axes now result in quite different loadings. Notice that prior to rotation, v_1 had a moderate loading (about .5) on both Factors 1 and 2. After rotation of the factors, v_1 has a fairly high loading on Factor 1 and virtually a zero loading on Factor 2. And this is the primary objective of factor rotation—to make the factors as distinctive as possible. If a variable loads high on one factor, we do not want it to also load high, or even moderately high, on another factor; otherwise how could we distinguish between the factors with respect to that variable?

Notice also in Figure 4 that variable v_5 had a high loading on Factor 1 and a moderate loading on Factor 2 prior to rotation. This is fairly good discrimination, but the rotation accentuates the difference. After rotation, v_5 still has a high loading on Factor 1, but now has a near zero loading on Factor 2. In the case of v_7, which prior to rotation had a moderate loading on both factors, winds up with a fairly high loading on Factor 2 and a relatively low loading on Factor 1 after the rotation is performed. Again, the result is a more unique definition of the respective factors.

In the preceding example we made no mention of Factor 3 from our original three-factor solution. In actuality all three factors would be rotated simultaneously in a three-dimensional space filled with all nine variables. This cannot be adequately portrayed on paper and the actual rotation is a task for the computer, as it must consider the ever-changing loadings of each of the nine variables on each of the three factors for every change in the orientation of the factor axes. This is a mind-boggling task at best. To accomplish it the computer employs an *iterative*, or repetitive, process of orienting and reorienting the factors until it can no longer improve upon a prespecified criterion of the spread of the loadings on the factors. Compare, for example, the *spread* of loadings on the factor axes in Figure 4 before and after rotation.

The results of our hypothetical rotation are shown in Figure 5. Prior to rotation, Factors 1, 2, and 3 accounted for 41%, 23%, and 14% of the total variance, respectively. After the rotation, the distribution is more or less equal among the three rotated factors—with 27%, 25%, and 26% of the variance being accounted for by the newly defined factors. Notice that in both cases, before and after rotation, the total variance explained by the three factors totals 78%. Thus, factor rotation does not result in a change in the number of

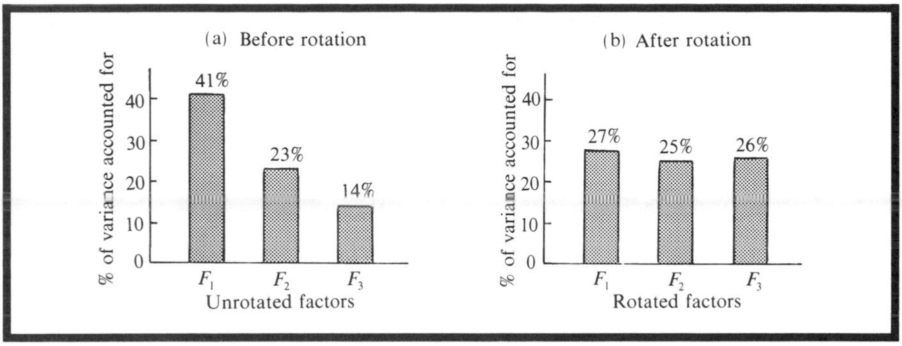

Figure 5 Distribution of variance accounted for by three factors before and after rotation.

factors, nor in the total variance explained. It only redefines the factors with respect to the manner in which the variables load on (correlate with) the factors. While successful in this example, factor rotation does not always overcome the lopsided distribution of explained variance.

A summary of the changes in the relative magnitudes of the factor loadings, and consequently the changes in the interpretation of the factors as a result of the rotation, is shown in Table 2. Notice the difference in the range of magnitudes of the factor loadings before and after rotation. Prior to rotation the variables tended to have either high or moderately high loadings on the first two factors and there were relatively few low loadings, except on Factor 3, which had *mostly* low and moderately high loadings.

After rotation, however, the loadings tend to be at the extremes, either high or low, with relatively few medium-sized loadings. Notice also that prior to rotation, eight of the nine variables had either a high or moderately high loading on Factor 1; the reason it accounted for so much of the variance at the start. After rotation, though, neither it nor any other factor has such a concentration of high loading variables.

To see how the factors differ from one another, the high loadings in Table 2 have been highlighted with circles. Variables v_1, v_4, v_5, and v_8 have high loadings on Factor 1; variables v_3 and v_7 have high loadings on Factor 2; and variables v_2, v_6, and v_9 have high loadings on Factor 3, just as portrayed at the start of the chapter in Figure 1.

As to what constitutes a *high loading* and what constitutes a *low loading*, it is not easy to generalize. There are no widely accepted statistical standards. To test a factor loading for statistical significance in the same way we would test any other correlation coefficient is not a very useful procedure, since the highest loadings on a factor are almost always well above the level of statistical significance—e.g., .2 to .3 for a sample of 100 objects. And even those loadings

Table 2 Factor loading matrix before and after factor rotation.

Relative magnitudes of factor loadings (H = High, M = Medium, L = Low)							
Before rotation				After rotation			
	Factors				Factors		
Variables	F_1	F_2	F_3	Variables	F_1	F_2	F_3
v_1	M	M	L	v_1	H	L	L
v_2	M	M	H	v_2	L	M	H
v_3	L	H	M	v_3	L	H	L
v_4	H	M	M	v_4	H	M	L
v_5	H	M	M	v_5	H	L	M
v_6	M	L	H	v_6	L	L	H
v_7	M	M	L	v_7	L	H	L
v_8	H	L	L	v_8	H	L	L
v_9	M	M	L	v_9	M	M	H
% of variance explained:	41 + 23 + 14 = 78%				27 + 25 + 26 = 78%		

which rank relatively low on a factor will still prove statistically different than zero. Consequently, the interpretation of the factor loading matrix is primarily dependent on the overall configuration of loadings.

The analyst must often use personal judgment as to what constitutes a meaningfully high loading, based on the distribution of loadings *within* a factor and *across* factors, as well as the absolute magnitudes of the loadings. In practice, loadings of .3, .4, or .5 are most often used as lower bounds for meaningful loadings. But again, these are only guidelines and there may be instances in which higher or lower bounds may be employed. As always, a degree of common sense must be used in the interpretation of the data.

9. The Naming of Factors

At this point in the analysis, an investigator will study the high-loading variables of each factor and attempt to give a *descriptive name* to the factors, names that represent a common element or abstraction of the individual

variables which load highly on the factors. Other investigators are less speculative and settle for a simple numerical label of the factors, because the act of naming the factors is perhaps, at the same time, one of the greatest benefits of the analysis and one of the biggest dangers.

On the one hand, if the investigator is especially insightful, he or she may detect the essence of the separate variables and identify a higher-order abstraction that provides a completely novel way of viewing the subject matter, and may contribute to the development of other hypotheses that could be tested in subsequent research studies.

On the other hand, a careless or casual naming of the factors might be completely misleading, jeopardizing the conclusions of the study. For it is a fact that the identity of the contributing variables is often foresaken once a factor has been named, and the label is then communicated to those who would hope to apply the results of the research. It is wise, therefore, to always study the variables which define a factor, and decide for oneself the meaning of the factors.

In the following section, where we will consider some specific examples, we will get a better idea of the difficulties and benefits inherent in the naming of factors.

10. Summary Presentation and Interpretation

Now that we have an idea of the key stages in the factor analysis procedure and the relationship between the input variables and the extracted factors, we can take a look at how the overall analysis is summarized and interpreted.

Tabular presentation. A useful way to summarize a factor analysis is to list the high loading variables for each of the named factors. Table 3 shows such a summary presentation. It is based on a factor analysis of the self-descriptions of a sample of individuals. Each person in the sample rated themselves on a 0 to 100 scale on each of the following personality variables: *active*, *aggressive*, *attractive*, *competitive*, *creative*, *generous*, *honest*, *humorous*, *influential*, *intelligent*, *intense*, *jealous*, *kind*, *lazy*, *loving*, *optimistic*, *persistent*, *sensitive*, *shy*, and *stubborn*; listed here alphabetically, but listed in Table 3 according to the factors upon which they loaded most highly.

As shown in the Table 3 summary, three factors were extracted in the analysis: the first, named Goodness, suggested by the high loading variables *kind*, *generous*, *loving*, and *honest*; the second, named Strength, suggested by the high loading variables *persistent*, *stubborn*, *influential*, and *intelligent*; and the third, named Dynamism, suggested by the high loading variables *creative*, *active*, *optimistic*, and *competitive*. A study of the variables with *moderately* high loadings on the respective factors further support the name labels.

Table 3 Summary of a factor analysis of self-descriptions.

Factor names and the high-loading variables	Factor loadings
Factor I: GOODNESS	
Kind	.71
Generous	.66
Loving	.59
Honest	.52
(Jealous)	−.51
Attractive	.43
Sensitive	.41
Factor II: STRENGTH	
Persistent	.75
Stubborn	.70
Influential	.67
Intelligent	.61
(Shy)	−.53
Intense	.42
Factor III: DYNAMISM	
Creative	.69
Active	.68
Optimistic	.57
Competitive	.55
Aggressive	.51
Humorous	.44
(Lazy)	−.43

A number of useful observations can be made from this example which illustrate the concepts discussed mostly in the abstract in the preceding sections. We learned, for example, that variables with high loadings on a particular factor would be those which are highly correlated with one another, but which have little or no correlation with the variables loading highly on the other factors. This is clear to see in the present example. Consider the three top-loading input variables on Factor I: *kind*, *generous*, and *loving*. It is not hard to imagine that persons rating themselves above average on one of these variables would be more likely than not to rate themselves above average on the others also. On the other hand, those individuals with below average ratings on one of the variables would probably have below average ratings on the other variables loading on that factor. In other words, there is a statistical dependency, a correlation, among these variables. The other self-images

loading on the Goodness factor will be seen to be similar in theme to the three listed above.

The variables loading high on Factor II—*persistent*, *stubborn*, *influential*, *intelligent*—reflect a Strength dimension, although other labels are possible. These self-images are all highly correlated with one another, but relatively independent of the set of inter-correlated images forming the Goodness factor. That Factors I and II are independent of one another means that one's "goodness" has nothing to do with one's "strength." If someone scores high on Factor I, they would be equally likely to score high or low on Factor II. One has nothing to do with the other. And that, as has been stressed often before, is the meaning of variables being independent of one another.

Factor III, named Dynamism, owes its label to the movement or non-static qualities of its high loading variables: *creative*, *active*, *optimistic*, and *competitive*. Again, they are correlated with one another, but uncorrelated with the images loading on the other two factors.

What the factor analysis has uncovered, then, is three relatively independent personality dimensions underlying the twenty individual input images. Someone can be good without being strong; someone can be dynamic without being good; and someone can be strong without being dynamic. It is interesting to note in this context that if we let .10 be the probability that an individual is in the top 10% on a given dimension, then, since the dimensions are independent, the probability of a person being in the top 10% on *all three* dimensions would be $.10 \times .10 \times .10 = .001$; that is, the probability of an individual being good, strong, and dynamic (or bad, weak, and sloth-like) is 1 in 1,000, based on the multiplication rule of independent probabilities.

Negative loadings. Another point to be made with this example is the issue of *negative factor loadings*. In Chapter 10 we learned that a negative correlation coefficient between two random variables meant that high scores on one tended to be associated with low scores on the other, and vice versa. The same thing is true with respect to factor loadings, inasmuch as a factor loading is a correlation coefficient between a variable and an abstract factor. So, when a variable loads negatively on a factor it simply means that a person or object scoring high on the variable will score low on the factor, and vice versa. In other words, such a negatively loading variable has a *meaning* opposite to that of the factor. Consider, for example, the *jealous* variable which loads $-.51$ on the Goodness factor. Immediately we know that the meaning of the factor will be the opposite of the meaning of the variable, and this is so, since Goodness is typically considered antithetical to jealousy. However, despite its negative loading, the variable has contributed to the interpretation of the factor. In terms of explanatory power, a $-.51$ loading is just as good as a $+.51$ loading.

Other negative loading variables can be found in Factors II and III. A *shy* self-image is negatively correlated with the Strength factor, and a *lazy* self-image

is negatively correlated with the Dynamism factor. Again, though, their presence in the analysis helped us to interpret the factors upon which they loaded, by contributing to the internal consistency of the set of variables loading on the respective factors.

Notice in Table 3 that we typically put a parentheses around a variable to highlight the fact that it has a negative loading on the factor, and that its opposite would conform to the meaning of the factor. For interpretation purposes, it is often helpful to reword such negatively loaded variables, by turning them into their opposite meaning. This can be done by placing the word "not" before the variable. For example, *not jealous* would load positively on the Goodness factor, *not shy* would load positively on the Strength factor, and *not lazy* would load positively on the Dynamism factor.

Magnitudes of loadings. Another point to be made with this example has to do with the *magnitudes* of the loadings. Recall that the size of the loading is an indication of the extent to which the variable correlates with the factor. It is not an indication of the *level* of endorsement of the various images, a common misinterpretation. As we learned in Chapter 10, the size of the correlation

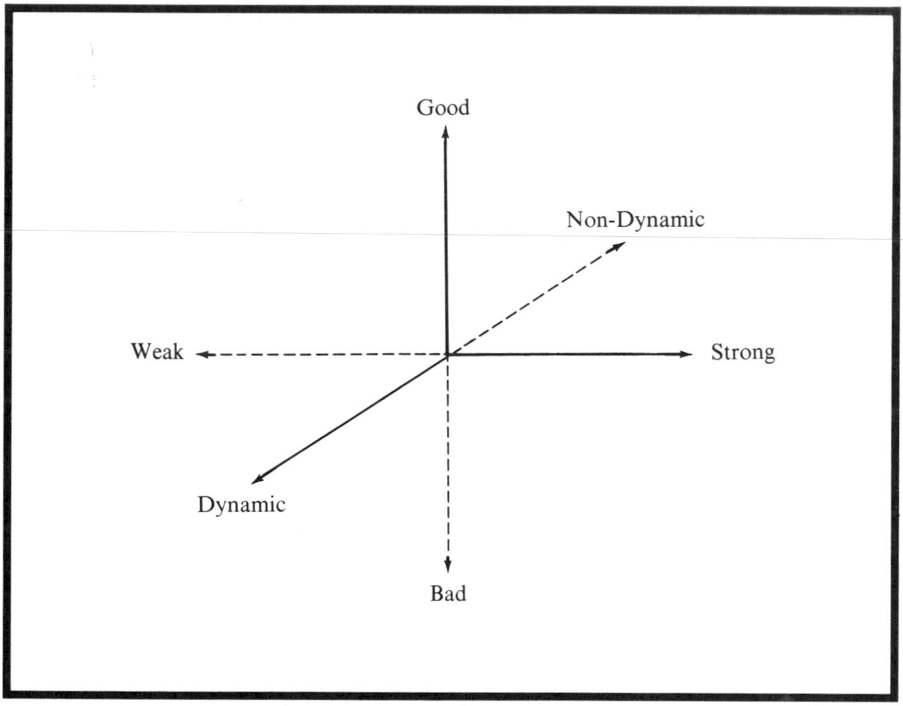

Figure 6 Graphical portrayal of three independent factors resulting from a factor analysis of a set of 20 self-image variables (See Table 3).

coefficient is not a function of the magnitude of the means of the variables. A variable with a low mean could correlate more highly with another variable or factor than a variable with a high mean. Thus, in the present example, the fact that *persistent* has the highest loading of any variable, it in no way implies that it was the highest rated image. To determine the mean ratings of the various variables we would have to consult the original data. It is an entirely separate analytical issue.

Graphical interpretation. Finally, a common misinterpretation of a factor analysis is that the various factors represent separate *groups* of people (or objects, more generally). For instance, in the present example, a misinterpretation would be that there are separate *non-overlapping* groups of good, strong, and dynamic individuals. This is not the case. Rather, the proper interpretation, as we have seen above, is that each factor represents a *continuum* along which people vary—some being above average, and some being below average—and that these continuums, dimensions, or factors are independent or orthogonal of one another, such that an individual's position on one is unrelated to their position on the others.

This interpretation is shown graphically in Figure 6, where the three factors correspond to three orthogonal axes in a three-dimensional space. It is

Table 4 Summary of a factor analysis of interests in TV programming.

Factor names and the high-loading variables	Factor loadings
Factor I: INFORMATION	
News programs	.81
Advances in science	.78
Social & cultural documentaries	.77
Animal documentaries	.71
Educational instruction	.69
Review of historical events*	.64
Talk shows*	.57
Factor II: ENTERTAINMENT	
Movies	.79
Variety shows	.74
Situation comedies	.72
Game shows	.65
Sports programs	.64
Talk shows*	.62
Review of historical events*	.50

*Listed on both factors.

interesting to note that in some psychological theories, the three dimensions of Goodness, Strength, and Dynamism might be interpreted as the Superego, Ego, and Id, respectively.

A two-factor example. Table 4 presents the summary of a factor analysis in which only two factors were extracted. It is based on interest ratings for various types of TV programming. The factors have been named Information and Entertainment, although alternative names are possible, and that task is left as an exercise.

Notice in this example that two of the variables—*talk shows* and *historical events*—are listed on both factors due to their moderately high loadings on both. In such situations we have the option of not listing those variables in the summary table, or listing them and indicating that they do appear on other factors also. It all depends upon whether or not they contribute to the meaning of the factors.

Although extracting only a few factors often results in "fuzzy" dimensions with meanings that are not always entirely obvious, other times they may lead to some fundamental understanding of the underlying dynamics of the subject matter. For example, in the present example of TV viewing interests, the two extracted factors might well be identified as *left-brain hemisphere* and *right-brain hemisphere* functions.

11. Criticisms of Factor Analysis

Despite the many obvious benefits to be derived from factor analysis, it has been the subject of a variety of criticisms. We have already discussed the danger of labelling the factors with misleading titles. Still, the insightful naming of the factors remains one of the greatest rewards of the analysis. The fact that the labelling is more of an art than a science should not lessen the overall appeal of the technique, as long as users of the research are aware that this aspect of the analysis involves a great deal of personal judgment.

A second criticism often levelled at the technique is that the derived factors are often so obvious that they could have been identified without the aid of a complicated computer analysis. Obviously, this criticism is made by individuals other than those who object to the factor-naming process, for if the factors are so intuitively clear then how could there be any difficulty in naming them appropriately. Also, it is one thing to make such claims of omniscience *after* an analysis has been performed, and quite another thing to identify the factors beforehand. Furthermore, this criticism does not show appreciation for the quantitative information resulting from the analysis, information that would not be available even if a successful "intuitive" factor analysis could be performed. Table 5 shows the factor-loading matrix for a factor analysis of consumer ratings of the importance of various variables in choosing an

Table 5 The rotated factor matrix for the factor analysis of the rated importance of 14 variables in choosing a car to buy; with the variables listed in random order.

	Extracted factors and their loadings			
Input variables	I	II	III	IV
1. Low-cost repairs	.31	.77	−.09	−.19
2. Comes in a variety of colors	.12	.29	.59	.15
3. Roomy interior	.86	−.11	.14	.35
4. Good gas mileage	.21	.81	.01	.14
5. Good handling	.29	−.07	.41	.49
6. Modern looking	.20	−.14	.71	.20
7. High resale value	.11	.61	.19	.05
8. Comfortable	.63	.19	.07	.38
9. Large engine	.35	−.57	.20	.11
10. Sleek appearance	.01	−.31	.65	.19
11. Easy to drive	.14	.16	.17	.73
12. Eye-catching	.31	−.21	.77	−.18
13. Large trunk space	.74	.15	.21	.29
14. Easy to park	.18	.11	.33	.65

automobile. It is left as an exercise to analyze, summarize, and interpret the analysis, and compare the results with one's *a priori* ideas as to the salient factors in choosing a car to buy.

A third criticism, one that has somewhat more validity than the preceding, is that the factor analysis procedure is subject to the *GIGO* principle—*Garbage In, Garbage Out*. But this is more a criticism of the nature of the input variables than it is of the factor analytic technique itself. Contrary to the expectations of some, factor analysis does not *create* new information. It merely organizes, summarizes, and quantifies information that is fed into the system. Consequently, if the input information is inadequate the final analysis will be inadequate. The resulting factors in the factor analytic technique, for example, could merely reflect the investigators' prejudices as to which variables are relevant in the particular area of study. If certain variables are overlooked, or only minimally represented, it will be no surprise if they do not appear as a factor in the end analysis. For this reason it is important that the input variables are obtained from as many different sources as possible, to ensure that all points of view are represented. Also, the initial compilation of variables should be subjected to a judgmental screening to eliminate variables which on face value, or logical grounds, are virtually identical. For it is a property of the factor analysis technique that the more variables loading highly on a factor, the more of the total variance it will be interpreted to represent. And this figure

will be artificially high if the analysis is "padded" with a large number of variables with only superficial differences in meaning or definition.

Then, there are the many theoretical controversies surrounding factor analysis. It is such a complicated sequence of procedures, with so many decisions to be made along the way, that there is not even agreement among authorities on factor analysis as to the optimum set of procedures to employ for a particular problem. But often these differences of opinion are more of theoretical than practical concern. The bottomline interpretation of factor analyses performed by the alternative procedures is not likely to differ to that great of an extent. What *is* important is that a specialist in the technique be a member of the research team in order to properly guide the analysis and answer questions from those who will eventually interpret and apply the research.

Finally, it has been argued that the factor analytic technique and its interpretation are so complicated that only individuals trained in advanced statistics could hope to understand it. But as we have seen in this chapter, it is possible to gain an appreciation of the system without a detailed knowledge of its theoretical underpinnings and computational intricacies.

In summary, it is safe to conclude that while the factor analysis procedure receives its share of criticism, it is still one of the most powerful data reduction techniques available, and when properly used can provide a deeper understanding of a wide range of problems and provide the necessary information for their solution.

12. Concluding Comments

While our introduction to factor analysis has been necessarily brief, overlooking many of the complexities and variations of the procedure, it has been designed to serve the needs of the non-specialist, who, most of all, needs to know of the *existence* of the technique, have an idea of the *key stages* in the analytical procedure, understand the *interpretation* of the results, and perhaps most importantly of all, to be able to spot the *types of problems* to which it can be applied.

Chapter 16

Cluster Analysis

1. Introduction

To this point we have been primarily concerned with the relationships that exist among variables. The objects upon which the variables were measured were assumed to be homogeneous in nature; that is, there was no reason to believe, or interest in the possibility, that a given set of objects could be divided into subsets which displayed reliable non-random differences. There are many situations, however, in which our main interest is in dividing a set of objects into subgroups which differ in meaningful ways.

Cluster analysis, also known by the names of segmentation analysis and taxonomy analysis, is a set of techniques for accomplishing the task of partitioning a set of objects into relatively homogeneous subsets based on the inter-object similarities. We can also view cluster analysis as encompassing the clustering of variables, as in the case of factor analysis, although it is usually thought of as being concerned with the clustering of objects.

The applications of cluster analysis are many. For example, we might be interested in clustering individuals based on their similarities with respect to social attitudes, self-images, blood chemistry, past ailments, or consumer needs. Plant or animal specimens, or entire predefined species, could be clustered based on various morphological, physiological, or environmental characteristics. Similarly, races, religions, cultures, ore samples, fossils, or archeological artifacts could be clustered into groupings based on their respective similarities. In short, any set of objects can be subjected to a cluster analysis.

2. Overview of Cluster Analysis

The overall cluster analysis procedure is shown in Figure 1. We typically begin by measuring each of a set of n objects on each of k variables. Next, a

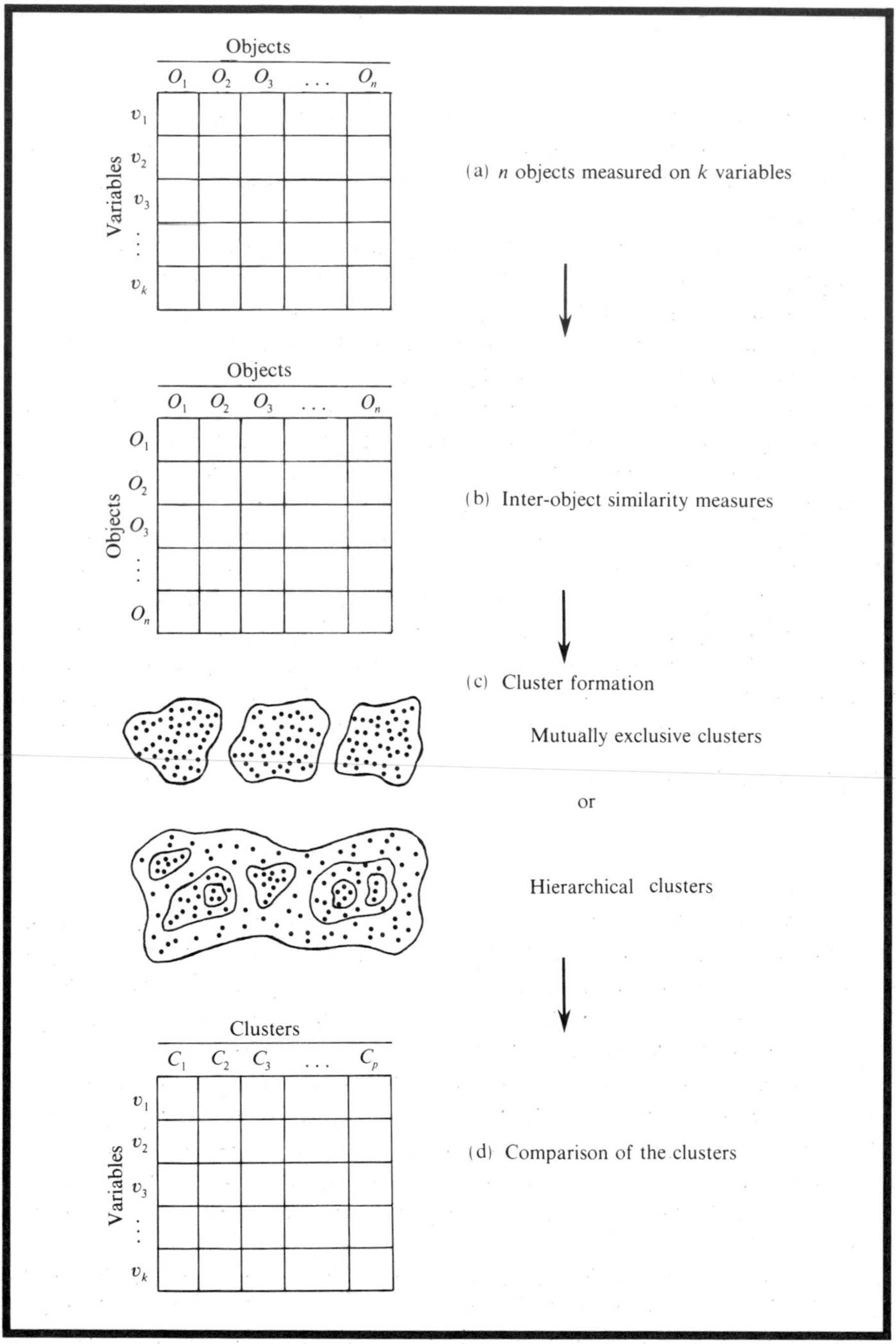

Figure 1 An outline of the cluster analysis procedure.

measure of the *similarity*—or, alternatively, the distance or difference—between each pair of objects must be obtained. Then some *algorithm*, or set of rules, must be employed to cluster the objects into subgroups based on the inter-object similarities. The ultimate goal is to arrive at clusters of objects which display *small within-cluster* variation, but *large between-cluster* variation. The differences between the resultant clusters can then be understood by comparing them with respect to their mean values on the input variables or other characteristics of interest.

The end benefit of forming and describing the clusters will depend upon the purpose of the particular analysis. For example, consumers who are clustered based on their needs can lead to alternative product formulations, as well as to customized advertising and marketing strategies. Citizens of a country who are clustered based on their social attitudes can result in new institutions and legislation to accommodate the alternative clusters. Patients clustered on the basis of their complaints or body chemistry can provide insights into the causes of the symptoms as well as for alternative therapies. At the very least, a cluster analysis will serve the heuristic function of generating hypotheses for further research.

While superficially similar, it is important to understand the distinction between cluster analysis and discriminant analysis. Whereas in discriminant analysis we begin with *a priori* well-defined groups in an attempt to identify the variables which distinguish the groups, in cluster analysis we begin with an undifferentiated group and attempt to form subgroups which differ on selected variables. That is, in discriminant analysis we essentially ask how the *given* groups differ. In cluster analysis we ask whether a given group can be partitioned *into subgroups* which differ.

Given the overall cluster analysis procedure it will be recognized that the two key problems involve (1) obtaining a measure of inter-object similarity, and (2) specifying a procedure for forming the clusters based on the similarity measures. There are literally dozens upon dozens of techniques for attacking these problems, making cluster analysis as much an art as a science, depending to a great degree on the limits of computer technology and the ingenuity of the researcher. As such, we will deal with the problems in a most general way, providing illustrations of the more commonly used techniques without getting bogged down with their practical and theoretical limitations, which admittedly are many.

3. Measures of Similarity

An essential step in the cluster analysis procedure is to obtain a measure of the *similarity* or "proximity" between each pair of objects under study. Alternatively, we can deal with the *distance* or difference between the pairs of

objects, similarities and distances being complements of one another. The following paragraphs present the most commonly used measures of similarity and distance. Throughout the discussion the technicality should be kept in mind that sometimes a large number reflects high similarity and sometimes low similarity, depending upon the particular measure employed. This is a detail which must be taken into account by the specific computer programs which process the similarity measures.

Correlation coefficients. One measure of inter-object similarity that we have already alluded to in the preceding chapter, in our discussion of inverse factor analysis, is the *correlation coefficient* between a pair of objects measured on several variables. The typical *objects* × *variables* matrix is inverted so that the columns represent the objects while the rows represent the variables, as shown in Figure 1*a*. In such a situation the correlation coefficient between two columns of numbers represents the correlation, or similarity of the profile, between two objects with respect to the set of variables. For example, a correlation coefficient of $r = .57$ between two given objects reflects a higher degree of similarity than a coefficient of $r = .12$ between two objects. Thus, two objects which are highly correlated can be interpreted analogously to two variables which are highly correlated.

Euclidean distances. Another commonly used measure of the similarity between two objects is the *Euclidean distance* based on the objects' values on each of the k variables under study. An example of how such a Euclidean distance is obtained is shown geometrically in Figure 2. Utilizing the Pythagorean Theorem, the Euclidean distance d_{12} between Objects 1 and 2 is found to be nothing more than the length of the hypotenuse of a right triangle as detailed in the figure.

The example shown in Figure 2 is restricted to the distance between two objects measured on only two variables, but the concept is easily generalized to k variables. Also, although the numerical values of the objects on the two variables shown in Figure 2*b* are in their raw score form for expositional purposes, in practice it is advisable to convert them into their z score form prior to calculating distances, so as to rule out spurious effects due to unequal variances of the respective empirical variables.

Matching-type measures of similarity. When objects are measured on dichotomous or *binary* variables, in which an object either *has*(1) or *does not have*(0) a given attribute, a matching-type measure of similarity is often used.

The procedure is outlined in Table 1 for a simple example involving only three objects measured on a set of nine binary variables. Objects 1 and 2 match on three of the nine attributes, compared to Objects 1 and 3 which match on five of the nine, while Objects 2 and 3 match on seven of the nine. By taking the ratio of the number of matches to the total number of attributes, an index of inter-object similarity is obtained. For example, Objects 2 and 3, with an

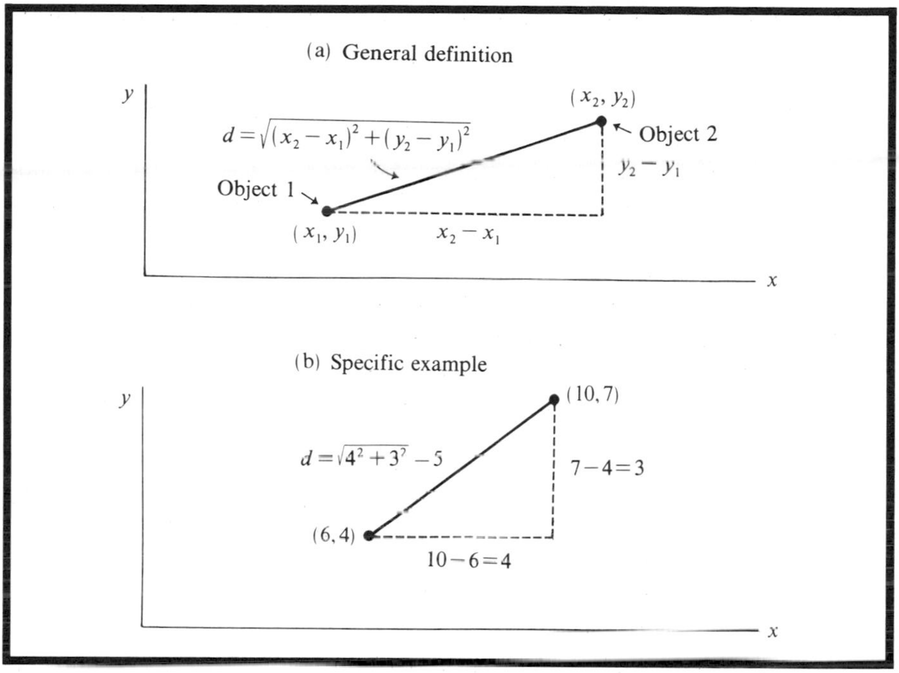

Figure 2 The Euclidean distance between two objects measured on two variables.

index of .78 are more similar than Objects 1 and 2 which have a similarity index of only .33. It is easy enough to see how this matching-count procedure can be extended to any number of objects.

A variation of the above procedure is to form a similarity index by taking the ratio of the number of matches between two objects, not to the total number of attributes, but to the number of attributes possessed by at least one or the other of the objects, discarding from the analysis the attributes possessed by *neither*. The rationale of this approach is that with a long enough list of characteristics, the majority of them will be possessed by neither of a pair of objects, yet if included in the denominator of the index would tend to "wash out" differences between object pairs as manifested in the numerators of the indices.

A more serious drawback of using binary variables to obtain inter-object similarities is that when only a handful of attributes are used, and which are completely orthogonal to one another, the subsequent clustering problem may be indeterminate. For example, an object may have the same number of matches with each of two other objects—hence, presumably equally

Table 1 A matching type measure of inter-object similarity: The number of matched attributes divided by the total number of attributes.

Binary variables	Objects				
	O_1	O_2	O_3	...	O_n
x_1	1	1	1		
x_2	1	0	0		
x_3	0	0	0		
x_4	1	0	1		
x_5	0	1	1		
x_6	0	0	0		
x_7	1	0	1		
x_8	0	1	1		
x_9	0	1	1		

Inter-object similarity measures:

$$S_{12} = \tfrac{3}{9} = .33$$
$$S_{13} = \tfrac{5}{9} = .56$$
$$S_{23} = \tfrac{7}{9} = .78$$

similar—yet the matches could have been based on opposite polarities of the attributes; e.g., (1,1) matches in one case, and (0,0) matches in the other. Probabilistically, this problem is less likely to arise if a large number of attributes are used. Also, it suggests that we should try to use continuous variables whenever possible, or at least those with several values.

Direct scaling of similarities. It is not always necessary to measure objects on a set of variables as a preliminary to obtaining the inter-object similarity measures. Rather, the similarities could be obtained directly by, say, having expert judges rate the similarity of each object to each of the other objects, or against some standard.

For example, to obtain the similarities among the images of a set of coffee brands (or any other stimulus objects) a sample of individuals could be presented the brands in a pair-wise fashion and asked to rate the similarity of the two on a 0 to 10 scale. The mean similarity ratings for the various pairs would then comprise the similarity matrix shown in Figure 1*b*.

This is a plausible approach if there is only a relatively small number of objects to be clustered, but if the number is large, the number of pair-wise comparisons becomes prohibitive. For example, with even 100 objects there are nearly 5,000 paired comparisons possible.

There are also instances of *naturally occurring* similarity matrices, which also allow us to bypass the initial step of measuring the objects on a set of

variables. Examples include the relative frequencies of consumer switching between brands in a given product category, the inter-readership of magazines, the trips between cities, the interactions between individuals, etc. In such situations, the concept of "similarity" takes on a more abstract meaning.

While all the above similarity measures have some practical and theoretical problems, they do represent the most commonly used approaches to assessing inter-object similarities. The most prudent course, then, is to use two different techniques to arrive at the similarity matrix and hope that the results are more or less the same.

4. Cluster Formation

Once the inter-object similarity matrix has been obtained, the task is to use that information to form clusters of objects such that the objects within a given cluster are similar to one another, but differ from objects in other clusters; that is, we wish to *maximize the between cluster variation relative to the within cluster variation*.

One major issue facing all clustering techniques is the *number* of clusters to form. There is a wide variety of criteria and guidelines for attacking that problem, too complicated in number and definition to discuss here, but a generally agreed upon approach is to solve for different numbers of clusters (e.g., 2, 3, 4, etc.) and then decide among the alternative solutions based on *a priori*, practical, common sense, or the more technical criteria which abound.

Perhaps the most difficult question is how to form the clusters in the first place. Again, the number of techniques is large, evidence that there is no simple solution to the problem. We will consider the two broadest types of techniques—inverse factor analysis vs. other clustering algorithms—with the understanding that each has many variations.

Inverse factor analysis. If the similarity matrix consists of correlation coefficients between objects based on an inverted *objects* × *variables* matrix, as shown in Figure 1*a*, factor analytic techniques can be used to form object clusters, as discussed briefly in the preceding chapter.

It will be recalled that in the typical factor analysis procedure, correlation coefficients are obtained between variables, and these variables are then found to load on the more abstract derived factors. In *inverse factor analysis* (sometimes called *Q*-type analysis), since the original objects × variables matrix is inverted, or tipped on its side, such that the role of objects and variables is interchanged, the resulting correlation coefficients represent the similarities between *objects* rather than the similarities between variables. Consequently, the factor matrix will not consist of variables loading on factors, but *objects loading on factors*, factors which in this context might more appropriately be

called *cluster dimensions*. Since each object will have a loading on each of these cluster dimensions, the objects can then be assigned to that cluster upon which they load most highly.

Among the problems with this approach is the interpretation we place upon an object which loads negatively on a cluster dimension. Whereas in conventional factor analysis, a variable with a negative loading on a factor is easily interpretable as its opposite meaning, this is not always possible in the case of objects. For example, whereas it is easy enough to interpret the opposite of a negative loading variable, say, "strong," as "not strong," it is not so easy to interpret the opposite of a negative loading object, say, "John Doe." Since such occurrences are not that common, one expedient is to discard those objects from the analysis, or place them in a cluster of their own.

Clustering algorithms. There are literally dozens upon dozens of *clustering algorithms* or sets of rules, other than factor analytic techniques, for forming clusters from a similarity matrix. Nearly all depend upon high-speed computer technology, and this is not the place to enumerate their many variations. Suffice it to say, that they all strive to meet some criterion which essentially maximizes the differences between clusters relative to the variation among the objects within the clusters, as depicted in Figure 3. Although the clusters in the figure are shown in two dimensions, in reality they exist in multidimensional space.

The ratio of the between cluster variation to the average within cluster variation is analogous, but not identical, to the *F* ratio discussed in conjunction with the analysis of variance procedure. To understand how such an *F* ratio might be maximized for a given number of clusters, imagine starting with

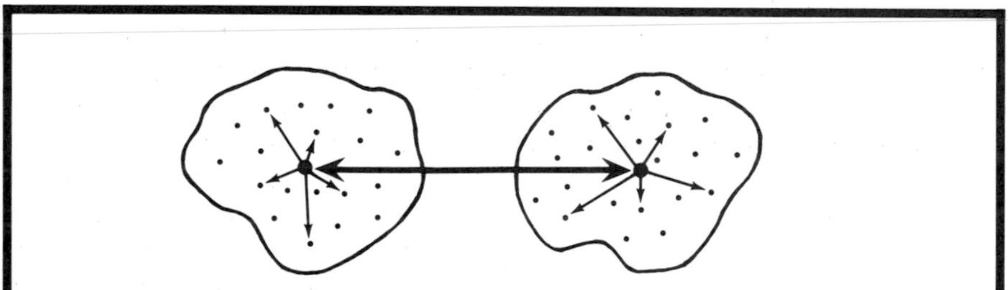

Figure 3 A schematic diagram of the variation between and within clusters. (In actuality, the clusters exist in multidimensional space.)

a given number of cluster centers chosen arbitrarily or on judgment, assigning objects to the nearest cluster center, computing the mean or center of gravity of the resulting clusters, and then juggling the objects back and forth between the clusters, each time recomputing the centers of gravity and the resultant between and within cluster variations, until the F ratio is maximized or sufficiently large.

A major drawback of this type of approach is that the end results may well differ depending upon the choice of the starting cluster centers. As such, it is advisable to repeat the procedure with different initial cluster centers and again hope that the resultant terminal clusters are more or less the same as before.

Because of the iterative trial and error nature of these clustering algorithms it is obvious that the amount and cost of computer time becomes a crucial factor in assessing the validity of any given approach; for with unlimited resources every conceivable cluster could be formed in the search for the optimum solution. Again, as with all statistical techniques, the purpose of the analysis and the cost involved must be taken into account when choosing one approach over another.

5. Cluster Comparisons

Once the object clusters have been formed, they must be compared in order to get some idea of how they differ. The most straightforward approach is to compare the clusters with respect to their means and variances on the various input variables. For example, we might find that Cluster 1 has an exceptionally high mean on variables x_2, x_5, and x_7; while Cluster 2 has an above average mean on variables x_1 and x_4; etc. From the comparative profiles of the clusters on the various input variables it is usually easy to draw a thumbnail sketch of the distinguishing qualities of the clusters.

The derived clusters can also be subjected to a discriminant analysis (see Chapter 14), either to determine which variables contributed most to the formation of the clusters, or to obtain a discriminant function for predicting cluster membership of a future sample of objects.

When the clusters have been based on a similarity matrix that has been scaled directly, without benefit of object measurements on input variables, as described in an earlier section, then the main recourse is to study the nature of the objects within each cluster and through this type of *content analysis* arrive at an intuitive or expert judgmental description of the clusters.

The above comparative procedures can be supplemented by a comparison of the clusters on external variables for which information is available on the cluster members, but which was not used in the clustering procedure. This

approach also allows for the testing of hypotheses about the deeper nature of the derived clusters.

6. Hierarchical Clustering

Instead of partitioning a set of objects into a given number of mutually exclusive clusters, clusters can be formed sequentially in a *hierarchical* or "nested" fashion in which smaller clusters occur within larger ones.

The essence of the hierarchical clustering approach is shown in Figure 4. The seven objects shown at the left of the tree diagram are merged into clusters at different stages depending upon their degree of similarity. We see, for example, that objects O_2 and O_7 are the most similar and merge into a cluster at the .9 level of the similarity index being used. Next, at a lower level of similarity, object O_1 joins that cluster, while objects O_3 and O_5 merge to form another cluster. This procedure of sequential clustering continues until all the objects merge into a single undifferentiated group.

Alternatively, the clustering can proceed in the opposite direction. Beginning at the right of Figure 4, the undifferentiated sample of objects splits into two groups, they in turn split into smaller clusters, and so on until each object stands by itself.

Unfortunately, the forward and backward clustering procedures are not likely to give exactly the same pattern of branching in the tree diagram. Still,

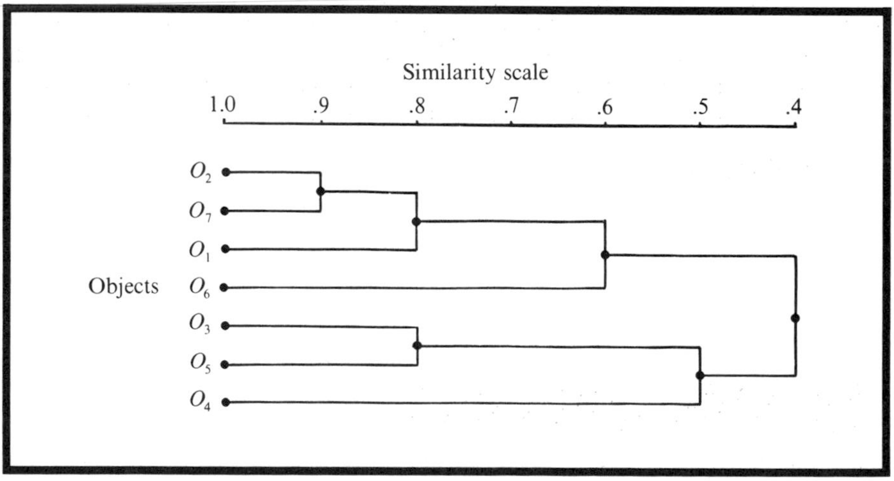

Figure 4 A sample tree diagram illustrating hierarchical clustering.

the differences that occur are likely to be informative, and the advantage of the approach is that one gets a better feel of how the clusters are formed and which particular objects are most similar to one another.

7. Concluding Comments

Cluster analysis is a rapidly developing area of study, highly dependent upon strides being made in computer technology, programming expertise, and analytical insight. We have touched upon the field only in its broadest aspects, emphasizing its rationales, general approaches, and applications.

The realization that broad populations can be segmented into subgroups with differences in their attitudes, aptitudes, and needs, cannot help but lead to changes in programs that will more adequately address those differences.

Chapter 17

Multidimensional Scaling

1. Introduction

An analysis that is often performed in conjunction with cluster analysis—inasmuch as it relies on the same inter-object similarity input data—is *multidimensional scaling*, often abbreviated MDS. The approach attempts to identify abstract variables (i.e., dimensions) which can be viewed as having generated the inter-object similarity or distance measures.

We have already seen one such approach in our study of factor analysis, in which a number of variables upon which a set of objects was measured were reduced to a smaller number of abstract dimensions or factors. Other multidimensional scaling techniques attempt the more challenging task of generating such abstract dimensions even when the objects have not been measured on a set of empirical variables; when only inter-object similarity measures are available, by any of the means discussed in the Cluster Analysis chapter.

2. One- and Two-Dimensional Representation

To understand the basic nature of the approach, we will consider a very simple example consisting of only three objects for which we have inter-object distance measures—call the objects A, B, and C, whether they represent students, crimes, species, product brands, diseases, or whatever.

We see in Figure 1 that depending on the given inter-object distance measures, the objects may or may not be able to be represented in either one or two dimensions.

In Figure 1*a*, for example, we see that the given inter-object distances allow us to portray the objects on a single dimension with no difficulty at all. Now if A, B, and C happened to correspond to "sales clerk," "librarian," and "attorney," we would not have to think long in order to give a meaningfully descriptive name to the created dimension upon which the occupations differ.

In Figure 1*b*, on the other hand, the given inter-object distances cannot be represented accurately on a single dimension, but only in an approximate way. Note that the problem is indeterminate in that the objects could not even be portrayed accurately in two dimensions.

In Figure 1*c*, we see yet another possibility, where the inter-object distances cannot be represented accurately in one dimension, but can be portrayed precisely in two dimensions. In Figure 1*d*, the same data is "compressed" to fit into a one-dimensional representation, sacrificing in the process some accuracy in portraying the given inter-object distances.

In parts *e* and *f* of Figure 1, we find another example of a set of objects alternatively represented in two dimensions vs. one dimension, respectively. Again, some accuracy is sacrificed as we move from two dimensions to one.

3. Three-Dimensional Representation

As would be suspected from the above discussion, the more diverse objects that we have, the more dimensions required to accurately represent them. Four objects, for example, although they could possibly be represented in one or two dimensions, depending upon their natures, can be portrayed in three-dimensional space.

In Figure 2*a*, four objects, all equidistant from each other, are represented precisely in three dimensions. The labels or meanings of the three dimensions upon which the four objects differ would require a content analysis of the objects themselves, which in this instance might represent alternative races or academic disciplines.

Figure 2*b* shows three alternative attempts at reducing the three-dimensional scaling to two dimensions. Here, as in cluster analysis, various algorithms or rules are required which specify which kinds of information we are willing to sacrifice as we move from a higher to a lower dimensional representation.

4. Multidimensional Representation

It is easy enough to extrapolate the above procedure to the many-object situation where many dimensions would be required to precisely represent the exploding number of inter-object distances. But anything beyond three dimensions is difficult to conceive and communicate, and also defeats our data reduction motive, so we strive to scale objects in two or three dimensions if possible.

Fortunately, empirical objects tend to contain a lot of redundancy and shared characteristics with other objects, so representing a large set of objects in two or three dimensions is not as difficult as it might seem.

(a)

Object Pair	Inter-Object Distance
A-B	2
A-C	6
B-C	4

A perfect representation in one dimension

(b)

Object Pair	Inter-Object Distance
A-B	2
A-C	5
B-C	2

An approximate representation in one dimension

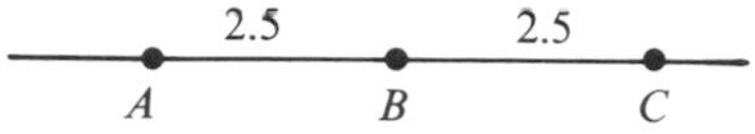

(c)

Object Pair	Inter-Object Distance
A-B	5
A-C	3
B-C	3

A perfect representation in two dimensions

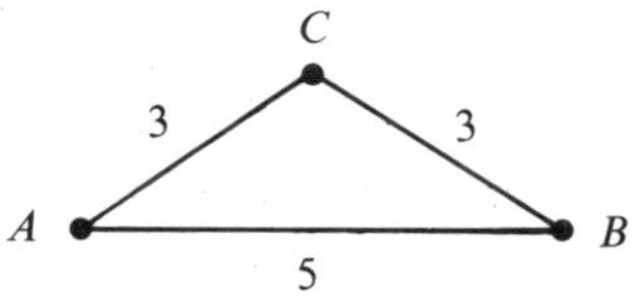

(d)

Object Pair	Inter-Object Distance
A-B	5
A-C	3
B-C	3

An approximate representation in one dimension

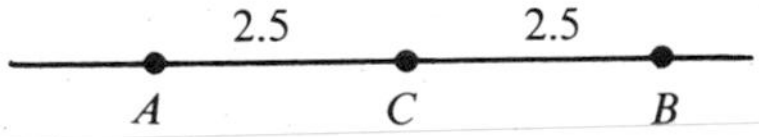

(e)

Object Pair	Inter-Object Distance
A-B	4
A-C	3
B-C	5

A perfect representation in two dimensions

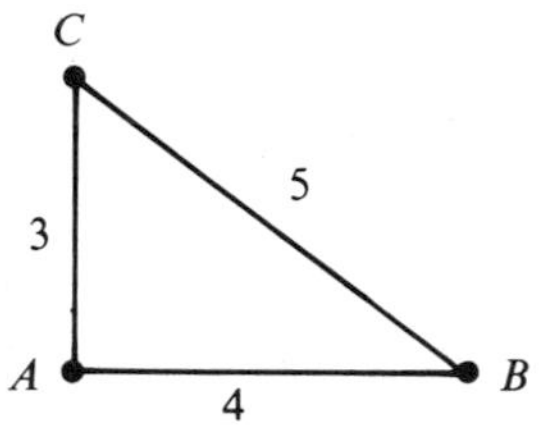

(f)

Object Pair	Inter-Object Distance
A-B	4
A-C	3
B-C	5

An approximate representation in one dimension

Figure 1 Three objects variously represented in one and two dimensions, based on the inter-object distances.

(a) Four objects represented perfectly in three dimensions.

Object Pair	Inter-Object Distance
A-B	3
A-C	3
A-D	3
B-C	3
B-D	3
C-D	3

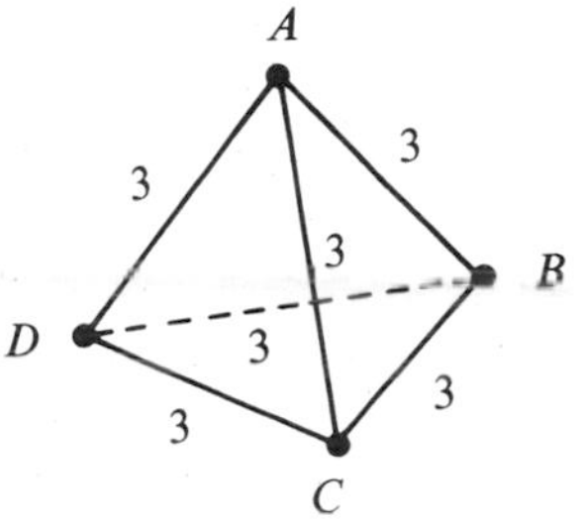

(b) Some alternative *approximate* representations in two dimensions.

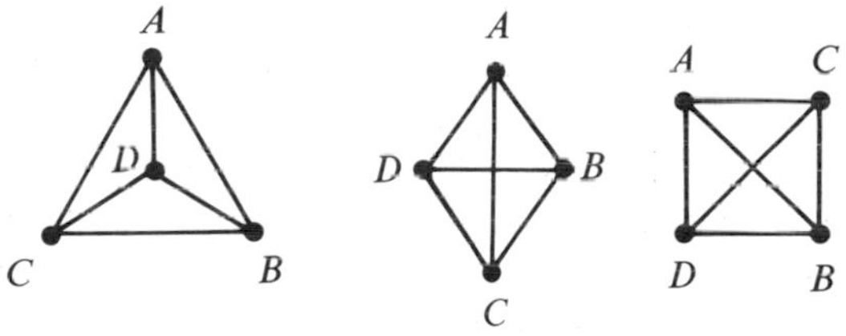

Figure 2 Representation of four objects in (a) three dimensions, and (b) alternatively in two dimensions.

If more than three dimensions are required to capture the inter-object data, some idea of the meaning of the dimensions can be had by studying the objects that score exceptionally high or low on each of the dimensions.

5. Perceptual Maps

Although the output of a multidimensional scaling procedure would consist of a set of coordinates for each object, representing their scaled value on each of the derived dimensions, the data is best presented as a *perceptual map*, which is a pictorial representation of the inter-object differences. Figures 1 and 2 show simple perceptual maps.

Figure 3 presents a more elaborate two-dimensional perceptual map of the perceived similarities of a wide variety of academic disciplines. Although the two dimensions have been given names, alternative labels are left as an exercise. Note, too, that the axes representing the derived dimensions can be

Figure 3 Perceptual map of academic disciplines.

rotated about the origin to any desired position, possibly yielding entirely different interpretations of the data.

A perceptual map can also be embellished by adding the results of a cluster analysis. Objects appearing in the map which belong to a particular cluster can be encircled, providing a broader basis on which to interpret the input data. If the objects happen to be measured on various empirical variables, those variables can be superimposed on the map and would appear as spokes or vectors emanating from the origin, their varying lengths reflecting their differential contribution to the map, and would help in the naming of the more abstract and interesting *orthogonal* (perpendicular) dimensions.

Such representations are often referred to as *joint-space* maps, and an example appears in Figure 4, portraying various cancer sites and their presumed etiologies. Notice that the four quadrants of a two-dimensional map represent another appearance of the very useful 2×2 contingency table.

6. Stress

We noticed in the earlier examples that moving from a *three*-dimensional to a *two*-dimensional, or from a *two*-dimensional to a *one*-dimensional representation of the given inter-object distances was sometimes difficult and always sacrificed information about the true inter-object distances. A statistical measure of this difficulty, or loss of information, in arriving at a particular dimensional representation be it one-, two-, three-, or *n*-dimensions—is appropriately named *stress*.

As we move from a higher to a lower number of dimensions, stress necessarily increases, for we lose more and more accuracy in representing the full matrix of ingoing inter-object data. Similarly, for a given dimensionality, stress will vary depending upon the configuration of the objects, as can be seen in the alternative arrangements in Figure 2*b*. The iterative trial and error scaling algorithm must try various configurations, attempting to hone in on one that minimizes stress.

A stress value of .15 or lower for a particular dimensional solution is usually considered satisfactory. Larger values suggest that there has been too much distortion of the input data. Stress values are routinely provided in the computer software packages which perform the multidimensional scaling.

To fully understand what stress is, try portraying the objects of Figures 2, 3, and 4 in *one* dimension.

7. Concluding Comments

Multidimensional scaling and perceptual mapping methods are useful data reduction techniques, often enabling us to make inferences about our empirical data that might otherwise escape our attention.

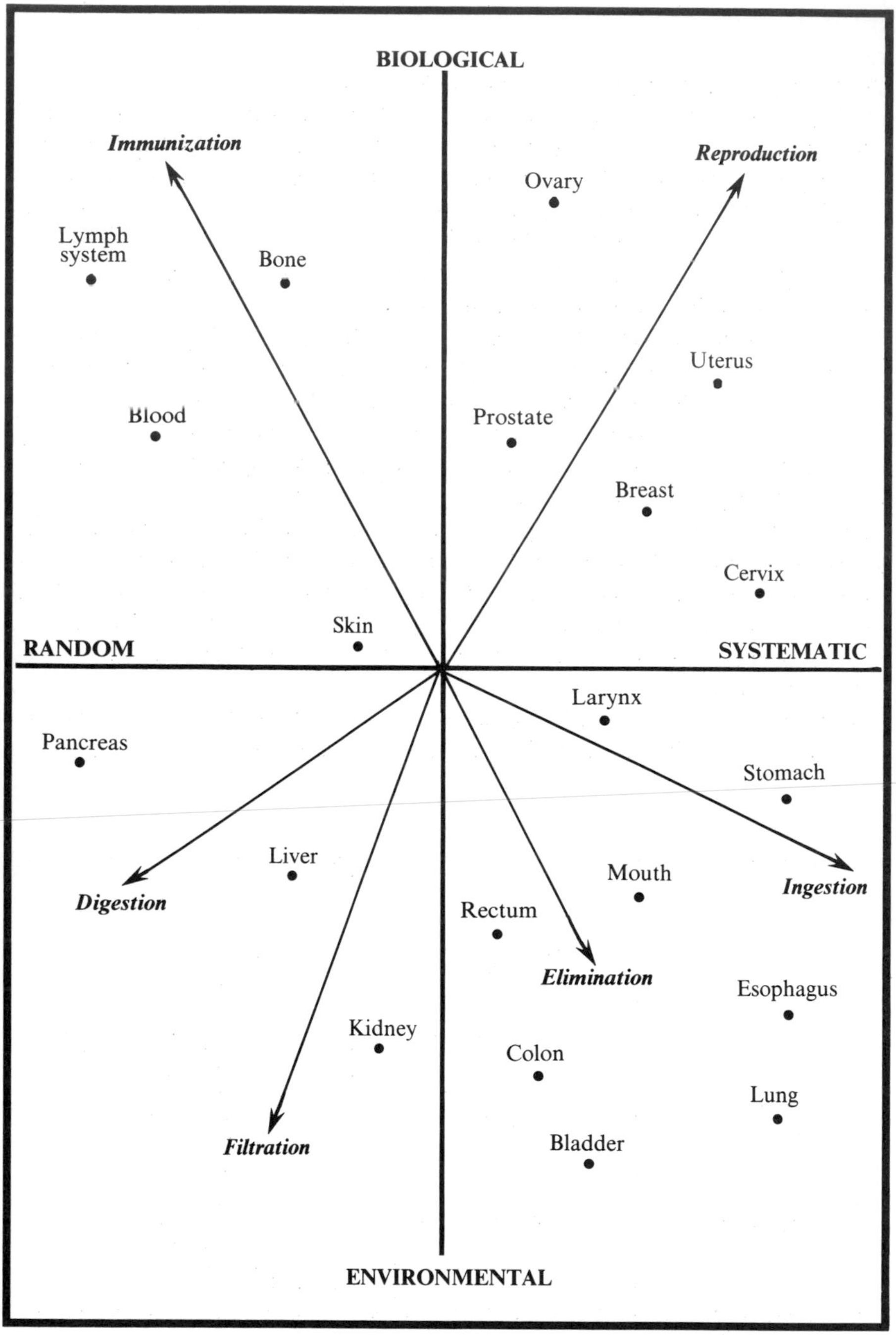

Figure 4 Perceptual joint-space map of cancer sites and causes.

Some additional issues which must be addressed in the multidimensional scaling approach, and which are the subject of much ongoing research, include whether to deal with the *individual's* perceptual data or to *aggregate* it, whether to use *non-metric* or *metric* similarity measures (e.g., obtaining object-pair similarities by ranking vs. magnitude estimation methods), and how to approach *non-symmetrical* inter-object relations such as preference or dominance.

Depending as it does to a large extent on computer technology, and the growing number of researchers in the field, multidimensional scaling is still in the early stages of development, and, like cluster analysis, holds promise for innovative approaches in the future, which should provide the basis for a better understanding of data originating in all disciplines.

Part V.
SELECTED SUBJECTS

Chapter 18

Time Series Analysis

1. Introduction

When we measure a criterion variable y repeatedly over time, the resulting sequence of values is referred to as a *time series*. For example, whether on a minute by minute, hourly, daily, monthly, yearly, decade, or geological time basis, we measure crime rate, sales levels, inflation rate, strike activity, social attitudes, crop yields, school enrollment, rainfall, hospital admissions, disease levels, pest populations, air pollutant levels, academic achievement, traffic accidents, commodity prices, the sun's temperature, etc.—in short, all variables of interest to us.

The major task of time series analysis is to describe the nature of the past variation of a variable in order that its future values can be predicted and acted upon accordingly. The situation can be portrayed as in Figure 1, in which we have a time-line consisting of the past, the present, and the future.

In this chapter we will introduce some of the techniques used to understand the nature of the most common types of time series, enabling us to describe the past and possibly predict or *forecast the future* based on the past. At the same time we will study some of the major problems inherent in trying to predict the future based on past observations.

Although we will be discussing *time* series, it should be kept in mind that the same techniques can be applied to *spatial* series; for example, measure-

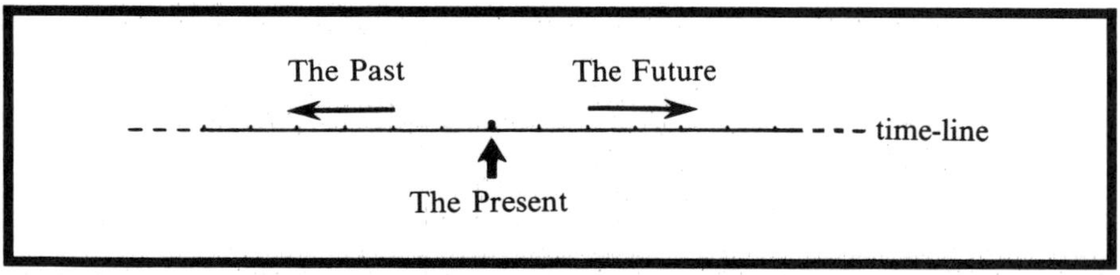

Figure 1 Schematic diagram of the time series analysis situation.

ments made along a city street, a tree boring, an earth probe, an artery, a genetic DNA strand, etc., where such lineal series are subdivided in analytical subunits of a convenient size, analogous to time periods. The goal again, is to understand past measurements in the series and forecast future ones.

2. Index Numbers

Since we often want to compare a given value of a time series with an earlier "benchmark" or reference value, it is convenient to use a quantity known as an *index number*, defined as the ratio of the one value to the other; i.e.,

$$\text{Index number} = \frac{\text{Comparison number}}{\text{Base number}} \times 100$$

The ratio of the two quantities is multiplied by 100 so that when the comparison number equals the base period number, the resulting index number will have a value of 100. This will be easily recognized as the familiar concept of a percentage, but with the percentage sign (%) omitted.

For example, if the population of a city in a specified base year was 50,000, and in the comparison year is 70,000, then in terms of an index number it is equal to

$$\text{Index number} = \frac{70{,}000}{50{,}000} \times 100 = 140$$

which tells us that the city's population in the comparison year is 40% higher than it was in the base period year.

In symbols we can define a time series index number as

$$\boxed{I_i = \frac{y_i}{y_o} \times 100} \qquad (1)$$

where y_o is the value of our criterion variable in the specified *base* period, y_i is the value of the variable for the *comparison* period i of interest, and I_i is the resulting index number for the comparison value in the ith time period.

Single variable vs. aggregate. Depending upon how the criterion variable y is defined, a number of different types of index numbers can be created. For instance, the criterion variable y could either equal a *single* quantity q or it could equal an *aggregate* of many quantities $q_1, q_2, \ldots, q_k$. The *simple single*

variable index number would be of the form:

$$I_i = \frac{q_i}{q_o} \times 100 \tag{2}$$

A *simple aggregate* index number, on the other hand, would take the form

$$I_i = \frac{q_{i1} + q_{i2} + \cdots + q_{ik}}{q_{o1} + q_{o2} + \cdots + q_{ok}} \times 100 \tag{3}$$

where the summation is across $1, 2, \ldots, k$ different variables—perhaps the populations of k cities, or the quantities consumed of k commodities, or the crime levels in k countries, or the levels of k diseases, or the prices of k services—and the aggregate in each case is compared for period i versus the designated base period.

Weighted vs. unweighted. In addition to index numbers being based on either a single variable or an aggregate of many variables, they can be based on quantities which are either *weighted* or *unweighted*. The index numbers defined in the above paragraph involve unweighted variables. However, if a variable q is weighted by another variable p, we would have a *weighted single variable* index number of the form:

$$I_i = \frac{p_i q_i}{p_o q_o} \times 100 \tag{4}$$

On the other hand, a *weighted aggregate* index number would take the form

$$I_i = \frac{p_{i1}q_{i1} + p_{i2}q_{i2} + \cdots + p_{ik}q_{ik}}{p_{o1}q_{o1} + p_{o2}q_{o2} + \cdots + p_{ok}q_{ok}} \times 100 \tag{5}$$

in which each quantity q has its respective weight p.

The above four basic types of index numbers are summarized in Table 1—the simple unweighted single variable case, the simple unweighted aggregate of many variables, the weighted single variable, and the weighted aggregate of many variables.

Applications. In the fields of business and economics, where index numbers find wide application, q and p typically stand for the quantity and price, respectively, for a particular commodity or service. Consequently, the product of the two, pq, represents the *value* of the object in question.

However, q could stand for any quantity, whether school enrollment, births, hospital admissions, rainfall level, crime rate, etc. Similarly, p can represent any kind of *importance weight* attached to each quantity. For

Table 1 Four basic types of index numbers, where the subscript i refers to the comparison time period and the subscript j refers to the objects or variables being aggregated.

Weights	Type of variable: *Single variable*	Type of variable: *Aggregate of many variables*
Unweighted ("*simple*")	$I_i = \frac{q_i}{q_o} \times 100$	$I_i = \frac{\sum_{j=1}^{k} q_{ij}}{\sum_{j=1}^{k} q_{oj}} \times 100$
Weighted	$I_i = \frac{p_i q_i}{p_o q_o} \times 100$	$I_i = \frac{\sum_{j=1}^{k} p_{ij} q_{ij}}{\sum_{j-1}^{k} p_{oj} q_{oj}} \times 100$

example, in a weighted aggregate crime index the weight attached to the quantity of murders could be different than the weight attached to the quantity of bank robberies.

Numerical examples of the main types of index numbers are presented in Table 2. A sample family's consumption of bread, butter, and eggs, and the respective prices paid during a base year and a later comparison year are shown. A careful study of the table will reveal all four types of index numbers introduced above. Also, it will be easy to see how such index numbers could just as well be based on a city or country as on a family.

Table 2 also presents three additional special types of index numbers used in business and economics—*Laspeyres aggregate price index*, the *Paasche aggregate price index*, and the *Fisher ideal aggregate price index*.

In the Laspeyres index, the combined value of a set of commodities or services is calculated based on the assumption that the quantities remained the same from base period to comparison period.

In the Paasche index, the base period value is calculated using the quantities consumed in the comparison period.

The Fisher ideal index, in turn, is defined as the square root of the product of the Laspeyres and Paasche index numbers. Although in the example of Table 2, all three index numbers are in close agreement, that will not necessarily be the case, depending upon interactions between price and quantity movements for the various goods or services involved.

Table 2 Numerical examples of common types of index numbers.

		Loaves of Bread		Pounds of Butter		Dozens of Eggs		Aggregates	
Base year		*Actual*	*Index*	*Actual*	*Index*	*Actual*	*Index*	*Actual*	*Index*
Quantity	q_o	157	(100)	11	(100)	117	(100)	$\Sigma q_o = 285$	(100)
Price	p_o	\$.96	(100)	\$2.71	(100)	\$1.55	(100)	$\Sigma p_o = 5.22$	(100)
Value	$p_o q_o$	\$150.7	(100)	\$29.8	(100)	\$181.4	(100)	$\Sigma p_o q_o = \$361.9$	(100)
Year *i*									
Quantity	q_i	169	(108)	10	(91)	121	(103)	$\Sigma q_i = 300$	(105)
Price	p_i	\$1.21	(126)	\$2.98	(110)	\$1.79	(115)	$\Sigma p_i = 5.98$	(115)
Value	$p_i q_i$	\$204.5	(136)	\$29.8	(100)	\$216.6	(119)	$\Sigma p_i q_i = 450.9$	(125)
Additional calculations for Laspeyres, Paasche, and Fisher index numbers:									
Current price × base quantity	$p_i q_o$	190.0		32.8		209.4		$\Sigma p_i q_o = 432.2$	
Base price × current quantity	$p_o q_i$	162.2		27.1		187.6		$\Sigma p_o q_i = 376.9$	

Laspeyres aggregate price index

$$= {}_L I_i = \frac{\Sigma p_i q_o}{\Sigma p_o q_o} = \frac{432.2}{361.9} = 119$$

Paasche aggregate price index

$$= {}_P I_i = \frac{\Sigma p_i q_i}{\Sigma p_o q_i} = \frac{450.9}{376.9} = 120$$

Fisher ideal aggregate price index

$$= {}_F I_i = \sqrt{{}_L I_i \cdot {}_P I_i} = \sqrt{(119)(120)} = 119.5$$

All the preceding index numbers were based on a comparison of a variable at two different points in time. However, we can also create index numbers by comparing a variable measured on two sets of objects at the *same point* in time. For example, we could index a given city's crime rate against the national average for the same year. Or, the average income of a minority group could be indexed against the national average. Or, the average score on an academic achievement test for a given school system could be indexed against a city or national average. Or, an individual's annual income increase could be indexed against the cost of living increase. Still, the basic concept remains the same—a *comparison number* is divided by a *base number* of interest. The advantage of such index numbers is that an index such as 125, for example, is more quickly interpreted than if we were provided with the two variable values upon which it was based.

3. The Forecasting Problem

As stated in the introduction, the ultimate purpose of describing the past performance of a variable is to predict its future performance. This is a problem of statistical inference, but it differs from the types of inferential

problems studied in Part III in which random samples were used to draw conclusions about the parent population. In the case of a time series, though, it is highly questionable whether we can think of the past time periods as being a random sample of all future time periods, much less of the immediately succeeding time periods.

Consequently, even though we might be able to precisely describe the past pattern of variation of our criterion variable, there is absolutely no theoretical assurance that it will recur in the immediate future, if at all. Fortunately, based on empirical evidence, we discover that things usually happen in the future as they have in the past. It is this regularity, then, that assures us of at least some success in forecasting the future. Such variables which vary over time in a probabilistic manner are referred to as *stochastic* variables, as contrasted with *deterministic* variables which do not have a probability component, if such truly exist.

The regularity or systematic variation of a variable can differ in degree. On the one hand, the variation can be so regular that no sophisticated techniques are needed to discern the pattern—it will be obvious to anyone. On the other hand, the variation may be so irregular as to defy description by any known means. We are left, then, with the intermediate situations in which the patterns of variation are evident but not so obvious as to preclude statistical analysis.

In the following sections we will study the most easily identified patterns of variation in a time series *seasonal*, *cyclical*, *trend*, and *irregular* patterns

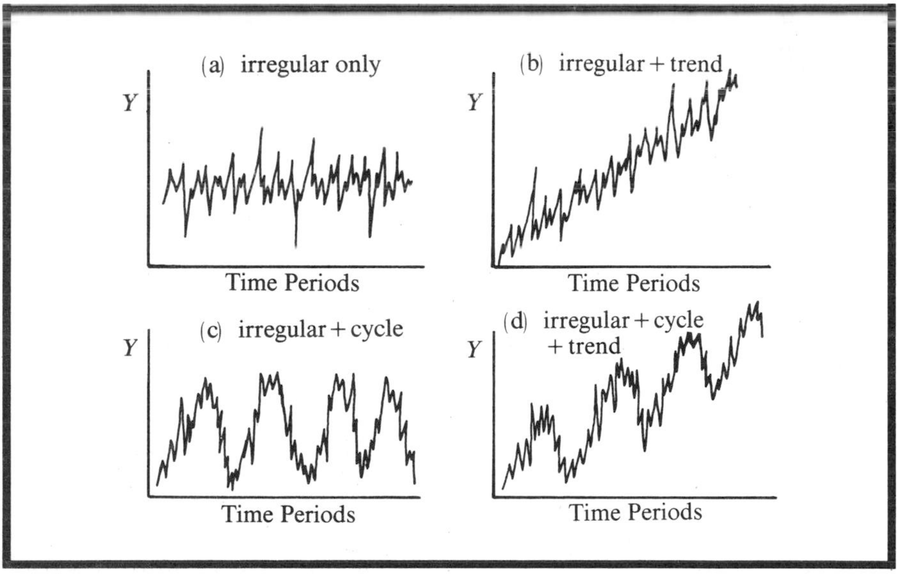

Figure 2 Common components of a time series.

—and see how they can be used to describe the past, and hopefully to forecast the future. As a preliminary to the discussion of these sources of variation, they are presented graphically in Figure 2 in some of the combinations in which they can occur.

4. Seasonal Variation

One of the most easily identified sources of variation in an extended time series is the *seasonal* variation, the regular up-and-down pattern of the variable within a well-defined recurring time period. For example, restaurant sales, body temperature, crime activity, and accident rates vary regularly within each day; retail customers, airline traffic, and absenteeism vary regularly within each week; department store sales, rainfall, and animal births vary regularly within each year; etc.

It will be noticed that the recurring time period upon which seasonal variation is based can be defined in natural terms—e.g., a day is an earth orbit, a month is an approximate moon orbit, a year is a sun orbit—while others can be defined in socio-cultural terms—e.g., weeks, decades, centuries. However, what distinguishes a *seasonal* cycle from other cycles is that the recurring time periods are well-defined and that they provide some kind of rationale for the variation. For example, we have a basis for expecting variation in ice cream sales throughout a year, or electricity usage throughout the day, or store sales throughout the week. In each case, the time interval is well-defined and provides the rationale for the observed variation in the criterion variable in question. In the case of sunspot activity, on the other hand, its supposedly cyclical behavior is not tied to any well-defined time period which provides us with a basis for interpreting the observed variation. As such, its variation is not labelled as seasonal. Similarly, apparent business or economic cycles extending over decades are not "seasonal" in nature, since the recurring time periods are usually defined *after* the limited data have been observed rather than before, and no accepted theory exists as to why the variation should occur regularly within a time period of a fixed duration.

Seasonal index numbers. As seen in the above examples of seasonal variation, one time period is always *nested* within another time period—hours are nested within days, days are nested within weeks, and months are nested within years. That is, the alternative time periods of one kind are exhausted within a single time period of another kind, the latter interval comprising a full seasonal cycle. By observing several of these cycles we hope to identify systematic variation over the nested time periods. For example, in searching for daily seasonal cycles we observe hourly data for several days, for weekly seasonal cycles we observe daily data for several weeks, and for yearly seasonal cycles we observe monthly or quarterly data for several years. By "several" we mean as many as possible.

Suppose, for example, we want to identify a yearly seasonal cycle for a given company's ice cream sales. The month-by-month sales over a three-year period are presented in Figure 3, and the seasonal variation in the sales is quite clear, and as expected, with sales cresting in the summer months.

To emphasize the similarity in monthly variation from one year to another, we can superimpose the three years of data on top of one another as in Figure 4. In addition to the monthly variation we also notice what appears to be an upward trend in sales from one year to the next. That aspect of the data will be addressed in a later section. For now, our concern is primarily with the month-by-month variation which *repeats* itself year after year.

The most common way of describing such seasonal variation is to index each month's sales against a meaningful base figure. The resulting index for each month should then communicate to us the extent to which that month's value is above or below the base figure. A reasonable base figure to choose would be one which represents *the average value of the variable throughout a complete cycle*, in this case a 12-month period.

As to which 12-month average should be used as a base for constructing the monthly index numbers, there is a wide variety of choices available. Table 3*a* presents the raw monthly data for the ice cream sales example, along with average monthly sales for each of the three *calandar* years—26.1, 35.6, and 43.6 tons, respectively.

Table 3*b*, on the other hand, presents a 12-month average for each month based on the *most recent* twelve months in each case. For example, for the twelve months ending in December of Year 1, the average sales were 26.1 tons. For the twelve months ending in July of Year 2, the average sales were 32.6 tons. Similarly, each month is associated with a *most recent* 12-month average. This type of *moving average* washes out all month-by-month seasonal effects, since a 12-month average contains both the ups and downs in the cycle.

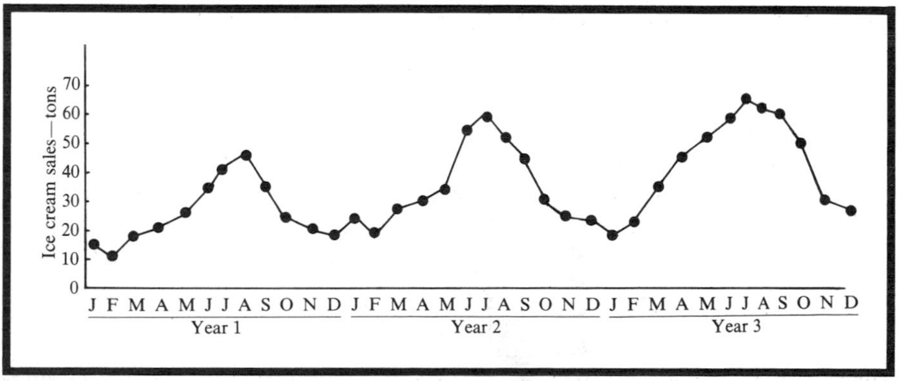

Figure 3 Monthly ice cream sales over three years.

Table 3 Monthly ice cream sales and selected 12-month averages.

Month	(a) Monthly sales				(b) Average monthly sales for the most recent 12-month period			(c) Average monthly sales for the 12-month period straddling the start of a given month (and the centered value)		
	Year 1	*Year 2*	*Year 3*	*Monthly Means*	*Year 1*	*Year 2*	*Year 3*	*Year 1*	*Year 2*	*Year 3*
Jan	15	24	19	19.3		26.8	35.2		31.1 (31.8)	38.8 (39.0)
Feb	12	19	23	18.0		27.4	35.5		32.6 (33.0)	39.2 (39.6)
Mar	19	27	33	26.3		28.1	36.0	NA	33.3 (33.8)	40.0 (40.6)
Apr	20	30	43	31.0		28.9	37.1		34.3 (34.6)	41.1 (42.0)
May	26	33	50	36.3	NA	29.5	38.5		34.8 (35.0)	42.8 (43.0)
Jun	36	55	58	49.7		31.1	38.8		35.2 (35.4)	43.2 (43.4)
Jul	42	60	66	56.0		32.6	39.2	26.1 (26.4)	35.6 (35.4)	43.6
Aug	44	53	62	53.0		33.3	40.0	26.8 (27.1)	35.2 (35.4)	
Sep	35	47	60	47.3		34.3	41.1	27.4 (27.8)	35.5 (35.8)	NA
Oct	25	30	50	35.0		34.8	42.8	28.1 (28.5)	36.0 (36.6)	
Nov	20	25	31	25.3		35.2	43.2	28.9 (29.2)	37.1 (37.8)	
Dec	19	24	28	23.7	26.1	35.6	43.6	29.5 (30.3)	38.5 (38.6)	
Yearly means:	26.1	35.6	43.6	35.1						

NA = not available

Table 3*c* presents yet another basis for forming a 12-month average upon which to index monthly figures. It too is a moving average, but rather than the most recent 12-months, it reflects a *straddled* 12-month period—six months on either side of the month in question. Each average appearing in parentheses in Table 3*c* represents the average of the 12-month averages above and below it. This minor adjustment is used so that the resulting average corresponds

Table 4 Alternative monthly index numbers for the ice cream sales reported in Table 3.

Month	(a) Sales indexed against the year's monthly average				(b) Sales indexed against average monthly sales for the most recent 12-month period				(c) Sales indexed against average monthly sales for the 12 month period straddling a given month			
	Year 1	*Year 2*	*Year 3*	*3-year Mean*	*Year 1*	*Year 2*	*Year 3*	*2-year Mean*	*Year 1*	*Year 2*	*Year 3*	*2-year Mean*
Jan	57	67	44	56		90	54	72		75	49	62
Feb	46	53	53	51		69	65	67		58	58	58
Mar	73	76	76	75		96	92	94	NA	80	81	80
Apr	77	84	99	87		104	116	110		87	102	94
May	100	93	115	103		112	130	121		94	116	105
Jun	138	154	133	142	NA	177	149	163		155	134	144
Jul	161	169	151	160		184	168	176	159	169		164*
Aug	169	149	142	153		159	155	157	162	150		156*
Sep	134	132	138	135		137	146	142	126	131	NA	128*
Oct	96	84	115	98		86	117	102	88	82		85*
Nov	77	70	71	73		71	72	72	68	66		67*
Dec	73	67	64	68	73	67	64	66	63	62		62*

NA = Not available

*Not directly comparable to the means for the first six months, due to their being based on different years.

approximately to the center of a month, rather than to the borderline between two months. Notice that the figures in Table 3*c* which are not in parentheses are exactly the same as those in part *b* of the table, only their location is shifted by six months.

Whether we create monthly index numbers based on 12-month averages from the calandar year, the most recent 12-month period, or a straddled 12-month period will largely be dictated by the purpose of our analysis and the nature of our data. For comparison purposes, monthly index numbers are presented in Table 4, based on all three methods.

The monthly index numbers in Table 4*a* were obtained by taking the ratio of the monthly sales figures in Table 3*a* to their respective calandar year monthly average. For example, for January of Year 1 we have

$$\frac{15}{26.1} \times 100 = 57$$

as its index number, suggesting that that month's sales were only about 57% of the calandar year average. Similarly, for May of Year 3 we have

$$\frac{50}{43.6} \times 100 = 115$$

as its index number, telling us that its sales were 15% above the calandar year's average monthly sales level. The same procedure is followed for each of the

other months, producing the complete set of monthly index numbers in Table 4*a*. This procedure for constructing index numbers is often referred to as the *ratio-to-current-period-average* method. Notice that the index numbers for any given month vary somewhat from year to year. To obtain a typical index number for a given month, the mean of the individual index numbers is taken. If they are badly skewed the median could be used instead.

In Table 4*b*, the monthly index numbers were obtained by taking the ratio of each month's sales against the average of the *most recent* 12-month period appearing in Table 3*b*. For example, for January of Year 2 we have

$$\frac{24}{26.8} \times 100 = 90$$

as its index number. Similarly, each of the other monthly index numbers is obtained by taking the ratio of the month's sales to the most recent 12-month average. Notice that index numbers for the first eleven months of Year 1 cannot be calculated since there was insufficient data to calculate the most recent 12-month average for those months. The two full years of index numbers, though, do allow us to obtain an average index number for each month. This procedure for forming seasonal index numbers is a *ratio-to-moving-average* method, where we must stipulate that the average is based on the most recent cycle.

Finally, in Table 4*c*, the monthly index numbers are obtained by taking the ratio of each month's sales to the 12-month average *straddling* the given month. For example, for July of Year 1, we have

$$\frac{42}{26.4} \times 100 = 159$$

as its index number, where 42 and 26.4 come from Table 3*a* and *c*, respectively. Notice that with this method we have insufficient data to calculate index numbers for the first six months of Year 1 and the last six months of Year 2. Although we have taken the averages of the monthly index numbers in Table 4*c*, it is ill-advised since the first six monthly means are based on different years than the last six monthly means. Consequently, if this technique is to be used to obtain a typical monthly index, many complete cycles of the data should be available, rather than just three. This, too, is a *ratio-to-moving-average* method, but where the average is based on the cycle *straddling* the *sub-period in question*.

A careful study of Table 4 shows that although the three alternative methods used to construct monthly index numbers yield different values, they are not all that different. If the data shows no overall trend from year to year, and if there is no interaction between the seasonal pattern and various years,

Table 5 Seasonal adjustment of ice cream sales.

(a) The raw sales data (from Table 3a):

	Month											
	J	*F*	*M*	*A*	*M*	*J*	*J*	*A*	*S*	*O*	*N*	*D*
Year 1	15	12	19	20	26	36	42	44	35	25	20	19
Year 2	24	19	27	30	33	55	60	53	47	30	25	24
Year 3	19	23	33	43	50	58	66	62	60	50	31	28

(b) Average monthly index numbers (from Table 4a):

J	*F*	*M*	*A*	*M*	*J*	*J*	*A*	*S*	*O*	*N*	*D*
56	51	75	87	103	142	160	153	135	98	73	68

(c) Seasonally adjusted sales (raw sales divided by monthly index × 100)

	J	*F*	*M*	*A*	*M*	*J*	*J*	*A*	*S*	*O*	*N*	*D*
Year 1	26.8	23.5	25.3	23.0	25.2	25.4	26.2	28.8	25.9	25.5	27.4	27.9
Year 2	42.9	37.3	36.0	34.5	32.0	38.7	37.5	34.6	34.8	30.6	34.2	35.3
Year 3	33.9	45.1	44.0	49.4	48.5	40.8	41.2	40.5	44.4	51.0	42.5	41.2

then the ratio-to-current-period-average is the simplest and most straight-forward approach. Otherwise, the ratio-to-moving-average approaches will yield more meaningful index numbers, especially when the moving average is of the straddled type.

Seasonal adjustments. Once a set of seasonal index numbers has been obtained, not only will they describe the nature of the particular seasonal variation, but they can be used to *seasonally adjust* future or past values of the variable in question. That is, they can be used to remove the seasonal source of variation in a time series, and then allow us to try to identify other sources in our quest for a full understanding of the time series pattern of variation.

Table 5 shows how the set of index numbers for the ice cream sales example are applied retroactively to the source data, in order to remove the seasonal variation. This is done simply by *dividing* an actual value of the variable by that time period's index number. For example, for January of Year 1, the actual sales were 15 tons, but since January's average index number (based on Table 4*a*) is 56, we have

$$\frac{15}{56} \times 100 = 26.8 \text{ tons}$$

as January's *adjusted* sales—i.e., if seasonal effects were absent. Similarly, adjusted sales are presented in Table 5*c* for each of the other months during the three-year period. As data becomes available for Year 4, the index numbers

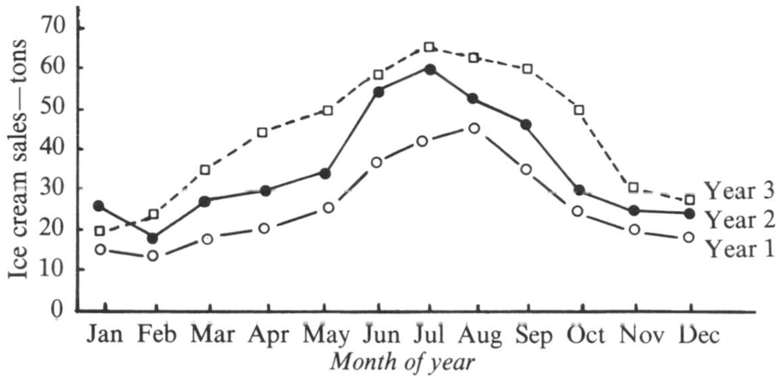

Figure 4 Monthly ice cream sales, by year.

could also be applied to them so that comparisons could be made with preceding adjusted values.

In the ice cream sales example we studied monthly sales were nested within years. It should be clear that the same type of analysis would apply for any seasonal period. In each case, however, we must be sure that the base figure for our index numbers is defined in terms of the time period over which the pattern of variation occurs. For example, if we want to determine day-of-the-week patterns, a 7-day period is needed for the base of the index numbers. For hour-of-the-day patterns, a 24-hour base period is needed for constructing the hourly index numbers.

Once we have identified and extracted seasonal patterns in a time series, we can proceed to search for other systematic sources of variation in the data, as discussed in the following sections.

5. Cyclical Variation

We learned in the preceding section that *seasonal variation is cyclical, but that all cycles are not seasonal.* Generally speaking, cycles tend to have a longer *period* than seasonals—the *period* being the length of time before the cycle repeats itself. More specifically, seasonal variation occurs within well-defined periods such as days, weeks, or years; periods defined in terms of physical and societal phenomenon—e.g., a day is an earth orbit, a week is a societal period. In the case of cycles, on the other hand, unless some theory exists which can predict the period of a hypothesized cycle, the period must be abstracted from the data itself.

There are several difficulties in identifying cycles in the values of a variable. The first, mentioned above, is the reliance on empirical data for estimating the cycle's period. Second, since several such periods must be observed in order to identify a reliable recurrence of the cyclical pattern, data must be collected over many years. For example, if an economic cycle is supposed to have a fifteen-year period, then it would take 45 years just to observe *three* such cycles, and three cycles is hardly enough to document all but the most clear-cut cyclical patterns, as seen in the earlier ice cream sales example. Finally, we rarely have a theory to substantiate such cycles, and consequently we are likely to interpret much random variation as being genuinely cyclical in nature, fitting a theory to data which is unlikely to hold up in the future.

This is not to say that cycles do not exist, but rather that the length of time needed to verify them is often beyond the limits of historical records. This is especially true of variables in the behavioral sciences, but less so in the biological and physical sciences. For example, clear-cut cycles in worker productivity have not been identified, but evidence exists for cycles in certain insect populations.

Supposing that a cycle has been identified, it can be analyzed in the same way as a seasonal. While a seasonal usually has a well-defined time period nested within the overall cycle—e.g., months within a year—a cycle is not likely to have such a natural calibration. Instead, the cycle would have to be arbitrarily divided into a convenient number of intervals for analysis of the pattern of variation. For example, if a cycle has a period of 7.5 years we could divide it into 15 time periods each .5 years in length, or into 5 periods each 1.5 years in length. Then, for several such cycles we could identify the systematic pattern of variation across the sub-periods in a manner analogous to the months-within-years analysis for the ice cream sales example of the preceding section.

Once cyclical patterns have been verified, a time series can be adjusted to remove those cyclical effects, just as seasonal effects were removed. The remaining variation could then be studied for other systematic sources of variation.

6. Trend Variation

In addition to seasonal and cyclical patterns of variation, a time series can show an overall *trend*, whether linear or curvilinear. Such long-term trends are often referred to as *secular trends* to distinguish them from short-term local trends due to cyclical patterns of variation.

The most obvious type of secular trend to be expected from a time series is a linear trend. Such a trend can be described by a linear equation of the form $y = a + bx$, where the parameters a and b can be determined from the data

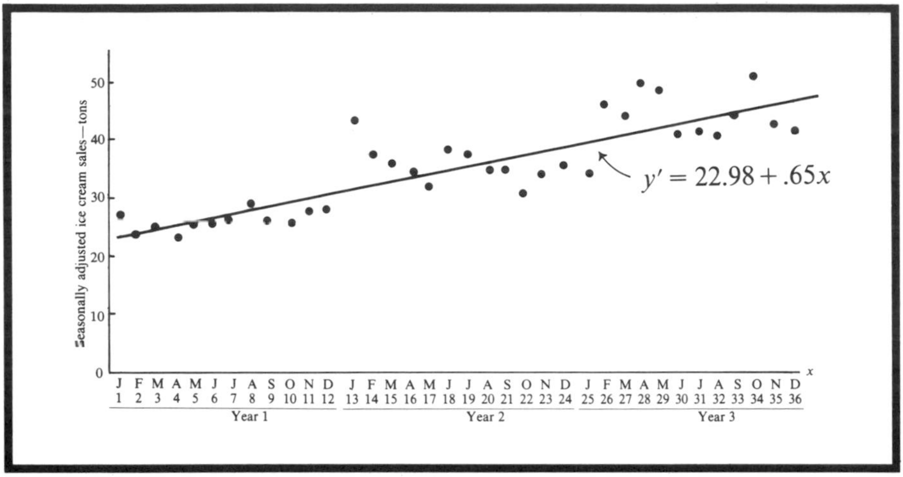

Figure 5 Seasonally adjusted monthly ice cream sales over three years.

using the least squares method introduced in the Regression Analysis chapter. Such an equation could be fitted either to the actual time series values or to the seasonally or cyclically adjusted values.

For example, Figure 5 presents the seasonally adjusted ice cream sales figures obtained from Table 5*c*. The 36 months for which the data is available are numbered from 1 to 36 and serve as the predictor variable x for the criterion variable y, the adjusted ice cream sales. Using the least squares criterion, the best-fitting line is found to be $y = 22.98 + .65x$. The .65 slope of the line tells us that for each increment in the value of x—i.e., each successive month—we can expect an increase in adjusted ice cream sales of .65 tons.

We can also use such an equation to forecast future values in the time series. For example, the forecast for January of Year 4, for which $x = 37$, would be $y = 22.98 + .65(37) = 47.0$ tons on a seasonally adjusted basis. Similarly, we could make forecasts for any other succeeding period, assuming the linear trend will continue. Each month's forecast of seasonally adjusted sales could then be converted back into actual sales by applying that month's seasonal index. For example, since January has a seasonal index of 56, the 47.0 ton *seasonally adjusted forecast* corresponds to a forecast of *actual* sales of $.56(47.0) = 26.3$ tons.

An alternative to the above forecasting procedure is to predict the time series value for a given period based on observed trends for that period. For example, we could directly forecast actual ice cream sales for January of Year 4 by using the trend observed over the preceding Januarys. That trend is shown in Figure 6*a* to be $y = 15.3 + 2.0x$. The forecast for January of Year 4 is then $y = 15.3 + 2.0(4) = 23.3$, which compares favorably with the 26.3 forecast ob-

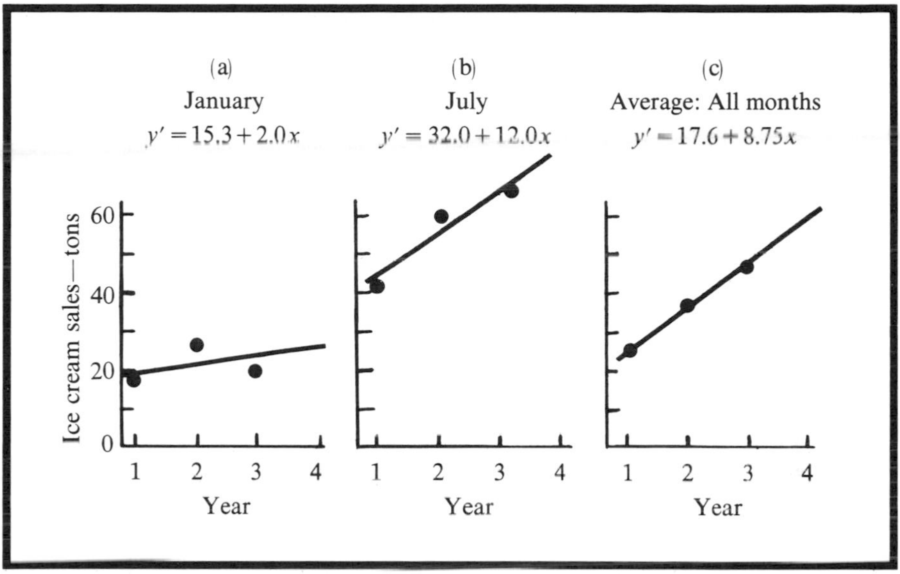

Figure 6 Yearly trend in ice cream sales, by selected months.

tained in the preceding paragraph. Compare, though, the January trend with the trend for the month of July shown in Figure 6*b*. These trends are based on the absolute minimum of three data points so they are very unreliable, but they do suggest that July sales are growing more rapidly than January sales. Similarly, trends for the other months could be identified, and it would be a matter of trial-and-error to determine if these monthly trends provide a better forecast than the overall trend discussed in the preceding paragraph and shown in Figure 5.

If the monthly trends were more or less alike, a forecast for Year 4 average monthly sales could be obtained from the trend line shown in Figure 6*c*, based on the average monthly sales for Years 1, 2, and 3. Again, whether the Year 4 forecast by this method is better than that obtained by averaging the 12 monthly forecasts that would result from the equation in Figure 5 would have to be determined from future data. The point is that we can look at our trend data from a number of different angles, and in the end what works best will be our choice.

If a time series shows a curvilinear rather than a linear trend, then instead of a best-fitting line of the form $y = a + bx$, we might try equations of the form $y = a + \log x$, $y = a + bx^2$, $y = a + b/x$, $y = a + bx + cx^2$ or any of a number of other data transformations which can be pursued in advanced studies of

time series analysis. Also, we have seen in Chapter 12 how analysis of variance techniques can be used to test hypotheses about contrasts of means reflecting linear, quadratic, cubic, and higher order trend sources of variation in a series of means.

7. Irregular Variation

After seasonal, cyclical, and trend variation have been identified in a time series, the left-over or residual variation is often referred to as *irregular*. It consists of two basic sources: (1) random variation, plus (2) systematic patterns of variation which are often too complex to readily identify and describe. However, to lump these two sources together is begging the question, for they are two distinct types of variation.

The random error component of the irregular variation is something that is inherent in all observations. By definition it is unpredictable. This error can be inherent in the measuring process or it may be a parametric characteristic of the criterion variable itself, or it could be a combination of the two. The best we can do in the analysis of this variation is to ignore it or to try and mask it in order to see if systematic variation remains. For example, in arriving at a best-fitting trend line we ignored deviations of the data points from the line as being chance error. A popular way to mask such random variation is to apply a moving-average to the time series. We have already encountered the concept of a moving-average in the ratio-to-moving-average method of determining seasonal index numbers. Table 6 shows a three-period moving-average applied to a series of thirty observations. We also see in the table that the placement of each average vis-a-vis the raw data is dictated by the purpose of the analysis—whether as a description of the past, a comparison for the present, or a forecast of the future.

A plot of the actual time series values appears in Figure 7*a*. In Figure 7*b*, however, the three-period moving average is superimposed upon the time series. Notice how the plot of the moving average is less variable than the original time series. If we used a five-period moving average it would be even less variable. We must be careful, though, not to interpret the slow undulating movement of the moving average as necessarily reflecting underlying cyclic patterns, since by the very definition of the moving average it cannot make abrupt swings; i.e., each successive average in the moving series contains values which appeared in the preceding average. This misleading feature of a moving average is especially evident in Figure 7*b*, since the time series shown is actually *completely random*, the successive values of the criterion variable having been drawn at random from a normal population with mean $\mu = 100$ and a standard deviation $\sigma = 10$. This should impress upon us that there is a

Table 6 Example of a three-period moving average.

Time period	Observation	Three-period moving average: As summary of the past	As comparison to current period	As forecast of next period
1	106			
2	124	108.7	—	—
3	96	106.3	108.7	—
4	99	99.3	106.3	108.7
5	103	99.3	99.3	106.3
6	96	104.3	99.3	99.3
7	114	104.0	104.3	99.3
8	102	102.3	104.0	104.3
9	91	97.7	102.3	104.0
10	100	88.7	97.7	102.3
11	75	90.0	88.7	97.7
12	95	89.7	90.0	88.7
13	99	99.7	89.7	90.0
14	105	96.0	99.7	89.7
15	84	97.0	96.0	99.7
16	102	99.0	97.0	96.0
17	111	104.7	99.0	97.0
18	101	105.7	104.7	99.0
19	105	107.0	105.7	104.7
20	115	105.3	107.0	105.7
21	96	101.7	105.3	107.0
22	94	98.7	101.7	105.3
23	106	103.0	98.7	101.7
24	109	109.3	103.0	98.7
25	113	110.0	109.3	103.0
26	108	104.0	110.0	109.3
27	91	97.3	104.0	110.0
28	93	88.3	97.3	104.0
29	81	94.3	88.3	97.3
30	109	—	94.3	88.3
31	?	—	—	94.3

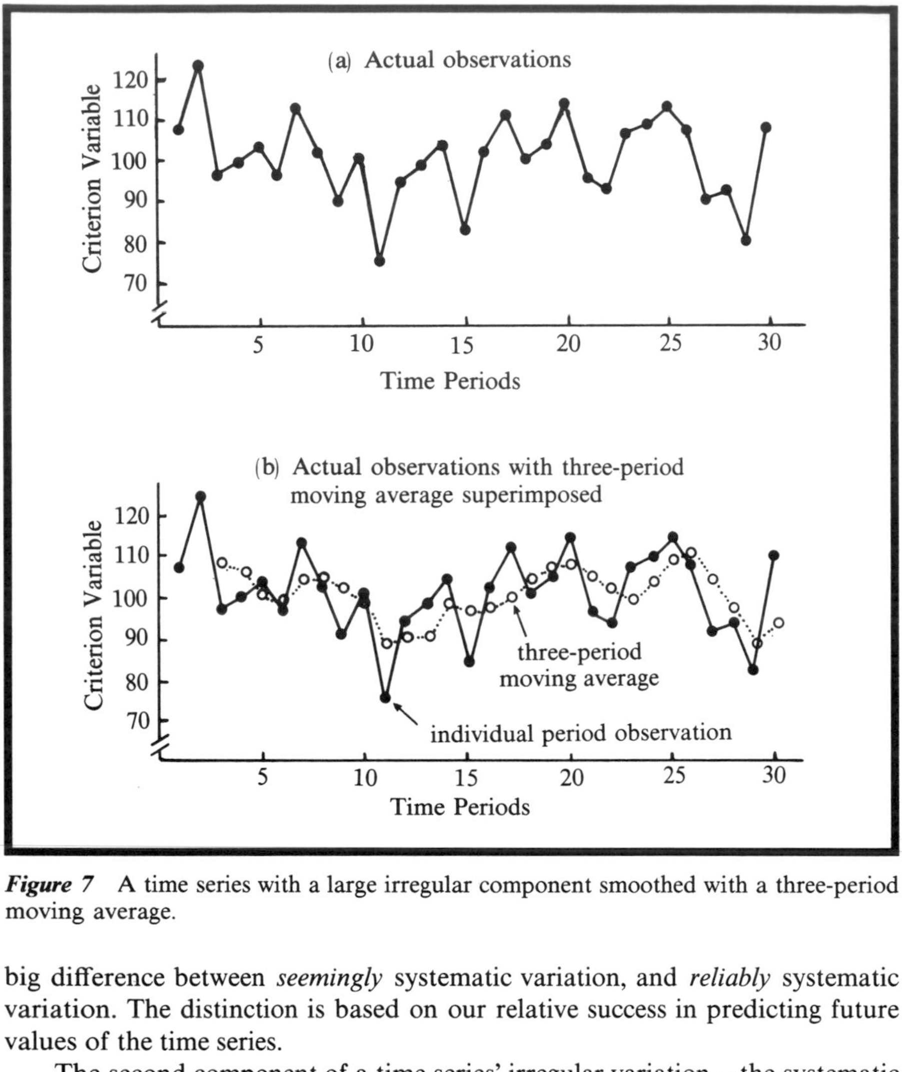

Figure 7 A time series with a large irregular component smoothed with a three-period moving average.

big difference between *seemingly* systematic variation, and *reliably* systematic variation. The distinction is based on our relative success in predicting future values of the time series.

The second component of a time series' irregular variation—the systematic variation which is too complex or subtle to identify as simple cyclical or secular trend variation—is perhaps the most challenging of all time series analyses. Since the variation is not entirely random, it is potentially describable; but since it is so complex or subtle it is not easily quantified such that future values can be forecast—a musical score and the stock market being two extreme examples. The following sections introduce some of the more promising techniques for tapping this type of variation. The techniques are all based on the fact that a time series' current value can be predicted better than chance

from a knowledge of one or more values earlier in the time series, actually the same premise upon which the preceding techniques of the chapter are based.

8. Autoregression

The concept of autoregression was first introduced in the Regression Analysis chapter. Whereas a conventional regression equation relates a criterion variable y to a predictor variable x, an autoregression equation relates the values of the criterion variable y to values of y itself observed earlier in the time series. For example, an autoregression equation of the form

$$y_i' = a + by_{i-1} \tag{6}$$

would predict the value of y in time period i, as a function of the value of y in period $i-1$, the immediately preceding value in the series.

More generally, we can express an autoregression equation as

$$y_i' = a + by_{i-k} \tag{7}$$

where the forecast for period i is now a function of y appearing k values back in the series. Also, *a multiple* autoregression equation can be formed

$$y_i' = a + b_1 y_{i-1} + b_2 y_{i-2} + \cdots + b_k y_{i-k} \tag{8}$$

in which the forecast for period i takes into account any number of its preceding values, each being weighted to a different degree.

The extent to which such autoregressive functions are successful in forecasting future values of a time series will depend upon the extent to which dependencies or non-randomness exist among the values in the time series. The existence of this necessary non-randomness can be verified by the serial correlation coefficient for various lagged values of the criterion variable, as discussed in the Correlation Analysis chapter. The value of r^2 will inform us of the amount of the variable's variance which can be accounted for by linear relationships with earlier values in the series.

While in many time series the amount of variance accounted for by an autoregressive function may be small, every little bit helps, provided the time and cost involved is not too great.

9. Exponential Smoothing

In the final column of Table 6 we saw how a moving average can be used to forecast future values of a time series. The logic behind such a forecasting system is that *recent* values of a time series are more likely than *remote* values

to be predictive of future values. An extension of this logic would suggest that a weighted moving average would be even better, since the most recent value in the series could be weighted more than the preceding one, and that one more than its preceding one, and so on. While the determination of such decreasing weights would be an analytical problem, the resulting equation would be similar to an autoregressive function in which a *fixed* number of preceding values are used to predict a future value.

The ultimate extension of the logic would be to weight every single preceding value of the time series from its very start, with the weights decreasing the further back we went. This procedure is the essence of a group of techniques known as *exponential smoothing*. In its simplest form the exponential smoothing system can be described as

$$\textit{Current forecast} = (\textit{weight})(\textit{last observation}) + (1 - \textit{weight})(\textit{last forecast})$$

where the weight applied to the last observation is a number between 0 and 1. That is, each forecast is a weighted average of the *last actual observation* and the *last forecast*. The relationship can be expressed symbolically as

$$\boxed{f_i = p y_{i-1} + (1-p) f_{i-1}} \qquad (9)$$

where

f_i is the forecast for period i,

y_{i-1} is the actual observation for the preceding period,

f_{i-1} is the forecast that was found for the preceding period,

p is a weight between 0 and 1 applied to the last observation,

and

$1 - p$ is the weight applied to the last forecast.

The weights p and $1 - p$ are not probabilities, although it is nice to think of them as such.

This type of *recursive* definition of a forecast, in which each successive forecast incorporates the preceding forecast, has the interesting feature that it takes into account *all* preceding values of the time series, although that may not be immediately obvious from the above equation. Since each forecast needs a preceding forecast, some initial forecast is needed to start the system going. That initial forecast can be choosen arbitrarily or on judgement. Since it is

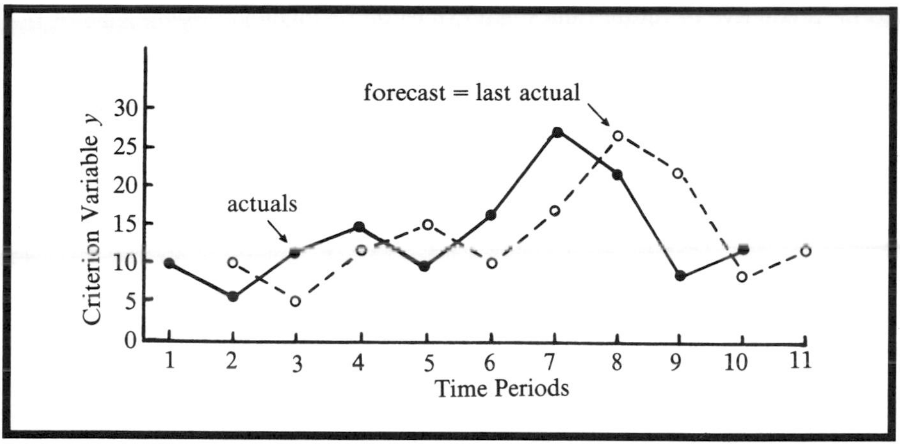

Figure 8 An example of using the latest observation as the forecast for the next observation.

intended to reflect past observations in the series, its choosen value will lose importance as the time series progresses, and therefore is not that crucial. In fact, exponential smoothing gets its name from the fact that values in a time series lose their importance for forecasting in an exponential manner as we travel further back in the series.

As for the respective weights that should be applied to the last observation and last forecast, no generalizations can be made as to their optimum size for minimizing forecasting errors. Every time series is different and therefore it is

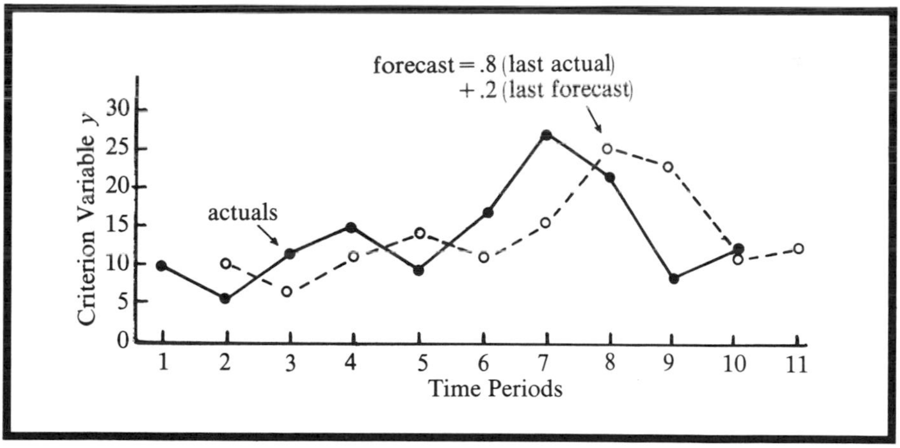

Figure 9 An example of forecasting by means of exponential smoothing.

largely a matter of trial-and-error and experience to determine the best weights. Figure 8, for example, shows an extreme situation in which a maximum weight of 1 is applied to the last actual observation in the series, in which case it becomes the forecast for the next period. That is, we forecast that what happened last will happen next. We see in Figure 8, that such a forecasting system is not all that bad. However, we may be able to improve upon it by not giving 100% importance to the last observation. Table 7 shows forecasts for the same time series in which weights of .8 and .2 are applied to the last observation and last forecast, respectively. A comparison of the actuals with the forecasts is shown in Figure 9.

In the preceding example a weight of .8 was applied to the last observation in the series. Such a high value will be very sensitive to short-term swings in the data. More often the weight for the last observation will be low, on the order of .2 or .1, or even lower. Again, the only way to determine the optimum weight is through trial-and-error, or through some insight into the dependencies that exist in the time series. A long segment of the time series would be tested with different weights to see which produced the smallest sum of squared prediction errors, our most common criterion for forecasting accuracy.

Once the weights have been decided upon, we no longer have to retain the historical time series, since each forecast contains all the information present in the earlier periods. This is a distinct advantage over the autoregression technique which requires a long history of the time series for each forecast and also

Table 7 Example of the exponential smoothing procedure.

Time periods	*Criterion variable (actual observations)*	*Forecast* = .8(*last actual*)+.2(*last forecast*)
1	10	.8(10)+.2(10*) = 10
2	6	.8(6)+.2(10) = 6.8
3	12	.8(12)+.2(6.8) = 10.96
4	15	.8(15)+.2(10.96) = 14.19
5	10	.8(10)+.2(14.19) = 10.84
6	16	.8(16)+.2(10.84) = 14.97
7	27	.8(27)+.2(14.97) = 24.59
8	21	.8(21)+.2(24.59) = 21.72
9	9	.8(9)+.2(21.72) = 11.54
10	12	.8(12)+.2(11.54) = 11.91
11	?	*The initial forecast is set equal to the last actual observation to get the system started.

requires considerably more computational time and expense than does the exponential smoothing method. However, as might be expected, autoregressive and smoothing techniques tap different features of the data, and consequently they are often used in weighted combination to attain forecasting accuracy better than with either technique alone.

10. Multivariate Time Series

In all the preceding techniques we relied solely on the sequence of values of the criterion variable y to forecast its future values, without recourse to other predictor variables, other than the numerical labels of the time periods themselves. However, if we can identify other variables $x_1, x_2, \ldots, x_k$ which are correlated with the criterion variable systematically over time, then we can use their values to help predict y. But to be practical, we must find predictor variables that anticipate the values of the criterion variable. Such a relationship, in its simplest form, would be

$$y_i' = a + bx_{i-k} \tag{10}$$

in which the forecast of the criterion y in period i is a function of the value of predictor variable x, occurring k periods earlier. That is, the lagged value of variable x is used to predict a current value of y. For example, steel production in a given time period could be used to predict automobile production for a later period. Bank interest rates could be used to predict future housing starts. The number of births during a given period could be used to predict future school and teacher needs.

In each of the above examples, a predictor variable x is a *leading indicator* or predictor of the criterion variable. This type of relationship is shown in Figure 10. The pattern of variation of the predictor variable or leading indicator mirrors and precedes that of the criterion variable.

Again, as with all correlation-type relationships, the predictability of y from a lagged value of x does not necessarily mean that the one causes the other. There may be many other variables which are confounded with the given predictor variable. At the very least, though, it can serve as a practical predictor of the criterion variable.

The above approach to forecasting a criterion variable y can be expanded to include any number of leading indicators, turning it into a multiple regression analysis. When used in conjunction with autoregressive and exponential smoothing techniques, an elaborate forecasting system results. But we must always be careful to validate a prediction system against a future time series, for although it may do well in describing a historical record of a variable, that success may be due largely to capitalizing on chance relationships in the data that are unlikely to recur in the future.

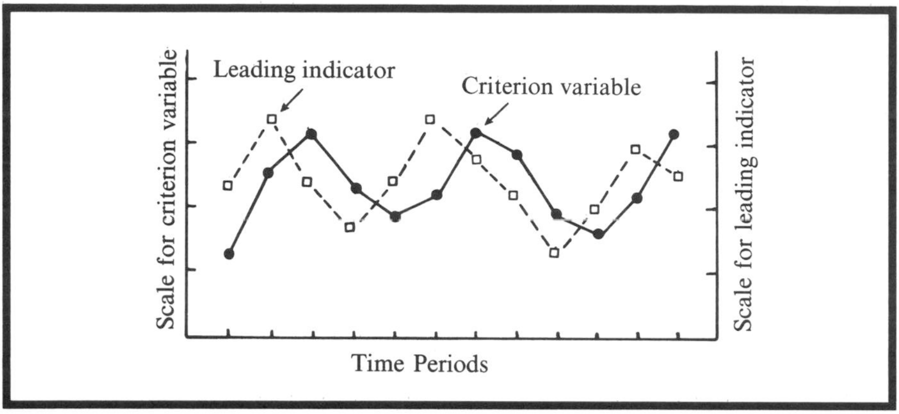

Figure 10 An example of a one-period leading indicator.

Perhaps the ultimate procedure for describing a complex time series is through the use of *Fourier analysis*, in which even the most irregular of time series can be potentially modelled as the sum of a large number of *sine*, *cosine*, and other trigonometric functions, varying in their periods and amplitudes such that when they are added together they can form any pattern we desire. But this approach again begs the issue, for it assumes that we have identified several underlying variables which vary in a regular trigonometric manner. Nonetheless, it can be an effective approach if the observed series is not too irregular, and it has the further benefit that it stimulates the search for underlying variables which contribute to our criterion variable of interest.

11. Contagion-Cure/Trial-Repeat Forecasting Models

We have considered various methods of forecasting a time series based on the past performance of the time series. An alternative method, which does not rely upon a past time series, is to construct a forecasting model or theory based on assumptions about the underlying "mechanism" or driving force which *generates* the time series. We will consider a simple type of such a forecasting model in order to understand the logic of the approach.

We are very often interested in the period-by-period growth or decline in the population incidence of a particular disease, drug usage, belief in a rumor or attitude, knowledge of a fact, use of a product or service, etc. The spread of these types of binary characteristics—in which each member of the population either has or does not have the characteristic at any given point in time—can often be described by a simple model which makes assumptions about (1) what percentage of the population will *acquire* the characteristic in a specified time period, and (2) what percentage of those who have acquired it will *continue to*

have it in the subsequent time period. Such models are often referred to as *contagion-cure models*, named after the two key parameters—the degree of *contagion* or *acquisition* of the characteristic in question, and the degree of *cure* or *loss* of the characteristic in subsequent periods. These terms are especially descriptive of the spread of a disease, but they apply generally to other characteristics as well that spread like a disease. An alternative name, especially descriptive of the spread of such characteristics as drug or consumer product usage, is *trial-repeat model*. The *trial* of a product being analogous to the *contagion* of a disease, and the *repeat usage* of the product in a subsequent time period being analogous to, not cure, but the *complement of cure*: i.e., repeat % = 100% − cure %. But this difference is just a technicality. If, instead, we called it a trial-dropout model it would be perfectly analogous to the contagion-cure model. Rather than duplicate our discussion for the two, we will refer to contagion-cure models, realizing that the discussion applies equally well to trial-repeat characteristics.

One of the simplest contagion-cure models is to assume that a fixed percentage of the virgin population will contract the characteristic in any given time period, and that another fixed percentage of those who have contracted it will be cured of it in the subsequent time period.

As an example, let us assume that a particular disease is contracted by 15% of the virgin population (those never having it before) each year. Next, let us assume that 60% of those who have the disease in a given time period will be cured of it in the subsequent time period.

Now we have all the information we need for our simplified model. First let us consider what percentage of the virgin population will contract the characteristic in each year. In the first year, 15% will contract it, leaving 85% of the virgin population. In the second year, 15% of those remaining 85% will contract it, or 15% × 85% = 12.75%. That leaves only 85% − 12.75% = 72.25% of the virgin population in the third year, when 15% of it will contract the attribute in question, and so on. The first column of Table 8a summarizes the percentages of the total population contracting the characteristic in each of the first seven years. This takes care of the contagion part of the model.

Next we must consider the cure part of the model which assumes that 60% of those with the characteristic will be cured of it (and 40% will not) in the subsequent time period. Thus, of the 15% who contracted it in the first year, only 15% × 40% = 6% (as % of total population) will have it in year two. But recall from the contagion part of the model that in the second year we had picked up 12.75% of the population that contracted the characteristic. Consequently, at the end of the second year there will be 6% + 12.75% = 18.75% of the population with the attribute in question—the 6% being the "non-cured" from the first year, and the 12.75% being the newly contracted members of the population.

This same series of calculations can then be made for each subsequent year, and the results are summarized in Table 8*a*. Notice that the percentage of the population with the characteristic initially rises and then tapers off as the years progress, reflecting the relatively high cure rate relative to the contagion rate *plus* the assumption that once an individual is cured they will not recontract the characteristic. It is left as an exercise to repeat the calculations assuming that the cured individuals are susceptible to contracting it in subsequent time periods.

Part *b* of Table 8 presents an alternative scenario to that in part *a*. The contagion rate of 15% is assumed again, but this time a cure rate of only 10% is postulated. Notice the rapid growth in the incidence of the characteristic.

Table 8 Contagion-cure rate forecasting model for an attribute *A*, with a contagion rate of 15% of the virgin population and a cure rate of (a) 60% vs. (b) 10%.

(a)

	% of population contracting attribute *A*	% of population with attribute *A*, given a 60% cure rate (40% repeat rate)						
		Year 1	*Year 2*	*Year 3*	*Year 4*	*Year 5*	*Year 6*	*Year 7*
Year 1	15.00	15.00	6.00	2.40	.96	.38	.15	.06
Year 2	12.75 →		12.75	5.10	2.04	.82	.33	.13
Year 3	10.84 →			10.84	4.34	1.73	.69	.28
Year 4	9.21 →				9.21	3.68	1.47	.59
Year 5	7.83 →					7.83	3.13	1.25
Year 6	6.66 →						6.66	2.66
Year 7	5.66 →							5.66
	Yearly totals:	15.00	18.75	18.34	16.55	14.44	12.43	10.63

(b)

	% of population contracting attribute *A*	% of population with attribute *A*, given a 10% cure rate (90% repeat rate)						
		Year 1	*Year 2*	*Year 3*	*Year 4*	*Year 5*	*Year 6*	*Year 7*
Year 1	15.00	15.00	13.50	12.15	10.94	9.84	8.86	7.97
Year 2	12.75 →		12.75	11.48	10.33	9.29	8.37	7.53
Year 3	10.84 →			10.84	9.76	8.78	7.90	7.11
Year 4	9.21 →				9.21	8.29	7.46	6.71
Year 5	7.83 →					7.83	7.05	6.34
Year 6	6.66 →						6.66	5.99
Year 7	5.66 →							5.66
	Yearly totals:	15.00	26.25	34.47	40.24	44.03	46.30	47.31

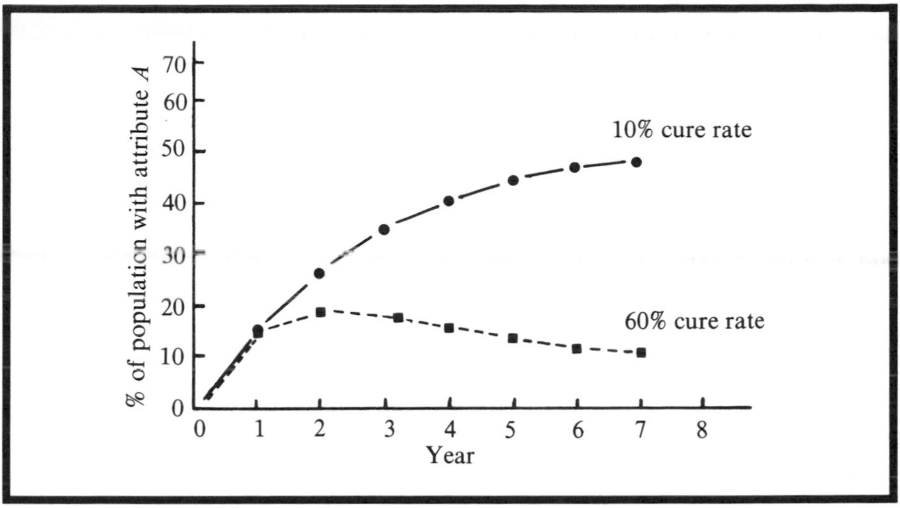

Figure 11 Yearly incidence of attribute *A* in the population, by cure rate.

Figure 11 presents a comparison of the growth rates for the attribute in question for the 60% vs. 10% cure rate conditions. As an exercise, graph the yearly *cumulative* percentage of the population that has *ever had* the characteristic under study.

The above type of contagion-cure or trial-repeat model is attractive in its simplicity. In practice, they are not likely to be so neat, since the two basic assumptions—the contagion and cure rates—are not likely to be constant from year to year, for any variety of reasons. Nonetheless, the basic model is a very useful starting point upon which to build refinements. Furthermore, departures of the data from the basic model will provide insights into the actual dynamics of the characteristic being studied.

12. Concluding Comments

We have seen in this chapter that the analysis of a time series is essentially an attempt to identify and model systematic sources of variation in the series, such as seasonal, cyclical, trend, and non-random irregular components. The goal of such analyses is to account for all of the time series variation except for the irreducible random component that is everpresent in all empirical observations.

The end benefit of understanding the nature of a time series is that its future values can be forecasted, allowing us to act accordingly. It is not an easy task, since we have no theoretical guarantee that the future will occur as the

past, and though much labor is expended in forecasting the future, it is a discipline in its early infancy, often being more of an art than a science, requiring an analyst who is intimately familiar with the data and with insights into its pattern of variation. Consequently, we should not expect any two analysts to approach a time series in exactly the same way nor arrive at the same result. Ultimately, the criterion for the validity of a given analytical approach will be its success in predicting the future, *time and time again*.

Chapter 19

Nonparametric Analysis

1. Introduction

There is a broad class of analytical techniques which are referred to as *nonparametric*, or *distribution-free*, in the sense that they do not involve assumptions about the defining characteristics of the sampled population distributions. This contrasts with the t and F ratio tests, for example, which require normally distributed populations with equal variances.

One of the most commonly used nonparametric methods is the application of the χ^2 test of independence to $r \times c$ contingency tables, as studied in Chapter 13, in which the expected frequencies are obtained from the set of data itself, rather than from some theory about the sampled population distributions. In this chapter we will look at a few additional nonparametric tests, those which can be considered counterparts to the main types of parametric testing techniques studied so far—a test of central tendency, a test of the difference between two sampled populations, a test of the association between two random variables, and a test of sequential randomness in a time series.

2. Sign Test of the Median

Suppose a mail-order marketer wants to test the sales response to its catalog among a particular list of prospects. A test mailing of 500 catalogs results in $n = 10$ orders in the following dollar amounts:

18 35 23 9 14 25 41 21 47 16

The research department wants to test the hypothesis that the median order size to be expected from the sampled list is 30 dollars; i.e.,

$$H_0: Mdn = \$30$$

Assigning a plus sign (+) to those sample observations above the hypothesized median, and a negative sign (−) to those below, yields

$$- + - - - - + - + -$$

for the $n=10$ sample orders, respectively. We find that there are 7 values below the hypothesized median, and 3 above it. (In the event that one of the observations coincided exactly with the hypothesized median, it could be discarded from the analysis and the sample size reduced by 1).

Since by definition the number of observations above and below the median should be equal, we can use the binomial distribution to determine the probability of the observed departure from the expected 5 : 5 split, if in fact the population median were as hypothesized. Referring to Appendix Table X we find that the probability of 7 or more −'s out of a possible 10 is .172, and similarly, the probability in the other direction of 7 or more +'s is .172, for an overall probability of .344 of observing at least 7 out of 10 outcomes of one kind (either +'s or −'s), a fairly likely event. Consequently, we cannot reject the working hypothesis that the true population median is 30.

A two-tailed test is appropriate in this situation since there was a probability that the median could have been either higher or lower than the hypothesized value, yielding a preponderance of either +'s or −'s. If it had been higher, the marketer would have been encouraged and proceeded with a full-scale mailing. If it had been lower, the mailing list would have been abandoned. As it turned out, the failure to reject the hypothesis that the true median was 30, resulted in cautious optimism and a follow-up test with a larger mailing.

The preceding analysis is an example of the *sign test* as applied to the median of a single sampled population. Note that no assumptions were made about the shape of the population. Note, also, that if we had cared to make the assumption that the population distribution was *symmetrical* in form, in which case the median would coincide with the mean μ, then the preceding analysis would also have been a test of the hypothesis $H_0: \mu = 30$. Such a test would be an alternative to the t test of the mean, which requires the more stringent assumption that the sampled population is normal in form.

The sign test, which essentially is an application of the binomial distribution, can be applied to any situation in which a set of independent observations can be classified as being one kind or another, or being above or below some designated value. If that designated value is the population median, then the binomial with $p=.5$ is used to evaluate the distribution of sample observations. If the designated value is other than the median, then the corresponding value of p must be used; e.g., to test the 25th percentile, we would use $p=.25$.

If the sample size is large, with both np and nq greater than 5, then the normal approximation to the binomial distribution can be used, as discussed in Chapter 9.

3. Rank Sum Test

The t test of the difference between two sample means requires that the sampled populations are normal in form and have equal variances. If these assumptions cannot be met, an alternative is to transfer the sample observations to an ordinal scale—i.e., ranks—and use the *rank sum* test to determine if the observations were drawn from the same population. In various equivalent forms the rank sum test is also known as the Wilcoxon or Mann-Whitney test.

As an example of the rank sum test, consider the researcher who assigns 14 laboratory rats at random to two groups of 7 rats each. Group I serves as a control, while Group II receives a dose of an experimental substance. The criterion variable is the number of minutes required to pass through a maze to food at the other end. The number of minutes required for the animals are as follows:

Group I:	21	56	38	17	43	71	63
Group II:	48	29	15	22	55	90+	18

The experiment was terminated for the second last rat in Group II since it still had not found its way through the maze after 90 minutes. Because of its unknown eventual score, and doubts about the normality of the population distributions, a t test was considered inappropriate.

Instead, all 14 scores were rank ordered, from lowest to highest, but retaining the group membership identities of the scores, as shown in part a of Table 1. Next, the scores were assigned the rank values 1 through 14, as shown in part b of Table 1. If there was no difference between the two sampled populations—the control vs. the experimental treatment—then the sum of the ranks for Group I would be expected to be the same as the sum for the equally sized Group II. In the event the two sample sizes differ, the sum of ranks is obtained for the *smaller* of the two groups. The rank sum test then evaluates the observed sum of ranks R for the smaller group, against an *expected* sum, given the assumption that the two populations are identical.

The sum of ranks R for Group I is shown in Table 1 to be $R = 57$. Appendix Table IX provides the range of values for accepting the null hypothesis of identical populations, for various pairs of sample sizes, and for

Table 1 Two groups of raw scores and their corresponding rank values.

(a) Raw data		(b) Rank values	
Group I: (*Control*)	*Group II*: (*Treatment*)	*Group I*: (*Control*)	*Group II*: (*Treatment*)
	15		1
17		2	
	18		3
21		4	
	22		5
	29		6
38		7	
43		8	
	48		9
	55		10
56		11	
63		12	
71		13	
	90		14
		Sum: $R = 57$	

various significance levels. For sample sizes of 7 and 7, we find that the probability of a sum of ranks outside of the interval extending from 37 to 68 is .05. Since the observed sum $R = 57$ falls within this interval, we cannot reject the hypothesis that the sampled populations are identical. That is, the experimental treatment does not appear to affect the time of traveling through the maze.

The reason the significance levels in Appendix Table IX are not exactly the conventional .05 and .01 values is that the sum of a given number of ranks is discrete in nature and therefore its sampling distribution is not continuous. For one-tailed tests the tabled significance levels should be halved.

Normal approximation. The expected or mean value of the rank sum R for the smaller of two samples is given by

$$\boxed{\mu_R = \frac{n_1(n_1 + n_2 + 1)}{2}} \tag{1}$$

and its standard error is

$$\boxed{\sigma_R = \sqrt{\frac{n_1 n_2(n_1 + n_2 + 1)}{12}}} \tag{2}$$

where n_1 and n_2 correspond to the smaller and larger sample sizes, respectively.

When both sample sizes exceed 10 or so, the distribution of the rank sum statistic R is very nearly normal in form, and the approximate normal deviate

$$z = \frac{R - \mu_R}{\sigma_R}$$

can be used to test the difference between the sampled population distributions.

For example, the overall ranking of student exam scores in two classes of sizes $n_1 = 11$ and $n_2 = 14$, was as follows:

Class I: 4 6 7 10 11 13 15 16 19 22 23 ($R = 146$)

Class II: 1 2 3 5 8 9 12 14 17 18 20 21 24 25

Using formula (1), the expected mean for the rank sum of the smaller sample, Class I, is

$$\mu_R = \frac{11(11+14+1)}{2} = 143$$

Using formula (2), the standard error of R is

$$\sigma_R = \sqrt{\frac{(11)(14)(11+14+1)}{12}} = 18.27$$

The approximate normal deviate is then

$$z = \frac{R - \mu_R}{\sigma_R} = \frac{146 - 143}{18.27} = .16$$

which does not allow us to reject the hypothesis that the performance of the students in the two classes represent identical population distributions.

4. Rank Correlation Coefficient

In Chapter 10 the product moment correlation coefficient r was used to measure the degree of linear relationship between the paired values of two continuous random variables. If we want to escape the burden of the population distribution assumptions accompanying the interpretation of r, we can

convert the raw scores on each variable into *ordinal* values—*ranks*—and correlate them instead. However, rather than assigning ranks based on the combined set of observations, as in the case of the preceding rank sum test, we assign ranks to the values of each variable separately. That is, the n values of one variable are ranked 1 through n, and the n values of the other variable are also ranked 1 through n. The result is n pairs of ranks. It is these ranks, then, instead of the raw scores, that are correlated, yielding the *rank* or *rank-order correlation coefficient*, defined as

$$r_o = 1 - \frac{6 \sum_{i=1}^{n} D_i^2}{n(n^2-1)} \tag{3}$$

where n is the number of pairs of ranks, D_i is the difference between the ith pair of ranks, and ΣD_i^2 is the sum of the squared differences between the paired ranks.

Consider the following application of the rank correlation coefficient r_o. A researcher interested in creating two equivalent forms of a test, administers initial drafts of the two forms to a sample of $n = 10$ individuals. The scores of the individuals on the two tests are then converted into rank values as shown in Table 2. For example, individual A ranked 2nd on Form I and 5th on Form II, while individual B ranked 7th and 6th on the two forms, respectively. It makes no difference whether rank 1 refers to the lowest or highest score in the distribution, so long as we are consistent for the two variables.

Next, the differences between the paired ranks are calculated, squared, and summed across all n individuals, yielding $\Sigma D_i^2 = 40$. Then, using formula (3), we have as the rank correlation coefficient

$$r_o = 1 - \frac{6(40)}{10(10^2-1)} = .76$$

To determine if an $r_o = .76$, based on $n = 10$ pairs of ranks, is larger than we would expect by chance, if in fact there were no relationship between the two variables, we can consult Appendix Table VIII. We find that an absolute value of r_o greater than .56 is needed at the $\alpha = .048$ significance level for a one-tailed test, so given the sample result $r_o = .76$ we can confidently reject the hypothesis of no relationship between the two test forms. That is, the preliminary research suggests that the alternate forms of the test are measuring the same thing, at least to *some* degree, and the researcher can proceed to evaluate the forms with larger samples. In this instance we were justified in using a

Table 2 Calculation of the rank correlation coefficient for two test forms.

Individuals	*Rank on Test I*	*Rank on Test II*	*Difference* (D_i)	D_i^2
A	2	5	−3	9
B	7	6	1	1
C	3	1	2	4
D	9	8	1	1
E	6	3	3	9
F	1	2	−1	1
G	10	9	1	1
H	4	7	−3	9
I	8	10	−2	4
J	5	4	1	1
				$\Sigma D_i^2 = 40$

$$r_o = 1 - \frac{6\Sigma D_i^2}{n(n^2-1)} = 1 - \frac{6(40)}{10(10^2-1)} = .76$$

one-tailed test of significance since it would be a rare test constructor indeed, who, despite all efforts, developed two forms that were *negatively* correlated.

If the number of paired observations n is sufficiently large, say, greater than 10, then

$$\boxed{t = \frac{r_o\sqrt{n-2}}{\sqrt{1-r_o^2}}} \tag{4}$$

with $df = n - 2$ can safely be used to test the hypothesis that r_o does not differ from zero. For example, if $n = 25$ pairs of ranks yielded an $r_o = .37$, we would have

$$t = \frac{.37\sqrt{25-2}}{\sqrt{1-.37^2}} = 1.91$$

which for $df = 23$ is not significant at the $\alpha = .05$ level. That is, if we tossed the ranks 1 through 25 into the air, and correlated their order of landing with another set similarly tossed, we could well get a value of $r_o = .37$ by chance alone.

It should be kept in mind that although the rank correlation coefficient is free of the assumptions of a joint bivariate normal population distribution, as

required in the case of the product moment correlation coefficient, the two coefficients are similar in the sense that neither necessarily reflects a causal relationship between the correlated variables.

5. Runs Test

We have learned in earlier chapters that the serial correlation coefficient can be used to test the randomness of a series of values of a continuous random variable. If the random variable is dichotomous in nature, then we can use the *runs test* to determine if the sequence of its values is randomly ordered.

Suppose, for example, that a roll-call vote of a committee of 26 individuals yielded the following sequence of yes (+) and no (−) votes on a particular issue:

$$+ + - + - - + + - - + - + + + - + + + - - + + - + +$$

In the above sequence there are 16 +'s and 10 −'s. The question here is not whether this result differs from a even split, but whether the *order* of the two outcomes is random. Since it is based on an *oral* roll-call vote we might suspect some sequential dependencies in which each person's vote was influenced by the preceding ones. To test the hypothesis of complete randomness we first count the number of *runs* in the sequence, where a run is defined as a sequence of identical values. For the given series we have a total of 15 runs

$$+ + / - / + / - - / + + / - - / + / - / + + + / - / + + + / - - / + + / - / + +$$

where the runs are separated by slashes. The question we must ask is whether 15 runs would be expected if the 26 dichotomous values—16 of one kind and 10 of another—were randomly arranged.

In general, the mean number of runs u to be expected from n_1 values of one type, and n_2 values of another, is given by

$$\boxed{\mu_u = \frac{2n_1n_2}{n_1 + n_2} + 1} \tag{5}$$

and the standard error of the number of runs is given by

$$\boxed{\sigma_u = \sqrt{\frac{2n_1n_2(2n_1n_2 - n_1 - n_2)}{(n_1 + n_2)^2(n_1 + n_2 - 1)}}} \tag{6}$$

If n_1 and n_2 are each greater than 10, then the number of runs u will be very nearly normally distributed. Consequently, we can use

$$z = \frac{u - \mu_u}{\sigma_u}$$

to test the hypothesis that the observed number of runs does not differ statistically from the expected number of runs.

In our voting example the expected number of runs is

$$\mu_u = \frac{2(16)(10)}{16+10} + 1 = 13.31$$

and the standard error is

$$\sigma_u = \sqrt{\frac{2(16)(10)[2(16)(10) - 16 - 10]}{(16+10)^2(16+10-1)}} = 2.36$$

The normal deviate is then

$$z = \frac{u - \mu_u}{\sigma_u} = \frac{15 - 13.31}{2.36} = .72$$

Since the obtained z does not exceed the critical value of 1.96 needed at the .05 significance level, we cannot reject the hypothesis that the sequence of yes and no votes is randomly ordered.

Although special tables exist for the runs test when either n_1 or n_2 is less than 10, it is very difficult to detect non-randomness in such short sequences. The runs test is best left for longer sequences of data.

6. Nonparametric vs. Parametric Tests

There are wide differences of opinion as to the practical value of nonparametric techniques. Proponents of the methods point out their ease of calculation, their generality to all distribution forms, the exact nature of their resulting probability statements, and that they sometimes have greater *power*—i.e., a greater ability to detect a real departure from the null hypothesis—than parametric tests when the population assumptions of the latter tests are not met.

Those who are not so enthusiastic about nonparametric techniques counter that with modern computing facilities the calculational ease of the tests is no real benefit. Also, since important decisions are seldom made based on small

samples, the use of large samples allows us to invoke the Central Limit Theorem as assurance that the sampling distributions of our statistics are nearly normal in form, even when the sampled populations are not.

Even when we must deal with small samples, the t test has been shown through extensive computer simulation work to be very *robust*, in the sense that violations of normality and equality of variance do not greatly affect the accuracy of the probability statements resulting from the tests. The major problem arises when the populations are severely skewed (e.g., J-shaped), or have extremely long tails in both directions. Such situations, though, can be handled by demanding a significance level at, say, the .01 or .025 level, but treating it practically as a .05 significance level. Also, adverse effects of heterogeneity of variance in the multiple sample situation can be minimized by using equal sample sizes whenever possible. In short, while violated assumptions of the t test reduce its power, it is still more powerful than the alternative nonparametric tests in the vast majority of situations encountered in the world at large.

Somewhere between these two positions is a reasonable medium, where each type of technique has its place, depending upon cost considerations and the purposes to which the analyses are to be put.

7. Concluding Comments

Virtually every parametric test has a nonparametric counterpart. We have looked at only a sampling of the more commonly used ones, primarily to see the distinction between the two approaches to data analysis. With these nonparametric techniques we have now studied analytical methods for variables measured on all four types of measurement scales: The z, t and F tests for interval and ratio scaled variables, the χ^2 test for nominally scaled variables, and nonparametric methods for ordinally scaled or ranked data. More than anything else, this variety of techniques should make us conscious of the inter-dependency between the nature of our observations and the analytical model used to evaluate them.

Chapter 20

Advanced Probability Topics

1. Introduction

We have learned in earlier chapters that in order to solve probability problems it is essential that we know the nature of the sample space for the probability experiment of interest. In some cases the sample space is simple and there is no difficulty in enumerating the alternative outcomes of an experiment—e.g., the tossing of a coin, the rolling of a die, or the inspection of a product for a major defect. Other times, the sample space is more complicated and the enumeration of the possible outcomes is a more challenging task, such as when we want to know the probability of finding 2 or more defective parts in a sample of 50, or the probability of at least 5 of 1000 mail solicitations resulting in a response, or the probability of no mutations among 100,000 offspring of a species, or the probability of a certain disease given a particular symptom.

In this chapter we will consider various approaches to solving some of the more common types of probability problems which rest on sample spaces that are not easily enumerated or which cannot be adequately modeled by the *normal*, t, F, or χ^2 distributions which we have largely depended upon to this point. We will begin by studying some efficient *counting* procedures, which are useful in and of themselves, as well as for creating sample spaces and probability distributions.

2. Permutations

Very often we want to know in how many ways we can arrange a set of n objects, either in space or time. For example, how many different ways can five choices be listed in a menu? How many ways can four negotiators be arranged

at a table? How many ways can five wine samples be presented to a panel of tasters? How many ways can ten schools finish in a competition?

Permuting *n* distinct objects. Each arrangement of n objects is referred to as a *permutation* of those objects. For example, *abc*, *acb*, *bac*, *bca*, *cab*, and *cba* are all permutations of the letters a, b, and c. If we are dealing with n *distinct* objects—i.e., no duplication among them—the determination of the number of possible permutations of the objects, without repetition of any one object, is a relatively simple matter.

We begin by mapping out the number of positions in time or space that must be occupied by the objects in question. For instance, if we are interested in the $n = 3$ objects represented by the letters a, b, and c, we begin by indicating the three positions they can fill:

——— ——— ———

Next we ask in how many ways each position can be occupied. The first position can be occupied by any of the $n = 3$ objects, either a, b, or c. Once the first position is filled, however, only two possibilities exist for the second position, and once it is filled, the remaining slot can only be filled by the left-over object. Consequently, we have

$$3 \times 2 \times 1 = 6$$

permutations of the letters a, b, c, which corresponds to the enumeration in the preceding paragraph. An alternative pictorial determination of the number of permutations for this example is presented in Figure 1 in the form of a tree diagram.

The above procedure can be extended to any number of distinct objects. If we want to know the number of permutations of $n = 4$ distinct objects, we again ask in how many ways each of the four positions can be occupied by the four objects, and we have

$$4 \times 3 \times 2 \times 1 = 24$$

different permutations possible. This, for example, would be the number of ways four people could be seated at a table.

It follows from the preceding examples that no matter how many distinct objects n we have, the number of permutations of all n objects can be expressed as

$$n \times (n-1) \times (n-2) \times \cdots \times 1$$

That is, beginning with the number of objects n, we multiply it first by *one* less than n, that result by *two* less than n, and so on until we reach the number 1.

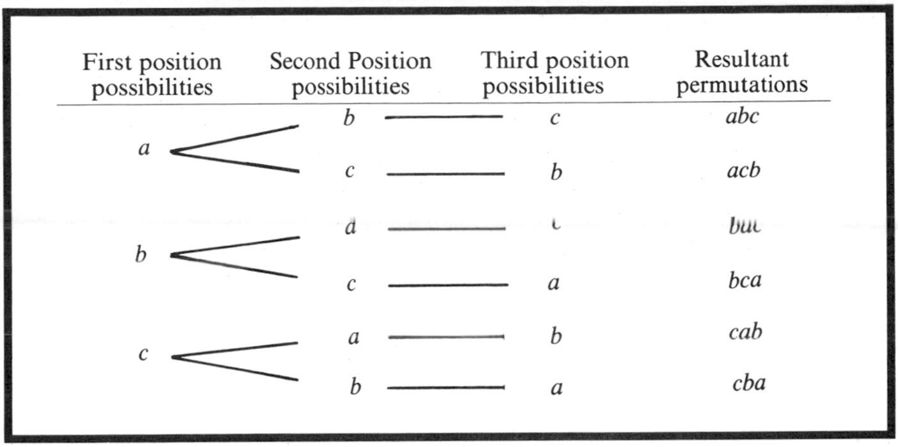

Figure 1 Generation of the $3 \times 2 \times 1 = 6$ permutations of the letters a, b, and c.

For convenience of expression this multiplication operation is designated with n followed by an exclamation mark

$$\boxed{n! = n \times (n-1) \times (n-2) \times \cdots \times 1} \tag{1}$$

and is read *n factorial*.

In summary, n distinct objects can be permuted, without repetition, in $n!$ ways—e.g., $5! = 5 \times 4 \times 3 \times 2 \times 1$, $4! = 4 \times 3 \times 2 \times 1$, $3! = 3 \times 2 \times 1$, $2! = 2 \times 1$, $1! = 1$, and, by definition, $0! = 1$, for reasons which will be apparent in the following sections.

Permuting *n* objects, *r* at a time. Sometimes we want to know how many ways we can permute some *subset* of n distinct objects; that is, instead of permuting all n objects, we want to know the number of permutations of r objects from among the total set of n. For example, in an election among $n = 7$ candidates, how many possibilities exist for the identities of the $r = 3$ top vote getters in rank order? Since three positions must be filled, we lay out a corresponding number of positions:

____ ____ ____

But now, any of the $n = 7$ candidates can fill the first slot, any of the remaining six could fill the second slot, and any of the remaining five could then fill the third slot, yielding

$$7 \times 6 \times 5 = 210$$

possible outcomes for the identities of the top three candidates, in order.

The above calculation can also be viewed as

$$\frac{7\times6\times5\times4\times3\times2\times1}{4\times3\times2\times1}=\frac{7!}{4!}=210$$

in which the 4! in the denominator cancels the last four terms of the numerator, leaving $7\times6\times5$, the original calculation. The advantage of viewing the above calculations in such a roundabout way is that it provides an understanding of the general formula for finding the *number of permutations of n distinct objects taken r at a time*, designated ${}_nP_r$, and defined as

$$\boxed{{}_nP_r=\frac{n!}{(n-r)!}} \tag{2}$$

For example, the number of ways of permuting *nine* distinct objects *five* at a time is given by

$${}_9P_5=\frac{9!}{(9-5)!}=\frac{9!}{4!}=\frac{9\times8\times7\times6\times5\times4\times3\times2\times1}{4\times3\times2\times1}=15{,}120$$

It is easily seen that the number of permutations accelerates rapidly as n increases.

Formula (2) is perfectly general for $r \le n$, and if we are interested in the number of permutations of n objects taken all n at a time we have

$${}_nP_n=\frac{n!}{(n-n)!}=\frac{n!}{0!}=\frac{n!}{1}=n!$$

which corresponds to our earlier definition of the number of ways in which n distinct objects can be permuted. We also see why it is convenient to define 0! as equal to 1.

Permutations with repetition. If there is no restriction on the number of times a given object can appear within a permutation, then n objects can form

$$n\times n\times\cdots\times n=n^r$$

permutations of length r.

For example, the letters of the alphabet, $n=26$, can create

$$26\times26\times26=26^3$$

or 17,576 permutations consisting of $r=3$ positions. This is to be compared

with the $26 \times 25 \times 24 = 15{,}600$ three-letter permutations in which a given letter cannot be used in more than one position, ruling out permutations such as *pup*, *did*, *see* or *mom*, which are possible in the first situation above.

Permuting *n* nondistinct objects. In the preceding sections we considered sets of distinct objects in which each object differed from every other one. Often, however, we must deal with sets of objects in which there is duplication among them; that is, the objects are not all distinct from one another. For example, the four letters of *work* are distinct, but the four letters of *wool* are not. The letters of *work* can be permuted in $4! = 24$ ways, but the four letters of *wool* can be permuted in something less than 24 arrangements, since the two *o*'s are indistinguishable. For instance, *work* and *wrok* are distinct permutations but wo_1o_2l and wo_2o_1l are not; where the subscripts have been added only to show the underlying identity of the two *o*'s.

In general, if a set of n objects consists of r_1 objects of one kind, r_2 of another kind, and so on to r_k of another, then the number of different permutations of all n objects is no longer $n!$, but

$$\boxed{{}_nP_{r_1, r_2, \ldots, r_k} = \frac{n!}{r_1! r_2! \ldots r_k!}} \tag{3}$$

where $r_1 + r_2 + \cdots + r_k = n$ For example, the number of different permutations of the ten letters of *statistics* is

$${}_{10}P_{3,3,2,1,1} = \frac{10!}{3!3!2!1!1!} = 50{,}400$$

where 3, 3, 2, 1, and 1 correspond to the number of *s*'s, *t*'s, *i*'s, *a*'s, and *c*'s, respectively. Compare this result with the over 3.6 million permutations of a word consisting of ten different letters.

It is easy to see, then, how an alphabet *abcdefghijklmnopqrstuvwxyz* can be permuted to generate libraries full of books. Also, it is not difficult to see how a small amount of genetic material can result in the wide variety within and between species.

3. Combinations

In contrast to the number of permutations of n objects taken r at a time is the number of *combinations* of n objects taken r at a time. With combinations the concern is only with the number of subsets of size r that can be constructed from the population of n objects, with *no concern* for the arrangement or order of the r objects. For example, *az* and *za* are two different

permutations of size $r = 2$ drawn from the alphabet, but they represent the same *combination*. Similarly, while 749, 794, 479, 497, 974, and 947 are all different permutations of the numbers 4, 7, and 9, all six represent the same combination, for which there is no concern for the arrangement of the objects.

Combining *n* objects, *r* at a time. Since any given set of r distinct objects can be permuted in $r!$ ways, it follows that the number of permutations of n objects taken r at a time must equal the corresponding number of combinations of those objects multiplied by a factor of $r!$, as given by the formula

$$ {}_nP_r = r!\,{}_nC_r \tag{4}$$

where ${}_nC_r$ stands for the number of combinations of n objects taken r at a time. That is, each combination of r objects generates $r!$ permutations. Alternatively, equation (4) can be written in its algebraically equivalent form

$$ {}_nC_r = \frac{{}_nP_r}{r!} \tag{5}$$

which yields

$$ \boxed{{}_nC_r = \frac{n!}{r!(n-r)!}} \tag{6}$$

when we substitute in (5) the definition of ${}_nP_r$ given by formula (2).

As an example of the application of the combinatorial formula, we can calculate the number of different committees of size $r = 3$ that can be formed from a group of $n = 7$ individuals:

$$ {}_7C_3 = \frac{7!}{3!(7-3)!} = \frac{7!}{3!4!} = 35$$

An alternative notation for ${}_nC_r$ that is commonly used is the placing of n over r and putting them in parentheses

$$\binom{n}{r} = {}_nC_r$$

Yet another notation is $C(n, r)$. Use of one notation or the other is purely a matter of preference.

Combinations of k classes of objects. Suppose there are n_1 distinct objects of one kind, n_2 of another kind, and so on to n_k of a kth kind, then

the number of combinations of size k will be:

$$n_1 \times n_2 \times \cdots \times n_k$$

For example, the number of different combinations of $n_1 = 50$ females with $n_2 = 40$ males is:

$$50 \times 40 = 2000$$

Or consider the number of combinations that exist for a package design based on one of $n_1 = 3$ sizes, one of $n_2 = 6$ colors, one of $n_3 = 2$ graphic designs, and one of $n_4 = 4$ shapes:

$$3 \times 6 \times 2 \times 4 = 144$$

Or suppose a researcher wants to study the effects of five different substances, singly and in combination. How many different preparations are possible. We can view this problem as consisting of $k = 5$ classes, with two possibilities per class—i.e., the prescense or absence of the substance. Then there will be

$$2 \times 2 \times 2 \times 2 \times 2 = 2^5 = 32$$

possible test combinations, but one of the 32 will consist of the absence of all five substances so there are really only 31 tangible possibilities. Notice that the result is the same as if we took the sum of ${}_5C_5$, ${}_5C_4$, ${}_5C_3$, ${}_5C_2$, ${}_5C_1$, and ${}_5C_0$. In general, then, if we have k distinct dichotomous classes, the number of combinations will be 2^k, with one of those combinations representing the "null" or empty set, in this instance a placebo condition.

4. Binomial Distribution

Suppose we have a binary population in which the probability of sampling one of the outcomes is p, and the probability of the other outcome is q. Since the two outcomes are mutually exclusive and exhaustive, the sum of p and q must necessarily equal 1, and again for the sake of convenience we can refer to the respective outcomes as a "success" and "failure".

If n independent samples are drawn from the above population, and if the value of p and q remain constant, the successive terms of the binomial

$$(p+q)^n$$

when expanded will yield the probabilities of n, $n-1$, $n-2, \ldots, 0$ successes; or equivalently, the probabilities of $0, 1, 2, \ldots, n$ failures.

Characteristics of the *binomial distribution* were discussed in Chapters 7, 8, and 9, and although the values of the individual terms of the binomial expansion were provided in tabular form, reproduced in Table 1, their derivation was never shown. Now that we have a background in permutations and combinations, that derivation can be accomplished.

If, for example, we are interested in how often the 1-dot face will turn up with $n = 5$ rolls of a die, the probabilities can be determined from terms of the expanded binomial

$$\left(\frac{1}{6}+\frac{5}{6}\right)^5 = \left(\frac{1}{6}\right)^5 + 5\left(\frac{1}{6}\right)^4\left(\frac{5}{6}\right) + 10\left(\frac{1}{6}\right)^3\left(\frac{5}{6}\right)^2 + 10\left(\frac{1}{6}\right)^2\left(\frac{5}{6}\right)^3 + 5\left(\frac{1}{6}\right)\left(\frac{5}{6}\right)^4 + \left(\frac{5}{6}\right)^5$$

where $p = 1/6$ is the probability of rolling a 1, and $q = 1 - p = 5/6$ is the probability of not rolling a 1 on a single roll or "experimental trial". The first term on the righthand side of the above equation represents the probability of all $n = 5$ rolls resulting in a 1; the second term the probability of exactly 4 of the 5 rolls turning up 1; the third term the probability of exactly 3 of the 5 rolls turning up 1; and so on to the final term which represents the probability of none of the rolls turning up 1.

To understand why the terms of the above binomial expansion should correspond to the probabilities in question, consider, for example, the

Table 1 The binomial distribution for selected numbers of trials.

Binomial expression	*Binomial expansion*
$(p+q)^1 =$	$p + q$
$(p+q)^2 =$	$p^2 + 2pq + q^2$
$(p+q)^3 =$	$p^3 + 3p^2q + 3pq^2 + q^3$
$(p+q)^4 =$	$p^4 + 4p^3q + 6p^2q^2 + 4pq^3 + q^4$
$(p+q)^5 =$	$p^5 + 5p^4q + 10p^3q^2 + 10p^2q^3 + 5pq^4 + q^5$
$(p+q)^6 =$	$p^6 + 6p^5q + 15p^4q^2 + 20p^3q^3 + 15p^2q^4 + 6pq^5 + q^6$
$(p+q)^7 =$	$p^7 + 7p^6q + \boxed{21p^5q^2} + 35p^4q^3 + 35p^3q^4 + \boxed{21p^2q^5 + 7pq^6 + q^7}$
↑ p: Probability of a success on a given trial. ↑ q: Probability of a failure on a given trial. ↑ exponent: Number of trials.	↑ $21p^5q^2$: This term is the probability of *exactly* 5 successes and 2 failures out of 7 trials. ↑ $21p^2q^5 + 7pq^6 + q^7$: The sum of these terms is the probability of getting 2 *or fewer* successes out of the 7 trials.

probability of getting exactly three "successes" out of five trials. One particular result corresponding to this outcome would be

$$SSSFF$$

where S stands for a success and F for a failure. Since the trials are independent, the probability of their joint occurrence will equal the product of their respective probabilities

$$pppqq = p^3q^2$$

But we know from our study of permutations, that five objects—three of which are one kind (success), and two of which are another kind (failure)—can be arranged in

$$_5P_{3,2} = \frac{5!}{3!2!} = 10$$

ways, based on formula (3). Alternatively, we can view this result as the number of ways three positions could be selected from the five, so as to be occupied by the three successes, and we have

$$_5C_3 = \frac{5!}{3!2!} = 10$$

possibilities, the same as before. So there are ten different ways that three successes and two failures can be arranged, each with a probability of p^3q^2. Consequently, the probability of three successes and two failures, disregarding their order of occurrence, will be

$$10p^3q^2$$

which corresponds to the third term of the binomial expansion for the die rolling example above.

In general, then, we can express a binomial probability as

$$\boxed{b(x;\, n, p) = \frac{n!}{x!(n-x)!}p^x q^{n-x}} \qquad (7)$$

where $b(x;\, n, p)$ is the binomial probability of getting *exactly* x successes from n trials in which the probability of success on a given trial is p, and the probability of failure is $q = 1 - p$. Now verify the terms in Table 1.

If the probability of x *or more* successes out of n binomial trials is desired, it is necessary to evaluate (7) not only for x but for each larger value up through n, and then to sum the respective probabilities. Appendix Table X saves this computational labor by providing the cumulative binomial probabilities for various values of x, n, and p.

Mean and variance. As developed in Chapter 7, the mean number of successes for a binomial distribution is

$$\mu = np \tag{8}$$

and the variance is

$$\sigma^2 = npq \tag{9}$$

For the die rolling example above, $\mu = 5(1/6) = .83$ and $\sigma^2 = .69$.

If we are interested in the proportion $\hat{p}$ rather than the number of successes, then the mean is

$$\mu_{\hat{p}} = p \tag{10}$$

and the variance is

$$\sigma_{\hat{p}}^2 = \frac{pq}{n} \tag{11}$$

For the die rolling example, $\mu_{\hat{p}} = 1/6$ and $\sigma_{\hat{p}}^2 = .028$.

Other applications of the binomial distribution, and the normal distribution approximation to it, are presented in Chapters 7, 8, and 9.

5. Hypergeometric Distribution

If we take repeated samples from a *finite* binary population without replacing the observations after each draw, the successive trials are no longer independent of one another as required by the binomial distribution. That is, a success on one trial will affect the probability of a success on a subsequent trial, since the population has only a finite number of successes to dispense.

Imagine, for example, a hat containing ten slips of paper numbered 1 though 10. Now consider two alternative ways of drawing a sample of say $n = 3$ numbers from the hat. One approach is *sampling with replacement* after each draw; that is, after a slip of paper is drawn and its number recorded, it is put back into the hat before the second draw is made, and then it too is replaced before the third draw is made. This sampling with replacement

procedure is equivalent to having an infinite population in which the probability of a success, say the drawing of a 7, remains constant from one sampling trial to the next, and the outcome of one trial in no way influences the outcome of another.

Compare the above with the procedure in which we sample *without replacement*. That is, we draw a number out of the hat, set it aside, draw a second number, set it aside, and draw a third number. Whereas in the sampling with replacement procedure, or with an infinite population, it would be possible to obtain three 7's in three trials, such a possibility would not exist when we sample without replacement, for there is only one 7 in our specified finite population. Similarly, the probabilities of any other outcomes would differ depending upon whether we sample with or without replacement after each trial draw.

When sampling without replacement from a finite population the probabilities can be determined by using the *hypergeometric distribution*, which takes into account the fixed number of "successes" present in the population. The following combinatorial formula provides the desired probabilities:

$$h(x;\, n, k, N) = \frac{\binom{k}{x}\binom{N-k}{n-x}}{\binom{N}{n}} \tag{12}$$

where $h(x;\, n, k, N)$ is the hypergeometric probability of drawing exactly x successes in a sample of size n, drawn without replacement from a population of size N in which there are k successes. The logic of the formula can be seen more clearly in Figure 2.

Consider, for example, the probability of drawing exactly $x = 3$ even numbers in a sample of $n = 4$ numbers drawn without replacement from the hat containing $N = 10$ numbers (1 through 10), $k = 5$ of which are even:

$$h(3;\, 4, 5, 10) = \frac{\binom{5}{3}\binom{10-5}{4-3}}{\binom{10}{4}} = \frac{\dfrac{5!}{3!2!}\cdot\dfrac{5!}{1!4!}}{\dfrac{10!}{4!6!}} = .238$$

Notice that the first term in the numerator is the number of ways that 3 successes could be drawn from the available 5, while the second term is the number of ways that 1 failure could be drawn from the available 5, and the denominator represents the number of different samples of size 4 that can be drawn from a population of size 10.

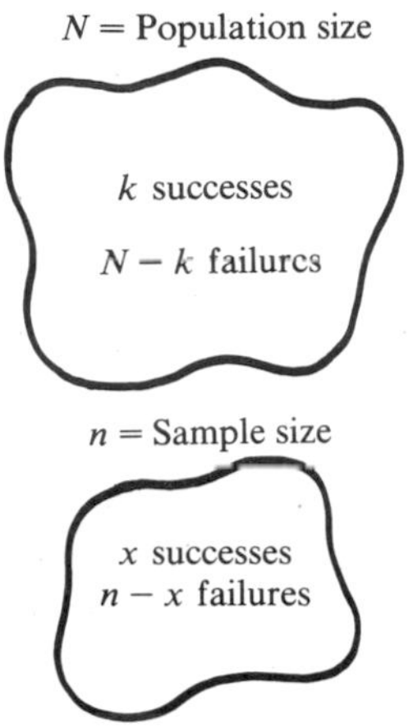

$\binom{k}{x} =$ ways of selecting x successes from among k available (1)

$\binom{N-k}{n-x} =$ ways of selecting $n-x$ failures from among $N-k$ available (2)

$\binom{k}{x}\binom{N-k}{n-x} =$ ways of selecting x successes and $n-x$ failures from the respective numbers available (1)×(2)

$\binom{N}{n} =$ number of samples of size n that can be selected from the population of size N (3)

$h(x; n, k, N) = \dfrac{\binom{k}{x}\binom{N-k}{n-x}}{\binom{N}{n}} =$ hypergeometric probability of getting x successes (and $n-x$ failures) in a sample of size n drawn from a population of size N containing k successes. $\dfrac{(1)\times(2)}{(3)}$

Figure 2 Derivation of the formula for determining hypergeometric probabilities.

The mean number of successes for the hypergeometric distribution is given by

$$\mu = n\left(\frac{k}{N}\right) \tag{13}$$

and the variance by

$$\sigma^2 = \frac{nk(N-k)(N-n)}{N^2(N-1)} \tag{14}$$

When the sample size n is small relative to the population size N, or as the population size approaches infinity, the hypergeometric distribution will approach the binomial distribution.

6. Poisson Distribution

The binomial and hypergeometric distributions are used to determine the probability of a certain number of successes in a set of n trials, in which either 0 or 1 successes can occur on each trial. There are situations, though, where we

are interested in the number of successes that occur over a given length of time, or throughout a given length, area, volume, or mass of material, in which it is possible for $0, 1, 2, \ldots, x$ successes to occur throughout each defined unit of measurement. Examples of this type of probability problem include the number of auto accidents in a given length of time, the number of defects in a given length of magnetic tape, the number of electrons emitted from a given area of a heated metallic plate, the number of fish in a given volume of sea water, the number of insect eggs in a given weight of wheat flour, or the number of murders on a given day.

If the occurrences of the event of interest are independent of each other, and if the probability of the event is proportional to the size of the interval of time, space, or mass in which it is observed, and if an interval exists with a low probability of more than one occurrence of the event, then we can use the *Poisson distribution* (pronounced *pwa-zahn*) to determine the probability of $0, 1, 2, \ldots, x$ successes in any given interval. Events which are likely to satisfy these conditions are rare events such as those named in the preceding paragraph—accidents, defects, murders, etc.

If we let x stand for the number of successes in which we are interested, and the Greek letter *lambda*, λ, stand for the expected or mean number of successes in the interval, then we have

$$\boxed{f_p(x;\ \lambda) = \frac{e^{-\lambda}\lambda^x}{x!}} \tag{15}$$

where $f_p(x;\ \lambda)$ is the Poisson probability of finding exactly x successes in the interval of measurement for which the expected number of successes is λ, and where e is an often encountered mathematical constant approximately equal to 2.7183...

As an example of the application of the Poisson distribution consider a governmental regulation that prohibits more than 2 fruitfly maggots per 100 grams of a particular consumer foodstuff. What is the probability that a random sample of 100 grams of the product will yield 3 or more such impurities, if in fact the population batch contains them at the rate of 2 per 100 grams. The Poisson probability

$$f_p(3;\ 2) = \frac{e^{-2}2^3}{3!}$$

would tell us the chances of finding exactly $x = 3$ maggots in the 100 gram sample, given that the expected number is $\lambda = 2$. But we want to know the probability for $x \geq 3$, so we would have to calculate the probability for values

of x equal to 3, 4, 5, etc., and sum those probabilities to obtain the overall probability that we desire. These calculations are unnecessary, though, since Appendix Table XI provides the cumulative Poisson probabilities for various values of x and λ. We find in this example that the probability of $x=3$ or more maggots is .323 for $\lambda=2$, so it follows that the probability of $x=2$ or fewer must be $1-.323$, or .677.

Implicit in the determination of λ in the preceding example was the evaluation of the relationship

$$\lambda = vu$$

where v is the rate of occurrence of the event per unit of measurement, and u is the number of such units being observed. For example, the presumed incidence of 2 maggots per 100 grams of the foodstuff corresponds to the rate v, and the observed sample of 100 grams corresponds to a u equal to 1 observational unit, yielding $\lambda = uv = (2)(1) = 2$. If, instead, we wanted to determine λ for a sample of 50 grams of the foodstuff, then $u=.5$, and we would have $\lambda = (2)(.5) = 1$. It is left as an exercise to verify that the probability of no maggots in a 50 gram sample is .368.

Binomial approximation. In addition to being used to find the probability of x successes within an observational unit of time, length, space, or mass, the Poisson distribution can also be used as a convenient approximation to the binomial distribution for determining the probability of x successes in a block of n independent trials, in which the block of n trials becomes analogous to an amount of time, space, or mass. The value of np is used for λ, and the resulting Poisson probability is a satisfactory approximation of the binomial probability provided np or nq are small, say less than 5—the condition for which the normal distribution is an unsatisfactory approximation to the binomial distribution.

For example, if the probability of success on one binomial trial is $p=.03$, what is the probability of 2 or fewer successes out of $n=100$ trials? A "success" in this example could represent a product with a defect, a response to a direct mail solicitation, a death from a routine medical operation, etc. Letting $np = \lambda$, we have

$$\lambda = np = 100(.03) = 3$$

expected successes for 100 trials. Substituting into the Poisson formula (15), we have

$$\frac{e^{-3}3^0}{0!}, \quad \frac{e^{-3}3^1}{1!}, \text{ and } \frac{e^{-3}3^2}{2!}$$

as the respective probabilities of exactly 0, 1, or 2 successes; which when summed provide the desired probability. Referring, instead, to Appendix Table XI we find the probability to be .423. Compare the ease of this procedure with the calculation of the binomial probabilities given by

$$\frac{100!}{x!(100-x)!}(.03)^x(.97)^{100-x}$$

which must be evaluated for $x = 0$, 1, and 2, since tabulated tables for values as high as $n = 100$ are not readily available.

An interesting feature of the Poisson distribution is that its mean

$$\mu = \lambda \tag{16}$$

and variance

$$\sigma^2 = \lambda \tag{17}$$

are equal to one another, each equal to the defining parameter λ.

In summary, the Poisson distribution is useful for studying the probabilities of rare events occurring over time, space, or mass; or for approximating binomial probabilities of rare events occurring over blocks of trials when the values of np or nq are less than 5, a condition for which the normal distribution approximation technique of Chapters 8 and 9 is unsatisfactory.

7. Bayes Formula

In Chapter 6 we learned that a conditional probability can be expressed

$$P(A|B) = \frac{P(A \text{ and } B)}{P(B)} \tag{18}$$

where $P(A|B)$ is the probability of outcome A, given outcome B. By way of review, Figure 3 presents a graphical example of the basic logic underlying the formula.

The importance of conditional probabilities lies in the fact that they can reflect the association between two random variables. We know, for example, that if two events, A and B, are independent of one another, then $P(A|B)$ will equal $P(A)$; that is, information about the occurrence of B in no way alters the probability of A. On the other hand, the extent to which $P(A|B)$ differs from $P(A)$ reflects the association or dependence between A and B, and information about the occurrence of B will help us to more accurately assess the probability of A under those conditions.

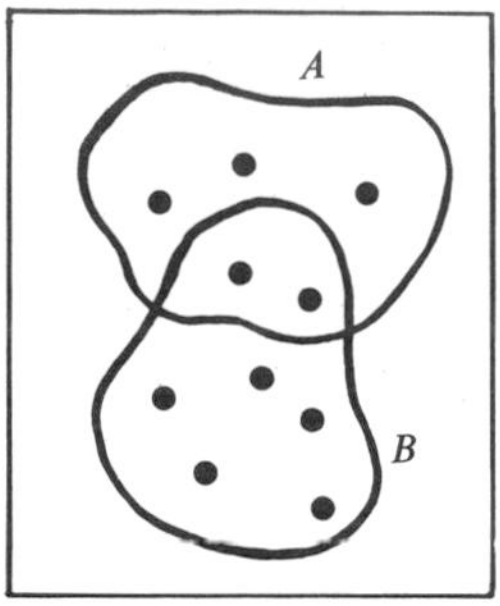

$$P(A) = \frac{5}{10}$$

$$P(B) = \frac{7}{10}$$

$$P(A \text{ and } B) = \frac{2}{10}$$

$$P(A|B) = \frac{P(A \text{ and } B)}{P(B)} = \frac{2/10}{7/10} = \frac{2}{7}$$

Figure 3 An illustration of the formula for a conditional probability, $P(A|B)$.

To more fully appreciate the usefulness of conditional probabilities, we will consider an example involving the association between color preference and personality. A large sample of individuals were asked two questions:

- Overall, which color do you like better: *blue* or *red*?
- Overall, which description fits you better: *introvert* or *extrovert*?

These two questions can be thought of as probability experiments, in which the answers are values of a random variable. Let us think of color preference as a predictor variable, call it v, with levels v_1 = "prefer blue" and v_2 = "prefer red." And let the personality self-description be a criterion variable, call it c, with levels c_1 = "introvert" and c_2 = "extrovert."

Now let us suppose that we want to know the probability that someone will describe themself as c_1 = introverted, given that their color preference is v_1 = blue. Simply changing the letter designations of formula (18) we have

$$P(c_1|v_1) = \frac{P(c_1 \textit{ and } v_1)}{P(v_1)} \tag{19}$$

as our desired probability. Now if we knew the probability of individuals describing themselves as introverted *and* preferring blue—i.e., $P(c_1 \textit{ and } v_1)$, the numerator of (19)—and if we also knew the overall probability of individuals preferring blue—i.e., $P(v_1)$, the denominator of (19)—then it would be a simple division operation to obtain the desired conditional probability. However, these probabilities may not be directly available. Rather,

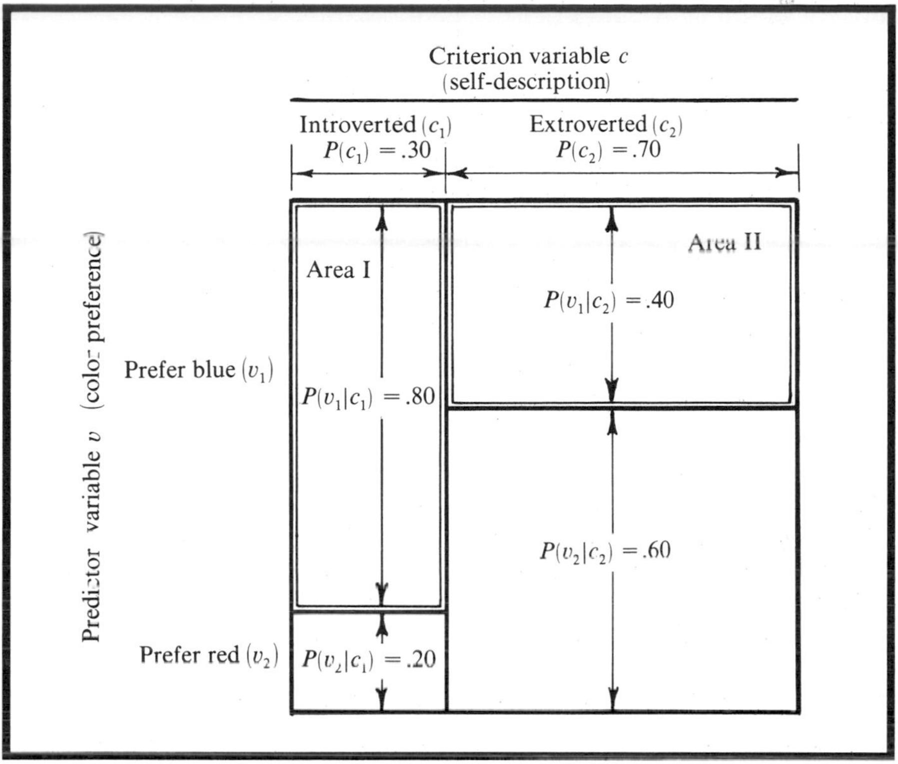

Figure 4 An example of input probabilities for the application of Bayes formula.

other probabilities which are available will have to be used to calculate the numerator and denominator of (19).

For the present example, suppose the only probabilities that are available are those presented in Figure 4; namely, the *unconditional* probabilities of being introverted vs. extroverted, and the *conditional* probabilities of preferring blue vs. red *given that the person is introverted vs. extroverted*. But the conditional probabilities in which we are interested are in the opposite direction—that is, we want to know the conditional probability of being introverted vs. extroverted given *a preference for blue vs. red*, not the other way around. The task is to convert the available probabilities into the desired probabilities.

Inspection of Figure 4 reveals that the upper left-hand cell, labeled Area I, represents $P(c_1 \textit{ and } v_1)$—the probability of being introverted *and* preferring blue—and is the value needed for the numerator of (19). Also, the sum of Area I and Area II represents $P(v_1)$, the overall probability of preferring blue, and is the value needed for the denominator of (19). Consequently, our desired

probability can be expressed in terms of the areas in Figure 4, where

$$P(c_1|v_1) = \frac{\text{Area } I}{\text{Area } I + \text{Area } II}$$

The areas of the rectangles in Figure 4 are easily calculated since their widths and heights correspond to the available probabilities, so we have

$$P(c_1|v_1) = \frac{P(c_1)P(v_1|c_1)}{P(c_1)P(v_1|c_1) + P(c_2)P(v_1|c_2)}$$

Substituting the numerical values from Figure 4 yields

$$P(c_1|v_1) = \frac{(.30)(.80)}{(.30)(.80)+(.70)(.40)} = .46$$

as the probability that a person is introverted, *given that they prefer blue to red*. Compare this conditional probability of .46 with the unconditional probability $P(c_1) = .30$ that a person will describe themself as introverted.

Similarly, we could calculate $P(c_1|v_2)$, $P(c_2|v_1)$, and $P(c_2|v_2)$. Thus, information about one variable (color preference) provides us with added information about another variable (personality). This technique will be recognized as similar in spirit to correlation and regression analysis involving the relationship between two continuous random variables.

The preceding example involved variables with only two values each, but this is not a limitation. In general, we can state

$$\boxed{P(c_l|v_k) = \frac{P(c_l)P(v_k|c_l)}{\sum_{i=1}^{n} P(c_i)P(v_k|c_i)}} \qquad (20)$$

where $P(c_l|v_k)$ is the probability of the lth value of variable c, given the occurrence of the kth value of variable v, and where c_i $(i = 1, 2, \ldots, n)$ represents the alternative values of c. This version of formula (19) is known as *Bayes Formula*, and is given more tangible meaning in Figure 5. When applied to the probabilities of Figure 4 it yields the probabilities given in Figure 6. Essentially, Figures 4 and 6 represent two alternative ways of looking at exactly the same data, with Bayes formula (20) being the means of transforming one set of probabilities to the other.

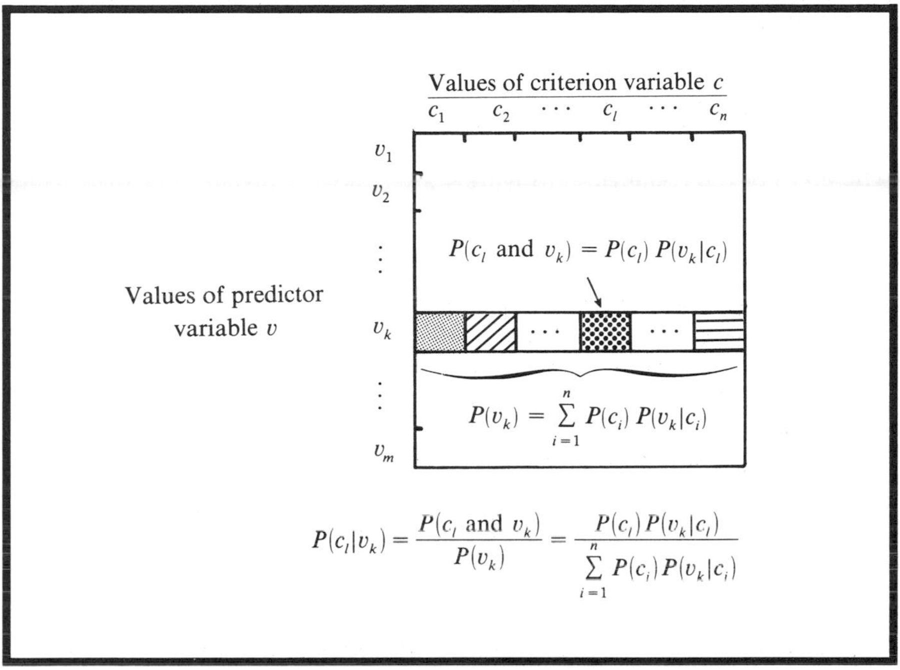

Figure 5 Depiction of the numerator and denominator of Bayes formula.

The conditional probabilities in Figure 4—i.e., the right-hand side of Bayes formula—are often referred to as *prior* probabilities; while those in Figure 6—i.e., the left-hand side of Bayes formula—are referred to as *posterior* probabilities, suggesting the order in which they are available.

Mathematical derivation. Bayes theorem (20) can also be derived from the definitional formula for conditional probabilities

$$P(c_l|v_k) = \frac{P(c_l \text{ and } v_k)}{P(v_k)} \tag{21}$$

by purely mathematical means, without recourse to the graphical device of Figure 4.

From (21) we can write

$$P(c_l \text{ and } v_k) = P(v_k)P(c_l|v_k) \tag{22}$$

or, simply exchanging the letter designations, we also have

$$P(v_k \text{ and } c_l) = P(c_l)P(v_k|c_l) \tag{23}$$

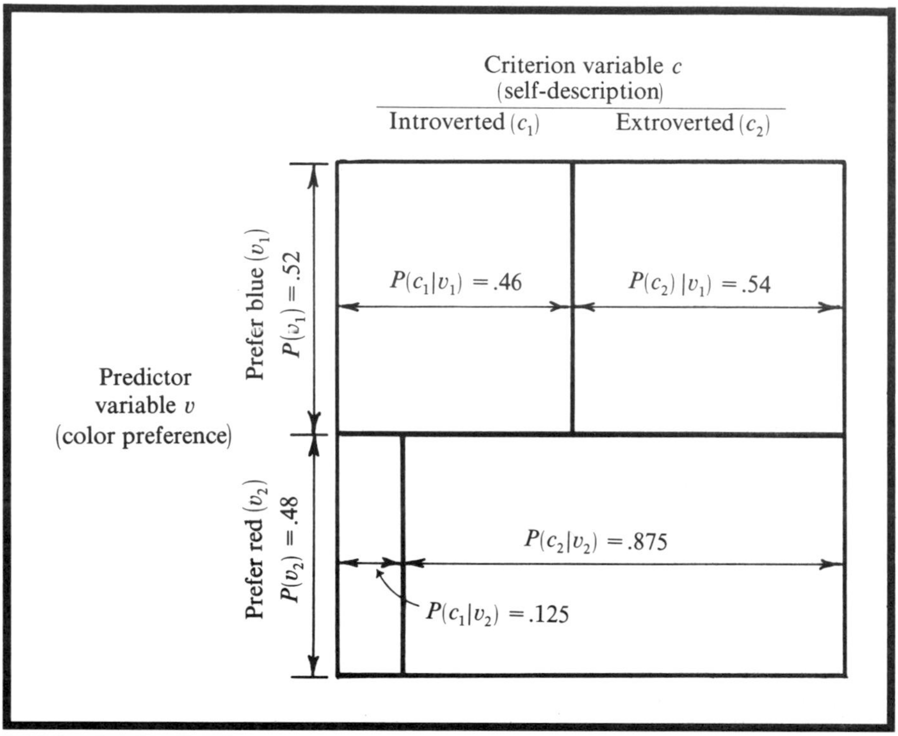

Figure 6 An example of output probabilities from the application of Bayes formula to the input probabilities of Figure 4.

But since

$$P(c_l \text{ and } v_k) = P(v_k \text{ and } c_l) \tag{24}$$

the right hand side of (23) can be placed in the numerator of (21), yielding

$$P(c_l|v_k) = \frac{P(c_l)P(v_k|c_l)}{P(v_k)} \tag{25}$$

Further, we know that the denominator $P(v_k)$ can be expressed as the sum of its joint probabilities of occurrence with the mutually exclusive and exhaustive values of c,

$$P(v_k) = P(v_k \text{ and } c_1) + P(v_k \text{ and } c_2) + \cdots + P(v_k \text{ and } c_l) + \cdots + P(v_k \text{ and } c_n) \tag{26}$$

Using (23) again, we can express the joint probabilities in (26) in their equivalent form

$$P(v_k) = P(c_1)P(v_k|c_1) + P(c_2)P(v_k|c_2) + \cdots + P(c_l)P(v_k|c_l) + \cdots + P(c_n)P(v_k|c_n) \tag{27}$$

or

$$P(v_k) = \sum_{i=1}^{n} P(c_i)P(v_k|c_i) \tag{28}$$

Substituting (28) into the denominator of (25) then yields

$$P(c_l|v_k) = \frac{P(c_l)P(v_k|c_l)}{\sum_{i=1}^{n} P(c_i)P(v_k|c_i)} \tag{29}$$

as stated earlier in (20).

It should be noted that v_k is only one of m values of variable v that could be entered into (29), as can be seen in Figure 5.

8. Concluding Comments

We have now expanded upon our understanding of the notion of probability first introduced in Chapter 6, addressing ourselves to the binomial, hypergeometric, Poisson, and Bayesian probabilities. Again, we learn that the nature of our data dictates the particular type of probability distribution that can be used to model the observations. While measures of central tendency, variation, and association can be viewed as the building blocks of statistical analysis, probability is the mortar that gives structure and meaning to them, singly and in combination, allowing inferences about our sampled populations and providing a deeper understanding of the world around us, the ultimate objective of statistical analysis.

Chapter 21

Decision Analysis

1. Introduction

Suppose we are travelling along, when suddenly we reach a fork in the road. We must decide whether to turn right or to turn left, uncertain of the fortunes or calamities that might befall us as a consequence of our decision. In the absence of any cues at the fork in the road informing us as to what we might expect upon either turn, we have no systematic basis on which to make a choice. If, however, we have some idea of the alternative consequences we might encounter, along with their respective values to us, and their probabilities of occurrence, then we can make a wise choice.

The above, in essence, is the paradigm or format for *decision analysis*, an analytical technique based on the theory that choice behavior can be modelled as a gambling game—that is, that the choice between alternative courses of action can be dictated by the values and probabilities of the alternative outcomes of our actions. In this chapter we will study the basics of the theory and try to assess its practical value.

2. Fundamental Concepts

Life is full of decisions—we must decide whether or not to modernize a factory; whether or not to build a new school; whether or not to perform a medical operation; whether to drill a well at site *A*, *B*, or *C*; whether to advertise a new product on TV, in magazines, or in newspapers; whether to choose mate *A* or *B*; whether to invest in stocks, bonds, or real estate; etc.

To the extent that the above types of decisions follow the format of gambling games—e.g., whether to bet red or black at the roulette wheel, whether to bet heads or tails on the flip of a coin, or whether to bet for or against a seven in a game of craps—then decision analysis provides a means of making optimal decisions.

There are basically *five* key concepts which form the foundation of decision analysis—the *decision alternatives*, the *outcomes* of those alternatives, the *outcome probabilities*, the *outcome payoff values*, and the *expected payoff values* of the decision alternatives. We will look at each of these concepts in turn.

Decision alternatives. Firstly, there are the *decision alternatives* or actions that we can take; for example, to bet heads or tails, to buy fire insurance or not, to invest in stocks or real estate, etc. For any given decision situation, there will be a minimum of *two* decision alternatives, but no limit on the maximum number. *To be or not to be*, that is a decision.

Outcomes. Secondly, each decision alternative is associated with a set of possible *outcomes* or future states. For example, whether we decide to buy fire insurance or not, each alternative is associated with possible outcomes of "fire" or "no fire." In fact, these outcomes will be recognized as the outcomes comprising the *sample space of a probability experiment*, as discussed in Chapter 6, where the decision we make corresponds to the probability experiment.

Outcome probabilities. Thirdly, each outcome has a *probability* of occurrence for each decision alternative. The probabilities of the various outcomes could, but need not be, the same for the various decision alternatives. For example, the probability of fire is the same whether we choose to insure or not. However, the probability of doubling our money is not necessarily the same if we choose to invest in stocks as opposed to real estate.

Outcome payoff values. Fourthly, each outcome, for each decision alternative, has a numerical *payoff value* to us. For example, if we choose not to buy fire insurance (a decision alternative) and it turns out that there is a fire (an outcome),we must assign a numerical value to that circumstance, just as we assign a numerical value to the outcome of a head when betting on a coin toss. The payoff value can be either positive or negative in value, and it may or may not be measured on a monetary scale. In gambling games and in many business settings the outcomes are measured in monetary terms, but this is not necessarily so. The payoff values could just as well be measured in numbers which represent prestige, health, frequency of smiling, number of friends, peace of mind, or whatever.

The important point is that the values of the outcomes are *ratio-scaled* for the individual(s) making the decision. For example, if the payoff value is measured in monetary terms, then a loss of $50,000 should be equivalent to ten times a loss of $5,000, which in turn should be equivalent to ten times a loss of $500. If not, some numerical scale which is meaningful to the decision maker in a ratio sense must be used.

The term *utility* is often used in place of payoff value so as not to suggest a monetary scale. However, since utility has a variety of other meanings, we will

use payoff value in a most general sense, in which it can be measured on a monetary or non-monetary scale.

Expected payoff value. Finally, each decision alternative, a_j, is associated with an *expected payoff value*

$$E(a_j) = \sum_{i=1}^{n} p_i x_i \tag{1}$$

which is the sum of the products of the payoff values x_i, and their respective probabilities p_i, for the $i = 1, 2, \ldots, n$ possible outcomes for a given decision alternative. This expected payoff value will be recognized as the mean of a probability distribution comprised of the various payoff values associated with the outcomes for a given decision alternative.

A decision, then, can be thought of as being a choice among probability distributions which differ in their expected value or mean, variance, range, high value, low value, or shape, any of which could serve as the criterion for choosing one distribution—decision alternative—over another. Some of the various strategies (and their names) include minimizing our maximum loss (*minimax*), maximizing our minimum gain (*maximin*), minimizing our minimum loss (*minimin*), and maximizing our maximum gain (*maximax*). However, as we will see, choosing the alternative with the highest *expected payoff value* is the most *rational* choice.

In the following sections we will look at these five basic concepts of decision analysis—decision alternatives, outcomes, outcome probabilities, outcome payoff values, and expected payoff values—in the context of some specific examples to see how they find practical application.

3. Decision Trees

It is often convenient to graphically portray a decision problem in the form of a branching *decision tree* diagram as in Figure 1. There, the decision is whether or not to pay $2,000 for a fire insurance policy on a remote and vacant warehouse valued at $100,000.

Looking at Figure 1 from left to right, the first branch formation represents the two decision alternatives—*insure* or *do not insure*. The next branch formation represents the uncertain outcomes—*fire* or *no fire*—and their respective probabilities which we have come to believe to be .01 and .99, respectively.

Beside each outcome is listed its numerical payoff value measured in dollars. For the "insure" decision alternative, the payoff value of a "fire" outcome and a "no fire" outcome are each a negative $2,000—the cost of the

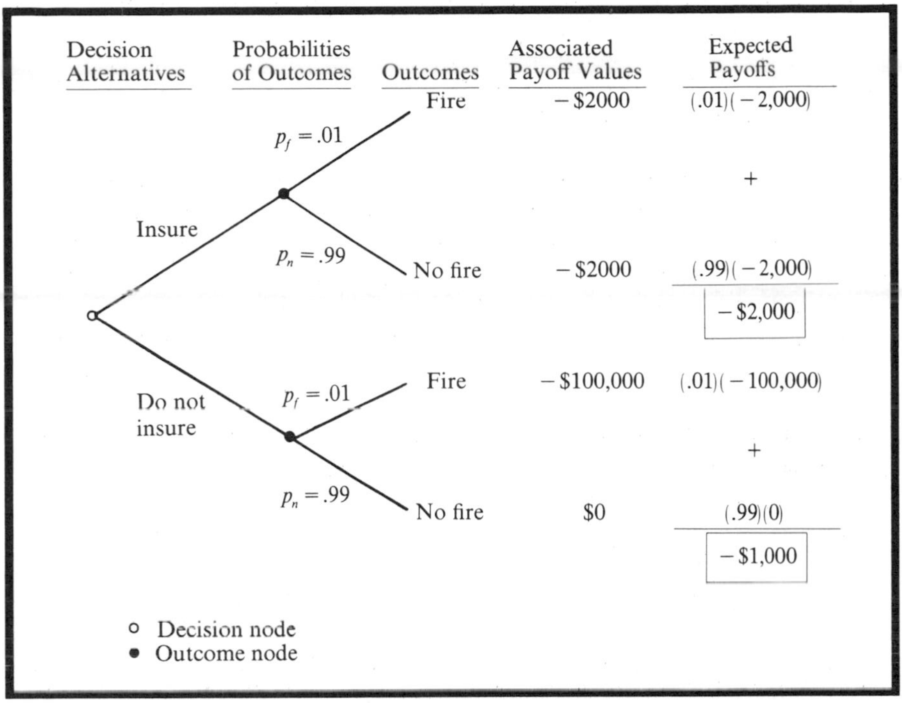

Figure 1 Decision tree for deciding whether to pay $2,000 for a fire insurance policy on a $100,000 ware house.

insurance policy—since the insurance company will replace our warehouse in the event of a fire. However, for the "do not insure" decision alternative, the payoff value of a "fire" outcome is a negative $100,000, the value of the warehouse. The payoff value for the "no fire" outcome, on the other hand, is $0; that is, we have spent no money for an insurance policy and we have lost no money due to fire. Notice that we use an open circle to represent the node of a *decision branch* and a solid circle to represent the node of an *outcome branch*.

Given the above scenario, the decision is whether to pay $2,000 for an insurance policy or not. In the long run, the optimum decision will be the one with the higher *expected payoff value*. The calculation of these expected payoffs is shown in the extreme right of Figure 1. For the "insure" decision alternative, multiplying the payoff values of the two possible outcomes by their respective probabilities, we have

$$.01(-2{,}000) + .99(-2{,}000) = -\$2{,}000$$

as the expected payoff—a two thousand dollar loss. For the "do not insure" decision alternative, the payoff values of a fire and no fire, and their respective probabilities, yield

$$.01(-100{,}000)+.99(0)=-\$1{,}000$$

as the expected payoff—a one thousand dollar loss. Consequently, since we prefer to lose $1,000 rather than $2,000, our wise choice would be to *not insure* the warehouse. While there is a chance that we could lose $100,000 due to a fire, our best bet in the *long run* is to be guided by the expected value of a distribution and not by its variance or highest or lowest possible value, which some decision makers choose to do, and in this case would lead them to pay for the insurance policy. In so doing, they could lose a maximum of $2,000, but in the long run, over many decisions, such a strategy will lose out to the strategy of choosing the alternative with the highest expected payoff value. Of course, the person with one warehouse might see things differently than the person with 1,000 warehouses.

The above example was relatively simple in that there were only two possible outcomes—fire or no fire—for each decision alternative, and their probabilities were independent of the decision. A somewhat more complex situation would be the decision between investing $50,000 in stocks or real estate. In this instance there are an *infinite* rather than a discrete number of outcomes for each decision alternative, in that the investments could decrease or increase by any number of amounts. The resulting continuous probability distributions of the payoffs for each decision alternative would be virtually impossible to ascertain. Rather, we could create a discrete probability distributions by partitioning the possible returns on investment into a convenient number of intervals for which we could estimate probabilities. The decision problem is shown in the decision tree of Figure 2. By one means or another we have come to believe that there is a .60 probability of doubling our investment by investing in stocks, a .30 probability of losing half our investment, and a .10 probability of staying even. For the real estate investment, however, we believe there is only a .30 probability of doubling our investment, but also only a .10 probability of losing half our investment, and .60 probability that it remains the same. Given the choice between the two, should we invest in the stocks or in the real estate? A comparison of the expected payoff values of the two alternative investments is shown in the right hand portion of Figure 2. We see that the expected payoff value from investing in the stocks is $22,500 compared to only $12,500 for the real estate investment, so given that our assessed probabilities are accurate, the stocks represent the best investment in this situation.

It should be apparent from the above examples that the validity of a decision analysis rests squarely on the accuracy with which we can estimate the

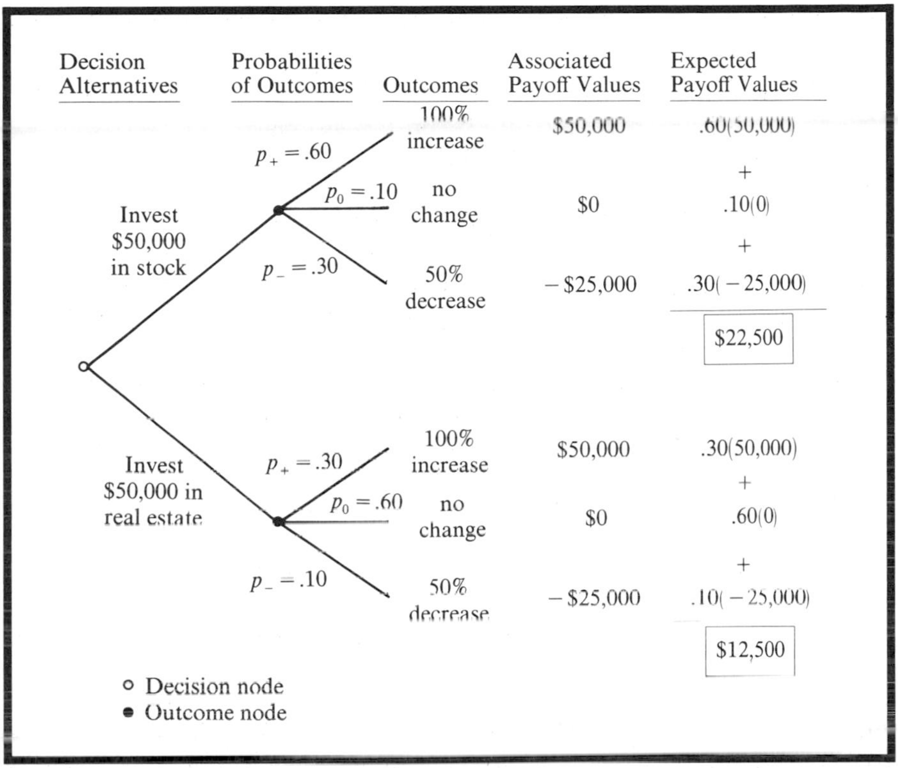

Figure 2 Decision tree for deciding whether to invest in a stock or real estate offering.

probabilities of the various outcomes and their respective payoff values. This is easy enough in most gambling games, but once we depart from such structured situations the probabilities and payoff values of the outcomes are not likely to be so well-defined.

4. Payoff Tables

Another way to portray a decision analysis problem is in the form of a *payoff table*, which, like a decision tree, shows the particular decision alternatives, the possible outcomes, the payoff values of the outcomes, and their respective probabilities.

Table 1 is the payoff table for the fire insurance decision problem portrayed in Figure 1. The rows of the table represent the various decision alternatives—insure vs. do not insure; the columns represent the alternative outcomes—fire vs. no fire; the cell entries represent the payoff values of the

Table 1 Payoff table for deciding whether to pay $2,000 for a fire insurance policy on a $100,000 warehouse. (See Figure 1).

Decision alternatives	Outcomes: *Fire*		Outcomes: *No fire*		*Expected payoff*
	payoff	probability	payoff	probability	
Insure	$2,000	.01	– $2,000	.99	– $2,000
Do not insure	– $100,000	.01	$0	.99	– $1,000

outcomes for the various decision alternatives; and the upper right-hand corner of each cell shows the probability of the given payoff.

Similarly, Table 2 is the payoff table for the stock vs. real estate investment problem portrayed in Figure 2. Notice that in this example the set of payoff values are the same for the two decision alternatives, whereas the probabilities of the payoffs differ. Compare this with Table 1 in which the probabilities of the outcomes are the same for the two decision alternatives, but the payoffs differ.

In the most general payoff table, as portrayed in Table 3, the sets of payoff values and their associated probabilities can differ from one decision alternative to another. Also, we see that the number of decision alternatives and outcomes is unlimited, provided there are at least two of each. In the above examples we were restricted to choosing only one of the available decision alternatives, but this need not be the case. In a problem involving many decision alternatives we might want to choose *all* those with a positive

Table 2 Payoff table for deciding whether to invest in a stock or real estate offering. (See Figure 2).

Decision alternatives:	Outcomes: *100 % increase*		Outcomes: *No change*		Outcomes: *50 % decrease*		*Expected Payoff Values*
	payoff	probability	payoff	probability	payoff	probability	
Invest $50,000 in stock	$50,000	.60	$0	.10	– $25,000	.30	$22,500
Invest $50,000 in real estate	$50,000	.30	$0	.60	– $25,000	.10	$12,500

Table 3 General format of a payoff table.

Decision alternatives	The uncertain *outcomes* (o), and their respective *probabilities* (p) and *payoff values* (x), for each *decision alternative* (a). o_1	o_2	$\cdots$	o_n	*Expected payoff values*
alternative a_1	x_{11} (p_{11})	x_{12} (p_{12})	$\cdots$	x_{1n} (p_{1n})	$p_{11}x_{11} + p_{12}x_{12} + \cdots + p_{1n}x_{1n}$
alternative a_2	x_{21} (p_{21})	x_{22} (p_{22})	$\cdots$	x_{2n} (p_{2n})	$p_{21}x_{21} + p_{22}x_{22} + \cdots + p_{2n}x_{2n}$
$\vdots$		$\vdots$			$\vdots$
alternative a_k	x_{k1} (p_{k1})	x_{k2} (p_{k2})	$\cdots$	x_{kn} (p_{kn})	$p_{k1}x_{k1} + p_{k2}x_{k2} + \cdots + p_{kn}x_{kn}$

expected value. For example, in the stock vs. real estate example, both alternatives yielded a positive expected value, albeit one larger than the other, so although choosing both alternatives would water down our expected gain from choosing the better investment, at least we would know that our expected payoff was positive. On the other hand, we might confront situations in which all alternatives have a negative expected payoff, a "no-win" situation, and if we had the option we would choose none of the alternatives.

5. Practical Value of Decision Theory

For games of chance, where the payoff values of the outcomes and their respective probabilities are well-defined, or potentially so, the principles of decision theory are indispensible. However, once we depart from well-structured gambling games, the *direct* benefit and validity of decision analysis becomes very tenuous.

The main reason for the limited practical value of decision analysis is our difficulty in accurately assessing the *probabilities* of the outcomes of interest. In gambling games the probabilities are usually *objective* in nature, based on logical considerations. In non-gambling risk situations, however, the probabilities are *subjective* in nature, their derivation known only to the decision maker and not replicable by impartial outside observers. As such, decision analysis is more of an art than a science, with the accurate assessment of probabilities varying from one individual to another.

There is the further problem of accurately assessing the *payoff values* of the outcomes. When betting on the color red at a roulette wheel, the gambler

knows ahead of time the exact payoff if red should occur. When buying a corporate stock, on the other hand, its value after a designated time period cannot be specified beforehand.

The above problems are compounded by the fact that the outcome payoff values and their respective probabilities form a *time series*, and their values could well differ between the time the decision is made and the time for the outcome to occur. And we know from our study of Time Series Analysis that forecasting any but the most regular time series is itself another art.

Finally, there is the problem that some individuals *will not act* upon the dictates of a decision analysis, choosing instead alternatives that do not maximize the expected payoff, but rather minimize one's losses in the short run, or perhaps are unable to make any decision at all. Thus it takes "guts" to maximize one's long term return on investment, a human parameter not accounted for by the theory.

There is, however, an *indirect* benefit of decision analysis which lies in its *heuristic* function—serving as a goad to thought, hypothesis construction, and exploratory behavior, which in the end probably will help to solve our problems in a round-about way. To expect more from the approach is being overly optimistic.

The above heuristic value of decision analysis can best be seen in its application to non-economic forms of behavior. For example, we might model various forms of criminal behavior as a decision analysis problem, in which the probabilities of the outcomes and/or their payoff values for the criminal differ from the cultural norm. Similarly, paranoia, phobias, and other forms of psychotic and neurotic behavior might be explained as being based on inaccurate perceptions of outcome probabilities and/or inordinate valuations of the outcomes. For example, the fear of flying might be based on an inaccurate assessment of the probability of a plane crash, or on a low valuation placed on the speed of arriving at one's destination, etc.

That the theory can be applied to virtually any type of behavior that you can name, should give us pause in evaluating its practical utility. More relevant, perhaps, is the study of how individuals differ in their assessment of probabilities and payoff values, and their ability to *act* on those evaluations. Who among us has not suddenly changed our assessed probabilities and payoff values of alternative outcomes just prior to the time for acting on alternative *A* vs. *B*. And why do two individuals equally adept at assessing probabilities and outcome values, differ in their ability to act on those assessments?

So, it is a long way from applying decision analysis to structured gambling games to its application to non-structured everyday situations. Still, the essential concept of decision analysis is appealing, and provided it is not taken too seriously when we are not absolutely certain of the probability and payoff values, it is a convenient framework upon which to organize one's thoughts prior to making decisions.

6. Concluding Comments

We have only looked at the very basic concepts of decision analysis—decision alternatives, outcomes, outcome payoff values, outcome probabilities, and expected payoff values. There are literally dozens of other extensions to the technique, which, while intellectually stimulating, do not always add to our basic understanding of the approach or its practical value. Also, it is difficult to do justice to the advanced techniques with a simple introduction. Rather, they should be pursued in extended and advanced courses of study. For our present purpose, we recognize it as a logical method for making decisions, but difficult to implement, its main benefit lying in its heuristic function of making us gather information and generating various scenarios which we can then evaluate more systematically, if only on an informed judgmental basis. As always, there is no substitute for common sense.

This concludes our introduction to statistical analysis, and above all we should leave with an appreciation that all statistical techniques revolve around the primary concepts of *objects*, *variables*, and *scales*; the building block concepts of *central tendency*, *variation*, and *association*; the integrating notion of *probability*; and the objectives of *data reduction*, *inference*, and the *identification of relations among objects and variables*. And, if through the application of these concepts and principles, the world is viewed differently and more meaningfully than before, ultimately leading to more fulfilling lives for all, then our original objectives will have been achieved.

APPENDIX

Exercises

Chapter 1: THE NATURE OF STATISTICAL ANALYSIS

1. State a general definition of the field of statistical analysis.
2. Name as many academic disciplines as you can, in which statistical analysis could be used.
3. Choose one of the disciplines named in exercise 2, and name three different examples of data that might be collected.
4. a. Name three broad objectives of statistical analysis. Do the objectives differ by discipline? Give examples.
   b. What is the ultimate objective of statistical analysis?
5. Explain the distinction between the following pairs of concepts, and give an example of each:
   a. population and sample
   b. parameter and statistic
   c. descriptive and inferential statistical analysis
6. What is meant by a random sample?
7. Explain the importance of drawing a random sample in an inferential statistical analysis.
8. Search the daily newspaper for an article which reports the results of a statistical analysis. Explain whether the sample used was a random one. Which of the objectives of statistical analysis did the study seek to fulfill?
9. What do we mean when we say a procedure is "well-defined"? Is the selection of a jury a well-defined procedure? Explain. Is it a random selection?
10. Discuss the key advantages and disadvantages of using computers to analyze data.
11. a. Do the voters in a national election represent a random sample of eligible voters? Explain.
    b. Are the winners of a lottery a random sample of the entrants?
12. It has been said by some that *all* statistical analyses are ultimately inferential in nature. How do you interpret this statement? Do you agree with it? Explain, giving examples.

Chapter 2: OBJECTS, VARIABLES, AND SCALES

1. a. Explain the meaning of the term "objects" as used in the text.
   b. Name five major classes of objects, and give a specific well-defined example of each.
2. For each of the following disciplines, give an example of the objects that might be studied.

Business	Psychology
Government	Agriculture
Medicine	Communications
Advertising	Economics
Literature	Sociology
Political Science	Anthropology
Archaeology	Markct Research
Engineering	Physics
Military	Education
Biology	Chemistry

3. a. In 25 words or less, give a general definition of a variable.
   b. What are some synonyms for a variable?
4. a. What do we mean by the values of a variable? Give an example.
   b. What is a synonym for the values of a variable?
5. Is there a limit on the number of values for a variable? If so, explain.
6. For each of the disciplines in exercise 2, give an example of a variable that might be of interest to a researcher. In doing so, also name the class of objects upon which the variable is measured.
7. What is the distinction between a variable and a scale?
8. a. Name and define the four types of scales discussed in the chapter.
   b. Give an example of each type of scale.
   c. Name a variable that can be defined on more than one scale, and identify two of the scales.
9. Distinguish between "metric" and "non-metric" scales, and give examples.
10. Define the property of transitivity, and name which types of scales have the property. Give specific examples of a scale which does, and one which does not, have the property.
11. Name the type of scale used to measure each of the following variables:
    a. Temperature (in °F)
    b. Temperature (in °C)
    c. Book sales (ranked on best-seller list)
    d. Book sales (in terms of dollar sales)
    e. Book sales (in terms of number sold)
    f. Toothpaste usage (brand used last)

g. Toothpaste usage (tubes per year)
h. Political affiliation (party registration)
i. Academic standing in a class (ranking)
j. Academic standing in a class (grade point average)

12. What is the distinction between a qualitative and quantitative variable? Give an example of each.

13. What is the distinction between a discrete and continuous variable? Give an example of each.

14. Give an example of a discrete quantitative variable.

15. What distinction have we placed between a dichotomous and a binary variable? Give examples of each.

16. What would you guess a "trichotomous" variable to be? Give an example.

17. Discuss the meaning of a dummy variable, giving examples.

18. What is the distinction between a variable and a constant? Give examples of each.

19. If x_i refers to the ith observation in the following set of values of variable x:

15 29 8 46 31 19 25

a. What is the value of x_3?
b. What is the value of x_7?
c. What is the value of $x_1 + x_4 + x_7$?
d. What is the value of $(x_4 - x_3) \div 2$?
e. What is the sum of all x_i, where i ranges from $i = 1$ to $i = 7$?

20. If x_{ij} designates the value in the ith row and jth column of the following matrix:

15	3	21	10
6	5	17	3
24	11	29	9

a. What is the maximum value of i?
b. What is the maximum value of j?
c. What is the value of x_{22}?
d. What is the value of x_{24}?
e. What is the value of x_{31}?
f. What is the value of $x_{12} + x_{34}$?
g. What is the value of $(x_{11} + x_{21} + x_{31}) \div 3$?
h. What is the value of $(x_{23})(x_{32})$?
i. What is the sum of all x_{ij}, such that $j = 2$ and i ranges from $i = 1$ to $i = 3$?
j. What is the sum of all x_{ij}, such that $i = 3$ and j ranges from $j = 1$ to $j = 4$?
k. What is the maximum value of x_{2j}?
l. What is the maximum value of x_{i2}?

21. Discuss the relative merits of using a 0 to 10 rating scale as opposed to a 1 to 10 scale.

Chapter 3: FREQUENCY DISTRIBUTIONS

1. Below are record high temperatures (in °F) for fifty weather stations:

112	112	109	110	111
100	108	106	116	120
127	118	112	108	113
120	117	114	109	120
134	116	115	121	116
118	118	118	113	105
105	121	117	120	110
110	114	118	119	118
106	114	122	111	112
109	105	106	102	114

   a. Construct a frequency distribution for the data (ungrouped).
   b. Draw a histogram of the distribution, labeling the axes and giving the figure an appropriate title.
   c. In a separate figure, draw a frequency polygon of the same data.
   d. Which record temperature occurred most often?
   e. What is the spread between the highest and lowest record temperature?

2. a. Using the data of exercise 1, construct a frequency distribution with intervals two units wide.
   b. Draw a histogram of the distribution created in part *a*.
   c. Superimpose a frequency polygon onto the histogram formed in part *b*.
   d. Which interval contains the most record temperatures? What is the midpoint of that interval? How does it compare with the answer to exercise 1*d*?

3. a. Using the ungrouped frequency distribution formed in exercise 1*a*, create a cumulative frequency distribution.
   b. Graph the cumulative frequency distribution formed in *a*.
   c. Which record temperature was exceeded by 50% of the stations?
   d. Which record temperature was exceeded by 25% of the stations?

4. Is the analysis of the data in exercise 1 an instance of descriptive or inferential statistical analysis? Explain.

5. a. Using the ungrouped frequency distribution created in exercise 1*a*, form the relative frequency distribution.
   b. Graph the distribution found in *a*.
   c. What percentage of the stations had a record temperature of 111° F?
   d. What percentage of the stations had a record temperature of 115° F?

6. a. Create the cumulative relative frequency distribution of the data in exercise 1 (ungrouped).
   b. Graph the distribution formed in *a*.
   c. What percentage of the stations had a record temperature exceeding 111° F?
   d. What percentage of the stations had a record temperature exceeding 115° F?
   e. Which record temperature was exceeded by 90% of the stations?
   f. Which record temperature was exceeded by 10% of the stations?

7. a. Using the data of exercise 1, create a grouped frequency distribution with intervals three units wide.
   b. Graph the distribution found in *a* as a histogram.
   c. Superimpose a frequency polygon on the histogram found in *b*.
   d. Using the same intervals as in *a*, form the relative frequency distribution and the cumulative relative frequency distribution.
   e. What is the midpoint of the interval containing the most record temperatures? How does this value compare with the answers to exercises 1*d* and 2*d*?

8. How would you describe the shape of the distribution of record temperatures?

9. The following is part of a grouped frequency distribution:

Interval	*Frequency*
⋮	⋮
60 – 63	15
64 – 67	19
68 – 71	22
⋮	⋮

   a. What size intervals are being used?
   b. In which interval would a value of 63.2 fall?
   c. What are the real boundaries of these intervals?
   d. What are the midpoints of these intervals?
   e. What would the next interval be?
   f. How would you redefine the interval scheme so that they had a midpoint that was a whole number?
   g. Assuming that the 19 values that fall in the 64 – 67 interval are evenly spread throughout the interval, what would you estimate the value of the 11th largest observation in the interval to be? What about the 5th largest value? (Hint: It will help to sketch the interval and imagine that the 19 observations are spread throughout it).

10. Below are the enrollment figures for a sample of colleges:

3527	310	400	992	760	1751
600	1225	1726	1275	1101	3299
148	105	1250	1833	336	2188
2041	7910	2583	1300	1363	2081
6516	1571	971	72	454	942
900	25,694	1888	1314	5987	1142
566	510	1850	260	1500	4207
3601	6627	700	1648	1831	1058
3932	2520	801	772	2071	1946
575	390	230	3222	4934	1206
2166	3159	1173	16,002	617	5507

   a. Decide upon an interval scheme and form a grouped frequency distribution of the data.
   b. Make a histogram of the distribution.
   c. How would you describe the distribution in terms of its shape?

11. a. Using the distribution formed in exercise 10*a*, form the relative frequency distribution, and the cumulative relative frequency distribution.
    b. What percentage of the colleges have enrollments greater than 1500?
    c. What enrollment figure is exceeded by 10% of the schools in the sample?
    d. What enrollment figure is exceeded by 50% of the schools?
    e. Which interval contains the most school enrollments? What are the boundaries of the interval? What is the midpoint of the interval?

12. If you were told that the school enrollments shown in exercise 10 were for United States colleges beginning with the letter *A*, would this be a random sample of all United States colleges? Explain.

13. Sketch a smooth curve which approximates the shape of the distribution found in exercise 10. How would you describe this distribution?

14. Sketch distributions fitting these descriptions:
    a. skewed to the right
    b. skewed to the left
    c. symmetrical
    d. rectangular
    e. bell-shaped
    f. *J*-shaped (skewed right)
    g. *J*-shaped (skewed left)
    h. *U*-shaped
    i. triangular

15. A sample of individuals was asked to name their primary concern in national affairs. The responses were coded into the following frequency distribution. Make a pie chart of the results, reporting the percentage of each response category.

Primary concern	*Frequency of mention*
Economy	533
Crime	319
Nuclear war	243
Declining morals	210
Drug use	127
Education system	87
Other	105

16. Explain why a pie chart is more appropriate than a frequency polygon for the data above.

Chapter 4: CENTRAL TENDENCY

1. What is meant by the notion of central tendency? Location?
2. Name and define three alternative measures of central tendency.
3. Use mathematical notation to define the mean of a sample of n values of a variable x.
4. What symbol is typically used to designate the mean of a population of values of a variable x?

5. For the following sample of data:

16 11 19 8 7 12 10 18 10

a. What is the modal value?
b. What is the median value?
c. What is the mean value?

6. For the following values of variable x:

2.3 2.9 1.7 2.9 3.3 1.4 2.6

a. What is the modal value?
b. What is the median value?
c. What is the mean value?

7. Find the mean of these binary values:

0 0 1 0 1 1 0 1 0 0 0 0

8. What proportion of the values in exercise 7 are equal to 1? How does this compare to your answer to exercise 7?

9. Given the following frequency distribution:

x	f
0	15
1	22
2	29
3	37
4	34
5	27
6	19
7	9

a. What is the modal value?
b. What is the median (to the nearest whole number)?
c. What is the median (to the nearest tenths decimal place)?
d. Calculate the mean.

10. Calculate the mean of the following binary variable x.

x	f
0	129
1	62

11. What proportion of the binary values in exercise 10 are equal to 1? How does this compare with the answer to exercise 10?

12. Make a sketch of a bimodal distribution.

13. If the median is greater than the mean of a distribution, what would you guess the shape of the distribution to be?

14. Given the following values of x and their associated weights w calculate the weighted mean.

x	14	9	3	11	10	6
w	1.1	.7	.9	1.0	1.4	1.2

15. Given the following values of y and their respective weights w calculate the weighted mean of y.

y	56	74	69	48
w	97	109	111	90

16. If $\bar{x} = 15.5$ is based on $n = 50$ observations, what is the value of Σx?

17. If $\bar{y} = 147.6$ is based on $n = 95$ observations, what is the value of Σy?

18. If a sample of observations yield $\bar{x} = 26.3$ and we know that $\Sigma x = 1236.1$, what must the sample size n equal?

19. Given the following values of variable x:

11 14 9 16 5

a. Calculate the mean.
b. Add 3 to each value and recalculate the mean. How does it compare with the answer to a?
c. Subtract 7 from each of the values, and recalculate the mean. How does it compare with the answer to a?
d. Multiply each value by 3, and recalculate the mean. How does it compare with the answer to a?
e. Divide each value by 5, and recalculate the mean. How does it compare with the answer to a?

20. Given the following values of x:

22 38 29 31

a. Calculate the mean.
b. Now subtract the mean from each value and recompute the mean.
c. Would the answer to b be the same regardless of the set of data we used? If not, provide an example where the answer is different.

21. For each of the following distributions, which would you expect to be greater in value, the mean or the median? Explain.
a. Scores on a difficult exam.
b. Scores on an easy exam.
c. Personal income.
d. House prices.

22. Sketch a distribution skewed to the right, and indicate the approximate locations of the mode, median, and mean.

23. Given the following grouped frequency distribution:

x	f
20-24	1
25-29	2
30-34	4
35-39	10
40-44	16
45-49	19
50-54	17
55-59	14
60-64	9
65-69	6
70-74	3
75-79	1

a. What is the modal value?
b. In which interval does the median fall?
c. Interpolate to find the value of the median to one decimal place.
d. What is the midpoint of the 20-24 interval? The 25-29 interval?
e. Calculate the mean, assuming the values of the variables are evenly distributed throughout each interval.

24. Letting M_x stand for the mean of a variable x, and c stand for any given constant value, complete the following expressions:
a. $M_{x+c} =$
b. $M_{x-c} =$
c. $M_{cx} =$
d. $M_{x/c} =$

25. If the mean of a variable x is equal to 27, what is the mean of a variable z defined as follows:
a. $z = x - 3$
b. $z = x + 7$
c. $z = x \div 9$
d. $z = 5x$
e. $z = 5x - 3$
f. $z = 5(x - 3)$
g. $z = x - \bar{x}$
h. $z = \dfrac{x - \bar{x}}{5}$
i. $z = \dfrac{x - \bar{x}}{5000}$

Chapter 5: VARIATION

1. What is meant by the concept of variation?
2. Name and define in words four alternative measures of variation.

3. Use mathematical notation to define the variance of a sample of data.
4. Use mathematical notation to define the standard deviation of a sample of data.
5. Use mathematical notation to define the variance of a finite population.
6. Define the standard deviation of a finite population, using mathematical notation.
7. Given the following sample of observations:

16 11 19 8 7 12 10 18 10

   a. What is the range of the data?
   b. What is the mean of the deviations from the mean?
   c. What is the mean of the absolute values of the deviations from the mean?
   d. What is the sample variance equal to?
   e. What is the sample standard deviation equal to?

8. For the following sample values of variable x:

23 29 17 29 33 14 26

   a. Calculate s^2
   b. Calculate s

9. If the observations in exercise 8 comprised a finite population instead of a sample, what would be the values of the population variance and standard deviation?
10. If a sample of 6 observations yield a sample variance of 15 what must the sum of the squared deviations about the mean be equal to?
11. If $\Sigma(x - \bar{x})^2 = 205$ for a sample size of $n = 15$, what is the value of s^2?
12. Calculate s^2 for the following sample:

3 9 6

13. Calculate s for the following sample:

16 29

14. Calculate σ^2 and σ for the following population of values:

30 21 46 29

15. Calculate σ^2 and σ for the following population of binary values:

0 0 1 0 1 1 0 0 1

16. Express the variance of a binary population as a function of the mean p of that population.

17. Given the following frequency distribution:

x	f
0	15
1	22
2	29
3	37
4	34
5	27
6	19
7	9

a. What is the sample size n?
b. What is the value of $\bar{x}$?
c. Calculate the sum of the squared deviations from the mean.
d. What is the value of s^2?
e. What is the value of s?

18. If the data in exercise 17 comprised a population rather than a sample, what would the values of σ^2 and σ be?

19. Suppose the definition of σ was incorrectly used to obtain the standard deviation of a sample of $n = 10$ observations, yielding a value of $s = 147$. What would the correct value of the sample standard deviation s be equal to?

20. Suppose the definition of s^2 was incorrectly used to obtain the variance of a population of 5 observations, yielding a value of $\sigma^2 = 46$. What would the correct value of σ^2 be?

21. Given the following sample of data:

4 10 8 14

a. Calculate s^2 and s.
b. Subtract 3 from each sample value and recalculate s^2 and s. How do they compare with the answers to a?
c. Add 10 to each sample value and recalculate s^2 and s. Compare the results to those found in a.
d. Multiply each sample value by 3 and recalculate s^2 and s.
e. Divide each sample value by the value of s found in a, and then recalculate s.

22. a. What happens to the variance of a set of scores when a constant value is subtracted from each score?
b. What happens to the variance of a set of scores when a constant value is divided into each score?
c. What happens to the standard deviation of a set of scores when a constant value is divided into each score?

23. If a distribution of fifty observations has a range of 25, what would you estimate the value of s to be?

24. Letting s_x stand for the standard deviation of a variable x, and c for any given constant, complete the following expressions:
a. $s_{x+c} =$
b. $s_{x-c} =$
c. $s_{cx} =$
d. $s_{x/c} =$
e. Repeat a through d for s^2.

25. If the standard deviation of a variable x is equal to 5, what is the standard deviation of a variable z defined as follows:
a. $z = x - 3$
b. $z = x + 3$
c. $z = 7x$
d. $z = x/5$
e. $z = x - 5$
f. $z = x - \overline{x}$
g. $z = \frac{x - \overline{x}}{5}$

26. Practice making a free-hand sketch of the normal distribution.

27. In a normal distribution, what percentage of the values are:
a. Within 1 standard deviation of the mean.
b. Above 2 standard deviations from the mean.
c. Below 1.96 standard deviations from the mean.
d. Between 2.58 and -2.58 standard deviations from the mean.
e. Above 1.28 standard deviations from the mean.

28. A population measured on variable x has $\sigma = 7$ and $\mu = 35$. Describe the procedure for standardizing the data so that it has $\sigma = 1$ and $\mu = 0$.

29. If the population described in exercise 28 were normal in form, what percentage of the values would you expect to exceed 25?

30. For a normal population with $\sigma = 16$ and $\mu = 100$ what percentage of the values would fall between 76 and 124?

31. Given a normal distribution with $\sigma = 25$ and $\mu = 75$, convert the following raw scores into standardized z scores (i.e., having $\mu = 0$ and $\sigma = 1$). Also, sketch the normal distribution in each case and indicate the approximate location of the z score.
a. $x = 50$
b. $x = 75$
c. $x = 25$
d. $x = 38$
e. $x = 121$
f. $x = 39$
g. $x = 100$
h. $x = 77$
i. $x = 54$
j. $x = 145$

32. Find the proportion of area under the normal curve meeting the following conditions. Also, sketch the normal distribution, shading the areas in question.
a. Above $z = 1.0$
b. Below $z = -1.0$
c. Between $z = -1.5$ and $z = 1.5$
d. Above $z = 1.96$
e. Between $z = -1.96$ and $z = 1.96$
f. Above $z = 1.65$
g. Below $z = -1.77$
h. Between $z = 0$ and $z = 2.58$
i. Below $z = -2.58$
j. Between $z = 1$ and $z = 2$

33. Find the standard deviation of a binary population with $p = .7$.

34. Sketch a binary population with $p = .4$. What is the variance of this population?

35. One theory of evolutionary change assumes gradual changes occur over time, while another assumes sudden large changes. Based on your understanding of the normal distribution, discuss whether these views are inconsistent with each other, and whether they could both be subsumed under the same statistical theory of variation.

36. If farm production is largely a function of random variation in yearly weather conditions, discuss the responsibility of the small farmer to society and vice versa. What is the difference between the small farmer and the large (corporate or collective) farmer in this equation? What are the implications for government involvement?

Chapter 6: BASIC PROBABILITY

1. "The probability of an event is equal to the likelihood of its occurrence." Explain why this is a circular definition of probability.

2. What is the difference between a theoretical and empirically based probability? Give an example of each.

3. Explain what is meant when we say that the probability is .5 that a tossed coin will land heads up.

4. Identify each of the following as a theoretical or empirically based probability.
   a. Drawing a king from a full poker deck.
   b. Rolling a 7 with a pair of dice.
   c. A rise in tomorrow's stock market value.
   d. The successful market penetration of a new product.
   e. Getting an *A* in Statistics.
   f. A red ace being drawn from a poker deck.
   g. The public's favorite winning the Kentucky Derby.
   h. Rain tomorrow.
   i. A well-drilling hitting water within 200 feet.
   j. Candidate *A* getting elected.
   k. Getting hit by a car when crossing the street.

5. Give a definition and example of a probability experiment.

6. Define a random variable and give two examples.

7. What is the sample space of a probability experiment?

8. What is meant by mutually exclusive outcomes of a probability experiment?

9. What constitutes an exhaustive set of outcomes of a probability experiment?

10. Use a tree diagram to determine the sample space for the probability experiment of

answering two exam questions in which the outcome of interest is "correct" or "wrong" in each case.

11. How does a composite outcome differ from a simple outcome?

12. Are composite outcomes defined on a sample space of simple outcomes necessarily mutually exclusive? Explain.

13. If a set of outcomes of a probability experiment are exhaustive, does this guarantee that they are mutually exclusive? Explain.

14. Suppose we conduct the probability experiment of tossing three coins into the air, with an interest in how many land heads up.
    a. Identify the random variable.
    b. Define a sample space of simple outcomes.
    c. What is the probability of no heads landing face up?
    d. What is the probability of at least one head landing face up?
    e. What is the probability of exactly one head landing face up?
    f. What is the probability of at least two heads landing face up?

15. What restrictions are there on the numbers we use for the probabilities of the simple outcomes of a probability experiment?

16. If $P(A)=.35$ and $P(B)=.75$ what do we know for sure about outcomes A and B?

17. If $P(A \text{ and } B)=.50$ and $P(B)=.60$, what is the value of $P(A|B)$?

18. If $P(B)=.20$ and $P(A|B)=.40$, what is the value of $P(A \text{ and } B)$?

19. If $P(A|B)=P(A)$, what can we say about outcomes A and B?

20. If $P(A \text{ and } B)=.80$ and $P(A|B)=.90$, determine $P(B)$.

21. Under what condition will $P(A)P(B)=P(A \text{ and } B)$?

22. State the multiplication rule of probabilities in words and in symbols.

23. State the addition rule of probabilities in words and in symbols.

24. Under what condition will $P(A \text{ or } B)=P(A)+P(B)$?

25. If $P(A \text{ and } B)=.12$, $P(A \text{ or } B)=.20$, and $P(B)=.06$, then what is the value of $P(A)$?

26. Define in words and symbols the expected value of a discrete random variable x.

27. Find the expected value of x:

x	0	1	2	3	4	5
$P(x)$	.50	.21	.11	.08	.06	.04

28. Given the probability distribution in exercise 27, what are the odds against $x=1$?

29. If the probability of a given new product succeeding in the marketplace is $p=.15$, what are the odds against it succeeding?

30. A red die and a blue die are rolled. We are interested in the number of spots facing up on each die. Create the sample space for this experiment, using a matrix in which the columns stand or the number of dots showing on the red die, and the rows stand for the number of dots showing the blue die. The cell entries represent the simple outcomes, consisting of the pair of faces showing on the respective dice —e.g., (2, 4) means that the red die showed 2 dots and the blue die showed 4 dots.

31. Referring to the sample space created in the preceding exercise, determine the following probabilities:
    a. The sum of the dots equals 7
    b. The sum of the dots equals 7, given that one of the dies shows a 2.
    c. The red and blue die show the same number of dots.
    d. The red and blue die each show 3 dots.
    e. The red or blue die shows 3 dots.
    f. The red and blue die sum to 7 or more.
    g. The red die shows more dots than the blue die.
    h. The dice sum to 12.
    i. The dice sum to 8.
    j. The dice sum to 8, given that the blue die shows more dots than the red die.
    k. The dice sum to 8 or more.

32. Again referring to the sample space created in exercise 30, find the following probabilities.
    a. The dots on the red die exceed those on the blue die, given that the blue die shows at least 3 dots.
    b. The blue die shows an odd number of dots, and the red die shows an even number.
    c. The red die shows two or more dots, and the blue die shows two or fewer dots.
    d. The dice sum to 5.
    e. The dice sum to 5, given that the blue die shows 2 or more dots.
    f. The dice sum to 5, given that either the blue or red die show exactly 3 dots.

33. What are the odds against rolling the following sums with a pair of dice:

a. 2	g. 8
b. 3	h. 9
c. 4	i. 10
d. 5	j. 11
e. 6	k. 12
f. 7	l. Repeat *a* through *k* for odds in favor of the sums.

34. Draw the probability distribution for the outcomes of the die rolling experiment listed in the preceding exercise. How would you describe the shape of this distribution?

35. Define the logit of p.

36. a. Use a scientific calculator to compute the logit of the following values of p.

    0 .01 .1 .2 .3 .4 .5 .6 .7 .8 .9 .99 1.0

    b. Convert the above values of p into odds.
    c. Graph odds vs. p.
    d. Graph the logit of p vs. p.

Chapter 7: SAMPLING DISTRIBUTIONS

1. a. Is every sampling distribution a probability distribution? If not, explain, giving an example.
   b. Is every probability distribution a sampling distribution? If not, explain, giving an example.
2. What constitutes a random sample?
3. Explain how you would go about drawing a random sample of size $n = 5$ names from your local telephone directory.
4. Would drawing every nth name (e.g., every 100th name) from a telephone directory constitute a random sample of the names in the directory? If not, explain what bias the sample might be subject to. (Note: This is called a *systematic* sample.)
5. Would all the names that start with the letter A in the phone directory constitute a random sample of the directory's names? If not, what bias might be introduced with this sampling procedure.
6. Devise a procedure for entering Appendix Table I, the table of random digits, at a random location.
7. Enter Appendix Table I at a random location, and then make a frequency distribution of the next 75 digits that you encounter (reading across the page as you would this sentence, moving on to the next lower line when you reach the end of the one you started with).
8. Repeat exercise 7, but this time enter Appendix Table II, the table of random normal deviates. Graph the distribution. What percentage of the 75 observations fall between -1.96 and $+1.96$? What percentage fall between -2.58 and $+2.58$? How do these percentages compare with those found in a normal distribution based on an infinite number of observations? By inspection, what is the approximate value of the mean of the distribution?
9. Enter Appendix Table I (random digits) at a random location again. This time find the sum of each successive triplet of digits (e.g., 74964864... yields 20, 18,...). Continue through the table as described in exercise 7, until you have fifty sums. Make a grouped frequency distribution of these sums. How does its shape compare with those found in exercises 7 and 8? By inspection, what is the approximate value of the mean of the distribution?
10. Divide the sums found in exercise 9 by the sample size $n = 3$. What is the smallest mean you encountered? The largest? How many means were larger than 7? If the sample size were increased from $n = 3$ to $n = 5$, would you expect the percentage of means which exceed 7 to increase or decrease? Explain.
11. a. Define the standard error of the mean in words.
    b. Define the standard error of the mean in mathematical notation.
12. If a normal population has $\mu = 50$ and $\sigma = 10$, what is the mean and standard deviation of the sampling distribution of the mean based on a sample size $n = 4$?
13. If the sample size in the preceding exercise were $n = 16$, what would be the value of the mean and standard deviation of the sampling distribution of the mean?

14. If $\mu = 110$ and $\sigma^2 = 25$ for a normal distribution, what is the value of $\sigma_{\bar{x}}$ if the sample size is $n = 100$?

15. Explain the difference between $\sigma_{\bar{x}}$ and $s_{\bar{x}}$.

16. If $s = 3$ what is the value of $s_{\bar{x}}$ for a sample size $n = 25$?

17. If we draw samples from a non-normal population, will the sampling distribution of the mean be normal? Explain.

18. Estimate the standard error of the mean given that $s = 50$ and $n = 36$.

19. How large must the sample size be before we can be confident that the sampling distribution of the mean will be nearly normal in form, even when the parent population is not?

20. Use mathematical notation to define the standard error of the difference between two independent means.

21. Under what conditions will the standard error of the difference between two independent means be normal in form.

22. Explain the difference between $\sigma_{\bar{x}_1 - \bar{x}_2}$ and $s_{\bar{x}_1 - \bar{x}_2}$.

23. a. If the means of two populations are $\mu_1 = 47$ and $\mu_2 = 36$, respectively, what is the mean of the sampling distribution of the difference between two sample means drawn from the respective populations when the respective sample sizes are $n_1 = 10$ and $n_2 = 20$?
    b. What would it be if the sample sizes were $n_1 = 5$ and $n_2 = 6$?

24. a. If the standard deviations of two populations are $\sigma_1 = 4$ and $\sigma_2 = 7$, respectively, what is the standard error of difference between two means drawn from the respective populations, when the respective sample sizes are $n_1 = 25$ and $n_2 = 30$?
    b. What would it be if the sample sizes were $n_1 = 50$ and $n_2 = 60$?

25. Given the following statistics from two samples, estimate the value of $\sigma_{\bar{x}_1 - \bar{x}_2}$.

	sample 1	sample 2
n	50	60
s	7	6
$\bar{x}$	73	79

26. Under what condition do we use $s_{\bar{x}}$ instead of $\sigma_{\bar{x}}$?

27. Under what conditions will the sampling distribution of the difference between two independent means be normal in form?

28. Given a sample of $n = 50$ paired measures (x, y), we obtain the difference between them $(D = x - y)$. If the standard deviation of the D measures is $s_D = 8$, what is the value of $s_{\bar{D}}$?

29. In the situation described in the preceding exercise, is it possible to obtain the following statistics: $\bar{D} = 0$ and $s_D = 8$? Explain, sketching the situation.

30. Under what conditions will the sampling distribution of $\bar{D}$, as described in the two preceding exercises, be normal in form?

31. Use mathematical notation to define the standard error of a sample proportion $\hat{p}$.

32. If the mean of a binary population is $p=.7$, what is the standard error of the sampling distribution of $\hat{p}$, when $\hat{p}$ is based on a sample of $n=100$?

33. Under what conditions will the sampling distribution of a proportion approach a normal distribution?

34. Write the expansion of the expression $(p+q)^9$.

35. An oil company drilled 125 wells one year. They struck oil in 15 of those wells.
   a. What was the proportion of successful drillings?
   b. What is the approximate standard error of the sampling distribution of the proportion of successes?
   c. Why is the answer to *c* only approximate?
   d. What is the approximate standard error of the sampling distribution of the number of successful drillings?

36. In the preceding exercise we must assume that the 125 drillings are independent tries. What does this mean?

37. A sample of 1000 registered voters are asked their preference between candidates *A* and *B*. Candidate *A* is preferred by 55% of the sample.
   a. Estimate the standard error of the *proportion* choosing candidate *A*.
   b. Estimate the standard error of the *number* choosing candidate *A*.
   c. Estimate the standard error of the *percentage* choosing candidate *A*.
   d. Why are the answers to *a*, *b*, and *c* only approximations of the true sampling distribution standard deviation?

38. Use mathematical notation to define the standard error of the difference between two independent proportions.

39. Given that the means of two independent binary populations are $p_1=.35$ and $p_2=.28$, find the value of $\sigma_{\hat{p}_1-\hat{p}_2}$, when the sample sizes are $n_1=100$ and $n_2=150$, respectively.

40. A sample of 400 men and 600 women were asked their preference between package design *A* and *B*. Package *A* was picked by 190 of the men, and by 315 of the women.
   a. Estimate the standard error of the difference between the male and female proportions favoring package *A*.
   b. Estimate the standard error of the difference between the number of males and females favoring package *A*.
   c. Estimate the standard error of the difference between the percentage of males and females favoring package *A*.

41. An advertising agency is making pitches for six new accounts. In the past, their success rate has been 15%. Use the binomial distribution to find the probabilities of the following outcomes.
   a. Getting two or more accounts.
   b. Getting all six accounts.
   c. Getting no accounts.
   d. Getting exactly one account.

42. Discuss the applicability of the binomial distribution to the problem in the preceding exercise.

43. What rule of thumb would you use to determine if the normal distribution is a good approximation of the binomial distribution.

44. Use Appendix Table X to determine the values of the successive terms of $(.7+.3)^{15}$.
   a. Graph the probability distribution of the number of "successes"; i.e., 0,1,2,...,15.
   b. What is the standard error of the number of successes?
   c. Along the baseline of the graph, under the number of successes, write the proportion of successes.
   d. What is the standard error of the proportion of successes?
   e. What percentage of the successes are more than 1.96 standard errors above the mean? How does this compare with the normal distribution?

Chapter 8: PARAMETER ESTIMATION

1. What is the distinction between a point and interval estimate of a parameter?

2. Name and explain the properties which we would like a parameter estimate to have.

3. What is meant by the 95% confidence interval for a parameter?

4. A sample mean $\bar{x} = 23$ is based on a random sample of $n = 50$ observations drawn from a normally distributed population with variance $\sigma^2 = 16$.
   a. What is the 95% confidence interval for μ?
   b. What is the 99% confidence interval for μ?
   c. Sketch the sampling distribution of $\bar{x}$, indicating the confidence intervals found in *a* and *b*.

5. A random sample of $n = 25$ from a normally distributed population with $\sigma = 7$ yields a sample mean $\bar{x} = 67.3$.
   a. What is the 99% confidence interval for μ?
   b. What is the 95% confidence interval for μ?
   c. Sketch the sampling distribution of $\bar{x}$, indicating the confidence intervals found in *a* and *b*.

6. A sample mean $\bar{x} = 123.7$ is based on a random sample of $n = 80$ observations from a normal population with unknown σ^2, but we have the sample variance $s^2 = 9$.
   a. What is the approximate 95% confidence interval for the mean μ?
   b. What is the approximate 99% confidence interval for the mean μ?
   c. Why are the answers to *a* and *b* only approximately correct intervals?
   d. Sketch the sampling distribution of $\bar{x}$, indicating the confidence intervals found in *a* and *b*.

7. A sample of size $n = 70$ is drawn from a normal population, yielding a mean of $\bar{x} = 4.37$ and a variance $s^2 = 1.97$.
   a. What is the 95% confidence interval for μ?
   b. What is the 99% confidence interval for μ?

c. If the sampled population had not been normal in form, would the intervals found in *a* and *b* still be approximately correct? Explain.

8. A sample of $n = 400$ consumers rate their purchase interest in product *A* before and after viewing a commercial. The mean of the differences between the before and after ratings is $\overline{D} = 2.4$ and the standard deviation is $s_D = 3$.
   a. What is the 95% confidence interval for the true mean difference μ_D?
   b. Find the 99% confidence interval for μ_D.
   c. Explain why the answers to *a* and *b* must necessarily be approximate in nature.

9. A sample of $n = 200$ consumers rate their liking of two soft drink formulations, *A* and *B*. The mean of the *A* minus *B* ratings is $\overline{D} = -1.4$, and the standard deviation is $s_D = 3.1$.
   a. Find the approximate 95% confidence interval of μ_D and indicate it on a sketch of the sampling distribution of $\overline{D}$.
   b. Do the same for the 99% confidence interval.

10. Two random samples of consumers of size $n_1 - 150$ and $n_2 = 160$, respectively, are drawn from a normally distributed population. One sample is shown commercial *A* and the other sample is shown commercial *B*, for the same product. Each person then rates their purchase interest in the product on a 0 to 100 scale. The mean ratings for commercial *A* and *B* are $\overline{x}_1 = 79.4$ and $\overline{x}_2 = 85.7$, respectively. The respective variances are $s_1^2 = 121$ and $s_2^2 = 144$, respectively.
    a. Find the approximate 95% confidence interval for $\mu_1 - \mu_2$.
    b. Find the approximate 99% confidence interval for $\mu_1 - \mu_2$.

11. Samples of size $n_1 = 75$ and $n_2 = 60$ are drawn from normally distributed populations with $\sigma_1^2 = 25$ and $\sigma_2^2 - 36$, respectively. The means of the samples are $\overline{x}_1 = 17.5$ and $\overline{x}_2 = 23.4$, respectively.
    a. Find the 95% confidence interval for $\mu_1 - \mu_2$.
    b. Find the 99% confidence interval for $\mu_1 - \mu_2$.
    c. Why are these confidence intervals exact, while those in the preceding exercise were only approximate?
    d. Sketch the sampling distribution of $\overline{x}_1 - \overline{x}_2$, indicating the confidence intervals found in *a* and *b*.

12. Samples of size $n_1 = 8$ and $n_2 = 9$ are drawn from normally distributed populations. The respective sample means and standard deviations are $\overline{x}_1 = 15.8$, $\overline{x}_2 = 12.3$, $s_1 = 3.2$, and $s_2 = 3.5$.
    a. Use Students *t* to determine the 95% confidence interval for $\mu_1 - \mu_2$.
    b. What is the 99% confidence interval for $\mu_1 - \mu_2$?

13. Since the populations described in the preceding exercise were normal, why did we have to resort to Student's *t* to determine the confidence intervals? What assumption did we have to make about the two population variances?

14. A sample of $n = 15$ is drawn from a normal population. The sample statistics are $\overline{x} = 673$ and $s^2 = 100$. What is the exact 95% confidence interval for μ?

15. A sample proportion $\hat{p} = .46$ is based on a sample $n = 100$.
    a. What is the approximate 95% confidence interval for p?
    b. What is the approximate 99% confidence interval for p?
    c. Why are these only approximately correct intervals?

16. A sample of $n = 200$ is drawn from a binary population with $p = .75$.
a. What is the 95% confidence interval for p?
b. If the sample size were $n = 100$, what would the 95% confidence interval be?
c. Are these intervals exact or approximate? Explain.

17. A sample proportion $\hat{p} = .1$ is based on a sample of $n = 1500$.
a. Find the 95% confidence interval for p.
b. Find the 99% confidence interval for p.
c. Why are these only approximate intervals?

18. Samples of $n_1 = 500$ and $n_2 = 400$ are drawn from two binary populations, yielding sample proportions $\hat{p}_1 = .29$ and $\hat{p}_2 = .21$, respectively.
a. Find the 95% confidence interval for $p_1 - p_2$.
b. Find the 99% confidence interval for $p_1 - p_2$.
c. Explain why these are approximately correct intervals.

19. A binary population has a mean $p = .7$.
a. If a sample of $n = 1000$ is drawn from the population, what is the 95% confidence interval for p?
b. If a sample of $n = 500$ is drawn from the population, what is the 95% confidence interval for the number of successes np?
c. If a sample of $n = 100$ is drawn from the population, what is the 95% confidence interval for the percentage of successes?

20. Two binary populations have means of $p_1 = .60$ and $p_2 = .65$, respectively.
a. Find the 95% confidence interval for $p_1 - p_2$, based on sample sizes of $n_1 = 100$ and $n_2 = 150$, respectively.
b. What would be the answer if $n_1 = 200$ and $n_2 = 200$?

21. Samples of size $n_1 = 200$ and $n_2 = 300$ from two populations yield sample proportions $\hat{p}_1 = .33$ and $\hat{p}_2 = .41$, respectively.
a. What is the weighted average of the sample $\hat{p}$'s?
b. Determine the approximate 95% confidence interval for $p_1 - p_2$, if the two populations are assumed to be different.
c. Determine the approximate 95% "confidence interval" for $\hat{p}_1 - \hat{p}_2$, if the two populations are assumed to be equal. Discuss the distinction with *b*.

22. Consider the advertiser that wishes to determine the difference in memorability of two alternative magazine advertisements for its product. Three hundred consumers are given a magazine, half of which have Ad 1 inserted, and half of which have Ad 2. They are asked to browse through the magazine under the pretense of evaluating its overall contents. A half hour later, after an intervening task, the consumers are asked to recall all the products they remember seeing advertised in the magazine. Of the 150 consumers exposed to Ad 1, 12% recall the advertiser's brand; of the 150 exposed to Ad 2, 16% recall the brand. Create the 95% confidence interval for the difference in recall levels for the two ads. Would you conclude that the two ads differ? Explain.

23. Suppose we want to know the average price consumers are willing to pay for a new product, but we do not want our estimate to be in error by more than \$2.50, with a probability of .95. A pilot research project provides an estimate $s = 9.8$ of the true standard deviation σ of the price prospective consumers are willing to pay for the product. What sample size should be used to meet the stated requirements?

Chapter 9: HYPOTHESIS TESTING

1. Discuss the similarity between parameter estimation and hypothesis testing.
2. Suppose we want to test the hypothesis that the mean of a given population is equal to 21. Express the hypothesis set for this problem using mathematical notation.
3. We hypothesize that if the mean of a given population is not equal to 61, then it must necessarily be greater than 61.
   a. Express the appropriate hypothesis set for this problem in mathematical notation.
   b. What special name do we give to this type of hypothesis set?
4. We hypothesize that the means of two populations are equal. Express the appropriate hypothesis set for this problem.
5. If we hypothesize that the difference between the means of two populations is equal to 17, what is the appropriate hypothesis set?
6. We hypothesize that if the difference between two population means is not equal to 17, then the only conceivable alternative is that the difference is less than 17. Express the relevant hypothesis set for this situation.
7. We want to test the hypothesis that the mean of a population is $\mu = 46$. We draw a sample of size $n = 100$ and obtain a sample mean $\bar{x} = 51$ and the population variance estimate $s^2 = 400$.
   a. State the hypothesis set.
   b. What is the value of the estimated standard error of the mean?
   c. What is the value of the relevant z ratio for testing the hypothesis?
   d. Do you reject the hypothesis at the $\alpha = .05$ significance level?
   e. Do you reject the hypothesis at the $\alpha = .01$ significance level?
   f. Is this test exact or approximate? Explain.
8. "The hypothesis that $\mu = 46$ is rejected at the $\alpha = .05$ significance level." What does this statement mean?
9. We want to test the hypothesis that the mean of a normal population with $\sigma^2 = 70$ has a mean $\mu = 100$. A sample of $n = 200$ yields a mean $\bar{x} = 99.4$ and variance $s^2 = 79.2$.
   a. State the hypothesis set.
   b. Do you reject the hypothesis at the $\alpha = .05$ level of significance?
   c. Sketch the sampling distribution of the mean, indicating the value of the z ratio obtained, and the critical rejection areas.
   d. How does this test differ from the one in exercise 7?
10. What is the distinction between a one- and two-tailed test?
11. Redo exercise 9, but using a one-tail test (the left tail).
12. What is the distinction between a type I and type II error?
13. What letters are used to designate the probabilities of type I and type II errors?
14. If a jury convicts an innocent person, what type of error has been committed?

15. Name the factor(s) upon which the probability of a type I error depends, in a statistical hypothesis test.

16. Identify the factors which influence the probability of a type II error.

17. A sample of $n = 80$ pairs is drawn from a population of paired measurements. We want to test the hypothesis that the mean difference of these paired measurements is equal to zero—i.e., $\mu_D = 0$. The mean of our eighty sample differences is $\bar{D} = -1.7$ and the standard deviation is $s_D = 7.1$.
   a. Express the hypothesis set.
   b. Sketch the relevant sampling distribution, shading the critical rejection areas. What are the critical values of the z ratio for the $\alpha = .05$ significance level?
   c. Calculate the z ratio.
   d. Do you reject the hypothesis $\mu_D = 0$ at the $\alpha = .05$ level?
   e. What is the probability of a type I error when using the $\alpha = .05$ significance level?
   f. Are the answers to the above questions exact or approximate? Explain.

18. The mean difference of a sample of 100 paired measures is $\bar{D} = 7.3$, and $s_D = 50.7$. Test the hypothesis that the observed mean difference is merely a chance deviation from a population mean difference of zero. Use the $\alpha = .05$ significance level for the test and state any qualifications in your conclusions.

19. Two normal populations have respective variances $\sigma_1^2 = 100$ and $\sigma_2^2 = 200$. Samples of size $n_1 = 8$ and $n_2 = 11$ are drawn from the respective populations, yielding means $\bar{x}_1 = 645$ and $\bar{x}_2 = 596$, respectively. We want to test the null hypothesis that there is no difference in the population means μ_1 and μ_2.
   a. Express the hypothesis set.
   b. Sketch the relevant sampling distribution indicating the critical rejection regions.
   c. Calculate the appropriate z ratio to test the null hypothesis.
   d. Do you reject the hypothesis at the $\alpha = .05$ significance level?

20. Samples of size $n_1 = 50$ and $n_2 = 60$ are drawn from two populations, yielding means $\bar{x}_1 = 26.7$ and $\bar{x}_2 = 38.5$, and standard deviations $s_1 = 11.1$ and $s_2 = 9.8$, respectively.
   a. Test the hypothesis that the two populations have identical means.
   b. Sketch the appropriate sampling distribution indicating the location of the z ratio.

21. A sample of $n = 15$ is drawn from a normally distributed population, yielding a mean $\bar{x} = 45.6$ and a standard deviation $s = 9.6$. Use Student's t to test the hypothesis that the population mean is $\mu = 50.0$.
   a. What are the degrees of freedom for the t ratio?
   b. What value of t is needed to reject the null hypothesis at the $\alpha = .05$ significance level?
   c. Calculate the value of t.
   d. Do you reject the hypothesis that $\mu = 50.0$?

22. Samples of size $n_1 = 7$ and $n_2 = 5$, respectively, are drawn from two normally distributed populations, yielding sample means $\bar{x}_1 = 81.9$ and $\bar{x}_2 = 67.2$, and standard deviations $s_1 = 19.2$ and $s_2 = 26.7$. Test the hypothesis that there is no difference between the population means, using the $\alpha = .05$ significance level.

23. Discuss the difference in the approach to testing the hypotheses in exercises 19 and 22, explaining the reasons for the differences.

24. What are the assumptions involved in using Student's t to test the difference between two sample means?

25. A sample of size $n = 100$ from a binary population yields a proportion $\hat{p} = .59$. Test the hypothesis that the population proportion is really $p = .50$, using the $\alpha = .05$ significance level.

26. A sample of $n = 1500$ from a binary population yields a proportion of $\hat{p} = .47$. Test the hypothesis that the mean of the population is $p = .50$, using the $\alpha = .05$ significance level.

27. Redo the preceding exercise using the $\alpha = .01$ significance level.

28. Samples of $n_1 = 100$ and $n_2 = 200$, respectively, are drawn from two binary populations, yielding sample proportions of $\hat{p}_1 = .21$ and $\hat{p}_2 = .26$. Test the null hypothesis that there is no difference in the means of the two binary populations. Use the $\alpha = .05$ significance level.

29. A sample of $n_1 = 1000$ from one binary population yields a mean $\hat{p}_1 = .73$, and another sample $n_2 = 1000$ from another binary population yields a mean $\hat{p}_2 = .65$. Test the hypothesis that the two populations have identical means. Test it at both the $\alpha = .05$ and $\alpha = .01$ significance levels.

30. Define the power of a statistical test, and name the factors upon which it depends.

31. Name two ways to increase the power of a statistical test.

32. What is the relationship between the power of a statistical test and the probability of a type II error?

33. Refer to exercise 9. What is the power of that statistical test if the true population mean is $\mu = 101.0$? Sketch the problem, indicating the power by shading the appropriate part of the relevant sampling distribution. Use the $\alpha = .05$ significance level.

34. What is the power of the test in exercise 9 if the true population mean is $\mu = 102.0$?

35. Refer to exercise 19. What is the power of that statistical test if the true difference between the population means is $\mu_1 - \mu_2 = 20$? Use the $\alpha = .05$ significance level.

36. What is the power of the test in exercise 19 if the true difference between population means is $\mu_1 - \mu_2 = 5$?

37. Refer to exercise 20. What is the approximate power of this test if the true difference between population means is $\mu_1 - \mu_2 = 3.2$? Use the $\alpha = .01$ significance level. Why is this answer only approximately correct?

38. The standard deviation of a given normal population is $\sigma = 10$. We hypothesize that the mean is $\mu = 65$, and test the hypothesis at the $\alpha = .05$ significance level. If in fact the true population mean is $\mu = 70$, we want to have a probability of .80 of rejecting the false hypothesis $\mu = 65$. What sample size should we use?

39. What sample size would we need in the preceding exercise if we only demanded that the power of our test be .40?

40. The respective variances of two populations are $\sigma_1^2 = 50$ and $\sigma_2^2 = 80$. We want to draw equally-sized samples from the two populations to test the null hypothesis that their means do not differ. If we use the $\alpha = .05$ significance level, what size samples should we draw so that the power of our test is .75 in the event the true population difference is $\mu_1 - \mu_2 = 5$? First define $\sigma_{\bar{x}_1 - \bar{x}_2}$.

41. What size samples should we use in the preceding exercise if we want the power of our test to be .90?

Chapter 10: CORRELATION ANALYSIS

1. What do we mean when we say two variables are associated, related, or "correlated"?

2. a. Name a pair of variables that are probably correlated, specifying the objects and scales upon which the variables are measured.
   b. Name a pair of variables that are probably not correlated, again specifying the objects and scales upon which they are measured.

3. A group of 200 students are administered two tests—one of math aptitude and one of verbal aptitude. Forty-five of the students scored above the respective medians on both tests. Portray the results of this analysis in a 2×2 contingency table. How many students scored above the median on the verbal test, but below the median on the math test? Does a simple inspection of the contingency table suggest that the variables of math and verbal aptitude are related to one another? Explain.

4. Given the following sample of $n = 9$ pairs of values on variables x and y:

x	4	7	1	5	8	5	2	4	3
y	80	92	52	76	106	100	69	71	65

   a. Make a scatter diagram of the paired measures.
   b. By inspection alone, do x and y appear to be correlated?
   c. Calculate r.
   d. Can we reject the hypothesis that the population correlation coefficient is zero, at the $\alpha = .05$ significance level?

5. Given the following sample of paired values on variables x and y:

x	15	11	14	12	20	11	13	10	16	15	17	14
y	27	12	18	21	28	17	29	7	22	24	34	10

   a. Make a scatter diagram of the data.
   b. Calculate r.
   c. Can we reject the hypothesis that the population correlation coefficient is zero, at the $\alpha = .05$ significance level?

6. What are the assumptions underlying the product moment correlation coefficient? Make a sketch of the assumption whenever possible.

7. Explain why the existence of a correlation between two variables does not ensure that one causes the other. Give examples to demonstrate your argument.

8. An investigator found a negative correlation between the cholesterol level in the blood and the amount of exercise the individuals reported that they performed in a typical week. Explain why a cause and effect interpretation cannot be placed on this finding. Is it conceivable that the blood chemistry influenced the individuals' activity, rather than vice versa? Are there any other explanations?

9. Assuming equal sample sizes in each instance, indicate which correlation coefficient in each of the following pairs indicates a stronger degree of relationship.
   a. $r = .63$ or $r = -.12$
   b. $r = .72$ or $r = -.72$
   c. $r = -.35$ or $r = -.53$
   d. $r = .59$ or $r = -.67$
   e. $r = 0$ or $r = -.27$

10. Think of two variables that are negatively correlated, specifying the objects being measured.

11. What is the distinction between a criterion and a predictor variable?

12. What is an alternative pair of names for criterion and predictor variables?

13. What is a correlation matrix?

14. Why are the diagonal elements of a correlation matrix equal to 1?

15. Describe in words how a multiple R is obtained.

16. Explain the advantage of converting the criterion and predictor variables into standardized score form with equal variances and equal means.

17. Explain the concept of partial correlation.

18. Explain the concept of canonical correlation.

19. What do we mean by the reliability of our measurements? Explain how the correlation coefficient can be used to assess reliability.

20. What do we mean by the validity of our measurements? Explain how the correlation coefficient can be used to assess validity.

21. Is it possible for our measurements to be reliable without being valid? Explain, giving an example.

22. Is it possible for our measurements to be valid without being reliable? Explain, giving an example to support your answer.

23. Sketch an instance in which two variables are related to one another, but for which the correlation coefficient is zero.

24. In addition to assessing the reliability and validity of our measurements, what other applications does correlation analysis lend itself to?

25. Study the scatter diagrams of exercises 4 and 5. Can you think of a procedure that would allow you to make a fairly good prediction of an object's y value, when given its x value?

26. It was pointed out in the chapter that the standard deviation of the difference between two *non-independent* variables, x and y, is given by $s_{x-y} = \sqrt{s_x^2 + s_y^2 - 2rs_xs_y}$. The same holds for σ. Discuss the geometry of this relationship in terms of triangles, for the condition when r is positive vs. when it is negative. Compare with Figure 6 of Chapter 7.

27. Find an equivalent expression for the quantity rs_xs_y, in terms of observed scores.

28. The sample *covariance* of two random variables, x and y, is defined

$$\text{Sample covariance}\,(x, y) = \sum_{i=1}^{n} \frac{(x_i - \bar{x})(y_i - \bar{y})}{n-1}$$

where n is the number of pairs of observations.

a. Calculate the sample covariance for the following pairs of observations

x_i	y_i
5	18
1	12
6	18
4	16

b. Discuss the advantage of r over the covariance.

29. Prove mathematically (or demonstrate with a numerical example) the equivalence of the following formulas for r.

$$\frac{\sum (x_i - \bar{x})(y_i - \bar{y})}{(n-1)s_xs_y} = \frac{\sum (x_i - \bar{x})y_i}{(n-1)s_xs_y}$$

Chapter 11: REGRESSION ANALYSIS

1. Compare correlation analysis and regression analysis.

2. Write the mathematical equation for the general form of a linear function, and identify each term.

3. Sketch three different lines that have the same slope but different y-intercepts.

4. Sketch three different lines that have the same y-intecept but different slopes.

5. Identify the slope and y-intercept for each of the following linear functions.
   a. $y = 7 + 3x$
   b. $y = -45 + 20x$
   c. $x + y = 16$
   d. $3y - 2x = -15$

6. Graph the following linear functions.
   a. $y = 10 + 2x$
   b. $y = 5 - 3x$
   c. $y = -9 + x$
   d. $y = -3 + .7x$
   e. $y = 12 - 1.5x$

7. Given the following associated values of variables x and y:

x	4	7	1	5	8	5	2	4	3
y	80	92	52	76	106	100	69	71	65

   a. Graph the data points.
   b. Using your judgment, draw what you think is a 'best-fitting' line through the points.
   c. Based on the least squares criterion, calculate the slope of the best-fitting line.
   d. Based on the least squares criterion, calculate the y-intercept of the best-fitting line.
   e. Draw the best-fitting line, based on the least squares criterion, through the data points. How do the slope and y-intercept of this objectively defined line compare with the values of the line you sketched in b?

8. Using the best-fitting line found in the preceding exercise, calculate the predicted values y' for each of the following values of x.
   a. $x = 6$
   b. $x = 3$
   c. $x = 5$
   d. $x = 8$

9. Calculate the standard error of estimate for the data in exercise 7. Use it to construct the approximate 95% confidence band for predicting y from x, in the figure from exercise 7e.

10. Annual income and expenditure figures were obtained from a sample of five households. The following figures are in thousands of dollars.

Income	*Expenditure*
9.9	9.7
12.5	13.7
15.5	15.0
17.4	14.9
22.3	21.2

   a. Graph the data points, with Income being the x variable.
   b. Calculate the slope of the best-fitting line, according to the least-squares criterion.
   c. Calculate the y-intercept.
   d. What is the equation of the best-fitting line? Graph it.
   e. What level of expenditure would you predict for a household with \$20,000 income?
   f. Calculate the standard error of estimate for predicting expenditure from income.

11. Discuss the regression equation found in the preceding exercise in terms of the real-life meanings of the slope and y-intercept values.

12. What are the assumptions underlying the linear regression model?

13. In exercise 10, why would it be inappropriate to use the obtained regression equation to make a prediction of a household's expenditures when its income is $30,000?

14. Explain how the correlation coefficient r can help us to interpret the accuracy of prediction achieved by a regression equation.

15. What is the distinction between a simple and multiple regression equation?

16. Write an example of a multiple regression equation, identifying the terms.

17. What is the distinction between b-coefficients and *beta* coefficients?

18. How does the square of the multiple correlation coefficient, R^2, help us to interpret a multiple regression equation?

19. A regression analysis using five predictor variables results in an adjusted R^2 of .75. Interpret this finding.

20. One of the five predictor variables in the preceding example is eliminated from the analysis and the value of R^2 drops from .75 to .71. Does this mean that the dropped variable is only marginally related to the criterion variable? Explain.

21. What interpretation can we place upon the *beta* coefficients of a regression analysis? Can the same be done for the b-coefficients? Explain.

22. Why are the coefficients of a multiple regression equation referred to as partial regression coefficients?

23. Briefly describe the stepwise regression procedure. Why is it used?

24. What is the multi-collinearity problem?

25. Describe how a qualitative variable, such as a person's gender, can be used as a predictor variable in a regression analysis.

26. An investigator entered three dummy variables into a regression analysis based on the three levels of political affiliation—Democrat, Republican, Independent. A multi-collinearity problem was encountered. Explain why.

27. A researcher wants to use a person's favorite type of music, from among a list of seven types, as a predictor of heaviness of record and tape buying. How can this be accomplished using regression analysis?

28. As a rule, what is the maximum number of dummy variables that can be entered into a regression analysis based on a qualitative variable with n levels? Why?

29. Briefly describe autoregression.

30. Discuss the regression to the mean phenomenon, giving an example from everyday life.

31. Describe how you would go about validating a regression equation.

32. A researcher used three years of data to develop a multiple regression equation that was able to do a fairly good job of predicting the stock market's value. When the

equation was applied during the next three years, the investigator did no better than if the buy and sell decisions had been made with the flip of a coin. Are you surprised? Explain.

33. Describe a situation in which the failure of a regression equation with a high R^2 to accurately predict the future is really a *practical success* of the regression analysis *per se*.

Chapter 12: ANALYSIS OF VARIANCE

1. Discuss the problem of using the z or t ratios to test the hypothesis that three or more sample means were drawn from the same population.
2. Define the F ratio in words and in mathematical notation.
3. What is the expected value of F?
4. How are the degrees of freedom for an F ratio determined?
5. For $df = 5, 15$ what value of F will be exceeded with a probability $\alpha = .05$?
6. For $df = 3, 25$ what value of F will be exceeded with a probability $\alpha = .01$?
7. Using our criterion of $\alpha = .05$ as an unlikely event, is it likely that an $F = 5.67$ with $df = 2, 35$ would occur?
8. Suppose we have five independent samples of seven observations each.
   a. Describe in words how the within groups variance would be obtained.
   b. Describe in words how the between groups variance would be obtained.
   c. What are the degrees of freedom for the within groups variance?
   d. What are the degrees of freedom for the between groups variance?
   e. What value of F for these degrees of freedom would be exceeded by chance with a probability $\alpha = .05$?
9. In a simple one-way analysis of variance is it necessary that there be an equal number of observations per sample?
10. What is assumed about the sampled populations when using the F test?
11. Name as many different reasons as you can to account for an F ratio turning out to be significantly higher than its expected value.
12. Four alternative programmed instruction manuals were randomly assigned to three students each. Below are the number of errors committed by the students on a subsequent achievement test.

Instruction manual used			
A	*B*	*C*	*D*
1	2	5	1
0	3	4	2
1	4	2	4

a. Calculate the mean number of errors for each sample.
b. Calculate the mean number of errors for all twelve students.
c. Calculate the sum of the squared deviations of the errors in group A about the mean for that group. Do the same for the other groups.
d. What is the value of the pooled within groups sum of squared deviations?
e. Calculate the between groups sum of squared deviations that is needed to determine the appropriate between groups estimate of the population variance.
f. What is the population variance estimate based on the within groups variation?
g. What is the population variance estimate based on the between groups variation?
h. What is the value of the F ratio for testing the hypothesis that there is no difference in the effectiveness of the alternative instruction manuals?
i. What are the degrees of freedom associated with the F ratio?
j. Do you reject the null hypothesis at the $\alpha = .05$ significance level? What value of F had to be exceeded in order to reject the hypothesis.
k. Arrange your results in a summary analysis of variance table.

13. What was the probability of a type I error in the preceding exercise?

14. Name as many factors as you can that would influence the probability of a type II error in an analysis such as the one in exercise 12.

15. Do you think the analysis of variance procedure could be used in place of the t test for testing the difference between two sample means? Explain why or why not.

16. What is the alternative to the null hypothesis in a simple analysis of variance. Is there one alternative or many? Explain.

17. In a survey of product usage, attitudes, and aptitudes, a short form of an intelligence test was included. A one-way analysis of variance was performed for four groups of yogurt eaters representing users of brand A vs. B vs. C vs. all other brands. A significant F ratio showed that they differed in their mean intelligence score. A follow-up analysis showed that users of brand A yogurt scored significantly higher than users of each of the other brands. The advertising manager wants to use a campaign slogan, "Our yogurt makes you smarter."
a. How would you advise this senior executive?
b. How does this analysis of variance differ from the one in exercise 12?
c. Suppose the F of exercise 12 had proven significant. How would the interpretation of that analysis differ from the present one? Explain why.

18. Suppose you have five means: 7, 11, 10, 6, 15.
a. Show the coefficient weights for a contrast of means comparing the average of the first four means against the fifth.
b. Do the same for a contrast comparing the mean of the first two against the mean of the last three.
c. What property must the coefficients for a contrast of means possess?
d. What property must the coefficients for two contrasts of means possess in order for them to be orthogonal to one another?
e. How many contrasts of means are possible for the five means above?
f. How many contrasts of means can be constructed such that they are all orthogonal to one another, for a given set of five means?
g. Create a full set of orthogonal contrasts for the five means above, and calculate the value of each contrast.

19. Three types of fine fiber were tested for their breaking point using five specimens for each type. The breaking points, measured in pounds, are presented below.

Type of fiber		
I	II	III
10	7	15
14	11	7
14	8	10
12	9	14
16	11	9

a. Perform an analysis of variance and summarize your results in a summary table. What is your conclusion with regard to the null hypothesis that there is no difference between strengths of the three fiber types?
b. Suppose the three types of fiber varied on a number of dimensions—e.g., type of material, type of manufacturing process, diameter, etc. If the analysis of variance suggested that the thread types did differ in strength, could we attribute that difference to any one particular variable? Explain your answer.
c. Design a factorial experiment to test the effects of four types of material and three construction methods on fiber strength.
d. Suppose the analysis of the experiment revealed an interaction effect between type of material and construction method. What would this mean? Give an example of such an effect with these particular variables.

20. An analysis of variance of the achievement test scores for three classes of statistics students at different colleges resulted in a significant F.
a. List as many factors as you can that could have accounted for the difference.
b. Design an experiment which could test the efficacies of the various factors you named above, in order that cause and effect conclusions could be made.

21. Below is the data for a randomized block experiment.

	Treatments		
	A	*B*	*C*
Block 1:	5	8	3
Block 2:	9	12	5

a. From your area of interest, identify what the treatments, blocks, and criterion measure could represent.
b. Create the summary analysis of variance table.
c. Evaluate the following contrasts of means, first by the interval estimation approach and then by the hypothesis testing approach. Use $\alpha = .05$.

A	*B*	*C*
+1	+1	−2
+1	−1	0

d. Discuss the results of the analysis.
e. What is it about the data that resulted in a significant F when only two blocks were used?

22. A completely randomized 2-factor 3×4 factorial experiment was conducted with $n = 5$ observations per cell. Name the various sources of variation and show their respective degrees of freedom.

23. Perform the analysis of variance of the following 2-factor 3×3 factorial experiment with $n = 2$ observations per cell.

Factor A

Factor B:		A_1	A_2	A_2
	B_1	10 12	11 9	14 18
	B_2	9 5	8 8	10 8
	B_3	7 7	6 10	9 7

24. Distinguish between a set of orthogonal and non-orthogonal comparisons.

25. Distinguish between a comparison-wise and set-wise significance level.

26. a. Perform the Table 13 type of analysis for the data of Table 11.
    b. To obtain a set-wise significance level α for a set of unplanned non-orthogonal contrasts, using the F test instead of the interval estimate approach, the critical variance ratio is $(m-1)F_c$ instead of F_c. Redo the Table 12 analysis using the variance ratio significance test. Does this affect your conclusions?

27. Perform the Table 11 type of analysis for the data of Table 13.

28. Analyze the data of Table 11 for a comparison-wise significance level of $\alpha = .01$.

29. The following five criterion means were obtained for equally spaced values of a predictor variable, in a one-way analysis of variance with an error variance estimate of $s^2 = 21$, based on 4 observations per group:

    10.2 12.5 15.9 16.3 17.1

    Graph the means and perform significance tests for the various possible trend components, reporting p values.

30. Distinguish between a repeated measures design and a randomized block design.

31. Discuss the conjoint measurement procedure and its goal.

32. Show the sources of variation and their associated degrees of freedom for a randomized block design with five treatments and seven blocks.

33. Under what conditions is a randomized block design preferable to a completely randomized design.

34. Discuss the notion of interaction among variables.

35. Distinguish between *anova* and *manova*.

36. Distinguish between the univariate, bivariate, and multivariate normal distributions.

37. Name the analogs to Student's t and the univariate F ratio in the multivariate analysis of variance.

38. Discuss the underlying principle or mechanism of analysis of covariance.

39. Discuss the variety of applications of *ancova*, identifying which of the uses are controversial and why.

40. From your area of interest, design research studies conforming to each of the following designs, describing the relevant variables (including levels used) and sample selection, indicating whether experimental or in-tact groups are being analyzed.
    a. One-way design.
    b. Randomized block design.
    c. Two-factor factorial design.
    d. Three-factor factorial design.
    e. Repeated measures design, with one predictor variable.
    f. Conjoint measurement design.
    g. Multivariate two-factor design (i.e., with multiple criterion variables).
    h. Analysis of covariance design.

Chapter 13: ANALYSIS OF CATEGORY DATA

1. Define in words and in symbols the χ^2 statistic.

2. Under what conditions does the χ^2 statistic closely follow the continuous chi-squared distribution.

3. To test k independent proportions with χ^2, what are the relevant degrees of freedom for the χ^2 statistic?

4. In an $r \times c$ contingency table in which the expected frequencies are determined from the data itself, what are the degrees of freedom for the χ^2 statistic?

5. What happens to the expected value of χ^2 as the degrees of freedom increase?

6. What value of χ^2 is needed for the $\alpha = .05$ significance level, when
    a. $df = 1$
    b. $df = 4$
    c. $df = 7$

7. Determine whether the following values of χ^2 are significant at the $\alpha = .01$ level.
    a. $\chi^2 = 4.07$, $df = 1$
    b. $\chi^2 = 12.16$, $df = 4$
    c. $\chi^2 = 17.54$, $df = 6$

8. The set of frequency cells subjected to a χ^2 analysis must be mutually exclusive and exhaustive. What does this mean?

9. The observations falling into the various cells subjected to a χ^2 analysis must be independent. What does this mean?

10. A researcher applied the χ^2 test to the number of employee absences by day of the week. Do you find anything wrong with this application of χ^2? Explain.

11. Determine if the following set of three mutually exclusive and exhaustive frequencies of independent observations differ from an equal split: 7, 12, 9.
    a. What are the expected frequencies?
    b. What is the value of χ^2?
    c. What are the degrees of freedom?
    d. Is the value of χ^2 significant at the $\alpha = .05$ level?

12. Repeat parts *a* through *d* of the preceding exercise, but this time test the hypothesis that the frequencies occur in the ratio 2 : 3 : 3.

13. A coin is tossed 20 times and lands Heads 7 times. Use χ^2 to determine if this outcome supports the hypothesis that it is an unfair coin.

14. An academic achievement test is intended to distinguish between students at different grade levels. For one particular item in the test, 37 of 50 fifth-graders and 34 of 40 sixth-graders answered the item correctly. Use χ^2 to determine if the response to the item is different for the two grade levels.

15. A survey of 105 Democrats, 72 Republicans, and 33 Independents, found that 76, 41, and 16, respectively, favored a particular referendum. Determine if the attitude toward the referendum is dependent upon political affiliation.

16. Three judges in a municipality are assigned cases on a random basis. Determine if the chance of being found guilty is dependent on the judge presiding over the case; based on the following information:

		Judge		
		A	*B*	*C*
Verdict:	Guilty:	2	8	9
	Not Guilty:	8	6	3

17. A sample of 150 consumers of product category *X* were asked their preference among three new product concepts. The preferences were tabulated by the brand they currently used most often. Is reaction to the new product concepts dependent on current brand usage?

		Current brand		
		A	*B*	Other
Preferred Concept:	I:	17	23	19
	II:	20	15	11
	III:	23	12	10

18. A sample of 100 coffee drinkers were asked if they could tell a difference between two cups of coffee. For half the sample the two contained exactly the same brew;

for the other half of the sample the two cups represented alternative brews. Analyze the data below to determine if the brews are distinguishable.

		Stimulus material	
		Two identical brews	Two different brews
Reaction:	Perceived difference:	32	28
	No perceived difference:	18	22

19. A personnel manager sought to develop a test which would predict job success. The predictions based on the test and the supervisor evaluations of the employees were as indicated below. Is there a relationship between the test's predictions and the supervisor's evaluations?

		Supervisor evaluation	
		Success	Failure
Prediction:	Success:	30	5
	Failure:	10	5

20. In the situation described in the preceding exercise, would the supervisor's knowledge of the prediction for a particular employee influence the results? Explain.

21. The responses of a sample of voters to two referendums are shown below. Are the referendums independent of one another?

		Vote on referendum B	
		Yes	No
Vote on referendum A	Yes:	145	15
	No:	5	35

22. In the preceding exercise 80% of the voters endorsed referendum *A* and 75% endorsed referendum *B*. Do the referendums differ in appeal? (Caution: These are not independent percentages).

23. Below are the responses to two test items made by a sample of 200 individuals. Are the items independent of one another?

		Answer to item 2	
		Correct	Incorrect
Answer to item 1:	Correct:	115	45
	Incorrect:	35	5

24. In the preceding exercise, determine if the level of correct response for the two items differs.

25. Compare the data of exercises 21 and 23. What is the fundamental difference in the nature of the relationships exhibited in each?

26. Give verbal descriptions of the concepts of loglinear models, log odds, the logit of p, and logistic regression.

27. Discuss stereotypes in terms of the 2×2 contingency table.

28. Explain why χ^2 is not sensitive to ordinal data.

Chapter 14: DISCRIMINANT ANALYSIS

1. Discuss the similarity and difference between regression analysis and discriminant analysis.

2. Think of five different criterion variables that are qualitative in nature.

3. Explain how the analysis of variance procedure is involved in determining the extent of relationship between the predictor variables and criterion variable in a discriminant analysis.

4. What two pieces of output from a discriminant analysis are essential for classifying objects on the criterion variable?

5. Define in words and in mathematical notation the discriminant function.

6. For a given set of data what determines how many discriminant functions are necessary?

7. Explain the cutoff score associated with a discriminant function in terms of the criterion for its selection.

8. What is the distinction between *b*-coefficients and *beta* coefficients in a discriminant analysis?

9. What is the advantage of expressing the discriminant function in terms of the standardized variables?

10. A discriminant function $L = 9x_1 + 3x_2 - 5x_3$ is associated with a cutoff score of $L_c = 36.1$ as the maximum value for assignment to level 1 of the dichotomous criterion variable. If objects have the following scores on predictor variables x_1, x_2, and x_3, respectively, to which level of the criterion variable would they be assigned?
    a. 3, 6, 1.4
    b. 5, 1, 2.1
    c. 2, 3, 1.1
    d. 2, 7, .8

11. Explain why the coefficients of the discriminant function in the preceding exercise give no indication of the relative importance of the three predictor variables in determining an object's value on the criterion variable.

12. Below are two discriminant functions; one in raw score form and the other in standardized score form. Is it possible that they are based on the same set of data? Explain.

$$L = 57x_1 - 17x_2 + .32x_3$$

$$L_z = .14z_1 - .81z_2 + .57z_3$$

13. a. In the preceding exercise, which predictor variables are most important in classifying objects on the criterion variable?
    b. Which is the least important? Explain.

14. Explain why it is necessary to take into account the intercorrelations of the predictor variables themselves in order to arrive at the weights in a discriminant function.

15. The cutoff score for the discriminant function $L = 6x_1 - 2x_2 + 5x_3 - 100x_4$ is $L_c = 14$. Objects with discriminant function values exceeding L_c are assigned to the second level of the dichotomous criterion variable. The values for several objects on variables x_1, x_2, x_3, and x_4 are given below. Assign each to a level of the criterion variable.
    a. 1, 4, 2, .02
    b. 1, 3, 1, .01
    c. 3, 5, 2, .03

16. Describe a procedure for validating a discriminant function.

17. Explain why a discriminant function will rarely perform as well on a set of data independent of the one upon which it was formed.

18. For each of the following discriminant functions based on standardized scores, identify which variable is most important in correctly classifying objects on the criterion variable.
    a. $L_z = .11z_1 - .63z_2 + .21z_3$
    b. $L_z = .31z_1 - .42z_2 - .27z_3$
    c. $L_z = .24z_1 - .33z_2$

19. Suppose we have a criterion variable with four levels, and three predictor variables. How many discriminant functions are needed? Explain how you arrived at your answer.

20. How many discriminant functions are needed when we have five predictor variables and a criterion variable with four levels? Explain.

21. Think of a criterion variable with three or more levels. Name the objects upon which it would be measured and develop a list of at least five predictor variables that might be successful in classifying the objects on the criterion variable.

22. What is meant by a derived variable? Give an example.

23. Sketch a scatter diagram showing the relationship between a specific dichotomous criterion variable of your choosing and a continuous predictor variable.

24. Discuss the common aspects of correlation analysis, regression analysis, analysis of variance, and discriminant analysis.

Chapter 15: FACTOR ANALYSIS

1. Discuss the key difference between factor analysis and other associative techniques such as correlation, regression and discriminant analysis.

2. Name and sketch the three basic matrices involved in the factor analysis procedure, identifying the rows and columns of each matrix.

3. Select a research area of particular interest to you, then list as many variables as you can that might usefully be subjected to a factor analysis. Specify the objects upon which the variables would be measured, and decide how many objects should be measured for the number of variables you have specified.

4. Discuss alternative criteria for deciding how many factors to extract from a given set of variables.

5. "Objects ________ on factors, variables ________ on factors." Fill in the missing words.

6. Discuss the meaning of a factor loading. What is its maximum and minimum value? Explain.

7. In general, what is meant by a factor? What is meant by a factor in the context of factor analysis?

8. What is accomplished by rotating a factor loading matrix?

9. a. If a factor analysis involves 65 variables, what percentage of the data's total variation would be expected to be accounted for by a single variable selected at random?
   b. What if 150 variables were involved in the analysis?

10. What is the meaning of an eigenvalue?

11. Assume a factor analysis involving 35 variables generates factors accounting for the percentages of total variance listed below. Translate each into its eigenvalue equivalent.
    a. 17% b. 45% c. 3% d. 61% e. 35%

12. Suppose a factor analysis involving 63 variables generates factors with the eigenvalues listed below. What percentage of the total variance of the data set is accounted for by each of the factors?
    a. 7 b. 5.3 c. 20 d. 1.1 e. .8

13. A factor analysis involves forty variables. What is the sum of the eigenvalues associated with the extracted factors?

14. Following are the eigenvalues associated with the ten factors of a factor analysis: 3.1, 2.0, 1.2, .8, .6, .6, .5, .5, .4, .3. How many factors would you be inclined to retain? Explain.

15. In the preceding exercise, what percentage of the total variance is accounted for by the first three extracted factors?

16. As a rule of thumb, what constitutes a high factor loading?

17. Interpret the meaning of a negative factor loading.

18. The twelve factors of a factor analysis account for 22%, 16%, 12%, 10%, 8%, 7%, 6%, 5%, 4%, 4%, 3%, and 3% of the total variance, respectively.
    a. Draw the scree curve for the data.
    b. How many factors would you retain based on this information alone? Explain.

19. Think of some alternative names for the factors presented in Table 3.

20. Define and explain the GIGO principle.

21. Study the factor loading matrix presented in Table 5 then summarize it in an intelligible manner, and discuss the results.

22. Explain how factor analysis might be used as a preliminary to a regression analysis.

23. In inverse factor analysis, inter-relationships are found between the objects rather than the input variables, with the intent of forming mutually exclusive and exhaustive groups of objects. However, since every object will have a loading on every factor, just as every variable has a loading on every factor in the conventional factor analysis, how might you go about assigning each object to one and only one factor?

24. Is the interpretation of a factor analysis a well-defined procedure? Explain.

25. It has been said that the various departments within a university represent a factor analysis of every conceivable variable of interest in the world. Explain what is meant by this observation. Do you agree with it? Why or why not.

Chapter 16: CLUSTER ANALYSIS

1. a. Specify a specific set of objects that could be subjected to a cluster analysis.
   b. Describe three alternative methods you would consider for obtaining the inter-object similarity measures.

2. The ordered pairs below represent the values on two variables for objects *A* through *F*. Plot the objects on graph paper.

A (3, 7)	*D*(1, 0)
B (6, 1)	*E* (7, 6)
C (5, 8)	*F* (4, 5)

3. From the data in the above exercise, calculate the Euclidean distance between the pairs of objects below.

a. *A-B*	d. *C-D*
b. *B-E*	e. *E-F*
c. *B-F*	f. *A-D*

4. a. From the data of Exercise 2, which two objects are most similar?
   b. Which two objects are most dissimilar?

5. Consider the occupations below:

Truck driver	Architect
Physician	Janitor
Cook	Cashier

Sales clerk	Waiter/waitress
Dentist	Carpenter
Police officer	Scientist
Auto mechanic	Skilled machinist
Real estate agent	Night watchguard
Secondary school teacher	Mailroom clerk
Forest ranger	Garbage pickup person
Freight handler	Telephone operator
Garment sewer	Bank executive
Sales manager	Billing clerk
Fireperson	Factory supervisor

a. Based on your perceptions of these occupations, sort them into two clusters. Discuss the within cluster similarities and the between cluster differences.
b. Repeat the above exercise, but this time sort the occupations into three clusters.
c. Now sort them into four clusters. Which clustering solution seems to make the most sense? Discuss.

Chapter 17: MULTIDIMENSIONAL SCALING

1. Compare the cluster analysis and multidimensional scaling procedures.
2. Suppose you have the following inter-object distances for objects A, B, and C.

A-B: 4 units   A-C: 4 units   B-C: 5 units

   a. Can the three objects be represented in one dimension? Sketch and discuss.
   b. Can the three objects be represented in two dimensions? Sketch and discuss.
3. Discuss the concept of stress, and illustrate it.
4. a. Referring to the occupations listed in exercise 5 of the preceding chapter, create a two-dimensional preceptual map based on your own perceptions.
   b. Name the perpendicular axes in the map that you created.
   c. If you were to create a third dimension, coming out of the paper toward you, what would you name it and which occupations would come along with it?
5. Consider the following sports:

Soccer	Basketball
Baseball	Pole vaulting
Hockey	Golf
Football	Boxing
Swimming	Speed skating
Tennis	Archery
Bowling	Weight lifting
Wrestling	Swimming
Horse racing	Marathon racing
Gymnastics	Karate

   a. Describe a method for obtaining inter-object similarity measures.
   b. Create a two-dimensional perceptual map based on your own perceptions.

c. Name the perpendicular axes in your map.
d. How would you cluster the objects? Encircle your clusters on the map.

6. Improve upon Figure 3.
7. Improve upon Figure 4.
8. Discuss war among nations as *stress*, in terms of the two-dimensional mapping of nationalities, religions and natural resources onto the surface of the globe.

Chapter 18: TIME SERIES ANALYSIS

1. Discuss the objectives of time series analysis and the major problems involved in achieving those objects.
2. Name five variables not mentioned in the chapter which could be subjected to a time series analysis, and sketch what you think the series would look like.
3. Distinguish between seasonal, cyclical, trend, and irregular patterns of variation in a time series.
4. What is the distinction between a local and secular trend?
5. Name a technique for identifying dependencies among the values of a time series showing a large irregular component.
6. Discuss the problems inherent in trying to identify cyclical patterns in a time series.
7. Define the concept of a time series index number.
8. Discuss the similarities and differences among the simple moving average, autoregressive, and exponential smoothing forecasting techniques.
9. Given the following time series for variable y:

x	1	2	3	4	5	6	7	8	9	10	11
y	43	51	59	48	56	63	69	59	71	66	61

   a. Graph the time series.
   b. Calculate the four-period moving average for the series, and superimpose the values on the graph of part *a* (Allign each average with the most recent value involved in the average).

10. Given the following time series:

x	1	2	3	4	5
y	11	14	16	20	23

   a. Graph the time series.
   b. Using the least squares method of Chapter 11, find the best-fitting line for the series and superimpose it on the graph of part *a*.

c. What is the forecast for period 6?
d. What is the forecast for period 9?

11. Given the time series in exercise 9, index the values using period $x = 5$ as the base period.

12. During a given base and comparison period, the price p, and quantity of traded shares q, for two stocks were as follows:

	Base period		Comparison period	
	p	q	p	q
Stock *A*	\$15	12,500	\$18	15,000
Stock *B*	\$41	9,000	\$37	11,000

Perform an index number analysis of the data similar to that shown in Table 2.

13. Name an advantage and disadvantage of using index numbers instead of the actual raw observations.

14. Explain the meaning of a seasonal index number.

15. Given the following time series for variable y measured quarterly over a five-year period:

	Quarter of the year			
	I	II	III	IV
Year 1	10	14	12	20
Year 2	8	17	10	25
Year 3	11	12	14	17
Year 4	9	12	10	22
Year 5	12	18	15	18

a. Graph the time series, and name a variable that might follow the given pattern.
b. Calculate seasonal index numbers for each observation using the ratio-to-current-period-average method, and then find the average index number for each quarter.
c. Calculate seasonal index numbers for each observation using the ratio-to-moving-average method (straddled method), and then find the average index number for each quarter.
d. Compare the results of b and c above. Under what conditions might the two approaches result in markedly different average index numbers?
e. Using the average index numbers found in b, seasonally adjust the data and graph it. By inspection, does there appear to be a secular trend present?

16. A variable y is measured daily over a 4-week period:

	Day of week						
	M	Tu	W	Th	F	Sa	Su
Week 1	2	2	5	4	7	11	6
Week 2	1	2	4	3	6	10	4
Week 3	2	0	4	5	10	12	5
Week 4	3	2	3	6	8	9	5

Repeat the analysis of exercise 15 for the above data. (Note that because of an odd number of intervals within the cycle, no "centering" step is needed for the ratio-to-moving-average method.)

17. Using the time series presented in Table 7, apply the exponential smoothing forecasting system in which the most recent observation is weighted with a value .3. (Let the first forecast equal 10.)

18. For the time series in Table 7:
    a. Find the autoregressive forecasting equation for a lag of $k=1$.
    b. For a lag of $k=2$.

19. Explain how you would go about comparing the relative success of two or more forecasting systems for the same set of data.

20. Forecasting accuracy aside, what are the respective advantages of exponential smoothing and the autoregression technique.

21. Define the concept of a leading indicator, explaining how it can be used as a forecasting tool. Provide examples.

22. Forecast the percentage of the population that will be using a new product for each of the first 5 years of its life. Assume 10% of the population will try it for the first year, and 10% of the non-triers will try it each succeeding year. Also, 50% of each year's users will re-use it in the subsequent year. Assume that the drop-outs will not re-use it, and that the population size is stable from year to year.

23. Repeat the above type of analysis for a new disease with:
    a. A contagion rate of 10% and cure rate of 40%.
    b. A contagion rate of 25% and cure rate of 60%, but where the cured can recontact the disease in subsequent years.

Chapter 19: NONPARAMETRIC METHODS

1. What is the distinction between parametric and nonparametric methods? Give examples of each, pointing out their distinguishing features.

2. Use the sign test to evaluate the hypothesis that the following sample of observations was drawn from a population with a median equal to 10.

   11 6 9 4 15 7 7 12 8 4 3 8

3. Each of a sample of 16 children was asked their preference between toys A and B: 12 picked B. Use the sign test to evaluate the hypothesis that the two toys are equally preferred in the sampled population.

4. An advertiser wants to know which of two commercials is more appealing to consumers. For a two-tailed test at the $\alpha=.05$ significance level, how many consumers out of a sample of 20 would have to choose one of the commercials to conclude that they differ in appeal?

5. Use the sign test to evaluate the hypothesis that the following sample of difference scores represent a population with a median of 0.

$$+3 \quad +7 \quad -1 \quad -10 \quad -4 \quad +6 \quad +9 \quad +2 \quad -6 \quad +9 \quad +8 \quad +1 \quad +3 \quad +3 \quad +1$$

6. Study exercises 2 and 5 and interpret the statement that the sign test, in these applications, discards information about the observations. Is the same true for the applications in exercises 3 and 4?

7. A comparison of the ages of samples of shoppers at two retail outlets suggested the populations were nonnormal with unequal variances. Consequently, a comparison was made of the number of members of each sample that were above or below the median of the two combined samples, as shown in the contingency table below. Use χ^2 to determine if the median ages for the two sampled populations are equal.

		Store *A* shoppers	Store *B* shoppers	Totals
Age:	Above overall median:	36	24	60
	Below overall median:	14	46	60
	Totals:	50	70	120

8. Describe the procedure for conducting the rank sum test for two independent samples. Could the rank sum test have been applied to the raw data of the preceding exercise? Explain why or why not. The rank sum test is sometimes called a test for "slippage." Show this with a sketch of the parent populations.

9. Consider the following criterion measures for a control and experimental group:
   Control: 76 57 93 89 84 97
   Experimental: 85 72 75 67 80 79 63 83
   Use the rank sum test to evaluate the hypothesis that these are samples from the identical population.

10. Does the rank sum test, as used in the preceding exercise, discard any information inherent in the raw scores? Explain.

11. For samples of size $n_1 = 4$ and $n_2 = 5$, what value of the rank sum statistic R is needed for significance at the $\alpha = .05$ level?

12. In the preceding exercise, what would be the maximum possible value of R?

13. What is the expected value of R for samples of size $n_1 = 9$ and $n_2 = 14$?

14. What is the standard error of R for the pair of samples in the preceding exercise?

15. Use the normal distribution to evaluate $R = 190$ based on a pair of samples of sizes $n_1 = 12$ and $n_2 = 15$.

16. Under what conditions is the normal distribution, as used in the preceding exercise, a good approximation of the exact rank sum test?

17. Below are the shares of market and advertising expenditures for the brands in a particular product category:

	Brand						
	A	B	C	D	E	F	Others
Share of market (%):	15.7	3.9	10.6	9.6	12.3	26.2	21.7
Ad. $ (000,000):	7.6	3.5	6.1	6.8	8.3	10.1	7.1

a. Calculate the rank correlation coefficient for the data.
b. Are the two variables independent of one another?

18. Is it likely than an $r_0 = .55$, based on 9 pairs of observations, would occur by chance if the variables were truly unrelated? Explain.

19. Given an $r_0 = .47$, based on 20 pairs of observations, test the hypothesis that the two variables involved are independent. Use a one-tailed test with $\alpha = .01$.

20. One use of the correlation coefficient is to determine the similarity between individuals with respect to some criterion. Consider, for example, the rankings of 7 concepts in overall appeal, made by individuals 1, 2, and 3:

	Concepts						
	A	B	C	D	E	F	G
Indiv. 1:	3	7	2	5	6	1	4
Indiv. 2:	1	6	2	7	3	4	5
Indiv. 3:	5	4	1	3	2	7	6

a. Calculate r_0 between the rankings of individuals 1 and 2.
b. Calculate r_0 between the rankings of individuals 1 and 3.
c. Calculate r_0 between the rankings of individuals 2 and 3.
d. Which two individuals are most alike with respect to their reactions to the concepts.
e. Which two individuals differ most?

21. Using the procedure described in the preceding exercise, which pair of individuals would be more similar—a pair with an $r_0 = 0$ or a pair with an $r_0 = -.59$? Explain.

22. Discuss the advantages and disadvantages of nonparametric techniques vis-a-vis the alternative parametric methods.

Chapter 20: ADVANCED PROBABILITY TOPICS

1. What is the distinction between a permutation and combination? Provide examples of each.
2. In how many ways could seven concepts be arranged in order of preference?

3. If the tastes of two randomly selected individuals are independent of one another, what is the probability that they will rank 5 foods exactly the same in terms of preference? (Assume the foods will have equal overall preference levels).

4. In exercise 3, what is the probability that the two individuals will both rank the five foods in the specified order B, E, C, D, A?

5. Evaluate the following:

   a. $7!$
   b. $(9-2)!$
   c. $(3-2)!$
   d. $0!$
   e. $(6-3)!\,3!$
   f. $3!2!$
   g. $\frac{8!}{6!}$
   h. $\frac{10!}{(10-3)!}$
   i. $\frac{9!}{(9-6)!6!}$

6. Evaluate the following:

   a. ${}_7P_3$
   b. ${}_5P_3$
   c. ${}_7C_3$
   d. $\binom{6}{4}$
   e. ${}_9P_6$
   f. ${}_{10}C_5$
   g. ${}_9P_{5,2,2}$
   h. $\binom{8}{5}$

7. From among a list of 10 TV programs, viewers are asked their 3 favorites in order of preference. How many different responses are possible?

8. An advertising manager for a magazine must arrange 6 full-page color ads, 4 full-page black and white ads, 2 half-page color ads, and 3 half-page black and white ads in an upcoming issue (one ad per page).
   a. How many different sequences of the ads are possible?
   b. How many sequences are possible in terms of the page size only?
   c. How many sequences are possible in terms of ad color only?
   d. How many sequences are possible in terms of the page size and color combinations?

9. If the ranks 1 through 10 are randomly divided into two groups of size 3 and 7, how many such partitions are possible...
   a. If the order of the ranks within the groups is irrelevant?
   b. If the order of the ranks within the groups is relevant?

10. If 15 independent well drillings result in 5 "strikes" and 10 failures, how many possible sequences of the strikes and failures are possible?

11. A food processor wants to test a product with alternative levels of sugar, salt, pepper, and garlic. If three levels of each is being considered, how many test products could be developed?

12. How many different matings are possible in a species colony containing 15 females and 10 males?

13. In a "true-false" exam consisting of 20 items, how many different sequences of answers are possible?

14. In a multiple choice exam consisting of 20 items, each with four alternatives, how many sequences of answers are possible?

15. Given the following expressions, state each in words and then evaluate them:
   a. $b(x=2;\ 7,\ .50)$
   b. $b(x=2;\ 7,\ .20)$
   c. $b(x\geq 2;\ 7,\ .50)$
   d. $b(x\leq 2;\ 7,\ .20)$
   e. $b(x\geq 9;\ 10,\ .50)$
   f. $b(x\leq 3;\ 10,\ .75)$
   g. $b(x\geq 5;\ 7,\ .30)$
   h. $b(x\leq 5;\ 7,\ .30)$

16. Expand the expression $(p+q)^7$.

17. What is the 17th term of $(p+q)^{23}$?

18. What is the probability of exactly 9 successes in 15 binomial trials in which the probability of failure is .3 on any given trial?

19. What is the probability of no failures in 20 binomial trials if the probability of a failure on a given trial is $1/6$?

20. Write the expression that will provide the probability of exactly 3 successes in a set of 50 binomial trials in which the probability of success on a given trial is $1/4$.

21. Given the following expressions, state each in words and then evaluate them:
   a. $h(x=2;\ 7,\ 5,\ 15)$
   b. $h(x=0;\ 4,\ 2,\ 8)$
   c. $h(x=3;\ 3,\ 3,\ 9)$
   d. $h(x\leq 1;\ 3,\ 3,\ 9)$

22. If 3 out of 10 particular job applicants would prove to be failures, what is the probability that a random hiring of 5 of those 10 would contain no failures?

23. If the day's production of 20 automobiles at a particular factory contains 3 defective ones, what is the probability that a dealer receiving 5 automobiles from the 20 will have at least one defective unit?

24. If a bushel of 40 apples contains 3 with worms, what is the probability that a pie made with a random sample of 5 apples contains at least one infested apple? Just show the necessary calculations.

25. Evaluate the following Poisson probabilities, $f_p(x;\ \lambda)$:
   a. $f_p(x=2;\ 1.5)$
   b. $f_p(x=0;\ 2.0)$
   c. $f_p(x\leq 1;\ .7)$
   d. $f_p(x\leq 2;\ 3.4)$
   e. $f_p(x\geq 4;\ 1.5)$

26. If 15 of 1000 individuals are typically afflicted with a certain disease within a one-year period, what would be the probability of at least 20 afflictions in a year?

27. If 3 percent of the members of a particular species have a peculiar characteristic, what is the probability that a sample of 60 members of the species will contain at least one member with that characteristic.

28. A mail-order marketer receives 7 orders from a test mailing of 1000 solicitations. Test the hypothesis H_0: $\lambda = 4$.

29. If .5 percent of a species' offspring are mutations, what is the probability that a sample of 200 offspring contains no mutations?

30. Under what conditions does the Poisson distribution closely approximate the binomial distribution?

31. If a particular magnetic tape contains a specific type of defect on average once every 1500 feet, what is the probability of at least one defect in a 4500 foot length?

32. Write Bayes theorem in symbolic form and state it in words.

33. Factories *A* and *B* produce the same product with rates of defectives of 1.7% and 1.2%, respectively. If factory *A* produces 75% of the products on the market, what is the probability that a given defective was produced at factory *A*?

34. a. In the preceding exercise, what is the unconditional probability that a given product will be defective?
    b. What is the probability that a non-defective product was produced at factory *B*?

35. Over a long period of time, 10% of the admissions into a hospital have a particular disease. Of the diseased individuals, 60% manifested a certain symptom, while only 25% of those without the disease had the symptom. Given that an incoming patient has the symptom in question, what is the probability that the individual has the disease?

36. In the preceding exercise, what is the overall probability of an incoming patient having the symptom in question?

37. In exercise 35, what is the probability of the disease without the symptom?

38. Using the results of exercises 33 to 35, or the color preference vs. personality example in the text, discuss the meaning of the statement, "the exception proves the rule."

Chapter 21: DECISION ANALYSIS

1. Name and define the five basic components of a decision analysis problem.

2. Create a specific decision analysis problem of your choice, using hypothetical numerical values wherever necessary.
   a. Make the tree diagram for the problem.
   b. Make the payoff table for the problem.
   c. Determine the optimum decision and explain how you arrived at the conclusion.

3. Name a decision problem that involves discrete payoff values, and one that involves continuous payoff values. How can we cope with the latter situation?

4. What difference, if any, is there between buying medical insurance and betting on a race horse? What are the similarities?

5. Is it possible for both the buyer and seller of an insurance policy to have a positive expected payoff? Explain.

6. Discuss the problems in applying decision theory to non-gambling situations.

7. What is the difference, if any, between a gamble and an investment? Give examples to support your answer.

8. Will choosing the decision alternative with the largest expected payoff result in the largest payoff each and every time a decision is made? Give an example to support your answer.

9. What is meant by the heuristic function of decision analysis?

10. Create a decision analysis problem that does not involve monetary payoff values.

11. Repeat the fire insurance example of Table 1, but assuming the probability of a fire is .02.
    a. What is the expected payoff of insuring?
    b. What is the expected payoff of not insuring?
    c. Which is the optimum decision?

12. Repeat the fire insurance example of Table 1, but assuming the probability of fire is .03 and the cost of the policy is $3500.
    a. What is the expected payoff of insuring?
    b. What is the expected payoff of not insuring?
    c. Which is the optimum decision?

13. How is time series analysis relevant to decision analysis?

14. Below is the payoff table for two different $2 bets:

	Win	Lose
Bet *A*	(.50) $2	(.50) – $2
Bet *B*	(.01) $150	(.99) – $2

    a. What is the expected payoff for Bet *A*?
    b. What is the expected payoff for Bet *B*?
    c. Which is the better bet?
    d. Under what condition might an individual choose the bet with the lower expected payoff?

15. A communications firm must decide whether to launch a satellite. If the launch succeeds a net profit of $5 million will be realized, while if the launch fails a net loss of $30 million will result. Assuming the probability of a successful launch is .9, what is the best decision—to launch or not to launch?

16. Suppose a farmer has the option of planting corn or soybeans in a particular plot of land. Based on an analysis of commodity future prices, the following payoff table was constructed:

	Future Prices		
	Up	No change	Down
Plant corn	(.5) $20,000	(.3) $0	(.2) – $30,000
Plant soybeans	(.3) $5,000	(.2) $0	(.5) – $10,000

a. Sketch the decision tree for the problem.
b. What is the expected payoff from planting corn?
c. What is the expected payoff from planting soybeans?
d. What should the farmer plant?
e. What should be planted if the farmer's strategy is to minimize his maximum loss?
f. If the answers to part *d* and *e* differ, explain which approach is the more rational, or make a case for both decisions.

17. By filling in the following payoff table, show how decision analysis can be used to account for procrastination behavior:

	Success	Failure	Expected Payoff
Act now	( )	( )	
Act later	( )	( )	

18. In the above exercise, is it possible to get the same expected payoff figures using different values within the four cells? Give an example. What does this say about the number of ways we can justify procrastinating. How would you account for an individual who decides to act, but then at the last minute decides not to?

19. During a given interval of time, one individual acts on twice as many decisions as another individual, but with equal average accuracy of assessing the expected values of acting vs. not acting. How are the lives and personalities of these two individuals likely to differ?

20. Consider the following three investments:
a. You are guaranteed a 10% return after one year.
b. There is a 50% chance that you will double your money, and a 50% chance that you will lose half of it, after one year.
c. There is a 25% chance that you will quadruple your money, and a 75% chance that you will lose it all, after one year.

Which of these alternatives would you choose? Discuss the differences between individuals choosing the other alternatives, and yourself.

Answers to Selected Exercises

This section contains answers to selected exercises. Depending upon the order of operations and the extent of rounding, your answers may not agree exactly with those provided.

Chapter 1: THE NATURE OF STATISTICAL ANALYSIS

11. (a) No, (b) Yes.

Chapter 2: OBJECTS, VARIABLES, AND SCALES

5. A minimum of 2, but no maximum.
11. (f) and (h) are nominal; (c) and (i) are ordinal; (a) and (b) are interval; and (d), (e), (g) and (j) are ratio scales.
19. (a) 8, (b) 25, (c) 86, (d) 19, (e) 173

Chapter 3: FREQUENCY DISTRIBUTIONS

1. (d) 118° F (e) 34° F 3. (c) 113° F (d) 117° F 5. (c) 4% (d) 2%
7. (e) There is a tie between the 114–116 interval with a 115 midpoint, and the 117–119 interval with a 118 midpoint.
9. (a) 4, (b) 60–63, (c) 59.50–63.49, 63.50–67.49, 67.50–71.49 (or 59.5–63.5, 63.5–67.5, 67.5–71.5 and flip a coin for borderline cases, if rare).
   (d) 61.5, 65.5, 69.5, (e) 72–75, (f) 60.00–63.99, 64.00–67.99, etc.
   (g) 65.82 would be 11/19ths of the way into the interval;
   64.55 would be 5/19ths of the way into the interval.

Chapter 4: CENTRAL TENDENCY

5. (a) 10, (b) 11, (c) 12.3

7. .33
9. (a) 3, (b) 3, (c) 3.3, (d) 3.33
11. .325, which is the same as the answer to Exercise 10.
13. Skewed left.
15. (a) 62.6
17. 14,022
19. (a) 11.0, (b) 14.0, (c) 4.0, (d) 33, (e) 2.2
23. (a) It falls in the 45–49 interval, with a midpoint of 47. (b) 44.5–49.5, (c) 49.4, (d) 22, (e) 49.75.
25. (a) 24, (b) 34, (c) 3, (d) 135, (e) 132, (f) 120, (g) 0, (h) 0, (i) 0

Chapter 5: VARIATION

7. (a) 12, (b) 0, (c) 3.55, (d) $s^2 = 18.75$, (e) $s = 4.33$
9. $\sigma^2 = 40.53$, $\sigma = 6.37$
11. $s^2 = 14.64$
13. $s = 9.19$
15. $\sigma^2 = .2469$, $\sigma = .4969$
17. (a) 192, (b) $\bar{x} = 3.328$, (c) $\Sigma f_i(x_i - \bar{x})^2 = 688.33$, (d) 3.60, (e) 1.90
19. $s = 154.95$
21. (a) $s^2 = 17.33$, $s = 4.16$, (b) no change, (c) no change, (d) $s^2 = 156.0$, $s = 12.5$ (e) $s = 1.0$
23. Approximately 4
25. (a) 5, (b) 5, (c) 35, (d) 1, (e) 5, (f) 5, (g) 1
27. (a) 68.3%, (b) 2.3%, (c) 2.5%, (d) 99.0%, (e) 90.0%
29. $x = 25$ corresponds to $z = -1.43$, which would be exceeded by 92.36% of a normal distribution.
31. (a) $z = -1.00$, (b) $z = 0$, (c) $z = -2.00$, (d) $z = -1.48$, (e) $z = 1.84$, (f) $z = -1.44$, (g) $z = 1.00$, (h) $z = .08$, (i) $z = -.84$, (j) $z = 2.80$
33. $\sigma = .458$

Chapter 6: BASIC PROBABILITY

17. .83
19. They are independent outcomes.
25. .26
27. 1.11

29. 5.67 to 1 (or 17 to 3)

31. (a) 6/36, (b) 2/11, (c) 6/36, (d) 1/36, (e) 11/36, (f) 21/36, (g) 15/36, (h) 1/36, (i) 5/36, (j) 2/15, (k) 15/36.

33. (a) 35:1, (b) 17:1, (c) 11:1, (d) 8:1, (e) 31:5, (f) 5:1, (g) 31:5, (h) 8:1, (i) 11:1, (j) 17:1, (k) 35:1

Chapter 7: SAMPLING DISTRIBUTIONS

13. $\mu = 50$, $\sigma_{\bar{x}} = 2.5$

23. (a) 11, (b) 11

25. 1.26

35. (a) .12, (b) .0291, (d) 3.63

37. (a) .0157, (b) 15.73, (c) 1.573

39. .060

41. (a) .2235, (b) .0000114 or $(.15)^6$, (c) .3771, (d) .3994

Chapter 8: PARAMETER ESTIMATION

5. (a) 63.69 to 70.91, (b) 64.56 to 70.04

7. (a) 4.0345 to 4.7055, (b) 3.9238 to 4.8162

9. (a) −1.829 to −.971, (b) −1.965 to −.835

11. (a) −7.793 to −4.007, (b) −8.392 to −3.408

15. (a) .362 to .558, (b) .331 to .589

17. (a) .0848 to .1152, (b) .0800 to .1200

19. (a) .6716 to .7284, (b) 330 to 370, (c) 61.02% to 78.98%

21. (a) .378, (b) $-.08 \pm .0857$, (c) $0 \pm .0868$

Chapter 9: HYPOTHESIS TESTING

3. (a) $H_0: \mu = 61$, $H_1: \mu > 61$
   (b) a truncated or one-tailed hypothesis set

5. $H_0: \mu_1 - \mu_2 = 17$
   $H_1: \mu_1 - \mu_2 \neq 17$

7. (a) $H_0: \mu = 46$, $H_1: \mu \neq 46$, (b) $s_{\bar{x}} = 2.0$, (c) $z = 2.50$,
   (d) yes, since $z = 2.50$ exceeds the critical $z_c = 1.96$ for $\alpha = .05$,
   (e) no, since $z = 2.50$ does not exceed the critical $z_c = 2.58$ for $\alpha = .01$
   (f) approximate, since we have only estimated $\sigma_{\bar{x}}$

9. (a) $H_0: \mu = 100$, $H_1: \mu \neq 100$,
   (b) no, since $z = -1.01$ is not beyond the critical $z_c = \pm 1.96$ needed for $\alpha = .05$.
   (d) In this example, we were provided with the actual value of σ, allowing us to calculate $\sigma_{\bar{x}}$ precisely. The value of s^2 given is not needed to solve the problem.

11. (a) $H_0: \mu = 100$, $H_1: \mu < 100$
(b) no, since $z = -1.01$ is not beyond the critical $z = -1.65$ needed for $\alpha = .05$ of a one-tailed test.

17. (a) $H_0: \mu_D = 0$, $H_1: \mu_D \neq 0$
(b) $z_c = 1.96$, (c) $z = -2.14$, (d) yes, since $z = -2.14$ is beyond the critical $z_c = 1.96$ needed for $\alpha = .05$, (e) it is equal to $\alpha = .05$, (f) approximate, since the population standard deviation was estimated.

19. (a) $H_0: \mu_1 - \mu_2 = 0$, $H_1: \mu_1 - \mu_2 \neq 0$
(c) $z = 8.84$, (d) yes, since $z = 8.84$ is beyond the critical $z_c = 1.96$ needed for $\alpha = .05$.

21. (a) 14, (b) $t_c = 2.145$, (c) $t = -1.78$, (d) no, since $t = -1.78$ is not beyond the critical $t_c = 2.145$ needed for $\alpha = .05$.

25. We cannot reject the hypothesis since $z = 1.84$ is not beyond the critical $z_c = 1.96$ needed for $\alpha = .05$, when using the normal approximation to the binomial distribution.

27. The hypothesis cannot be rejected since $z = 2.33$ is not beyond the critical $z_c = 2.58$ needed for $\alpha = .01$, using the normal approximation to the binomial.

29. We can reject the hypothesis since $z = 3.87$ is beyond the critical $z_c = 2.58$ needed for $\alpha = .01$, using the normal approximation to the binomial.

33. .39 35. .95 37. .16 39. $n = 12$ 41. $n = 55$

Chapter 10: CORRELATION ANALYSIS

3. The relevant contingency table:

		Verbal score		
		Below median	Above median	Totals
Math score:	Above median:	55	45	100
	Below median:	45	55	100
	Totals:	100	100	200

5. (b) $r = .689$, (c) yes, since $r = .689$ is beyond the critical $r_c = .576$ needed for $\alpha = .05$ for a two-tailed test.

9. (a) .63, (b) same degree of correlation, (c) $-.53$, (d) $-.67$, (e) $-.27$.

Chapter 11: REGRESSION ANALYSIS

5. (a) $b = 3$, $a = 7$, (b) $b = 20$, $a = -45$, (c) $b = -1$, $a = 16$, (d) $b = 2/3$, $a = -5$

7. (c) $b = 7.00$, (d) $a = 48.67$, (e) $y = 48.67 + 7.00x$

9. $s_{y \cdot x} = 8.26$

Chapter 12: ANALYSIS OF VARIANCE

3. 1.00

5. 2.90

7. It is unlikely, since for $df = 2, 30$, the critical value for F is 3.32 for $\alpha = .05$.

13. The significance level $\alpha = .05$

19.

Source of Variation	Sum of Squared Deviations	*df*	Variance Estimate	*F*
Between groups (Fiber type)	40.13	2	20.07	3.03
Within groups (error)	79.60	12	6.63	
Totals	119.73	14		

We cannot reject the null hypothesis that the fiber types do not differ, since $F = 3.03$ with $df = 2, 12$ does not exceed the critical $F_c = 3.89$ needed for $\alpha = .05$

21.

Source of variation	Sum of squared deviations	*df*	Variance estimate	*F* ratio
Treatments	36.000	2	18.000	26.99*
Blocks	16.667	1	16.667	24.99*
Residual error	1.333	2	.667	—
Totals	54.000	5		

$^*p < .05$

23.

Source of variation	Sum of squared deviations	*df*	Variance estimate	*F* ratio
Between groups	128.00	8	16.00	4.49*
Factor *A*	25.38	2	12.69	3.56
Factor *B*	81.18	2	40.59	11.40**
$A \times B$ interaction	21.44	4	5.36	1.51
Within groups (error)	32.00	9	3.56	—
Totals	160.00	17		

$^*p < .05$ $\quad ^{**}p < .01$

Chapter 13: ANALYSIS OF CATEGORY DATA

3. $k - 1$

5. It also increases.

7. (a) No, $\chi^2 = 4.07$ does not exceed $\chi_c^2 = 6.64$
(b) No, $\chi^2 = 12.16$ does not exceed $\chi_c^2 = 13.28$
(c) Yes, $\chi^2 = 17.54$ exceeds $\chi_c^2 = 16.81$

11. (a) 9.33, 9.33, 9.33
    (b) $\chi^2 = 1.36$
    (c) $df = 2$
    (d) No, $\chi^2 = 1.36$ does not exceed $\chi_c^2 = 5.99$

13. We cannot reject the hypothesis that it is a fair coin. The 7:13 Heads: Tails split yields a $\chi^2 = 1.80$ which does not exceed $\chi_c^2 = 3.84$ needed for $df = 1$ at $\alpha = .05$. This test is equivalent to the normal curve approximation to the binomial distribution and therefore is not exact.

15. We can reject the hypothesis that the attitude toward the referendum is independent of party affiliation, since the obtained value of $\chi^2 = 8.10$ with $df = 2$ exceeds $\chi_c^2 = 5.99$ for $\alpha = .05$.

17. We cannot reject the hypothesis that concept preference is independent of current brand used, since the obtained value of $\chi^2 = 5.69$ with $df = 4$ does not exceed $\chi_c^2 = 9.49$ for $\alpha = .05$.

19. The obtained $\chi^2 = 2.38$ with $df = 1$ does not exceed $\chi_c^2 = 3.84$ needed for $\alpha = .05$. We cannot reject the hypothesis that the predictions and evaluations are independent of one another.

21. Since the obtained $\chi^2 = 104.17$ with $df = 1$ exceeds $\chi_c^2 = 6.64$ needed for $\alpha = .01$, we can reject the hypothesis that the responses to the two referendums are independent of one another.

23. Since the obtained $\chi^2 = 4.17$ with $df = 1$ exceeds $\chi_c^2 = 3.84$ needed for $\alpha = .05$, we can reject the hypothesis that the responses to the two items are independent of one another.

Chapter 14: DISCRIMINANT ANALYSIS

13. (a) Variable 2, (b) Variable 1

15. (a) $L = 6$, assign to first level (i.e., Group I)
    (b) $L = 4$, assign to first level (i.e., Group I)
    (c) $L = 15$, assign to second level (i.e., Group II)

19. 3

Chapter 15: FACTOR ANALYSIS

5. Score, load

9. (a) 1.54%, (b) .67%

11. (a) 5.95, (b) 15.75, (c) 1.05, (d) 21.35, (e) 12.25

13. 40

15. 63%

23. Typically, the object is assigned to the factor "group" upon which it has the highest loading. Objects with high negative loadings can be discarded from the analysis, since they cannot easily be interpreted in the same way as *variables* with negative loadings in the conventional form of factor analysis.

Chapter 16: CLUSTER ANALYSIS

3. (a) 6.7, (b) 5.1, (c) 4.5

Chapter 17: MULTIDIMENSIONAL SCALING

5. The sports are probably best portrayed in the three dimensions of accuracy, speed, and strength.

Chapter 18: TIME SERIES ANALYSIS

9. (b) 50.25, 53.50, 56.5, 59.00, 61.75, 65.50, 66.25, 64.25

11. 77, 91, 105, 86, 100, 112, 123, 105, 127, 118, 109

15. (b)

	Quarterly Index Numbers			
	I	II	III	IV
Year 1	71	100	86	143
Year 2	53	113	67	167
Year 3	81	89	104	126
Year 4	68	91	75	166
Year 5	76	114	95	114
Averages:	70	101	85	143

(c) These are the 4 quarter moving averages (and the centered values) needed for calculating the ratio-to-moving-average index numbers:

	I		II		III		IV	
Year 1				14.00		13.50		14.25
					(13.75)		(13.88)	
Year 2		13.75		15.00		15.75		14.50
	(14.00)		(14.38)		(15.38)		(15.12)	
Year 3		15.50		13.50		13.00		13.00
	(14.00)		(14.50)		(13.25)		(13.00)	
Year 4		12.00		13.25		14.00		15.50
	(13.50)		(12.62)		(13.62)		(14.75)	
Year 5		16.75		15.75				
	(16.12)		(16.25)					

The ratio-to-moving-average index numbers

	I	II	III	IV
Year 1	—	—	87	144
Year 2	57	118	65	165
Year 3	79	83	106	131
Year 4	67	95	73	149
Year 5	74	111	—	—
Average of Years 2-4:	68	99	81	148

(e)

Seasonally adjusted data

	I	II	III	IV
Year 1	14.3	13.9	14.1	14.0
Year 2	11.4	16.8	11.8	17.5
Year 3	15.7	11.9	16.5	11.9
Year 4	12.9	11.9	11.8	15.4
Year 5	17.1	17.8	17.6	12.6

Chapter 19: NONPARAMETRIC ANALYSIS

3. Based on the binomial distribution, the probability of 12 or more out of 16 preferences in favor of one toy or the other is $2 \times .0384 = .0768$, which is not an unlikely outcome vis-a-vis an $\alpha = .05$ significance level. Therefore, we cannot reject the hypothesis that the two toys are equally preferred.

5. Given the 15 difference scores, the probability of 11 or more of either sign is $2 \times .0592 = .1184$, which is not an unlikely outcome vis-a-vis an $\alpha = .05$ significance level. Therefore, we cannot reject the hypothesis that the median difference is 0.

7. The obtained value of $x^2 = 16.59$ with $df = 1$ exceeds $x_c^2 = 6.64$ needed for $\alpha = .01$. Therefore, we can reject the hypothesis that the two stores draw shoppers of equal median age.

9. The rank sum for the smaller group is $R = 56$. Since it does not fall outside the 30-60 "acceptance" interval for an $\alpha = .06$ when the sample sizes are 6 and 8, we cannot reject the hypothesis that these two samples represent the same population.

11. R should be outside the interval from 12 to 28 in order to reject the null hypothesis at the $\alpha = .06$ significance level.

13. $\mu_R = 108$

15. $z = 1.07$ is not beyond $z_c = \pm 1.96$ therefore we cannot reject the hypothesis that the two samples represent the same population.

17. (a) $r_o = .82$
    (b) Since $r_o = .82$ exceeds the critical value of $r_o = .79$ for an $\alpha = .048$ we can reject the hypothesis that the two variables are independent of one another.
19. The obtained value of $t = 2.26$ with $df = 18$ does not exceed $t_c = 2.55$ needed for $\alpha = .01$ for a one-tailed test, therefore we cannot reject the hypothesis that the two variables are independent of one another.

Chapter 20: ADVANCED PROBABILITY TOPICS

3. 1/120
5. (a) 5,040, (b) 5,040, (c) 1, (d) 1, (e) 36, (f) 12, (g) 56, (h) 720, (i) 84
7. 720
9. (a) 120, (b) ${}_{10}P_3 \times 7! = 10! = 3{,}628{,}800$
11. 81
13. $2^{20} = 1{,}048{,}576$
15. (a) .1641, (b) .2753, (c) .9375, (d) .8520
    (e) .0107, (f) .0035, (g) .0288, (h) .9962
17. ${}_{23}C_7 p^7 q^{16}$
19. $\left(\frac{5}{6}\right)^{20} = .026$
21. (a) .392, (b) .214, (c) .012
23. .601
25. (a) .2510, (b) .1353, (c) .8442, (d) .3397, (e) .0656
29. .3679, 31. .9502, 33. .8095, 35. .2105, 37. 0559

Chapter 21: DECISION ANALYSIS

11. (a) − \$2,000, (b) − \$2,000, (c) it's a toss-up
15. The launch decision is best since its expected value is \$1.5 million, while the no-launch alternative has a \$0 expected value.

Statistical Tables

Table I Random Digits

Each location in the table is equally likely to be the digit 0, 1, 2, 3, 4, 5, 6, 7, 8, or 9.

39203	59841	91168	32021	82081	60164	37385	52925	91004	71887
39965	79079	97829	95836	26651	12495	68275	20281	73978	07258
17752	87652	07004	95860	89325	56997	70904	91993	13209	50274
04284	63927	07533	60557	41339	16728	96512	11116	92345	04612
03440	97786	37416	24541	36408	63936	36480	87028	05094	95318
07466	12899	31434	06525	81175	38234	24468	30891	89620	50129
83343	72721	52695	36309	67961	73792	63300	89222	10618	24229
03745	48015	85373	77206	76214	85412	83510	73998	13500	65084
27975	70407	56983	07913	38682	89173	40739	40168	95705	46872
54284	28109	48080	80215	85753	64411	27938	56201	16005	49409
79521	93795	56291	03839	16098	44436	22678	37566	45822	26879
17817	48797	59971	28104	68171	05068	98190	33721	13991	73487
56213	82716	77356	91791	31267	19598	25159	28785	57736	72346
75194	03658	65212	50828	73031	12498	30153	80522	30866	05307
44549	28479	49939	43539	66337	61547	25104	27361	27060	17720
11543	45735	21121	46119	96548	48237	30815	01082	00715	18213
27327	47369	72686	74153	67849	91820	22255	91564	28009	19796
65332	83444	40231	84229	48713	46748	54693	63440	03439	97497
45214	30409	35466	73494	39421	86061	88928	55676	68453	66827
77929	36175	61017	71350	93393	32687	29040	74575	45306	22552
54366	88887	16301	19105	51147	31217	41907	42982	64904	63597
08535	65466	48869	58315	23905	24696	66332	22822	37808	78375
36947	67802	81864	59051	52076	34784	06530	51015	39540	61780
28323	33789	56413	16652	28571	53781	63579	42659	53203	29708
16748	41349	75175	66405	75745	33003	32043	01747	49361	61584
33178	69744	11252	49458	86585	85536	92257	24864	48761	31924
26466	93243	88962	31547	05650	29480	92795	39219	22342	60169
36535	14197	72029	40094	61100	17633	38541	08250	04353	13417
66835	93340	09121	97179	24446	47809	87930	83677	46036	07924
09357	02826	35480	92998	35244	39454	50956	36244	31511	40640
07296	75285	29833	78926	48012	97299	56635	57142	00203	77302
01106	48819	40679	96311	90666	91712	16907	65802	94408	76429
15742	99837	87999	36431	96530	84598	62879	82602	57911	18505
16523	51356	37907	65491	39889	49415	97503	09430	39471	12136
03536	42548	50478	54022	18614	03129	68513	08643	91870	93123
73445	35057	97928	83183	57729	35701	70757	28092	97686	90810
52017	99654	63051	87131	87755	29329	52001	24808	54075	48002
63724	57039	06679	46472	92762	75952	54470	88720	57702	61299
16675	01990	38803	84706	24066	41937	26551	58381	04810	35915
01377	36919	49327	24518	61098	25962	04427	33234	04480	02438
49752	61849	05823	84198	18174	74419	10322	95196	47893	77825
40734	81595	96763	68282	34155	29452	94005	23972	66115	40478
64213	91973	62604	00789	21825	25568	00981	89250	24446	86013
24505	41214	03031	34756	31600	84374	36871	83645	80482	22081
34248	31337	78109	49077	10187	84757	45754	51435	52726	24296
60229	06451	61294	53777	17640	85533	10178	23212	02002	08264
36712	16560	35055	99750	53169	58659	37377	53580	16829	10472
94150	42762	54989	58564	12434	81297	36197	84099	55629	03717
36402	94992	51794	59245	87178	84460	58370	34416	75064	07568
15853	95261	90876	66395	72788	66605	08718	96740	45414	81015

Selected from the Rand Corporation's, *A Million Random Digits with 100,000 Normal Deviates*, Free Press, Glencoe, IL, 1955. Used with the permission of the Rand Corporation.

Table I Random Digits (Continued)

84807	71928	78331	51465	39259	63729	32989	80330	57238	98955
98408	62427	04782	69732	83461	01420	68618	11575	24972	14040
61825	69602	11652	56412	22210	03517	40796	29470	49044	10343
39883	29540	45090	05811	62559	50967	66031	48501	05426	82446
68403	57420	50632	05400	81552	91661	37190	95155	26634	01135
58917	60176	48503	14559	18274	45809	09748	19716	15081	84704
72565	19292	16976	41309	04164	94000	19939	55374	26109	58722
58272	12730	89732	49176	14281	57181	02887	84072	91832	97489
92754	47117	98296	74972	38940	45352	58711	43014	95376	57402
34520	96779	25092	96327	05785	76439	10332	07534	79067	27126
18388	17135	08468	31149	82568	96509	32335	65895	64362	01431
06578	34257	67618	62744	93422	89236	53124	85750	98015	00038
67183	75783	54437	58890	02256	53920	61369	65913	65478	62319
26942	92564	92010	95670	75547	20940	06219	28040	10050	05974
06345	01152	49596	02064	85321	59627	28489	88186	74006	18320
24221	12108	16037	99857	73773	42506	60530	96317	29918	16918
83975	61251	82471	06941	48817	76078	68930	39693	87372	09600
86232	01398	50258	22868	71052	10127	48729	67613	59400	65886
04912	01051	33687	03296	17112	23843	16796	22332	91570	47197
15455	88237	91026	36454	18765	97891	11022	98774	00321	10386
88430	09861	45098	66176	59598	98527	11059	31626	10798	50313
48849	11583	63654	55670	89474	75232	14186	52377	19129	67166
33659	59617	40920	30295	07463	79923	83393	77120	38862	75503
60198	41729	19897	04805	09351	76734	10333	87776	36947	88618
55868	53145	66232	52007	81206	89543	66226	45709	37114	78075
22011	71396	95174	43043	68304	56773	83931	43631	50995	68130
90301	54934	08008	00565	67790	84760	82229	64147	28031	11609
07586	90936	21021	54066	87281	63574	41155	01740	29025	19909
09973	76136	87904	54419	34370	75071	56201	16768	61934	12083
59750	42528	19864	31595	72097	17005	24682	43560	74423	59197
74492	19327	17812	63897	65708	07709	13817	95943	07909	75504
69042	57646	38606	30549	34351	21432	50312	10566	43842	70046
16054	32268	29828	73413	53819	39324	13581	71841	94894	64223
17930	78622	70578	23048	73730	73507	69602	77174	32593	45565
46812	93896	65639	73905	45396	71653	01490	33674	16888	53434
04590	07459	04096	15216	56633	69845	85550	15141	56349	56117
99618	63788	86396	37564	12962	96090	70358	23378	63441	36828
34545	32273	45427	30693	49369	27427	28362	17307	45092	08302
04337	00565	27718	67942	19284	69126	51649	03469	88009	41916
73810	70135	72055	90111	71202	08210	76424	66364	63081	37784
60555	94102	39146	67795	05985	43280	97202	35613	25369	47959
58261	16861	39080	22820	46555	32213	38440	32662	48259	61197
98765	65802	44467	03358	38894	34290	31107	25519	26585	34852
39157	58231	30710	09394	04012	49122	26283	34946	23590	25663
08143	91252	23181	51183	52102	85298	52008	48688	86779	21722
66806	72352	64500	89120	13493	85813	93999	12558	24852	04575
08289	82806	36490	96421	81718	63075	54178	39209	03050	47089
12989	31280	71466	72234	26922	04753	61943	86149	26938	53736
44154	63471	30657	62298	56461	48879	54108	97126	43219	95349
63788	18000	10049	49041	28807	64190	39753	17397	48026	76947

Table I Random Digits (Continued)

78430	41928	49878	82180	02986	03277	85181	96077	00659	41871
07863	13414	76306	51980	84228	12883	38365	91592	74379	13866
32321	91324	37880	95452	31527	80901	29889	36203	27743	74489
95070	36517	67584	35751	11060	40353	15332	23997	27336	24279
87206	96890	55249	33560	74471	17002	10975	09199	76527	72317
26440	51243	24554	69642	11187	72190	82285	13794	53720	51748
41891	15361	93112	36080	73068	99598	30775	70331	61749	67411
42439	64699	17263	88392	67665	25475	69981	20671	63027	52722
61731	30743	54385	11838	23577	86165	61587	94306	32708	69154
35304	31248	88424	04778	26859	91464	00841	52430	96223	58288
84433	96803	71684	87226	77023	16202	29315	85191	13093	70332
21496	34638	04723	20855	83112	79093	76176	38943	62595	81904
68751	82350	61143	98260	54514	66945	48129	53944	07455	51565
64026	13539	26282	04818	24657	64247	49185	53643	29526	80568
08939	79721	83524	39276	26649	36611	54565	23827	52909	52226
25672	69980	01220	95872	19009	01449	66189	70539	76013	81897
19311	20843	95775	66661	28488	84824	26080	16946	76397	35886
07118	48379	27498	35264	54995	25028	82089	53554	93380	80369
61263	71051	71076	78586	88453	46606	98191	44612	21977	35776
72910	75684	78411	35006	53246	85908	62244	28599	64430	99888
41024	02668	35409	19690	55644	40781	38977	76619	98111	97436
54167	46600	68243	93639	64428	98790	06482	52846	71149	49165
86088	72610	29566	83019	43069	07015	96383	57419	43639	54244
62571	11846	64555	66501	43015	99060	16426	04954	80824	50812
25646	14826	71348	21171	75901	10486	66603	44924	77359	32672
48611	75706	55419	91766	54313	48282	71170	18666	49490	42359
87418	85663	76456	76570	82005	25071	87983	52144	44550	74109
28553	39599	78675	88219	66706	31909	16243	48626	89548	26693
59608	15311	58459	79330	63554	76310	37805	76808	05963	17714
25170	19298	89788	89687	43673	42621	03495	18881	62148	21806
27114	26753	18141	48890	58595	13628	88469	80769	80900	73242
02025	97477	24129	39861	58284	58802	89928	04248	77091	03082
35900	13901	21381	85427	28397	71885	82843	99179	14383	25330
66850	38514	91121	52626	26152	59741	52321	41858	05735	22684
29636	90501	41942	95824	05949	81648	21484	53382	88500	61323
20309	98582	70822	06613	74978	52668	80347	16925	51369	22102
25897	63995	09393	40461	28505	50114	07949	28388	18587	52538
64469	86768	17259	60715	40689	65318	09903	60867	71068	77075
50729	37461	03846	57091	67102	06772	55886	74325	15545	19660
42757	29440	04464	33800	08192	25420	07797	52746	42703	12244
84778	48723	24130	10402	02742	99232	86292	61863	53266	84796
24683	23733	23722	26571	80500	62377	27457	30711	37785	18135
42672	25838	02376	73951	11594	85636	56949	79720	10272	75359
50071	55670	45351	85848	67239	48381	47206	25461	56847	63347
41007	99628	29089	87080	68711	76148	71171	89335	21804	54064
23088	47607	57453	73795	06625	56375	30879	87792	70489	86525
46285	37426	29643	84872	74009	39626	24893	56387	72107	47316
33801	28058	53708	69838	19070	44239	59639	70558	65223	07143
36591	85097	79603	50138	40202	02530	88912	20779	86754	20299
75742	18166	00221	88682	80750	49621	07774	75575	28919	16130

Table I Random Digits (Continued)

69646	90312	78612	16226	82083	37253	17871	27639	41481	80595
73399	81179	95187	23079	11664	13781	35118	74347	48120	24831
16198	72047	61633	09005	45888	72153	37394	69180	98988	43413
65695	03685	46983	71608	51419	73164	43124	60401	12347	15386
01347	05357	39655	58816	73829	10065	33905	39625	22764	27502
71911	76315	70232	55220	91833	35366	94688	15851	16604	11079
91425	65220	35977	80393	35445	13778	95878	59019	74597	52285
99256	76783	24094	23600	72289	37715	60961	40281	97409	14587
18661	12968	66351	60111	21967	98959	89030	89684	28851	92787
33211	97675	46427	11911	38810	96933	74118	58983	11672	31864
65605	93505	64129	54327	48180	29604	40944	74432	43025	69354
86085	61911	90089	43839	81259	13881	18847	01323	70806	60059
94626	59218	31003	97933	10236	20929	47742	95381	57787	22939
39080	18210	03809	79751	15908	66319	40766	23123	49434	34303
38585	01695	60797	01636	22957	72990	66578	57143	64013	92452
84199	87855	26041	11720	68742	04843	38617	41346	63668	21253
45180	32395	41294	57527	31188	29114	28020	28160	23466	69822
91449	59545	66646	03526	03206	51770	07826	03855	18683	96534
03484	91441	50638	91952	26842	87203	57145	06388	71702	06274
81946	40983	02379	96024	68861	41027	50271	68712	23220	01538
58928	65252	99959	02703	94744	95844	29531	23663	28175	76699
30751	41453	46591	42356	36416	32338	21419	80531	61696	44754
36852	13204	63023	59316	49928	09697	76504	24774	67908	61289
70631	46050	12059	46115	36953	44908	20851	08902	17496	14404
05897	25533	60963	06811	77582	05636	37683	38348	80953	33703
21826	18286	14524	58214	71924	13487	09601	34575	47012	58758
38881	45465	01785	66584	58209	29981	04935	82855	16799	04066
32412	11942	33275	04011	56956	52944	00602	84617	16765	35087
66969	52749	74039	29799	04644	45823	10845	48994	99909	33370
98773	25122	02663	62295	76587	77167	88451	79691	57467	15464
03682	52876	21049	23878	15192	80767	49345	47174	11631	50849
11303	82571	32812	72023	11268	70705	12155	44815	30742	06489
81730	86862	98124	69512	68243	78662	79943	84693	89574	46209
26847	91358	19666	90971	08113	31567	19927	23081	77086	43837
49345	01536	67740	99734	34278	24006	85755	17842	16231	04575
17822	70450	88628	89492	43980	39317	63772	79086	25930	56648
29492	20296	54952	08632	38798	41328	46686	96846	27757	97779
65059	50866	95747	90465	54326	96444	44177	84764	08279	04466
57272	87429	91897	94691	44809	98801	29265	15266	47307	47032
89660	14268	50266	20169	55054	61066	56564	20871	97664	26144
82194	63818	47242	93127	99372	61336	47044	93350	71512	91277
32105	39637	21349	39020	60340	90158	53182	95230	65548	07303
14065	28934	55969	79322	57601	54258	87271	68671	13174	23287
90808	35485	37035	72678	97131	66561	03799	04276	43998	17070
67235	27057	36931	04192	79649	01732	92747	47509	26616	01251
33593	57002	38286	98741	27351	62198	90080	45578	03468	42995
86815	68708	44773	97944	92051	76161	88301	78783	37331	66909
98855	33644	52625	87163	29139	61193	77546	41111	80116	93077
55256	96796	47169	05378	33453	77568	58119	86531	63289	73287
87810	65903	75686	38388	12946	75624	05512	61744	74345	62173

Table II   Random Normal Deviates

Each entry in the table is a random selection from a normal population with a mean $\mu = 0$ and standard deviation $\sigma = 1$.

.464	.137	2.455	−.323	−.068	.296	−.288	1.298	.241	−.957
.060	−2.526	−.531	−.194	.543	−1.558	.187	−1.190	.022	.525
1.486	−.354	−.634	.697	.926	1.375	.785	−.963	−.853	−1.865
1.022	−.472	1.279	3.521	.571	−1.851	.194	1.192	−.501	−.273
1.394	−.555	.046	.321	2.945	1.974	−.258	.412	.439	−.035
.906	−.513	−.525	.595	.881	−.934	1.579	.161	−1.885	.371
1.179	−1.055	.007	.769	.971	.712	1.090	−.631	−.255	−.702
−1.501	−.488	−.162	−.136	1.033	.203	.448	.748	−.423	−.432
−.690	.756	−1.618	−.345	−.511	−2.051	−.457	−.218	.857	−.465
1.372	.225	.378	.761	.181	−.736	.960	−1.530	−.260	.120
−.482	1.678	−.057	−1.229	−.486	.856	−.491	−1.983	−2.830	−.238
−1.376	−.150	1.356	−.561	−.256	−.212	.219	.779	.953	−.869
−1.010	.598	−.918	1.598	.065	.415	.169	−.313	−.973	−1.016
−.005	−.899	.012	−.725	1.147	−.121	1.096	.481	−1.691	.417
1.393	−1.163	−.911	1.231	−.199	−.246	1.239	−2.574	−.558	.056
−1.787	−.261	1.237	1.046	−.508	−1.630	−.146	−.392	−.627	.561
−.105	−.357	−1.384	.360	−.992	−.116	−1.698	−2.832	−1.108	−2.357
−1.339	1.827	−.959	.424	.969	−1.141	−1.041	.362	−1.726	1.956
1.041	.535	.731	1.377	.983	−1.330	1.620	−1.040	.524	−.281
.279	−2.056	.717	−.873	−1.096	−1.396	1.047	.089	−.573	.932
−1.805	−2.008	−1.633	.542	.250	−.166	.032	.079	.471	−1.029
−1.186	1.180	1.114	.882	1.265	−.202	.151	−.376	−.310	.479
.658	1.141	1.151	1.210	.927	.425	.290	−.902	.610	2.709
−.439	.358	−1.939	.891	−.227	.602	.873	−.437	−.220	−.057
−1.399	−.230	.385	−.649	−.577	.237	−.289	.513	.738	−.300
.199	.208	−1.083	−.219	−.291	1.221	1.119	.004	−2.015	−.594
.159	.272	−.313	.084	−2.828	−.439	−.792	1.275	−.623	−1.047
2.273	−.606	.606	−.747	.247	1.291	.063	−1.793	−.699	−1.347
.041	−.307	.121	.790	−.584	.541	.484	−.986	.481	.996
1.132	2.098	.921	.145	.446	−1.661	1.045	−1.363	−.586	−1.023
.768	.079	−1.473	.034	2.127	.665	.084	−.880	−.579	.551
.375	−1.658	−.851	.234	−.656	.340	−.086	−.158	−.120	.418
−.513	−.344	.210	−.736	1.041	.008	.427	−.831	.191	.074
.292	−.521	1.266	−1.206	−.899	.110	−.528	−.813	.071	.524
1.026	2.990	−.574	−.491	−1.114	1.297	−1.433	−1.345	−3.001	.479
−1.334	1.278	−.568	−.109	−.515	.566	2.923	.500	.359	.326
.287	−.144	−.254	.574	−.451	−1.181	−1.190	−.318	−.094	1.114
.161	−.886	−.921	−.509	1.410	−.518	.192	−.432	1.501	1.068
−1.346	.193	−1.202	.394	−1.045	.843	.942	1.045	.031	.772
1.250	−.199	−.288	1.810	1.378	.584	1.216	.733	.402	.226
−.128	1.463	−.436	−.239	−1.443	.732	.168	−.144	−.392	.989
1.879	−2.456	.029	.429	.618	−1.683	−2.262	.034	−.002	1.914
.680	.252	.130	1.658	−1.023	.407	−.235	−.224	−.434	.253
−.631	.225	−.951	−1.072	−.285	−1.731	−.427	−1.446	−.873	.619
−1.273	.723	.201	.505	−.370	−.421	−.015	−.463	.288	1.734
−.643	−1.485	.403	.003	−.243	.000	.964	−.703	.844	−.686
−.435	−2.162	−.169	−1.311	−1.639	.193	2.692	−1.994	.326	.562
−1.706	.119	−1.566	.637	−1.948	−1.068	.935	.738	.650	.491
−.498	1.640	.384	−.945	−1.272	.945	−1.013	−.913	−.469	2.250
−.065	−.005	.618	−.523	−.055	1.071	.758	−.736	−.959	.598

Selected from the Rand Corporation's, *A Million Random Digits with 100,000 Normal Deviates*, Free Press, Glencoe, IL, 1955. Used with the permission of the Rand Corporation.

Table II Random Normal Deviates (Continued)

-.013	-1.115	.070	.478	-.563	-1.282	-.307	.076	-1.166	1.033
-.539	-.770	-1.375	1.287	.092	-2.508	1.028	.305	-1.083	.688
.667	-1.461	2.003	1.746	.709	-.272	1.221	-.595	-1.120	.198
1.972	-1.242	.197	-.333	-1.875	1.017	-.855	-.364	.548	-.677
-.508	1.157	1.246	1.775	-.545	2.385	.465	.775	-.732	-.165
.748	1.376	.380	-.390	1.408	.752	-.456	.329	-.459	.340
.072	.458	-1.203	-1.653	-.579	.597	-.398	-.945	.499	2.242
-2.238	.234	-2.065	.277	-.519	-.861	-1.288	-.057	-.898	-.632
-1.037	1.200	.221	.177	-1.168	-.975	.381	-.244	-.308	.444
.924	-.048	1.770	-.405	.785	-.880	-.158	.491	1.395	-.678
.909	.073	-1.510	1.131	.196	1.884	1.609	1.350	1.341	.309
-.353	1.966	-.125	.686	.435	-.287	1.231	-2.472	.244	-.260
.301	.108	.906	.420	.279	.730	1.150	-.134	-.569	.983
-.196	-.574	1.273	-.275	-.810	-.254	1.334	.016	1.003	-.406
-.120	.198	-1.333	-.356	.076	.449	-1.162	-.619	-2.233	-.345
.245	-.687	.190	-1.874	-.800	.841	.351	-.096	.584	2.537
.101	.222	.561	.620	-1.215	.719	-2.098	.699	.126	.385
.021	-.453	.810	-1.180	-.129	-1.083	.551	-.187	.711	2.010
.876	2.001	-1.773	-.292	-.980	.291	-.900	-.370	1.820	.885
-1.430	.264	-1.657	-.326	-.281	1.007	.191	.266	-.508	2.185
.326	-1.290	-.869	.240	-1.388	-.996	-1.888	1.699	-.265	.119
2.549	1.373	.103	-2.214	.110	-1.991	.218	-.806	.376	-.127
-.260	.190	-.701	.464	.324	-1.037	-.418	-.111	.192	-1.201
-1.652	.646	-.810	.666	-1.145	.478	.554	-.707	2.005	.850
.808	-1.994	-2.024	.638	.625	-.254	.446	-1.065	-.038	.278
-.511	-.459	.020	-.609	-.315	1.098	-.214	-.596	.523	-.408
-.282	-.787	1.358	.725	-.817	.316	-.025	-.709	.938	-1.667
-1.818	1.012	1.009	-.131	1.288	.649	1.533	.071	-1.092	-.793
-.774	-.554	-.573	-.409	2.461	-.734	1.235	1.531	-1.754	.472
-.334	-.195	.581	.318	-.341	.351	-2.334	-.240	.463	-.376
.031	-.405	-.060	-.004	.558	-.475	-.572	-1.684	-.270	1.204
.453	-1.077	-1.405	1.817	.237	.130	-.994	1.066	-.250	.150
-.699	1.055	-1.128	-1.291	-.545	-1.988	-1.201	-.239	-.475	-.764
-1.411	-1.681	-2.006	1.140	-1.257	.129	.599	-.267	.670	-1.587
-1.275	.846	-1.835	-.486	.736	1.567	-1.488	.042	-.157	.706
.509	-.694	-2.568	.566	-.686	.442	1.571	.340	-.443	-.930
.480	.775	.131	.713	.414	-.354	.189	-.258	.571	-1.706
.759	-.734	.169	.281	-1.202	.588	1.094	-.988	.126	1.139
-.545	.543	-.334	-1.091	-.031	.265	-.052	.218	.173	-.832
.519	.512	1.765	.427	-.193	-.723	.569	-.894	.808	.751
-1.074	1.323	-1.659	-.186	-.612	-1.612	2.159	-1.210	-.596	1.421
-1.518	2.101	.397	.516	1.169	-1.821	-1.346	2.435	1.165	.428
.935	.206	-1.117	.241	-.963	-.099	-.412	1.344	-.411	.583
-1.360	.380	-.031	1.066	.893	.431	.081	-.099	.500	2.441
-.115	.211	-1.471	.332	.750	.652	.812	-1.383	.355	-.638
-.082	.309	-.355	-.402	-.774	.150	.015	2.539	.756	-1.049
-1.492	-.259	.323	.697	.509	-.968	.053	-1.033	.220	-2.322
.203	-.548	1.494	1.185	.083	1.196	-.749	-1.105	-1.324	.689
1.857	.167	-1.531	-1.551	.848	.120	.415	.317	-1.446	1.002
.669	1.017	-2.437	-.558	-.657	-.940	.985	.483	.361	-.095

Table II Random Normal Deviates (Continued)

-.615	-1.137	1.495	1.395	1.030	.569	.347	-1.814	-.085	-.354
1.544	-1.102	-.023	1.939	-.182	1.014	1.417	.332	-1.449	.413
-.637	-.463	-1.289	.660	-.535	-.599	-.624	-.987	-.813	-.739
-.480	-.621	-1.248	-2.158	2.187	-.356	.264	-1.651	.130	.225
-.309	.212	.743	-.904	-.629	-1.476	-1.897	2.078	.464	-1.159
.065	-.842	.360	-.983	-.970	-.081	1.611	-1.552	1.387	.363
1.417	-1.127	-.913	.299	-.168	-.622	-.914	.166	.280	-2.797
.243	.063	-1.297	-1.511	-.461	.806	.616	-.103	-.414	-.901
1.294	-.526	.927	1.232	-.870	-.468	-1.020	2.005	-.202	-1.099
-1.004	2.161	.696	2.914	-.389	.080	.135	.363	.710	-1.374
-.715	-.840	.531	.942	1.442	-.191	-.266	.472	-.337	-.538
1.540	-.201	.137	1.040	-.121	.377	-.687	.006	.713	-1.650
-1.565	-1.037	-2.592	.210	2.424	.309	-.915	-1.990	.356	1.736
-.560	1.326	1.207	-.245	-.076	-.592	-.268	-.519	-.063	-.336
-.519	-.060	1.270	.070	-2.245	1.581	.853	-1.034	-.369	-.810
-.693	-.512	.900	-.695	.978	-1.214	-.305	1.891	-1.916	-.602
.058	.558	.824	-1.855	-1.245	-.684	.329	-.026	-.616	.458
-1.116	.286	-.407	-.740	1.187	-.180	.872	.117	.809	-.347
.276	-.765	1.189	-.537	-.185	-1.009	-.722	.808	.357	-1.075
-.118	.376	-.349	-.969	1.541	-1.868	1.118	1.159	.401	-.667
.706	1.007	-2.128	-1.609	-1.743	-.681	.361	-.687	-.995	-.900
-.751	.713	.247	1.144	-.456	.291	-.427	.143	.378	-.997
-1.769	1.493	.167	-.835	.165	-.463	.141	.081	.070	-1.220
-1.589	.730	.618	-.381	1.442	1.894	.999	-.428	.274	1.357
-.565	.687	1.932	-.315	.602	1.597	-1.150	-.020	1.416	-.174
-.961	-.617	.235	.050	.678	-1.018	.789	-.496	-1.446	.039
.540	1.150	-1.850	-1.892	-.805	.560	-.023	.087	2.336	-.298
1.455	.480	-.260	-.567	.199	1.873	-.940	-1.148	.524	.334
-1.211	.613	-.516	-.442	.268	-.206	-.492	.131	1.302	-.700
1.305	2.710	.422	.795	1.964	.773	1.200	1.912	.615	1.106
1.207	-1.662	-1.325	.026	-2.026	-1.430	-.907	.123	-1.197	2.066
.774	.906	-1.426	-.282	-.739	-1.125	1.146	.001	-.519	.431
.260	1.338	.532	.091	1.635	-.407	-.849	-1.573	1.008	-.422
.188	.781	1.140	.317	-.324	-1.186	.310	-.498	-2.698	.461
1.056	.738	.967	-.451	1.457	1.241	.781	2.307	-.127	-.400
-1.717	.767	1.742	.489	-.312	.856	1.078	-1.280	-2.687	.870
.996	-.805	1.421	-.190	-.285	-.530	1.243	.829	.783	-.284
-.531	.458	1.139	-.137	-1.626	1.522	-1.105	-.558	-.178	.739
1.551	1.676	.680	-.428	.616	-.986	.201	-2.131	.442	-.380
1.781	-.768	1.806	1.604	.447	.735	-1.791	.404	-.361	-1.923
.181	-.583	1.478	-.181	-.281	.559	-1.985	1.122	-1.106	-1.441
1.549	1.183	-2.089	-1.997	-.343	-1.275	.676	.212	-1.252	.163
.978	1.067	-2.640	-.134	.328	.052	-.030	-.273	-.570	-1.026
-.596	-.420	-.318	-.057	-.695	-1.148	-.333	.531	-2.037	-1.587
-.440	-.032	.163	1.029	.079	1.148	.762	1.961	-.674	-.486
.443	1.100	-.728	2.397	-.543	-.872	.568	-.980	.174	-.728
2.401	-1.375	-1.332	-2.177	-2.064	-.245	-.039	-.585	1.344	1.386
.311	.322	.158	-.359	.103	.371	.735	.011	2.091	.490
1.209	-.241	1.488	-.667	-1.772	-.197	-.741	1.303	-1.149	-2.251
-.575	1.227	-1.674	-1.400	.289	.005	.185	1.072	-.431	1.096

Table II Random Normal Deviates (Continued)

.039	1.434	−.384	1.272	−.480	−.747	−1.005	−.309	−.872	−2.264
−1.818	.037	1.044	.513	.559	−.420	.064	−.606	−.176	.157
−1.715	−1.533	.763	.992	1.200	−1.673	−2.186	−1.185	−.040	−.944
−1.646	−.755	.240	−1.261	2.077	.567	.996	.829	−1.402	.968
.622	1.286	1.071	−2.153	1.345	−.525	.594	−.329	−.368	1.272
1.233	−.141	−.503	−.360	−1.034	−1.577	−.027	.614	1.008	−.945
−.532	−1.442	1.462	.911	−1.401	−.990	−.434	.956	.618	−1.026
−.266	.029	−.892	.310	.130	−1.362	1.203	−.015	−.623	1.858
−1.094	.818	.517	−.983	−.498	−.056	.805	−.622	.086	.123
1.619	.040	.667	−1.350	.239	−1.509	.086	1.787	.345	−1.280
−1.306	−.761	−.402	−.539	1.780	−.036	−.334	−.471	.991	.118
−1.729	1.511	.374	.193	−1.326	1.919	.655	−.627	.183	.096
.777	.162	−.310	−1.412	−1.305	.658	−.950	.160	1.506	−1.390
−.628	1.539	1.806	−.612	.765	−1.403	1.811	.174	.647	−.522
1.448	1.112	.100	1.275	1.474	−.162	−.996	.522	−.588	−.818
.401	−.283	−.245	−.984	−1.128	−.670	1.238	.603	.417	−1.262
−1.118	−.046	−.080	−.057	−.497	.027	.618	−.163	.467	−.237
−1.686	−.385	−2.686	.345	.782	−.652	.171	−.769	−.777	.487
1.004	−.040	−.913	.769	.078	−.108	1.350	−.064	.852	−.112
−1.187	1.767	−.243	−.330	−.771	−2.056	.940	2.003	−.470	−1.031
−.247	−.491	.110	.080	−.479	−1.296	−1.319	1.589	−1.370	−1.008
−1.403	1.498	−.305	1.326	1.379	−.603	1.843	.018	1.763	1.194
.268	1.128	−.973	.393	−.135	−1.329	.514	.683	.362	−.683
−.912	1.047	.306	−1.448	−.038	.882	.153	.257	.489	.116
−1.491	.221	−.679	.588	−.516	−.598	.990	−.964	.221	.066
−.013	−.278	1.122	1.277	−1.092	−.105	2.114	−.627	.304	−.280
−.034	−1.229	.617	.996	.474	1.378	1.045	.555	−.831	−.243
−1.480	−.633	−.645	.134	.636	.338	1.413	−.623	−.010	.340
.714	1.595	−.234	−.342	−1.286	−.757	.359	.328	1.832	1.511
−.400	−1.243	−.887	.466	−.892	.966	−.682	−.169	.074	.526
.551	−.116	.379	−2.348	−.370	−1.227	−1.341	−.203	.617	.027
−1.162	3.214	−.628	−.168	−.793	−.600	.402	1.604	2.491	−.118
.382	1.170	.090	−.688	1.436	.049	1.622	−2.939	2.087	1.396
−1.111	−.062	.031	1.678	.817	.614	−.569	−.636	.095	.049
−.782	.668	1.935	.509	.038	−.178	.737	.230	.155	−.568
.800	−.850	−2.020	−1.180	.646	.275	.045	1.295	−.226	−.314
−.718	−1.095	−1.264	−.373	.130	−1.113	−.814	.943	−.685	−1.044
−.698	−1.249	−1.579	.451	−.363	−.750	.191	.756	.390	.176
1.032	1.388	−1.136	−.416	.181	1.071	−.583	.745	.336	−.186
.473	1.410	−.127	−.151	.711	.750	.827	−.489	−.435	−.980
−.190	−.272	1.216	.227	1.358	.215	2.306	−1.301	−.597	−1.401
−.817	−.769	−.470	−.633	−.187	.517	−.888	−1.712	−1.774	.162
−.265	.676	−.244	1.897	.629	−.206	−1.419	−1.049	.266	.438
.221	−.678	2.149	1.486	1.361	−1.402	.028	−.493	.744	.195
.436	−.358	.602	−.107	.085	.573	.529	1.577	.239	1.898
.010	−.475	.655	.659	.029	−.029	−.126	1.335	−1.261	−2.036
−.244	−1.654	1.335	.610	−.617	.642	.371	.241	.001	1.799
−.932	−1.275	−1.134	−1.246	−1.508	−.949	1.743	.271	−1.333	−1.875
.199	−1.285	−.387	−.191	.726	.151	−.064	.803	−.062	−.780
.251	−.431	−.831	−.036	.464	−1.089	−.284	.451	−1.693	1.004

Table III Normal Distribution

The tabled entries represent the proportion p of area under the normal curve above the indicated values of z. (Example: .0694 or 6.94% of the area is above $z = 1.48$). For negative values of z, the tabled entries represent the area less than $-z$. (Example: .3015 or 30.15% of the area is beneath $z = -.52$).

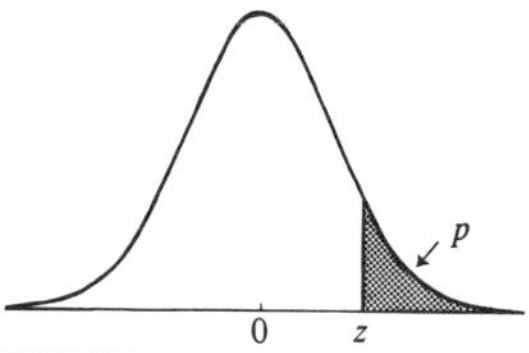

	Second decimal place of z									
z	.00	.01	.02	.03	.04	.05	.06	.07	.08	.09
0.0	.5000	.4960	.4920	.4880	.4840	.4801	.4761	.4721	.4681	.4641
0.1	.4602	.4562	.4522	.4483	.4443	.4404	.4364	.4325	.4286	.4247
0.2	.4207	.4168	.4129	.4090	.4052	.4013	.3974	.3936	.3897	.3859
0.3	.3821	.3783	.3745	.3707	.3669	.3632	.3594	.3557	.3520	.3483
0.4	.3446	.3409	.3372	.3336	.3300	.3264	.3228	.3192	.3156	.3121
0.5	.3085	.3050	.3015	.2981	.2946	.2912	.2877	.2843	.2810	.2776
0.6	.2743	.2709	.2676	.2643	.2611	.2578	.2546	.2514	.2483	.2451
0.7	.2420	.2389	.2358	.2327	.2297	.2266	.2236	.2206	.2177	.2148
0.8	.2119	.2090	.2061	.2033	.2005	.1977	.1949	.1922	.1894	.1867
0.9	.1841	.1814	.1788	.1762	.1736	.1711	.1685	.1660	.1635	.1611
1.0	.1587	.1562	.1539	.1515	.1492	.1469	.1446	.1423	.1401	.1379
1.1	.1357	.1335	.1314	.1292	.1271	.1251	.1230	.1210	.1190	.1170
1.2	.1151	.1131	.1112	.1093	.1075	.1056	.1038	.1020	.1003	.0985
1.3	.0968	.0951	.0934	.0918	.0901	.0885	.0869	.0853	.0838	.0823
1.4	.0808	.0793	.0778	.0764	.0749	.0735	.0721	.0708	.0694	.0681
1.5	.0668	.0655	.0643	.0630	.0618	.0606	.0594	.0582	.0571	.0559
1.6	.0548	.0537	.0526	.0516	.0505	.0495	.0485	.0475	.0465	.0455
1.7	.0446	.0436	.0427	.0418	.0409	.0401	.0392	.0384	.0375	.0367
1.8	.0359	.0351	.0344	.0336	.0329	.0322	.0314	.0307	.0301	.0294
1.9	.0287	.0281	.0274	.0268	.0262	.0256	.0250	.0244	.0239	.0233
2.0	.0228	.0222	.0217	.0212	.0207	.0202	.0197	.0192	.0188	.0183
2.1	.0179	.0174	.0170	.0166	.0162	.0158	.0154	.0150	.0146	.0143
2.2	.0139	.0136	.0132	.0129	.0125	.0122	.0119	.0116	.0113	.0110
2.3	.0107	.0104	.0102	.0099	.0096	.0094	.0091	.0089	.0087	.0084
2.4	.0082	.0080	.0078	.0075	.0073	.0071	.0069	.0068	.0066	.0064
2.5	.0062	.0060	.0059	.0057	.0055	.0054	.0052	.0051	.0049	.0048
2.6	.0047	.0045	.0044	.0043	.0041	.0040	.0039	.0038	.0037	.0036
2.7	.0035	.0034	.0033	.0032	.0031	.0030	.0029	.0028	.0027	.0026
2.8	.0026	.0025	.0024	.0023	.0023	.0022	.0021	.0021	.0020	.0019
2.9	.0019	.0018	.0018	.0017	.0016	.0016	.0015	.0015	.0014	.0014
3.0	.0013	.0013	.0013	.0012	.0012	.0011	.0011	.0011	.0010	.0010

Adapted with rounding from Table II of R. A. Fisher and F. Yates, *Statistical Tables for Biological, Agricultural, and Medical Research*, 6th Edition, Longman Group, Ltd., London, 1974. (Previously published by Oliver & Boyd, Ltd., Edinburgh). Used with permission of the authors and publishers.

Table IV Student's *t* Distribution

For various degrees of freedom (*df*), the tabled entries represent the critical values of *t* above which a specified proportion *p* of the *t* distribution falls. (Example: For $df = 9$, a *t* of 2.262 is surpassed by .025 or 2.5% of the total distribution). By symmetry, negative values of *t* cut off equal areas *p* in the left tail of the distribution. Double the *p* for two-tailed probabilities.

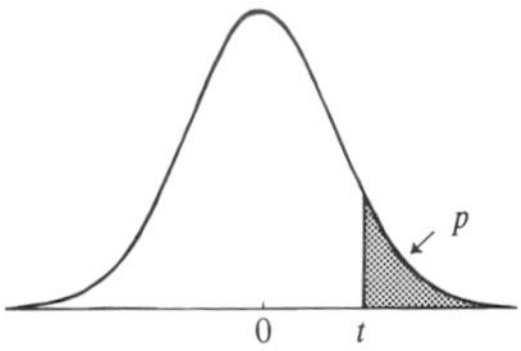

	p (*one-tailed probabilities*)				
df	.10	.05	.025	.01	.005
1	3.078	6.314	12.706	31.821	63.657
2	1.886	2.920	4.303	6.965	9.925
3	1.638	2.353	3.182	4.541	5.841
4	1.533	2.132	2.776	3.747	4.604
5	1.476	2.015	2.571	3.365	4.032
6	1.440	1.943	2.447	3.143	3.707
7	1.415	1.895	2.365	2.998	3.499
8	1.397	1.860	2.306	2.896	3.355
9	1.383	1.833	2.262	2.821	3.250
10	1.372	1.812	2.228	2.764	3.169
11	1.363	1.796	2.201	2.718	3.106
12	1.356	1.782	2.179	2.681	3.055
13	1.350	1.771	2.160	2.650	3.012
14	1.345	1.761	2.145	2.624	2.977
15	1.341	1.753	2.131	2.602	2.947
16	1.337	1.746	2.120	2.583	2.921
17	1.333	1.740	2.110	2.567	2.898
18	1.330	1.734	2.101	2.552	2.878
19	1.328	1.729	2.093	2.539	2.861
20	1.325	1.725	2.086	2.528	2.845
21	1.323	1.721	2.080	2.518	2.831
22	1.321	1.717	2.074	2.508	2.819
23	1.319	1.714	2.069	2.500	2.807
24	1.318	1.711	2.064	2.492	2.797
25	1.316	1.708	2.060	2.485	2.787
26	1.315	1.706	2.056	2.479	2.779
27	1.314	1.703	2.052	2.473	2.771
28	1.313	1.701	2.048	2.467	2.763
29	1.311	1.699	2.045	2.462	2.756
30	1.310	1.697	2.042	2.457	2.750
40	1.303	1.684	2.021	2.423	2.704
60	1.296	1.671	2.000	2.390	2.660
120	1.289	1.658	1.980	2.358	2.617
∞	1.282	1.645	1.960	2.326	2.576

Adapted from Table III of R. A. Fisher and F. Yates, *Statistical Tables for Biological, Agricultural, and Medical Research*, 6th Edition, Longman Group, Ltd., London, 1974. (Previously published by Oliver & Boyd, Ltd., Edinburgh). Used with permission of the authors and publishers.

Table V ***F* Distribution**

For various pairs of degrees of freedom (v_1, v_2), the tabled entries represent the critical values of F above which a proportion p of the distribution falls. (Example: For $df = 4{,}16$ an $F = 2.33$ is exceeded by $p = .10$ of the distribution). Tables are provided for values of p equal to .10, .05, .01 and .001.

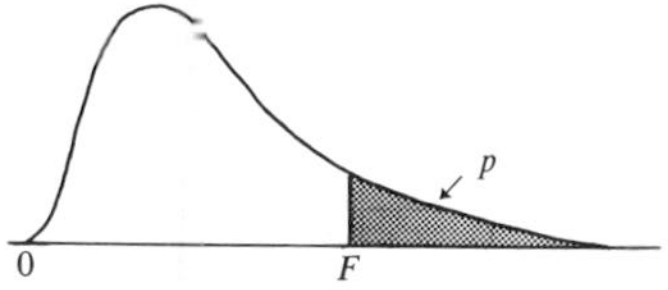

$p = .10$ values

Degrees of freedom for denominator v_2	Degrees of freedom for the numerator, v_1: 1	2	3	4	5	6	7	8	9	10	12	15	20	30	40	60	120	∞
1	39.86	49.50	53.59	55.83	57.24	58.20	58.91	59.44	59.86	60.19	60.71	61.22	61.74	62.26	62.53	62.79	63.06	63.33
2	8.53	9.00	9.16	9.24	9.29	9.33	9.35	9.37	9.38	9.39	9.41	9.42	9.44	9.46	9.47	9.47	9.48	9.49
3	5.54	5.46	5.39	5.34	5.31	5.28	5.27	5.25	5.24	5.23	5.22	5.20	5.18	5.17	5.16	5.15	5.14	5.13
4	4.54	4.32	4.19	4.11	4.05	4.01	3.98	3.95	3.94	3.92	3.90	3.87	3.84	3.82	3.80	3.79	3.78	3.76
5	4.06	3.78	3.62	3.52	3.45	3.40	3.37	3.34	3.32	3.30	3.27	3.24	3.21	3.17	3.16	3.14	3.12	3.10
6	3.78	3.46	3.29	3.18	3.11	3.05	3.01	2.98	2.96	2.94	2.90	2.87	2.84	2.80	2.78	2.76	2.74	2.72
7	3.59	3.26	3.07	2.96	2.88	2.83	2.78	2.75	2.72	2.70	2.67	2.63	2.59	2.56	2.54	2.51	2.49	2.47
8	3.46	3.11	2.92	2.81	2.73	2.67	2.62	2.59	2.56	2.54	2.50	2.46	2.42	2.38	2.36	2.34	2.32	2.29
9	3.36	3.01	2.81	2.69	2.61	2.55	2.51	2.47	2.44	2.42	2.38	2.34	2.30	2.25	2.23	2.21	2.18	2.16
10	3.29	2.92	2.73	2.61	2.52	2.46	2.41	2.38	2.35	2.32	2.28	2.24	2.20	2.16	2.13	2.11	2.08	2.06
11	3.23	2.86	2.66	2.54	2.45	2.39	2.34	2.30	2.27	2.25	2.21	2.17	2.12	2.08	2.05	2.03	2.00	1.97
12	3.18	2.81	2.61	2.48	2.39	2.33	2.28	2.24	2.21	2.19	2.15	2.10	2.06	2.01	1.99	1.96	1.93	1.90
13	3.14	2.76	2.56	2.43	2.35	2.28	2.23	2.20	2.16	2.14	2.10	2.05	2.01	1.96	1.93	1.90	1.88	1.85
14	3.10	2.73	2.52	2.39	2.31	2.24	2.19	2.15	2.12	2.10	2.05	2.01	1.96	1.91	1.89	1.86	1.83	1.80
15	3.07	2.70	2.49	2.36	2.27	2.21	2.16	2.12	2.09	2.06	2.02	1.97	1.92	1.87	1.85	1.82	1.79	1.76
16	3.05	2.67	2.46	2.33	2.24	2.18	2.13	2.09	2.06	2.03	1.99	1.94	1.89	1.84	1.81	1.78	1.75	1.72
17	3.03	2.64	2.44	2.31	2.22	2.15	2.10	2.06	2.03	2.00	1.96	1.91	1.86	1.81	1.78	1.75	1.72	1.69
18	3.01	2.62	2.42	2.29	2.20	2.13	2.08	2.04	2.00	1.98	1.93	1.89	1.84	1.78	1.75	1.72	1.69	1.66
19	2.99	2.61	2.40	2.27	2.18	2.11	2.06	2.02	1.98	1.96	1.91	1.86	1.81	1.76	1.73	1.70	1.67	1.63
20	2.97	2.59	2.38	2.25	2.16	2.09	2.04	2.00	1.96	1.94	1.89	1.84	1.79	1.74	1.71	1.68	1.64	1.61
21	2.96	2.57	2.36	2.23	2.14	2.08	2.02	1.98	1.95	1.92	1.87	1.83	1.78	1.72	1.69	1.66	1.62	1.59
22	2.95	2.56	2.35	2.22	2.13	2.06	2.01	1.97	1.93	1.90	1.86	1.81	1.76	1.70	1.67	1.64	1.60	1.57
23	2.94	2.55	2.34	2.21	2.11	2.05	1.99	1.95	1.92	1.89	1.84	1.80	1.74	1.69	1.66	1.62	1.59	1.55
24	2.93	2.54	2.33	2.19	2.10	2.04	1.98	1.94	1.91	1.88	1.83	1.78	1.73	1.67	1.64	1.61	1.57	1.53
30	2.88	2.49	2.28	2.14	2.05	1.98	1.93	1.88	1.85	1.82	1.77	1.72	1.67	1.61	1.57	1.54	1.50	1.46
40	2.84	2.44	2.23	2.09	2.00	1.93	1.87	1.83	1.79	1.76	1.71	1.66	1.61	1.54	1.51	1.47	1.42	1.38
60	2.79	2.39	2.18	2.04	1.95	1.87	1.82	1.77	1.74	1.71	1.66	1.60	1.54	1.48	1.44	1.40	1.35	1.29
120	2.75	2.35	2.13	1.99	1.90	1.82	1.77	1.72	1.68	1.65	1.60	1.55	1.48	1.41	1.37	1.32	1.26	1.19
∞	2.71	2.30	2.08	1.94	1.85	1.77	1.72	1.67	1.63	1.60	1.55	1.49	1.42	1.34	1.30	1.24	1.17	1.00

Adapted from Table 18 of E. S. Pearson and H. O. Hartley (Eds.), *Biometrika Tables for Statisticians*, Vol. 1, Cambridge University

Table V *F* Distribution (Continued)

***p* = .05 values**

Degrees of freedom for denominator v_2	Degrees of freedom for the numerator, v_1: 1	2	3	4	5	6	7	8	9	10	12	15	20	30	40	60	120	∞
1	161.4	199.5	215.7	224.6	230.2	234.0	236.8	238.9	240.5	241.9	243.9	245.9	248.0	250.1	251.1	252.2	253.3	254.3
2	18.51	19.00	19.16	19.25	19.30	19.33	19.35	19.37	19.38	19.40	19.41	19.43	19.45	19.46	19.47	19.48	19.49	19.50
3	10.13	9.55	9.28	9.12	9.01	8.94	8.89	8.85	8.81	8.79	8.74	8.70	8.66	8.62	8.59	8.57	8.55	8.53
4	7.71	6.94	6.59	6.39	6.26	6.16	6.09	6.04	6.00	5.96	5.91	5.86	5.80	5.75	5.72	5.69	5.66	5.63
5	6.61	5.79	5.41	5.19	5.05	4.95	4.88	4.82	4.77	4.74	4.68	4.62	4.56	4.50	4.46	4.43	4.40	4.36
6	5.99	5.14	4.76	4.53	4.39	4.28	4.21	4.15	4.10	4.06	4.00	3.94	3.87	3.81	3.77	3.74	3.70	3.67
7	5.59	4.74	4.35	4.12	3.97	3.87	3.79	3.73	3.68	3.64	3.57	3.51	3.44	3.38	3.34	3.30	3.27	3.23
8	5.32	4.46	4.07	3.84	3.69	3.58	3.50	3.44	3.39	3.35	3.28	3.22	3.15	3.08	3.04	3.01	2.97	2.93
9	5.12	4.26	3.86	3.63	3.48	3.37	3.29	3.23	3.18	3.14	3.07	3.01	2.94	2.86	2.83	2.79	2.75	2.71
10	4.96	4.10	3.71	3.48	3.33	3.22	3.14	3.07	3.02	2.98	2.91	2.85	2.77	2.70	2.66	2.62	2.58	2.54
11	4.84	3.98	3.59	3.36	3.20	3.09	3.01	2.95	2.90	2.85	2.79	2.72	2.65	2.57	2.53	2.49	2.45	2.40
12	4.75	3.89	3.49	3.26	3.11	3.00	2.91	2.85	2.80	2.75	2.69	2.62	2.54	2.47	2.43	2.38	2.34	2.30
13	4.67	3.81	3.41	3.18	3.03	2.92	2.83	2.77	2.71	2.67	2.60	2.53	2.46	2.38	2.34	2.30	2.25	2.21
14	4.60	3.74	3.34	3.11	2.96	2.85	2.76	2.70	2.65	2.60	2.53	2.46	2.39	2.31	2.27	2.22	2.18	2.13
15	4.54	3.68	3.29	3.06	2.90	2.79	2.71	2.64	2.59	2.54	2.48	2.40	2.33	2.25	2.20	2.16	2.11	2.07
16	4.49	3.63	3.24	3.01	2.85	2.74	2.66	2.59	2.54	2.49	2.42	2.35	2.28	2.19	2.15	2.11	2.06	2.01
17	4.45	3.59	3.20	2.96	2.81	2.70	2.61	2.55	2.49	2.45	2.38	2.31	2.23	2.15	2.10	2.06	2.01	1.96
18	4.41	3.55	3.16	2.93	2.77	2.66	2.58	2.51	2.46	2.41	2.34	2.27	2.19	2.11	2.06	2.02	1.97	1.92
19	4.38	3.52	3.13	2.90	2.74	2.63	2.54	2.48	2.42	2.38	2.31	2.23	2.16	2.07	2.03	1.98	1.93	1.88
20	4.35	3.49	3.10	2.87	2.71	2.60	2.51	2.45	2.39	2.35	2.28	2.20	2.12	2.04	1.99	1.95	1.90	1.84
21	4.32	3.47	3.07	2.84	2.68	2.57	2.49	2.42	2.37	2.32	2.25	2.18	2.10	2.01	1.96	1.92	1.87	1.81
22	4.30	3.44	3.05	2.82	2.66	2.55	2.46	2.40	2.34	2.30	2.23	2.15	2.07	1.98	1.94	1.89	1.84	1.78
23	4.28	3.42	3.03	2.80	2.64	2.53	2.44	2.37	2.32	2.27	2.20	2.13	2.05	1.96	1.91	1.86	1.81	1.76
24	4.26	3.40	3.01	2.78	2.62	2.51	2.42	2.36	2.30	2.25	2.18	2.11	2.03	1.94	1.89	1.84	1.79	1.73
30	4.17	3.32	2.92	2.69	2.53	2.42	2.33	2.27	2.21	2.16	2.09	2.01	1.93	1.84	1.79	1.74	1.68	1.62
40	4.08	3.23	2.84	2.61	2.45	2.34	2.25	2.18	2.12	2.08	2.00	1.92	1.84	1.74	1.69	1.64	1.58	1.51
60	4.00	3.15	2.76	2.53	2.37	2.25	2.17	2.10	2.04	1.99	1.92	1.84	1.75	1.65	1.59	1.53	1.47	1.39
120	3.92	3.07	2.68	2.45	2.29	2.17	2.09	2.02	1.96	1.91	1.83	1.75	1.66	1.55	1.50	1.43	1.35	1.25
∞	3.84	3.00	2.60	2.37	2.21	2.10	2.01	1.94	1.88	1.83	1.75	1.67	1.57	1.46	1.39	1.32	1.22	1.00

Table V *F* Distribution (Continued)

$p = .01$ values

Degrees of freedom for denominator υ_2	Degrees of freedom for the numerator, υ_1: 1	2	3	4	5	6	7	8	9	10	12	15	20	30	40	60	120	∞
1	4052	4999.5	5403	5625	5764	5859	5928	5981	6022	6056	6106	6157	6209	6261	6287	6313	6339	6366
2	98.50	99.00	99.17	99.25	99.30	99.33	99.36	99.37	99.39	99.40	99.42	99.43	99.45	99.47	99.47	99.48	99.49	99.50
3	34.12	30.82	29.46	28.71	28.24	27.91	27.67	27.49	27.35	27.23	27.05	26.87	26.69	26.50	26.41	26.32	26.22	26.13
4	21.20	18.00	16.69	15.98	15.52	15.21	14.98	14.80	14.66	14.55	14.37	14.20	14.02	13.84	13.75	13.65	13.56	13.46
5	16.26	13.27	12.06	11.39	10.97	10.67	10.46	10.29	10.16	10.05	9.89	9.72	9.55	9.38	9.29	9.20	9.11	9.02
6	13.75	10.92	9.78	9.15	8.75	8.47	8.26	8.10	7.98	7.87	7.72	7.56	7.40	7.23	7.14	7.06	6.97	6.88
7	12.25	9.55	8.45	7.85	7.46	7.19	6.99	6.84	6.72	6.62	6.47	6.31	6.16	5.99	5.91	5.82	5.74	5.65
8	11.26	8.65	7.59	7.01	6.63	6.37	6.18	6.03	5.91	5.81	5.67	5.52	5.36	5.20	5.12	5.03	4.95	4.86
9	10.56	8.02	6.99	6.42	6.06	5.80	5.61	5.47	5.35	5.26	5.11	4.96	4.81	4.65	4.57	4.48	4.40	4.31
10	10.04	7.56	6.55	5.99	5.64	5.39	5.20	5.06	4.94	4.85	4.71	4.56	4.41	4.25	4.17	4.08	4.00	3.91
11	9.65	7.21	6.22	5.67	5.32	5.07	4.89	4.74	4.63	4.54	4.40	4.25	4.10	3.94	3.86	3.78	3.69	3.60
12	9.33	6.93	5.95	5.41	5.06	4.82	4.64	4.50	4.39	4.30	4.16	4.01	3.86	3.70	3.62	3.54	3.45	3.36
13	9.07	6.70	5.74	5.21	4.86	4.62	4.44	4.30	4.19	4.10	3.96	3.82	3.66	3.51	3.43	3.34	3.25	3.17
14	8.86	6.51	5.56	5.04	4.69	4.46	4.28	4.14	4.03	3.94	3.80	3.66	3.51	3.35	3.27	3.18	3.09	3.00
15	8.68	6.36	5.42	4.89	4.56	4.32	4.14	4.00	3.89	3.80	3.67	3.52	3.37	3.21	3.13	3.05	2.96	2.87
16	8.53	6.23	5.29	4.77	4.44	4.20	4.03	3.89	3.78	3.69	3.55	3.41	3.26	3.10	3.02	2.93	2.84	2.75
17	8.40	6.11	5.18	4.67	4.34	4.10	3.93	3.79	3.68	3.59	3.46	3.31	3.16	3.00	2.92	2.83	2.75	2.65
18	8.29	6.01	5.09	4.58	4.25	4.01	3.84	3.71	3.60	3.51	3.37	3.23	3.08	2.92	2.84	2.75	2.66	2.57
19	8.18	5.93	5.01	4.50	4.17	3.94	3.77	3.63	3.52	3.43	3.30	3.15	3.00	2.84	2.76	2.67	2.58	2.49
20	8.10	5.85	4.94	4.43	4.10	3.87	3.70	3.56	3.46	3.37	3.23	3.09	2.94	2.78	2.69	2.61	2.52	2.42
21	8.02	5.78	4.87	4.37	4.04	3.81	3.64	3.51	3.40	3.31	3.17	3.03	2.88	2.72	2.64	2.55	2.46	2.36
22	7.95	5.72	4.82	4.31	3.99	3.76	3.59	3.45	3.35	3.26	3.12	2.98	2.83	2.67	2.58	2.50	2.40	2.31
23	7.88	5.66	4.76	4.26	3.94	3.71	3.54	3.41	3.30	3.21	3.07	2.93	2.78	2.62	2.54	2.45	2.35	2.26
24	7.82	5.61	4.72	4.22	3.90	3.67	3.50	3.36	3.26	3.17	3.03	2.89	2.74	2.58	2.49	2.40	2.31	2.21
30	7.56	5.39	4.51	4.02	3.70	3.47	3.30	3.17	3.07	2.98	2.84	2.70	2.55	2.39	2.30	2.21	2.11	2.01
40	7.31	5.18	4.31	3.83	3.51	3.29	3.12	2.99	2.89	2.80	2.66	2.52	2.37	2.20	2.11	2.02	1.92	1.80
60	7.08	4.98	4.13	3.65	3.34	3.12	2.95	2.82	2.72	2.63	2.50	2.35	2.20	2.03	1.94	1.84	1.73	1.60
120	6.85	4.79	3.95	3.48	3.17	2.96	2.79	2.66	2.56	2.47	2.34	2.19	2.03	1.86	1.76	1.66	1.53	1.38
∞	6.63	4.61	3.78	3.32	3.02	2.80	2.64	2.51	2.41	2.32	2.18	2.04	1.88	1.70	1.59	1.47	1.32	1.00

Table V *F* Distribution (Continued)

$p = .001$ values

Degrees of freedom for denominator	Degrees of freedom for the numerator, v_1																	
v_2	1	2	3	4	5	6	7	8	9	10	12	15	20	30	40	60	120	∞
1	4053*	5000*	5404*	5625*	5764*	5859*	5929*	5981*	6023*	6056*	6107*	6158*	6209*	6261*	6287*	6313*	6340*	6366*
2	998.5	999.0	999.2	999.2	999.3	999.3	999.4	999.4	999.4	999.4	999.4	999.4	999.4	999.5	999.5	999.5	999.5	999.5
3	167.0	148.5	141.1	137.1	134.6	132.8	131.6	130.6	129.9	129.2	128.3	127.4	126.4	125.4	125.0	124.5	124.0	123.5
4	74.14	61.25	56.18	53.44	51.71	50.53	49.66	49.00	48.47	48.05	47.41	46.76	46.10	45.43	45.09	44.75	44.40	44.05
5	47.18	37.12	33.20	31.09	29.75	28.84	28.16	27.64	27.24	26.92	26.42	25.91	25.39	24.87	24.60	24.33	24.06	23.79
6	35.51	27.00	23.70	21.92	20.81	20.03	19.46	19.03	18.69	18.41	17.99	17.56	17.12	16.67	16.44	16.21	15.99	15.75
7	29.25	21.69	18.77	17.19	16.21	15.52	15.02	14.63	14.33	14.08	13.71	13.32	12.93	12.53	12.33	12.12	11.91	11.70
8	25.42	18.49	15.83	14.39	13.49	12.86	12.40	12.04	11.77	11.54	11.19	10.84	10.48	10.11	9.92	9.73	9.53	9.33
9	22.86	16.39	13.90	12.56	11.71	11.13	10.70	10.37	10.11	9.89	9.57	9.24	8.90	8.55	8.37	8.19	8.00	7.81
10	21.04	14.91	12.55	11.28	10.48	9.92	9.52	9.20	8.96	8.75	8.45	8.13	7.80	7.47	7.30	7.12	6.94	6.76
11	19.69	13.81	11.56	10.35	9.58	9.05	8.66	8.35	8.12	7.92	7.63	7.32	7.01	6.68	6.52	6.35	6.17	6.00
12	18.64	12.97	10.80	9.63	8.89	8.38	8.00	7.71	7.48	7.29	7.00	6.71	6.40	6.09	5.93	5.76	5.59	5.42
13	17.81	12.31	10.21	9.07	8.35	7.86	7.49	7.21	6.98	6.80	6.52	6.23	5.93	5.63	5.47	5.30	5.14	4.97
14	17.14	11.78	9.73	8.62	7.92	7.43	7.08	6.80	6.58	6.40	6.13	5.85	5.56	5.25	5.10	4.94	4.77	4.60
15	16.59	11.34	9.34	8.25	7.57	7.09	6.74	6.47	6.26	6.08	5.81	5.54	5.25	4.95	4.80	4.64	4.47	4.31
16	16.12	10.97	9.00	7.94	7.27	6.81	6.46	6.19	5.98	5.81	5.55	5.27	4.99	4.70	4.54	4.39	4.23	4.06
17	15.72	10.66	8.73	7.68	7.02	6.56	6.22	5.96	5.75	5.58	5.32	5.05	4.78	4.48	4.33	4.18	4.02	3.85
18	15.38	10.39	8.49	7.46	6.81	6.35	6.02	5.76	5.56	5.39	5.13	4.87	4.59	4.30	4.15	4.00	3.84	3.67
19	15.08	10.16	8.28	7.26	6.62	6.18	5.85	5.59	5.39	5.22	4.97	4.70	4.43	4.14	3.99	3.84	3.68	3.51
20	14.82	9.95	8.10	7.10	6.46	6.02	5.69	5.44	5.24	5.08	4.82	4.56	4.29	4.00	3.86	3.70	3.54	3.38
21	14.59	9.77	7.94	6.95	6.32	5.88	5.56	5.31	5.11	4.95	4.70	4.44	4.17	3.88	3.74	3.58	3.42	3.26
22	14.38	9.61	7.80	6.81	6.19	5.76	5.44	5.19	4.99	4.83	4.58	4.33	4.06	3.78	3.63	3.48	3.32	3.15
23	14.19	9.47	7.67	6.69	6.08	5.65	5.33	5.09	4.89	4.73	4.48	4.23	3.96	3.68	3.53	3.38	3.22	3.05
24	14.03	9.34	7.55	8.59	5.98	5.55	5.23	4.99	4.80	4.64	4.39	4.14	3.87	3.59	3.45	3.29	3.14	2.97
30	13.29	8.77	7.05	6.12	5.53	5.12	4.82	4.58	4.39	4.24	4.00	3.75	3.49	3.22	3.07	2.92	2.76	2.59
40	12.61	8.25	6.60	5.70	5.13	4.73	4.44	4.21	4.02	3.87	3.64	3.40	3.15	2.87	2.73	2.57	2.41	2.23
60	11.97	7.76	6.17	5.31	4.76	4.37	4.09	3.87	3.69	3.54	3.31	3.08	2.83	2.55	2.41	2.25	2.08	1.89
120	11.38	7.32	5.79	4.95	4.42	4.04	3.77	3.55	3.38	3.24	3.02	2.78	2.53	2.26	2.11	1.95	1.76	1.54
∞	10.83	6.91	5.42	4.62	4.10	3.74	3.47	3.27	3.10	2.96	2.74	2.51	2.27	1.99	1.84	1.66	1.45	1.00

*Multiply these entries by 100.

Table VI χ^2 Distribution

For various degrees of freedom (*df*), the tabled entries represent the values of χ^2 above which a proportion p of the distribution falls. (Example: For $df = 5$, a $\chi^2 = 11.070$ is exceeded by $p = .05$ or 5% of the distribution).

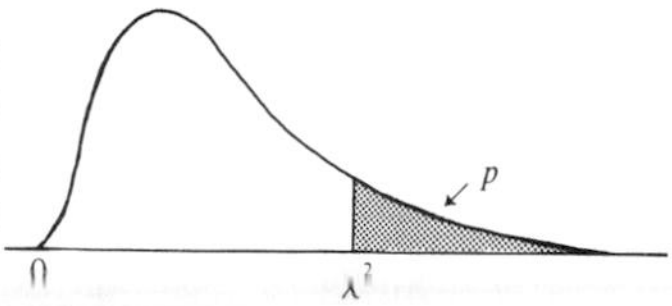

df	*p*						
	.99	.95	.90	.10	.05	.01	.001
1	$.0^3157$	.00393	.0158	2.706	3.841	6.635	10.827
2	.0201	.103	.211	4.605	5.991	9.210	13.815
3	.115	.352	.584	6.251	7.815	11.345	16.266
4	.297	.711	1.064	7.779	9.488	13.277	18.467
5	.554	1.145	1.610	9.236	11.070	15.086	20.515
6	.872	1.635	2.204	10.645	12.592	16.812	22.457
7	1.239	2.167	2.833	12.017	14.067	18.475	24.322
8	1.646	2.733	3.490	13.362	15.507	20.090	26.125
9	2.088	3.325	4.168	14.684	16.919	21.666	27.877
10	2.558	3.940	4.865	15.987	18.307	23.209	29.588
11	3.053	4.575	5.578	17.275	19.675	24.725	31.264
12	3.571	5.226	6.304	18.549	21.026	26.217	32.909
13	4.107	5.892	7.042	19.812	22.362	27.688	34.528
14	4.660	6.571	7.790	21.064	23.685	29.141	36.123
15	5.229	7.261	8.547	22.307	24.996	30.578	37.697
16	5.812	7.962	9.312	23.542	26.296	32.000	39.252
17	6.408	8.672	10.085	24.769	27.587	33.409	40.790
18	7.015	9.390	10.865	25.989	28.869	34.805	42.312
19	7.633	10.117	11.651	27.204	30.144	36.191	43.820
20	8.260	10.851	12.443	28.412	31.410	37.566	45.315
21	8.897	11.591	13.240	29.615	32.671	38.932	46.797
22	9.542	12.338	14.041	30.813	33.924	40.289	48.268
23	10.196	13.091	14.848	32.007	35.172	41.638	49.728
24	10.856	13.848	15.659	33.196	36.415	42.980	51.179
25	11.524	14.611	16.473	34.382	37.652	44.314	52.620
26	12.198	15.379	17.292	35.563	38.885	45.642	54.052
27	12.879	16.151	18.114	36.741	40.113	46.963	55.476
28	13.565	16.928	18.939	37.916	41.337	48.278	56.893
29	14.256	17.708	19.768	39.087	42.557	49.588	58.302
30	14.953	18.493	20.599	40.256	43.773	50.892	59.703

Adapted from Table IV of R. A. Fisher and F. Yates, *Statistical Tables for Biological, Agricultural, and Medical Research*, 6th Edition, Longman Group, Ltd., London, 1974. (Previously published by Oliver & Boyd, Ltd., Edinburgh). Used with permission of the authors and publishers.

Table VII Product Moment Correlation Coefficient

The tabled entries represent the critical values of r, based on n pairs of observations, for testing the hypothesis $\rho = 0$ at the .05 and .01 significance levels.

		Two-tailed p		One-tailed p	
n	df	.05	.01	.05	.01
4	2	.950	.990	.900	.980
5	3	.878	.959	.805	.934
6	4	.811	.917	.729	.882
7	5	.754	.875	.669	.833
8	6	.707	.834	.621	.789
9	7	.666	.798	.582	.750
10	8	.632	.765	.549	.715
11	9	.602	.735	.521	.685
12	10	.576	.708	.497	.658
13	11	.553	.684	.476	.634
14	12	.532	.661	.457	.612
15	13	.514	.641	.441	.592
16	14	.497	.623	.426	.574
17	15	.482	.606	.412	.558
18	16	.468	.590	.400	.542
19	17	.456	.575	.389	.529
20	18	.444	.561	.378	.515
25	23	.396	.505	.337	.462
30	28	.361	.463	.306	.423
40	38	.312	.402	.264	.367
60	58	.254	.330	.214	.300
120	118	.179	.234	.151	.212

For unlisted values of n, the critical value of r is given by $r = t_c/\sqrt{t_c^2 + n - 2}$, where t_c is the critical t value associated with a given significance level and has $df = n - 2$.

Table VIII Rank Correlation Coefficient

The tabled entries represent the probabilities of equalling or surpassing the critical values of r_o based on chance alone, assuming independent variables.

Number of pairs n	Critical r_o	Probabilities: *Two-tailed*	Probabilities: *One-tailed*
4	1.00	.083	.042
5	1.00	.017	.008
	.90	.083	.042
6	.94	.017	.008
	.83	.058	.029
	.77	.102	.051
7	.86	.024	.012
	.79	.048	.024
	.68	.110	.055
8	.81	.022	.011
	.74	.046	.023
	.64	.096	.048
9	.78	.018	.009
	.68	.050	.025
	.60	.096	.048
10	.73	.020	.010
	.65	.048	.024
	.56	.096	.048

For values of n greater than 10, the critical values for r listed in Table VII are also approximately correct for r_o.

Table IX Rank Sum Statistic (Critical Values)

The tabled entries, of the form $a, b(p)$, represent the probability p that the rank sum statistic R will be equal to or less than a *or* equal to or greater than b. (Example: For two samples of size 5 and 7, respectively, the probability is $p = .05$ that $R \leq 20$ *or* $R \geq 45$). For a one-tailed test, use one half of the indicated probabilities. (Note: For a more detailed table, see the primary source indicated at the foot of the table.)

Size of larger sample	Size of the smaller sample								
	2	3	4	5	6	7	8	9	10
3	3,9 (.20)	6,15 (.10)							
4	3,11 (.13)	6,18 (.06) 7,17 (.11)	10,26 (.03) 11,25 (.06) 12,24 (.11)						
5	3,13 (.10)	6,21 (.04) 7,20 (.07)	10,30 (.02) 12,28 (.06) 13,27 (.11)	15,40 (.01) 18,37 (.06) 19,36 (.10)					
6	3,15 (.07)	6,24 (.02) 7,23 (.05) 8,22 (.10)	10,34 (.01) 12,32 (.04) 14,30 (.11)	16,44 (.01) 19,41 (.05) 20,40 (.08)	23,55 (.01) 26,52 (.04) 28,50 (.09)				
7	3,17 (.06) 4,16 (.11)	6,27 (.02) 7,26 (.03) 9,24 (.12)	11,37 (.01) 13,35 (.04) 15,33 (.11)	17,48 (.01) 20,45 (.05) 22,43 (.11)	24,60 (.01) 28,56 (.05) 30,54 (.10)	33,72 (.01) 37,68 (.05) 39,66 (.10)			
8	3,19 (.04) 4,18 (.09)	6,30 (.01) 8,28 (.05) 9,27 (.08)	11,41 (.01) 14,38 (.05) 16,36 (.11)	18,52 (.01) 21,49 (.05) 23,47 (.09)	26,64 (.01) 30,60 (.06) 32,58 (.11)	34,78 (.01) 39,73 (.05) 41,71 (.09)	44,92 (.01) 49,87 (.05) 52,84 (.10)		
9	3,21 (.04) 4,20 (.07)	6,33 (.01) 8,31 (.04) 10,29 (.10)	12,44 (.01) 15,41 (.05) 17,39 (.11)	19,56 (.01) 23,52 (.06) 25,50 (.11)	27,69 (.01) 31,65 (.05) 33,63 (.09)	36,83 (.01) 41,78 (.05) 43,76 (.09)	46,98 (.01) 51,93 (.05) 54,90 (.09)	57,114 (.01) 63,108 (.05) 66,105 (.09)	
10	3,23 (.03) 4,22 (.06)	7,35 (.01) 9,33 (.05) 11,31 (.11)	12,48 (.01) 16,44 (.05) 18,42 (.11)	20,60 (.01) 23,57 (.04) 26,54 (.11)	28,74 (.01) 32,70 (.04) 35,67 (.09)	37,89 (.01) 42,84 (.04) 46,80 (.11)	48,104 (.01) 54,98 (.05) 57,95 (.10)	59,121 (.01) 66,114 (.05) 69,111 (.09)	72,138 (.01) 79,131 (.05) 82,128 (.09)

Table X Binomial Distribution (cumulative)
The table entries represent cumulative binomial probabilities $b(x \geq x_0; n, p)$. (Example: For $n = 7$ trials with $p = .20$, the probability of observing $x = 5$ *or more* successes is .0047). To determine the probability of *exactly* x_0 successes, subtract the two adjacent cumulative probabilities. (Example: For $n = 7$ trials with $p = .20$, the probability of exactly 5 successes is $.0047 - .0004 = .0043$).

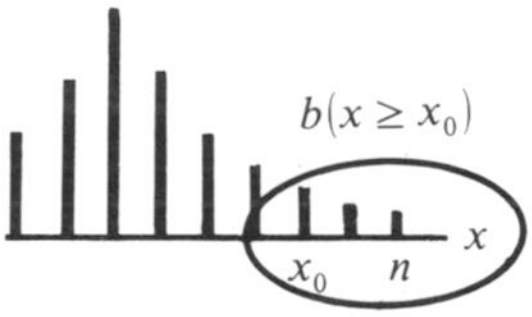

n	x_0 \ p	.05	.10	.15	.20	.25	.30	.35	.40	.45	.50
2	1	.0975	.1900	.2775	.3600	.4375	.5100	.5775	.6400	.6975	.7500
	2	.0025	.0100	.0225	.0400	.0625	.0900	.1225	.1600	.2025	.2500
3	1	.1426	.2710	.3859	.4880	.5781	.6570	.7254	.7840	.8336	.8750
	2	.0072	.0280	.0608	.1040	.1562	.2160	.2818	.3520	.4252	.5000
	3	.0001	.0010	.0034	.0080	.0156	.0270	.0429	.0640	.0911	.1250
4	1	.1855	.3439	.4780	.5904	.6836	.7599	.8215	.8704	.9085	.9375
	2	.0140	.0523	.1095	.1808	.2617	.3483	.4370	.5248	.6090	.6875
	3	.0005	.0037	.0120	.0272	.0508	.0837	.1265	.1792	.2415	.3125
	4	.0000	.0001	.0005	.0016	.0039	.0081	.0150	.0256	.0410	.0625
5	1	.2262	.4095	.5563	.6723	.7627	.8319	.8840	.9222	.9497	.9688
	2	.0226	.0815	.1648	.2627	.3672	.4718	.5716	.6630	.7438	.8125
	3	.0012	.0086	.0266	.0579	.1035	.1631	.2352	.3174	.4069	.5000
	4	.0000	.0005	.0022	.0067	.0156	.0308	.0540	.0870	.1312	.1875
	5	.0000	.0000	.0001	.0003	.0010	.0024	.0053	.0102	.0185	.0312
6	1	.2649	.4686	.6229	.7379	.8220	.8824	.9246	.9533	.9723	.9844
	2	.0328	.1143	.2235	.3447	.4661	.5798	.6809	.7667	.8364	.8906
	3	.0022	.0158	.0473	.0989	.1694	.2557	.3529	.4557	.5585	.6562
	4	.0001	.0013	.0059	.0170	.0376	.0705	.1174	.1792	.2553	.3438
	5	.0000	.0001	.0004	.0016	.0046	.0109	.0223	.0410	.0692	.1094
	6	.0000	.0000	.0000	.0001	.0002	.0007	.0018	.0041	.0083	.0156
7	1	.3017	.5217	.6794	.7903	.8665	.9176	.9510	.9720	.9848	.9922
	2	.0444	.1497	.2834	.4233	.5551	.6706	.7662	.8414	.8976	.9375
	3	.0038	.0257	.0738	.1480	.2436	.3529	.4677	.5801	.6836	.7734
	4	.0002	.0027	.0121	.0333	.0706	.1260	.1998	.2898	.3917	.5000
	5	.0000	.0002	.0012	.0047	.0129	.0288	.0556	.0963	.1529	.2266
	6	.0000	.0000	.0001	.0004	.0013	.0038	.0090	.0188	.0357	.0625
	7	.0000	.0000	.0000	.0000	.0001	.0002	.0006	.0016	.0037	.0078
8	1	.3366	.5695	.7275	.8322	.8999	.9424	.9681	.9832	.9916	.9961
	2	.0572	.1869	.3428	.4967	.6329	.7447	.8309	.8936	.9368	.9648
	3	.0058	.0381	.1052	.2031	.3215	.4482	.5722	.6846	.7799	.8555
	4	.0004	.0050	.0214	.0563	.1138	.1941	.2936	.4059	.5230	.6367
	5	.0000	.0004	.0029	.0104	.0273	.0580	.1061	.1737	.2604	.3633
	6	.0000	.0000	.0002	.0012	.0042	.0113	.0253	.0498	.0885	.1445
	7	.0000	.0000	.0000	.0001	.0004	.0013	.0036	.0085	.0181	.0352
	8	.0000	.0000	.0000	.0000	.0000	.0001	.0002	.0007	.0017	.0039

Adapted from *Handbook of Tables for Mathematics*, 4th Ed. Cleveland, OH: Chemical Rubber Company. Used with permission of CRC Press, Inc.

Table X Binomial Distribution (Continued)

n	x_0 \ p	.05	.10	.15	.20	.25	.30	.35	.40	.45	.50
9	1	.3698	.6126	.7684	.8658	.9249	.9596	.9793	.9899	.9954	.9980
	2	.0712	.2252	.4005	.5638	.6997	.8040	.8789	.9295	.9615	.9805
	3	.0084	.0530	.1409	.2618	.3993	.5372	.6627	.7682	.8505	.9102
	4	.0006	.0083	.0339	.0856	.1657	.2703	.3911	.5174	.6386	.7461
	5	.0000	.0009	.0056	.0196	.0489	.0988	.1717	.2666	.3786	.5000
	6	.0000	.0001	.0006	.0031	.0100	.0253	.0536	.0994	.1658	.2539
	7	.0000	.0000	.0000	.0003	.0013	.0043	.0112	.0250	.0498	.0898
	8	.0000	.0000	.0000	.0000	.0001	.0004	.0014	.0038	.0091	.0195
	9	.0000	.0000	.0000	.0000	.0000	.0000	.0001	.0003	.0008	.0020
10	1	.4013	.6513	.8031	.8926	.9437	.9718	.9865	.9940	.9975	.9990
	2	.0861	.2639	.4557	.6242	.7560	.8507	.9140	.9536	.9767	.9893
	3	.0115	.0702	.1798	.3222	.4744	.6172	.7384	.8327	.9004	.9453
	4	.0010	.0128	.0500	.1209	.2241	.3504	.4862	.6177	.7340	.8281
	5	.0001	.0016	.0099	.0328	.0781	.1503	.2485	.3669	.4956	.6230
	6	.0000	.0001	.0014	.0064	.0197	.0473	.0949	.1662	.2616	.3770
	7	.0000	.0000	.0001	.0009	.0035	.0106	.0260	.0548	.1020	.1719
	8	.0000	.0000	.0000	.0001	.0004	.0016	.0048	.0123	.0274	.0547
	9	.0000	.0000	.0000	.0000	.0000	.0001	.0005	.0017	.0045	.0107
	10	.0000	.0000	.0000	.0000	.0000	.0000	.0000	.0001	.0003	.0010
11	1	.4312	.6862	.8327	.9141	.9578	.9802	.9912	.9964	.9986	.9995
	2	.1019	.3026	.5078	.6779	.8029	.8870	.9394	.9698	.9861	.9941
	3	.0152	.0896	.2212	.3826	.5448	.6873	.7999	.8811	.9348	.9673
	4	.0016	.0185	.0694	.1611	.2867	.4304	.5744	.7037	.8089	.8867
	5	.0001	.0028	.0159	.0504	.1146	.2103	.3317	.4672	.6029	.7256
	6	.0000	.0003	.0027	.0117	.0343	.0782	.1487	.2465	.3669	.5000
	7	.0000	.0000	.0003	.0020	.0076	.0216	.0501	.0994	.1738	.2744
	8	.0000	.0000	.0000	.0002	.0012	.0043	.0122	.0293	.0610	.1133
	9	.0000	.0000	.0000	.0000	.0001	.0006	.0020	.0059	.0148	.0327
	10	.0000	.0000	.0000	.0000	.0000	.0000	.0002	.0007	.0022	.0059
	11	.0000	.0000	.0000	.0000	.0000	.0000	.0000	.0000	.0002	.0005
12	1	.4596	.7176	.8578	.9313	.9683	.9862	.9943	.9978	.9992	.9998
	2	.1184	.3410	.5565	.7251	.8416	.9150	.9576	.9804	.9917	.9968
	3	.0196	.1109	.2642	.4417	.6093	.7472	.8487	.9166	.9579	.9807
	4	.0022	.0256	.0922	.2054	.3512	.5075	.6533	.7747	.8655	.9270
	5	.0002	.0043	.0239	.0726	.1576	.2763	.4167	.5618	.6956	.8062
	6	.0000	.0005	.0046	.0194	.0544	.1178	.2127	.3348	.4731	.6128
	7	.0000	.0001	.0007	.0039	.0143	.0386	.0846	.1582	.2607	.3872
	8	.0000	.0000	.0001	.0006	.0028	.0095	.0255	.0573	.1117	.1938
	9	.0000	.0000	.0000	.0001	.0004	.0017	.0056	.0153	.0356	.0730
	10	.0000	.0000	.0000	.0000	.0000	.0002	.0008	.0028	.0079	.0193
	11	.0000	.0000	.0000	.0000	.0000	.0000	.0001	.0003	.0011	.0032
	12	.0000	.0000	.0000	.0000	.0000	.0000	.0000	.0000	.0001	.0002

Table X Binomial Distribution (Continued)

n	x_0 \ p	.05	.10	.15	.20	.25	.30	.35	.40	.45	.50
13	1	.4867	.7458	.8791	.9450	.9762	.9903	.9963	.9987	.9996	.9999
	2	.1354	.3787	.6017	.7664	.8733	.9363	.9704	.9874	.9951	.9983
	3	.0245	.1339	.2704	.4983	.6674	.7975	.8868	.9421	.9731	.9888
	4	.0031	.0342	.0967	.2527	.4157	.5794	.7217	.8314	.9071	.9539
	5	.0003	.0065	.0260	.0991	.2060	.3457	.4995	.6470	.7721	.8666
	6	.0000	.0009	.0053	.0300	.0802	.1654	.2841	.4256	.5732	.7095
	7	.0000	.0001	.0013	.0070	.0243	.0624	.1295	.2288	.3563	.5000
	8	.0000	.0000	.0002	.0012	.0056	.0182	.0462	.0977	.1788	.2905
	9	.0000	.0000	.0000	.0002	.0010	.0040	.0126	.0321	.0698	.1334
	10	.0000	.0000	.0000	.0000	.0001	.0007	.0025	.0078	.0203	.0461
	11	.0000	.0000	.0000	.0000	.0000	.0001	.0003	.0013	.0041	.0112
	12	.0000	.0000	.0000	.0000	.0000	.0000	.0000	.0001	.0005	.0017
	13	.0000	.0000	.0000	.0000	.0000	.0000	.0000	.0000	.0000	.0001
14	1	.5123	.7712	.8972	.9560	.9822	.9932	.9976	.9992	.9998	.9999
	2	.1530	.4154	.6433	.8021	.8990	.9525	.9795	.9919	.9971	.9991
	3	.0301	.1584	.3521	.5519	.7189	.8392	.9161	.9602	.9830	.9935
	4	.0042	.0441	.1465	.3018	.4787	.6448	.7795	.8757	.9368	.9713
	5	.0004	.0092	.0467	.1298	.2585	.4158	.5773	.7207	.8328	.9102
	6	.0000	.0015	.0115	.0439	.1117	.2195	.3595	.5141	.6627	.7880
	7	.0000	.0002	.0022	.0116	.0383	.0933	.1836	.3075	.4539	.6047
	8	.0000	.0000	.0003	.0024	.0103	.0315	.0753	.1501	.2586	.3953
	9	.0000	.0000	.0000	.0004	.0022	.0083	.0243	.0583	.1189	.2120
	10	.0000	.0000	.0000	.0000	.0003	.0017	.0060	.0175	.0426	.0898
	11	.0000	.0000	.0000	.0000	.0000	.0002	.0011	.0039	.0114	.0287
	12	.0000	.0000	.0000	.0000	.0000	.0000	.0001	.0006	.0022	.0065
	13	.0000	.0000	.0000	.0000	.0000	.0000	.0000	.0001	.0003	.0009
	14	.0000	.0000	.0000	.0000	.0000	.0000	.0000	.0000	.0000	.0001
15	1	.5367	.7941	.9126	.9648	.9866	.9953	.9984	.9995	.9999	1.0000
	2	.1710	.4510	.6814	.8329	.9198	.9647	.9858	.9948	.9983	.9995
	3	.0362	.1841	.3958	.6020	.7639	.8732	.9383	.9729	.9893	.9963
	4	.0055	.0556	.1773	.3518	.5387	.7031	.8273	.9095	.9576	.9824
	5	.0006	.0127	.0617	.1642	.3135	.4845	.6481	.7827	.8796	.9408
	6	.0001	.0022	.0168	.0611	.1484	.2784	.4357	.5968	.7392	.8491
	7	.0000	.0003	.0036	.0181	.0566	.1311	.2452	.3902	.5478	.6964
	8	.0000	.0000	.0006	.0042	.0173	.0500	.1132	.2131	.3465	.5000
	9	.0000	.0000	.0001	.0008	.0042	.0152	.0422	.0950	.1818	.3036
	10	.0000	.0000	.0000	.0001	.0008	.0037	.0124	.0338	.0769	.1509
	11	.0000	.0000	.0000	.0000	.0001	.0007	.0028	.0093	.0255	.0592
	12	.0000	.0000	.0000	.0000	.0000	.0001	.0005	.0019	.0063	.0176
	13	.0000	.0000	.0000	.0000	.0000	.0000	.0001	.0003	.0011	.0037
	14	.0000	.0000	.0000	.0000	.0000	.0000	.0000	.0000	.0001	.0005
	15	.0000	.0000	.0000	.0000	.0000	.0000	.0000	.0000	.0000	.0000

Table X Binomial Distribution (Continued)

n	x_0 \ p	.05	.10	.15	.20	.25	.30	.35	.40	.45	.50
16	1	.5599	.8147	.9257	.9719	.9900	.9967	.9990	.9997	.9999	1.0000
	2	.1892	.4853	.7161	.8593	.9365	.9739	.9902	.9967	.9990	.9997
	3	.0429	.2108	.4386	.6482	.8029	.9006	.9549	.9817	.9934	.9979
	4	.0070	.0684	.2101	.4019	.5950	.7541	.8661	.9349	.9719	.9894
	5	.0009	.0170	.0791	.2018	.3698	.5501	.7108	.8334	.9147	.9616
	6	.0001	.0033	.0235	.0817	.1897	.3402	.5100	.6712	.8024	.8949
	7	.0000	.0005	.0056	.0267	.0796	.1753	.3119	.4728	.6340	.7228
	8	.0000	.0001	.0011	.0070	.0271	.0744	.1594	.2839	.4371	.5982
	9	.0000	.0000	.0002	.0015	.0075	.0257	.0671	.1423	.2559	.4018
	10	.0000	.0000	.0000	.0002	.0016	.0071	.0229	.0583	.1241	.2272
	11	.0000	.0000	.0000	.0000	.0003	.0016	.0062	.0191	.0486	.1051
	12	.0000	.0000	.0000	.0000	.0000	.0003	.0013	.0049	.0149	.0384
	13	.0000	.0000	.0000	.0000	.0000	.0000	.0002	.0009	.0035	.0106
	14	.0000	.0000	.0000	.0000	.0000	.0000	.0000	.0001	.0006	.0021
	15	.0000	.0000	.0000	.0000	.0000	.0000	.0000	.0000	.0001	.0003
	16	.0000	.0000	.0000	.0000	.0000	.0000	.0000	.0000	.0000	.0000
17	1	.5819	.8332	.9369	.9775	.9925	.9977	.9993	.9998	1.0000	1.0000
	2	.2078	.5182	.7475	.8818	.9499	.9807	.9933	.9979	.9994	.9999
	3	.0503	.2382	.4802	.6904	.8363	.9226	.9673	.9877	.9959	.9988
	4	.0088	.0826	.2444	.4511	.6470	.7981	.8972	.9536	.9816	.9936
	5	.0012	.0221	.0987	.2418	.4261	.6113	.7652	.8740	.9404	.9755
	6	.0001	.0047	.0319	.1057	.2347	.4032	.5803	.7361	.8529	.9283
	7	.0000	.0008	.0083	.0377	.1071	.2248	.3812	.5522	.7098	.8338
	8	.0000	.0001	.0017	.0109	.0402	.1046	.2128	.3595	.5257	.6855
	9	.0000	.0000	.0003	.0026	.0124	.0403	.0994	.1989	.3374	.5000
	10	.0000	.0000	.0000	.0005	.0031	.0127	.0383	.0919	.1834	.3145
	11	.0000	.0000	.0000	.0001	.0006	.0032	.0120	.0348	.0826	.1662
	12	.0000	.0000	.0000	.0000	.0001	.0007	.0030	.0106	.0301	.0717
	13	.0000	.0000	.0000	.0000	.0000	.0001	.0006	.0025	.0086	.0245
	14	.0000	.0000	.0000	.0000	.0000	.0000	.0000	.0005	.0019	.0064
	15	.0000	.0000	.0000	.0000	.0000	.0000	.0000	.0001	.0003	.0012
	16	.0000	.0000	.0000	.0000	.0000	.0000	.0000	.0000	.0000	.0001
	17	.0000	.0000	.0000	.0000	.0000	.0000	.0000	.0000	.0000	.0000
18	1	.6028	.8499	.9464	.9820	.9944	.9984	.9996	.9999	1.0000	1.0000
	2	.2265	.5497	.7759	.9009	.9605	.9858	.9954	.9987	.9997	.9999
	3	.0581	.2662	.5203	.7287	.8647	.9400	.9764	.9918	.9975	.9993
	4	.0109	.0982	.2798	.4990	.6943	.8354	.9217	.9672	.9880	.9962
	5	.0015	.0282	.1206	.2836	.4813	.6673	.8114	.9058	.9589	.9846
	6	.0002	.0064	.0419	.1329	.2825	.4656	.6450	.7912	.8923	.9519
	7	.0000	.0012	.0118	.0513	.1390	.2783	.4509	.6257	.7742	.8811
	8	.0000	.0002	.0027	.0163	.0569	.1407	.2717	.4366	.6085	.7597
	9	.0000	.0000	.0005	.0043	.0193	.0596	.1391	.2632	.4222	.5927
	10	.0000	.0000	.0001	.0009	.0054	.0210	.0597	.1347	.2527	.4073

Table X Binomial Distribution (Continued)

n	x_0 \ p	.05	.10	.15	.20	.25	.30	.35	.40	.45	.50
18	11	.0000	.0000	.0000	.0002	.0012	.0061	.0212	.0576	.1280	.2403
	12	.0000	.0000	.0000	.0000	.0002	.0014	.0062	.0203	.0537	.1189
	13	.0000	.0000	.0000	.0000	.0000	.0003	.0014	.0058	.0183	.0481
	14	.0000	.0000	.0000	.0000	.0000	.0000	.0003	.0013	.0049	.0154
	15	.0000	.0000	.0000	.0000	.0000	.0000	.0000	.0002	.0010	.0038
	16	.0000	.0000	.0000	.0000	.0000	.0000	.0000	.0000	.0001	.0007
	17	.0000	.0000	.0000	.0000	.0000	.0000	.0000	.0000	.0000	.0001
19	1	.6226	.8649	.9544	.9856	.9958	.9989	.9997	.9999	1.0000	1.0000
	2	.2453	.5797	.8015	.9171	.9690	.9896	.9969	.9992	.9998	1.0000
	3	.0665	.2946	.5587	.7631	.8887	.9538	.9830	.9945	.9985	.9996
	4	.0132	.1150	.3159	.5449	.7369	.8668	.9409	.9770	.9923	.9978
	5	.0020	.0352	.1444	.3267	.5346	.7178	.8500	.9304	.9720	.9904
	6	.0002	.0086	.0537	.1631	.3322	.5261	.7032	.8371	.9223	.9682
	7	.0000	.0017	.0163	.0676	.1749	.3345	.5188	.6919	.8273	.9165
	8	.0000	.0003	.0041	.0233	.0775	.1820	.3344	.5122	.6831	.8204
	9	.0000	.0000	.0008	.0067	.0287	.0839	.1855	.3325	.5060	.6762
	10	.0000	.0000	.0001	.0016	.0089	.0326	.0875	.1861	.3290	.5000
	11	.0000	.0000	.0000	.0003	.0023	.0105	.0347	.0885	.1841	.3238
	12	.0000	.0000	.0000	.0000	.0005	.0028	.0114	.0352	.0871	.1796
	13	.0000	.0000	.0000	.0000	.0001	.0006	.0031	.0116	.0342	.0835
	14	.0000	.0000	.0000	.0000	.0000	.0001	.0007	.0031	.0109	.0318
	15	.0000	.0000	.0000	.0000	.0000	.0000	.0001	.0006	.0028	.0096
	16	.0000	.0000	.0000	.0000	.0000	.0000	.0000	.0001	.0005	.0022
	17	.0000	.0000	.0000	.0000	.0000	.0000	.0000	.0000	.0001	.0004
	18	.0000	.0000	.0000	.0000	.0000	.0000	.0000	.0000	.0000	.0000
20	1	.6415	.8784	.9612	.9885	.9968	.9992	.9998	1.0000	1.0000	1.0000
	2	.2642	.6083	.8244	.9308	.9757	.9924	.9979	.9995	.9999	1.0000
	3	.0755	.3231	.5951	.7939	.9087	.9645	.9879	.9964	.9991	.9998
	4	.0159	.1330	.3523	.5886	.7748	.8929	.9556	.9840	.9951	.9987
	5	.0026	.0432	.1702	.3704	.5852	.7625	.8818	.9490	.9811	.9941
	6	.0003	.0113	.0673	.1958	.3828	.5836	.7546	.8744	.9447	.9793
	7	.0000	.0024	.0219	.0867	.2142	.3920	.5834	.7500	.8701	.9423
	8	.0000	.0004	.0059	.0321	.1018	.2277	.3990	.5841	.7480	.8684
	9	.0000	.0001	.0013	.0100	.0409	.1133	.2376	.4044	.5857	.7483
	10	.0000	.0000	.0002	.0026	.0139	.0480	.1218	.2447	.4086	.5881
	11	.0000	.0000	.0000	.0006	.0039	.0171	.0532	.1275	.2493	.4119
	12	.0000	.0000	.0000	.0001	.0009	.0051	.0196	.0565	.1308	.2517
	13	.0000	.0000	.0000	.0000	.0002	.0013	.0060	.0210	.0580	.1316
	14	.0000	.0000	.0000	.0000	.0000	.0003	.0015	.0065	.0214	.0577
	15	.0000	.0000	.0000	.0000	.0000	.0000	.0003	.0016	.0064	.0207
	16	.0000	.0000	.0000	.0000	.0000	.0000	.0000	.0003	.0015	.0059
	17	.0000	.0000	.0000	.0000	.0000	.0000	.0000	.0000	.0003	.0013
	18	.0000	.0000	.0000	.0000	.0000	.0000	.0000	.0000	.0000	.0002

Table XI Poisson Distribution (cumulative)
The tabled entries represent the cumulative Poisson probabilities $f_p(x \geq x_0;\ \lambda)$. (Example: For $\lambda = 2.7$ the probability of $x = 6$ *or more* successes is $p = .0567$). To determine the probability of *exactly* x_0 successes, subtract the two adjacent cumulative probabilities. (Example: For $\lambda = 2.7$ the probability of exactly $x_0 = 6$ successes is $.0567 - .0206 = .0361$).

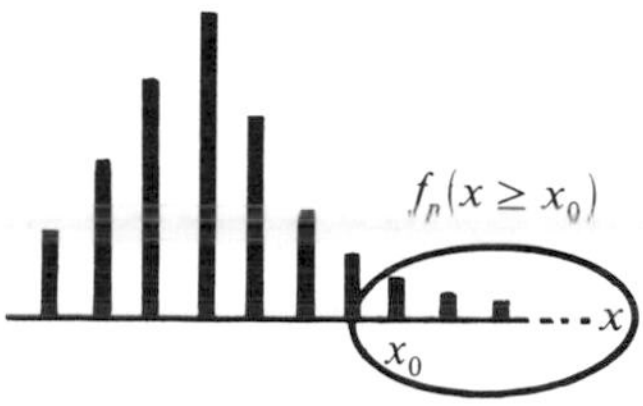

x_0 \ λ	0.1	0.2	0.3	0.4	0.5	0.6	0.7	0.8	0.9	1.0
1	.0952	.1813	.2592	.3297	.3935	.4512	.5034	.5507	.5934	.6321
2	.0047	.0175	.0369	.0616	.0902	.1219	.1558	.1912	.2275	.2642
3	.0002	.0011	.0036	.0079	.0144	.0231	.0341	.0474	.0629	.0803
4	.0000	.0001	.0003	.0008	.0018	.0034	.0058	.0091	.0135	.0190
5	.0000	.0000	.0000	.0001	.0002	.0004	.0008	.0014	.0023	.0037
6	.0000	.0000	.0000	.0000	.0000	.0000	.0001	.0002	.0003	.0006
7	.0000	.0000	.0000	.0000	.0000	.0000	.0000	.0000	.0000	.0001
x_0 \ λ	1.1	1.2	1.3	1.4	1.5	1.6	1.7	1.8	1.9	2.0
1	.6671	.6988	.7275	.7534	.7769	.7981	.8173	.8347	.8504	.8647
2	.3010	.3374	.3732	.4082	.4422	.4751	.5068	.5372	.5663	.5940
3	.0996	.1205	.1429	.1665	.1912	.2166	.2428	.2694	.2963	.3233
4	.0257	.0338	.0431	.0537	.0656	.0788	.0932	.1087	.1253	.1429
5	.0054	.0077	.0107	.0143	.0186	.0237	.0296	.0364	.0441	.0527
6	.0010	.0015	.0022	.0032	.0045	.0060	.0080	.0104	.0132	.0166
7	.0001	.0003	.0004	.0006	.0009	.0013	.0019	.0026	.0034	.0045
8	.0000	.0000	.0001	.0001	.0002	.0003	.0004	.0006	.0008	.0011
9	.0000	.0000	.0000	.0000	.0000	.0000	.0001	.0001	.0002	.0002
x_0 \ λ	2.1	2.2	2.3	2.4	2.5	2.6	2.7	2.8	2.9	3.0
1	.8775	.8892	.8897	.9093	.9179	.9257	.9328	.9392	.9450	.9502
2	.6204	.6454	.6691	.6916	.7127	.7326	.7513	.7689	.7854	.8009
3	.3504	.3773	.4040	.4303	.4562	.4816	.5064	.5305	.5540	.5768
4	.1614	.1806	.2007	.2213	.2424	.2640	.2859	.3081	.3304	.3528
5	.0621	.0725	.0838	.0959	.1088	.1226	.1371	.1523	.1682	.1847
6	.0204	.0249	.0300	.0357	.0420	.0490	.0567	.0651	.0742	.0839
7	.0059	.0075	.0094	.0116	.0142	.0172	.0206	.0244	.0287	.0335
8	.0015	.0020	.0026	.0033	.0042	.0053	.0066	.0081	.0099	.0119
9	.0003	.0005	.0006	.0009	.0011	.0015	.0019	.0024	.0031	.0038
10	.0001	.0001	.0001	.0002	.0003	.0004	.0005	.0007	.0009	.0011
11	.0000	.0000	.0000	.0000	.0001	.0001	.0001	.0002	.0002	.0003
12	.0000	.0000	.0000	.0000	.0000	.0000	.0000	.0000	.0001	.0001

Adapted from *Handbook of Tables for Mathematics*, 4th Ed. Cleveland, OH: Chemical Rubber Company. Used with permission of CRC Press, Inc.

Table XI Poisson Distribution (Continued)

x_0 \ λ	3.1	3.2	3.3	3.4	3.5	3.6	3.7	3.8	3.9	4.0
1	.9550	.9592	.9631	.9666	.9698	.9727	.9753	.9776	.9798	.9817
2	.8153	.8288	.8414	.8532	.8641	.8743	.8838	.8926	.9008	.9084
3	.5988	.6201	.6406	.6603	.6792	.6973	.7146	.7311	.7469	.7619
4	.3752	.3975	.4197	.4416	.4634	.4848	.5058	.5265	.5468	.5665
5	.2018	.2194	.2374	.2558	.2746	.2936	.3128	.3322	.3516	.3712
6	.0943	.1054	.1171	.1295	.1424	.1559	.1699	.1844	.1994	.2149
7	.0388	.0446	.0510	.0579	.0653	.0733	.0818	.0909	.1005	.1107
8	.0142	.0168	.0198	.0231	.0267	.0308	.0352	.0401	.0454	.0511
9	.0047	.0057	.0069	.0083	.0099	.0117	.0137	.0160	.0185	.0214
10	.0014	.0018	.0022	.0027	.0033	.0040	.0048	.0058	.0069	.0081
11	.0004	.0005	.0006	.0008	.0010	.0013	.0016	.0019	.0023	.0028
12	.0001	.0001	.0002	.0002	.0003	.0004	.0005	.0006	.0007	.0009
13	.0000	.0000	.0000	.0001	.0001	.0001	.0001	.0002	.0002	.0003
14	.0000	.0000	.0000	.0000	.0000	.0000	.0000	.0000	.0001	.0001

x_0 \ λ	4.1	4.2	4.3	4.4	4.5	4.6	4.7	4.8	4.9	5.0
1	.9834	.9850	.9864	.9877	.9889	.9899	.9909	.9918	.9926	.9933
2	.9155	.9220	.9281	.9337	.9389	.9437	.9482	.9523	.9561	.9596
3	.7762	.7898	.8026	.8149	.8264	.8374	.8477	.8575	.8667	.8753
4	.5858	.6046	.6228	.6406	.6577	.6743	.6903	.7058	.7207	.7350
5	.3907	.4102	.4296	.4488	.4679	.4868	.5054	.5237	.5418	.5595
6	.2307	.2469	.2633	.2801	.2971	.3142	.3316	.3490	.3665	.3840
7	.1214	.1325	.1442	.1564	.1689	.1820	.1954	.2092	.2233	.2378
8	.0573	.0639	.0710	.0786	.0866	.0951	.1040	.1133	.1231	.1334
9	.0245	.0279	.0317	.0358	.0403	.0451	.0503	.0558	.0618	.0681
10	.0095	.0111	.0129	.0149	.0171	.0195	.0222	.0251	.0283	.0318
11	.0034	.0041	.0048	.0057	.0067	.0078	.0090	.0104	.0120	.0137
12	.0011	.0014	.0017	.0020	.0024	.0029	.0034	.0040	.0047	.0055
13	.0003	.0004	.0005	.0007	.0008	.0010	.0012	.0014	.0017	.0020
14	.0001	.0001	.0002	.0002	.0003	.0003	.0004	.0005	.0006	.0007
15	.0000	.0000	.0000	.0001	.0001	.0001	.0001	.0001	.0002	.0002
16	.0000	.0000	.0000	.0000	.0000	.0000	.0000	.0000	.0001	.0001

Table XI Poisson Distribution (Continued)

x_0 \ λ	5.1	5.2	5.3	5.4	5.5	5.6	5.7	5.8	5.9	6.0
1	.9939	.9945	.9950	.9955	.9959	.9963	.9967	.9970	.9973	.9975
2	.9628	.9658	.9686	.9711	.9734	.9756	.9776	.9794	.9811	.9826
3	.8835	.8912	.8984	.9052	.9116	.9176	.9232	.9285	.9334	.9380
4	.7487	.7619	.7746	.7867	.7983	.8094	.8200	.8300	.8396	.8488
5	.5769	.5939	.6105	.6267	.6425	.6579	.6728	.6873	.7013	.7149
6	.4016	.4191	.4365	.4539	.4711	.4881	.5050	.5217	.5381	.5543
7	.2526	.2676	.2829	.2983	.3140	.3297	.3456	.3616	.3776	.3937
8	.1440	.1551	.1665	.1783	.1905	.2030	.2159	.2290	.2424	.2560
9	.0748	.0819	.0894	.0974	.1056	.1143	.1234	.1328	.1426	.1528
10	.0356	.0397	.0441	.0488	.0538	.0591	.0648	.0708	.0772	.0839
11	.0156	.0177	.0200	.0225	.0253	.0282	.0314	.0349	.0386	.0426
12	.0063	.0073	.0084	.0096	.0110	.0125	.0141	.0160	.0179	.0201
13	.0024	.0028	.0033	.0038	.0045	.0051	.0059	.0068	.0078	.0088
14	.0008	.0010	.0012	.0014	.0017	.0020	.0023	.0027	.0031	.0036
15	.0003	.0003	.0004	.0005	.0006	.0007	.0009	.0010	.0012	.0014
16	.0001	.0001	.0001	.0002	.0002	.0002	.0003	.0004	.0004	.0005
17	.0000	.0000	.0000	.0001	.0001	.0001	.0001	.0001	.0001	.0002
18	.0000	.0000	.0000	.0000	.0000	.0000	.0000	.0000	.0000	.0001

x_0 \ λ	6.1	6.2	6.3	6.4	6.5	6.6	6.7	6.8	6.9	7.0
1	.9978	.9980	.9982	.9983	.9985	.9986	.9988	.9989	.9990	.9991
2	.9841	.9854	.9866	.9877	.9887	.9897	.9905	.9913	.9920	.9927
3	.9423	.9464	.9502	.9537	.9570	.9600	.9629	.9656	.9680	.9704
4	.8575	.8658	.8736	.8811	.8882	.8948	.9012	.9072	.9129	.9182
5	.7281	.7408	.7531	.7649	.7763	.7873	.7978	.8080	.8177	.8270
6	.5702	.5859	.6012	.6163	.6310	.6453	.6594	.6730	.6863	.6993
7	.4098	.4258	.4418	.4577	.4735	.4892	.5047	.5201	.5353	.5503
8	.2699	.2840	.2983	.3127	.3272	.3419	.3567	.3715	.3864	.4013
9	.1633	.1741	.1852	.1967	.2084	.2204	.2327	.2452	.2580	.2709
10	.0910	.0984	.1061	.1142	.1226	.1314	.1404	.1498	.1505	.1695
11	.0469	.0514	.0563	.0614	.0668	.0726	.0786	.0849	.0916	.0985
12	.0224	.0250	.0227	.0307	.0339	.0373	.0409	.0448	.0490	.0534
13	.0100	.0113	.0127	.0143	.0160	.0179	.0199	.0221	.0245	.0270
14	.0042	.0048	.0055	.0063	.0071	.0080	.0091	.0102	.0115	.0128
15	.0016	.0019	.0022	.0026	.0030	.0034	.0039	.0044	.0050	.0057
16	.0006	.0007	.0008	.0010	.0012	.0014	.0016	.0018	.0021	.0024
17	.0002	.0003	.0003	.0004	.0004	.0005	.0006	.0007	.0008	.0010
18	.0001	.0001	.0001	.0001	.0002	.0002	.0002	.0003	.0003	.0004
19	.0000	.0000	.0000	.0000	.0001	.0001	.0001	.0001	.0001	.0001

Suggested Reading

For more information on the topics introduced in the text, consult the latest editions of the books listed below. They, in turn, will provide other references to pursue.

Anderson, T. W. *An Introduction to Multivariate Statistical Analysis, 2nd Ed.* Wiley, New York, 1984.

Bishop, Y. M. M., Fienberg, S. E., & Holland, P. W. *Discrete Multivariate Analysis.* MIT Press, Cambridge, 1975.

Box, G. E. P. & Jenkins, G. *Time Series Analysis, Forecasting, & Control.* Holden-Day, Oakland, 1976.

Cochran, W. G. & Cox, G. M. *Experimental Designs*, 2nd Ed. Wiley, New York, 1957.

Cooley, W. W. & Lohnes, P. R. *Multivariate Data Analysis.* Wiley, New York, 1971.

Deming, W. E. *Sample Design in Business Research.* Wiley, New York, 1960.

Dixon, W. J. (Ed.) *BMDP: Statistical Software.* University of California Press, Berkeley, 1983.

Dixon, W. J. & Massey, F. J. *Introduction to Statistical Analysis*, 3rd Ed. McGraw-Hill, New York, 1969.

Draper, N. R. & Smith, H. *Applied Regression Analysis*, 2nd Ed. Wiley, New York, 1981.

Edwards, A. L. *Multiple Regression and the Analysis of Variance and Covariance.* W. H. Freeman, San Francisco, 1979.

Feller, W. *An Introduction to Probability Theory and its Applications*, Vol. 1, 3rd Ed. Wiley, New York, 1968.

Fisher, R. A. *Statistical Methods for Research Workers*, 14th Ed. Hafner, New York, 1973.

Green, P. E. & Rao, V. R. *Applied Multidimensional Scaling.* Holt, Rinehart, & Winston, New York, 1972.

Harman, H. H. *Modern Factor Analysis*, 3rd Ed. University of Chicago Press, Chicago, 1976.

Hartigan, J. A. *Clustering Algorithms*. Wiley, New York, 1975.

Hoel, P. G. *Introduction to Mathematical Statistics*, 4th Ed. Wiley, New York, 1971.

Huitema, B. E. *The Analysis of Covariance and Alternatives*. Wiley, New York, 1980.

Johnson, R. A. & Wichern, D. W. *Applied Multivariate Analysis*. Prentice-Hall, Englewood Cliffs, 1982.

Kendall, M. *Multivariate Analysis*. Hafner, New York, 1980.

Lachenbruch, P. A. *Discriminant Analysis*. Hafner, New York, 1975.

McNemar, Q. *Psychological Statistics*, 4th Ed. Wiley, New York, 1969.

Morrison, D. F. *Multivariate Statistical Methods*, 2nd Ed. McGraw-Hill, New York, 1976.

Nie, N. et al. *Statistical Package for the Social Sciences*. McGraw-Hill, New York, 1981.

Rao, C. R. *Advanced Statistical Methods in Biometric Research*, Wiley, New York, 1952.

Romney, A. K., Shepard, R. N., & Nerlove, S. *Multidimensional Scaling: Theory and Applications in the Behavioral Sciences*. Academic Press, New York, 1972.

Siegel, S. *Nonparametric Statistics for the Behavioral Sciences*. McGraw Hill, New York, 1956.

Snedecor, G. W. & Cochran, W. G. *Statistical Methods*, 6th Ed. Iowa State University Press, Ames, Iowa, 1967.

Srivastava, M. S. & Carter, E. M. *An Introduction to Applied Multivariate Statistics*. North-Holland, New York, 1983.

Von Neumann, J. & Morgenstern, O. *Theory of Games & Economic Behavior, Rev. Ed.* Princeton University Press, Princeton, 1953.

Wilks, S. S. *Mathematical Statistics*. Wiley, New York, 1962.

Winer, B. J. *Statistical Principles in Experimental Design*, 2nd Ed. McGraw-Hill, New York, 1971.

Index

T

U

V

W